Sociological Odyssey

Contemporary Readings in Introductory Sociology

FOURTH EDITION

PATRICIA A. ADLER
University of Colorado

PETER ADLER
University of Denver

 WADSWORTH
CENGAGE Learning·

Australia • Brazil • Japan • Korea • Mexico • Singapore • Spain • United Kingdom • United States

WADSWORTH
CENGAGE Learning·

Sociological Odyssey: Contemporary Readings in Introductory Sociology, Fourth Edition
Patricia A. Adler and Peter Adler

Senior Publisher: Linda Ganster

Acquiring Sponsoring Editor: Erin Mitchell

Assistant Editor: John Chell

Editorial Assistant: Mallory Ortberg

Associate Media Editor: Melanie Cregger

Marketing Manager: Andrew Keay

Marketing Coordinator: Jack Ward

Marketing Communications Manager:
Laura Localio

Production Management, and Composition:
PreMediaGlobal

Manufacturing Planner: Judy Inouye

Rights Acquisitions Specialist: Roberta Broyer

Cover Designer: Ingalls Design

Cover Image: PBNJ Productions/Blend
Images/Corbis

For product information and technology assistance, contact us at **Cengage Learning Customer & Sales Support, 1-800-354-9706**

For permission to use material from this text or product, submit all requests online at **www.cengage.com/permissions**. Further permissions questions can be emailed to **permissionrequest@cengage.com**.

Library of Congress Control Number: 2011943245

ISBN-13: 978-1-111-82955-1

ISBN-10: 1-111-82955-1

Wadsworth
20 Davis Drive
Belmont, CA 94002-3098
USA

Cengage Learning is a leading provider of customized learning solutions with office locations around the globe, including Singapore, the United Kingdom, Australia, Mexico, Brazil and Japan. Locate your local office at **www.cengage.com/global**.

Cengage Learning products are represented in Canada by Nelson Education, Ltd.

For your course and learning solutions, visit **www.cengage.com**.

Purchase any of our products at your local college store or at our preferred online store **www.cengagebrain.com**.

Printed in the United States of America
1 2 3 4 5 6 7 16 15 14 13 12

To Our Protégés
Kathy, Angela, Laurie, Dana, Joanna, Jen, Katy I., Adina, Alice, Raegan, Kevin,
Katie, Kelly, Andrea, Kathi, Ross, Molly G., Maren, Elisabeth, Rob, Lori, Jessica,
Katy S., Robert, Molly J., Allison, Marc, Patrick, Jesse, Tracy, Chloe, and Tess

May all teachers be blessed with students such as these.

Table of Contents

Preface

…the times they are a-changin'

So sang Bob Dylan, perhaps America's most famous troubadour, more than fifty years ago. In fact, when he wrote that song, the civil rights movement was only in its infancy, women's liberation was still more than a decade away, the environmental movement was not even a twinkle in anyone's eyes, and social class in America was deeply divided and entrenched. Prophetic as Dylan was—and the times of the 1960s and 1970s certainly ushered in monumental social, political, legal, and economic change—the rapidity with which social change is occurring now, in the second decade of the new millennium, makes Dylan's notions of change seem glacial in speed. The United States in which you are living is experiencing a change to its landscape like few, if any, before. Beginning with the new immigrations of people from all over the world that began in the last quarter of the twentieth century, the demographics of our country have been dramatically altered. When this book was first published, not that long ago at the beginning of the century, Latinos (or Hispanics), for the first time, overtook African-Americans as our largest minority population, but just barely. In 2010, by the time of the next census, the percentage of African-Americans remained fairly constant (between 12% and 13%), while Latinos continued to grow to about 16%, with projections for even greater expansion in the next decade and beyond. These two groups, alone, represent about 30% of the entire population. Add to this the large numbers of people coming from Asian countries and other nations from around the world and the United States takes on a very different character, much more diverse, than in your parents' and certainly your grandparents' generations.

Changes in the types of people are only one sign of the speed with which social change is moving. In terms of how we communicate, it is estimated that almost 250 million Americans (84% of the population) subscribe to a cellular-phone service. This means that our idea of "place" and "home" has changed, as people are basically reachable at all times of day and in almost all venues they visit. About half of all Americans have access to a computer, yet approximately half of the world's population has not even *seen* a telephone. Similarly, more than three-quarters of North Americans use the Internet regularly, but only about 11% in Africa and 28% of the people worldwide have *ever* been exposed to it. Other innovations, such as texting, blogging, chat rooms, and Skyping, are part of the ways that North Americans communicate, but as shown above, these advances are mostly available only to people in Westernized countries. Clearly, there is a huge schism between how we live in North America and how people from around the globe are still mired in technology that is ancient to us now. This only further highlights the changes of our times.

Social changes are abundant in the world in which you live. Although merely a decade or so ago, it would have been apocryphal to speak about the advent of same-sex marriage, as of this writing five states and the District of Columbia (DC) have legalized this form of union since Massachusetts paved the way in 2004. No doubt more states will follow. To think that we would have a bi-racial president as early as 2008, too, would have been unheard of just years earlier. Licit drugs, such as tobacco, have become effectively banned from indoor use in more than half the states, while an illicit drug, marijuana, is now legal for those who can prove medical need in fifteen states and DC, and its decriminalization is widespread (prompting the Obama administration to say that prosecution of small amounts of marijuana is a lowest level priority, though it strongly opposed legalization). Women, although still not equally paid for the same jobs as men, make up over half of our law school students, about half of our medical school students, about one-third of engineering students, and over-all, are approaching nearly 60% of all undergraduates in our nation's colleges and universities. To imagine that wholesale changes about gender equality, already in the works, will not continue would be absurd. While we should not be ready to applaud total gender equality, the future is much brighter than even at the turn of the century. Bi-racial marriages represent about 7% of all marriages in the United States, still relatively small, but remember that miscegenation laws (banning interracial marriages) were made unconstitutional by the Supreme Court only as recently as 1967. In fact, some of America's most recognized figures, such as Barack Obama in politics, Tigers Woods in sports, and Halle Berry in entertainment, are bi-racial. Some White ethnic groups practice intermarriage to such a degree that they are practically disappearing, as is the case of Jews, who now represent less than 1% of the U.S. population. In religion, though this country remains dominated by Christian-based sects (almost 80%), the fastest growing "religious" group is "no religion," at just about 15% of the population.

yes, indeed, the times they are a-changing…

And sociology, the science of society, has been there to witness the impact of these profound transformations. Arguably, no other academic discipline is better suited to assess, analyze, evaluate, and predict these changes.

You are about to begin an odyssey, a journey into a new realm of thinking that, if taken seriously, can change the way you view the world. The "sociological imagination," as the mid-twentieth-century sociologist C. Wright Mills referred to it, can transform one's perspective on reality so that even the most mundane, seemingly trivial aspects of everyday life take on added significance. Further, the big issues that face us, such as social inequality, immigration, economic turmoil, political corruption, crime, and environmental decay, can also be looked at with fresh eyes to lend increasing clarity to why these exist and how we can improve conditions. Like the visionary filmmaker Stanley Kubrick predicted years ago, the early part of the twenty-first century has already been a space odyssey into domains we could not even have dreamed about a scant decade ago. Given the tremendous amount of change we have witnessed in the last decade, who knows what the future has to offer?

We are pleased to be your guides in these travels. We hope that at the end of this journey, you will have a better understanding of the social dynamics of society and the complexities that make up your world.

NEW TO THE FOURTH EDITION

To mark the incredible amount of change that has occurred in the time since this book was conceived in 2000, it became clear to us that some of the same old theories, research, facts, figures, and ways of looking at the world had to change as well. To that end, we have endeavored to replace some of the pieces that appeared in previous editions with the most contemporary research to reflect the shifting tides of North American societies. It is our belief that, due to the demographic changes mentioned above, a whole different type of college student is emerging, one who might likely be the first of his or her family to even attend college, and we feel that this book should represent them. Members of racial and ethnic group that had low college attendance rates are now making up a much larger percentage of students, as we would expect given population trends. At the same time, colleges and universities are still filled with "traditional" students, White, aged 18–22, and upper middle class, who have been the primary population in higher education for centuries. To this end, we have tried to select articles that are relevant to this wide panoply of people, as well as research that is "experience near" (familiar to most readers of this book) and "experience far" (perhaps giving readers insights to parts of culture that remain distant from them). We have added new articles on White privilege and identity and how it operates to continue to provide power to Euro-Americans; the simulation and construction of authentic ethnicity as portrayed in the ubiquitous Mexican restaurants in the United States; how Asian students (Cambodians, in this case) negotiate their ways through America's educational system; the success of Korean entrepreneurs in the banking industry; first- and second-generation Mexican-American gardeners in Southern California and how they are using their businesses successfully (or not) to gain a foothold in American society; how Latinos, some of whom are now comfortably ensconced as third-generation Americans, balance the identities of their ethnic group, bi-ethnic relations, and changing values from their families; the persistence of female-dominated homes in the underclass and how women, in particular, must fend for their families as they navigate the welfare system; the experiences of Black and White fraternity members; how gender is scripted at college dance parties; the changes wrought by the recession in the 2000s and its effects on family decisions, labor, and relations; the effects of the recession on working-class families and how they have had to adapt to changing economic times; new family types that are emerging, such as what happens when same-sex partners have children; fathers taking a greater role in their upbringing and care of their children; the movement away from religion altogether to the alternative of atheistic beliefs; and the explosion of online relations and the changes that occur when much of social interaction is no longer face-to-face.

Many of the articles we have kept also address issues of great import to the contemporary collegian: the gay subculture; the straight edge movement; fraternities and sororities; the Black middle class; fashion models and their bodies, male cheerleaders, cohabiters, and leisure seekers. Other topics that are inherently fascinating include the culture of homophobia among men; women who suffer with the stigma of sexually transmitted diseases; people's behavior in public bathrooms; the medicalization of American youth with drugs such as Adderall and Ritalin; the ritual of the engagement proposal; contrasting images of beauty among Blacks and Whites; criminality and race; the effects of September 11 on Muslims in the United States; the disparities in educational opportunities between the rich and the poor; the outsourcing of the American service industry; and children's clique behavior. The growing equality between the genders (or the illusion of such) has produced a great deal of research in sociology about gender.

To that end, this book tackles the gender divide in about a dozen different chapters. Race and ethnicity issues can be found in about 15 of the articles. Issues involving social class are found in about a dozen readings, although in all of these categories, race, class, and gender, there is considerable overlap. We have also bulked up the sections on the impact of technology and globalization, two phenomena that affect you on an everyday basis but that you may simply take for granted and not notice. Finally, in a departure from what we hear too often about sociology's irrelevance to everyday life, the last chapter of the book offers an optimistic solution for how sociology can be applied to make for a better, more sustainable universe.

In sum, it has been our hope to create a book that resonates with our likely readers. We understand that most people who are assigned *Sociological Odyssey* have never taken a sociology course before. We have tried to avoid articles that are excessively jargonized, preferring to select pieces that are accessible and readable. Although we have included some classics, our primary focus has been on contemporary works that reflect current societal interests. We have avoided abstraction where a firmly grounded, empirical piece can illustrate basic sociological concepts. The selections have been edited, avoiding repetition, unnecessary detail, and theoretical discussions that might be out of the reach of the novice sociologist. In all, we hope that you find the following chapters, both old and new to be of interest, and that the selections will spark debate, controversy, and discussion. If so, we have accomplished a major part of our task.

ACKNOWLEDGMENTS

We began the initial project for the first edition during the hectic months of the Fall, 1999 semester. For years we had toyed with the idea of creating an exciting, new introductory reader for students to enjoy, but it wasn't until Eve Howard, then stalwart acquisitions editor for Wadsworth, flew to Boulder to twist our arms that we finally agreed that the time was ripe. The first edition met with moderate success, but like many new books, it had its problems finding its niche. However, as the years have passed, we are honored that more professors and students have discovered some of the joyful tidbits of sociology we included. In the second edition, we tried to improve on the original version, replaced about half of the selections, reflecting changes in society that had occurred in the past half decade, what we thought college students wanted to read about, and with a finer eye toward the nuances in how students approach each of these topics. For the third go-round, we kept what our readers have told us works, we jettisoned some articles that did not resonate well, and we added research and sections to this book that continue to reflect the warp speed at which our society is changing. It used to be that three years was a short time to produce a new edition of a book such as this, but now length of time seems like an eternity when considering how profoundly each year brings new innovations and emerging social relationships. Thus, as described, for this fourth edition, we felt that we needed to be particularly sensitive to changes that were discovered through the results of the 2010 US Census Reports, as well as the rapid social changes we have described above. Throughout this process, Erin Mitchell and John Chell have been strong supporters of the book, urged us to keep the book as relevant and current as possible, and provided us with the trust and leeway we needed to make the decisions we made.

We have also had the support of two departments at the University of Colorado and University of Denver, respectively, which have given us the foundation to pursue a project such as this. Our colleagues have been most open to hearing about our various selections, suggesting possible readings, and understanding the value of creating such a text. Most particularly, Paul Colomy (University of Denver) provided critical feedback at a time when we were unsure of the precise direction we wanted to take. His intellectual guidance will forever be appreciated. Other friends and colleagues, such as Dan Cress, Gary Alan Fine, Andy Fontana, Chuck Gallmeier, Bob Granfield, Paul Harvey, Lori Hunter, Leslie Irvine, Rebecca Plante, Rik Scarce, D. Angus Vail, and Glenda Walden, gave us some ideas that were used in making our choices. Our children, Jori and Brye, offered us perspectives to evaluate what young adults want to hear, as well as a model for what parents want to see in their progeny. No thank-you would be complete without the inclusion of Dorene Miller, Administrative Assistant at the University of Denver, who time and again bailed us out, unselfishly did the tough work, and lent the type of support without which projects such as this could not flourish.

Finally, we are fortunate to have a cadre of comrades, our protégés, who have also given us advice and encouragement along the way. Kathy Fox, Angela Yancik Baldasare, Laurie Scarborough Voss, Dana Jones Hubbard, Joanna

Gregson, Jennifer Lois, Katy Irwin, Adina Nack, Alice Fothergill, Raegan Cormaci, Kevin Vryan, Katie Coroso Robinson, Kelly Teitelbaum, Andrea Schmitt, Kathi Schaff, Ross Haenfler, Molly George, Maren Scull, Rob Gardner, Lori Peek, Jessica Johnston, Katy Sirles, Robert Duran, Molly Jenkins, Allison Hicks, Marc Eaton, Patrick O'Brien, Jesse Smith, Tracy Bacon, Nini Adams, Tess Dougherty, and Chloe Grossman are all now full-fledged collaborators with us in the joint sociological enterprise we share. We lovingly dedicate this book to them.

A team of sociological reviewers were also an important cog in assuring that the selections in the first edition had the most relevance to students' lives: David Boden (Lake Forest College), John Bridges (Kutztown University), Kevin Early (Oakland University), Patti Guiffre (Southwest Texas State University), Jane Johnson (Southwest Texas State University), David Maines (Oakland University), and Alvar Nieves (Wheaton College). We are grateful for the quick turn-around they gave in reviewing the book, as well as the thoughtful comments about the numerous ways that introductory sociology can be taught.

Additionally, we would like to thank Amy D'Olivio (Centenary College), Ann M. Hickey (University of Kansas), Jennifer Karas (University of Denver), and Deborah J. Shepherd (Anoka-Ramsey Community College) for their valuable comments that contributed to the development of the second edition. For the third edition, we received excellent feedback and advice from another fine stable of reviewers, including Gary Laird (Florida Community College), Ronald L. Mize (Cornell University), Richard Perry (Wake Technical Community College), Rebecca Plante (Ithaca College), Daniel Renfrew (Pacific Lutheran University), and D. R. Wilson (Houston Baptist College).

For the fourth edition, we received insightful recommendations from several reviewers, including Donald Stewart (University of Nevada, Las Vegas), Elizabeth Wissinger (Borough of Manhattan Community College), Patricia H. O'Brien (Elgin Community College), and Rebecca Plante (Ithaca College).

At Wadsworth Cengage Publishing Company, in addition to the steadfast work of Eve Howard and Chris Caldeira (whose presence we miss sorely) in the early iterations of this book, there have been a number of individuals who have assisted us. Most especially, Dee Dee Zobian, assistant to Eve Howard during the production of the first edition, did more of the daily work than anyone else to assure the timeliness of the project. Dee Dee treated us as if we were her only project, when in reality she was simultaneously juggling a dozen or more ventures. Others, such as Matt Ballantyne, Rebecca Boorsma, Bobbie Broyer, Jerilyn Emori, Wendy Gordon, Bob Jucha, Bob Kauser, Kristin Marrs, Cheri Palmer, Erin Parkins, Jennifer Walsh, Michelle Williams, and Matt Wright made sure that the final product was produced with the professionalism that has become the hallmark of Wadsworth/Cengage's operation.

This list would not be complete without mentioning others in the publishing industry who either directly or indirectly helped us. Notably, Serina Beauparlant was literally the first person to try to cajole us into doing our first anthology, *Constructions of Deviance*, and presented the first germ of the idea for this book. Mitch Allen (Left Coast Press), Fran Benson (Cornell University Press), Dean Birkenkamp (Paradigm Press), Martha Heller (formerly of Rutgers University

Press), and more recently, Ilene Kalish (NYU Press) are all editors with whom we've worked who reminded us of the scholarly and intellectual fortunes, rather than merely economic ones, that are borne out through our books. They represent the best that publishing has to offer.

To our students, past, present, and future, and to our readers, we thank you for your patience, diligence, and understanding. We hope that we have been able to convey the sociological perspective in such a way as to make you as excited about applying it to your lives as it is for us to teach it to you.

About the Editors

Patricia A. Adler (Ph.D., University of California, San Diego) is Professor of Sociology at the University of Colorado, Boulder. In 1999, she was named Outstanding Teacher in the Faculty of Arts and Sciences and in 2005 she received the Outstanding Researcher Award for the Boulder campus. In 2004, Patti was awarded the Mentor Excellence Award from the Society for the Study of Symbolic Interaction. She has written and taught in the areas of deviance, social psychology, sociology of gender, and the sociology of children. A second edition of her book, *Wheeling and Dealing* (Columbia University Press), a study of upper-level drug traffickers, was published in 1993.

Peter Adler (Ph.D., University of California, San Diego) is Professor of Sociology and Criminology at the University of Denver, where he served as Chair from 1987 to 1993. In 1998, he was named University Lecturer, an award that represents outstanding achievement in scholarship and research and in 2005 he received the United Methodist Church Scholar-Teacher of the Year. Also in 2005, Peter was honored with the Mentor Excellence Award from the Society for the Study of Symbolic Interaction. His research interests include social psychology, qualitative methods, drugs and society, and sociology of work, sport, and leisure. His first book, *Momentum*, was published in 1981 by Sage.

Together, the Adlers served as Co-Presidents of the Midwest Sociological Society from 2006 to 2007. In 2010, they were the recipients of the George H. Mead Award for Lifetime Achievement, the first collaborative winners of this Award. They have edited the *Journal of Contemporary Ethnography* and were the founding editors of *Sociological Studies of Child Development*. They are also editors of another Cengage book, *Constructions of Deviance*, a collection of readings on deviant behavior, now in its seventh edition, as well as *Drugs and the American Dream* (Wiley 2012). Among their many books are *Membership Roles in Field Research*, a treatise on qualitative methods published by Sage in 1987, *Backboards and Blackboards*, a participant-observation study of college athletes that was published by Columbia University Press in 1991, *Peer Power*, a study of the culture of elementary schoolchildren that was published by Rutgers University Press in 1998, and *Paradise Laborers*, an ethnography of the subculture of resort workers, published in 2004 by Cornell University Press, that won the North Central Sociological Association's Outstanding Scholarship Award in 2005. Their most recent book, *The Tender Cut*, which focuses on the culture of self-injurers (cutters and burners), was released by NYU Press in 2011.

Peter and Patti have been researching and writing together for approximately forty years, ever since they were undergraduates at Washington University in St. Louis. They have two grown children, both of whom graduated from Emory University with a major in sociology. Peter and Patti divide their times between their homes in Boulder, Colorado, and Maui, Hawaii.

About the Contributors

Sine Anahita is Associate Professor in Sociology and Northern Studies at the University of Alaska Fairbanks. Her research and teaching interests focus on inequalities, especially inequalities based on race, gender, and sexuality. Recently, she has begun ethnographic work on the fiddle in Interior Alaska.

Eric Anderson is affiliated with the Department of Sport Studies at the University of Winchester, England. He is known for his research on sport, masculinities, sexualities, and homophobia. His work has been published in dozens of academic journals and is regularly featured in the popular press. His work shows an increasingly positive relationship between gay male athletes and sport, as well as a growing movement of young heterosexual men's masculinity becoming softer and more inclusive. He also researches matters related to men's monogamy/cheating, men's improving recognition of bisexuality, and the increased acceptance of young heterosexual men kissing.

Elizabeth Aries earned her Ph.D. at Harvard University in Social Psychology in 1973 and has been teaching at Amherst College since 1975. She is the author of *Men and Women in Interaction: Reconsidering the Differences* (Oxford University Press 1996), *Adolescent Behavior: Readings and Interpretations* (Dushkin/McGraw-Hill 2001), and *Race and Class Matters at an Elite College* (Temple University Press 2008). Her research examines the ways in which race- and class-based aspects of students' identities influence and are influenced by the college experience, and the extent to which students are learning from race- and class-based diversity at an elite college.

Peter Bain was Lecturer in the Department of Human Resource Management at the University of Strathclyde at the time of his death in 2007. His most recent work had examined the development of, and working conditions within, Scottish call centers.

Dana Berkowitz is Assistant Professor in the Department of Sociology and the Program in Women's and Gender Studies at Louisiana State University. Her research and teaching interests focus on the social construction of gender, sexualities, and families. She has published extensively on gay men's procreative, fathering, and family identities. She was the recipient of the National Council on Family Relations Jesse Bernard Award for Outstanding Research Paper from a feminist perspective. Her current project focuses on Botox, bodies, and the politics of non-surgical anti-aging interventions.

Nancy Berns is Associate Professor of Sociology at Drake University. She received her Ph.D. from the University of Illinois at Urbana-Champaign. Her teaching and research interests are in the areas of grief, death, violence, media, and social justice. She is the author of *Framing the Victim: Domestic Violence, Media and Social Problems* (Aldine de Gruyter 2004) and *Closure: Untangling the Rush to End Grief and What It Costs Us* (Temple University Press 2011).

A. Ayres Boswell was Director of the Intake Department for the Community Mental Health Foundation in Doylestown, Pennsylvania. She did evaluations, assessments, and made recommendations for treatment as well as conducted psychotherapy with clients. She currently lives in Doylestown, Pennsylvania and keeps busy raising her four children.

Spencer E. Cahill passed away in October 2006. At the time of his death, he was Outgoing Editor of *Social Psychology Quarterly* and Professor of Sociology at the University of South Florida. He was also Co-editor of *Journal of Contemporary Ethnography* and published articles on a range of topics, including gender identity acquisition, public life, childhood, disability, and professional socialization. His final research project examines adolescent culture and relations through notes and other personal documents that adolescents wrote and exchanged among themselves.

Joel Charon earned his Ph.D. in Sociology at the University of Minnesota. He worked in the Department of Sociology and Criminal Justice at Moorhead State University, which he chaired from 1993 to 2001. He has written *Symbolic Interactionism* (10th edition, Pearson/Prentice-Hall), *The Meaning of Sociology* (9th edition, Pearson/Prentice-Hall), *The Meaning of Sociology: A Reader* (9th edition, Pearson/Prentice-Hall), *Ten Questions* (7th edition, Cengage/Wadsworth), and *Social Problems* (Cengage/Wadsworth, 4th edition). Since 2001, he has devoted the rest of his professional life to research and writing.

Vichet Chhuon is Assistant Professor of Curriculum and Instruction with an affiliation in the Asian American Studies Program at the University of Minnesota. He received his Ph.D. in Education from the University of California, Santa Barbara. His work focuses on meeting the teaching and learning needs of immigrant and ethnic minority students; examining the relationships between educational institutions, families, and social context; and employing interdisciplinary frameworks and diverse methodologies to the study of education.

Randall Collins is Professor of Sociology at the University of Pennsylvania. His books include *Conflict Sociology* (1975), *The Credential Society* (1979), *Weberian Sociological Theory* (1986), *The Sociology of Philosophies* (1998), *Macro-History: Essays in Sociology of the Long Run* (1999), *Interaction Ritual Chains* (2004), and *Violence: A Microsociological Theory* (2008).

Paul Colomy is Professor of Sociology and Criminology at the University of Denver. He is Editor of *The Dynamics of Social Systems*, *Functionalist Sociology* and *Neofunctionalist Sociology*, and has contributed articles to *Sociological Theory*, *Social Problems*, *Sociological Forum* and *The Sociological Quarterly*. His current work examines the origins and transformation of juvenile justice.

Marie Cornwall is Professor of Sociology and the recipient of an FHSS College Professorship at Brigham Young University; she has also been the editor of the *Journal for the Scientific Study of Religion* (2008–2012). Her research interests include the sociology of religion, social movements, gender, family, and social change. She has published in the *JSSR, the Journal of Marriage and Family, Social Forces, Mobilization*, and *Gender & Society* and co-edited, with Tim Heaton and Larry Young, *Contemporary Mormonism: Social Science Perspectives* (2001). Her research has included the study of religious socialization, religion and family, the suffrage movement, unemployment and family well-being, and, most recently the sociology of reproduction.

William Finlay is Professor of Sociology and Department Head at the University of Georgia. He received his Ph.D. in Sociology from Northwestern University. His main area of research is the sociology of work and organizations and his most recent project examines the entry of South African wines into the U.S. market, focusing on the forging of ties between winemakers and importers. His latest book, *The Sociology of Work: Structures and Inequalities* (co-authored with Steven Vallas and Amy Wharton), was published by Oxford University Press in 2009.

Marie Sarita Gaytán is Assistant Professor of Sociology and Gender Studies at the University of Utah. She is currently working on a book manuscript that examines the relationship between commodity culture and national identity.

Kathleen Gerson is Professor of Sociology and Collegiate Professor of Arts and Science at New York University, where her research focuses on gender, work, and family change in contemporary societies. Her most recent book, *The Unfinished Revolution*, provides a first-hand account of the experiences, outlooks, and strategies of the generation who grew up amid the gender revolution in work and family patterns. After receiving her Ph.D. from the University of California, Berkeley, she joined the NYU faculty, where she has taught for several decades. During this time, she has also served as Chair of the American Sociological Association Section on the Family and as President of the Eastern Sociological Society. A recognized authority on work, gender, and family life, she has also contributed to many policy initiatives supporting gender equality and work-family balance and is a recipient of the Rosabeth Moss Kanter Award for excellence in Work-Family Research.

Patricia Greenfield, Distinguished Professor of Psychology at UCLA and Director of the Children's Digital Media Center@Los Angeles, is an expert on culture, and human development. She is the author of *Mind and Media: The Effects of Television, Video Games, and Computers* (1984), subsequently translated into nine languages; co-editor of *Effects of Interactive Entertainment Technologies on Development* (1994); co-editor of *Children, Adolescents, and the Internet: A New Field of Inquiry in Developmental Psychology* (2006); and co-editor of *Social Networking on the Internet: Developmental Implications* (2008). Her empirical research on the developmental implications of interactive media has included action video games, massive multiplayer online role-playing games, teen chat rooms, and social networking sites. In 2010, she received the Urie Bronfenbrenner Award for Lifetime Contribution to Developmental Psychology in the Service of Science and Society from the American Psychological Association.

Ross Haenfler is Associate Professor of Sociology at the University of Mississippi. His interests revolve around how people pursue social change in their daily lives. He is the author of *Straight Edge: Clean Living Youth, Hardcore Punk, and Social Change* (Rutgers University Press), *Goths, Gamers, and Grrrls: Deviance and Youth Subcultures* (Oxford University Press), and co-author of *The Better World Handbook: Small Changes That Make a Big Difference* (New Society). Ross teaches courses on social movements, youth subcultures, men and masculinities, and political sociology. In his spare time he enjoys hiking, backpacking, and traveling.

David R. Hodge is Associate Professor in Social Work at Arizona State University and a Senior Nonresident Fellow at the University of Pennsylvania's Program for Research on Religion and Urban Civil Society. His scholarly interests include the study of spirituality as a protective factor, particularly among minority and disenfranchised populations. He is currently conducting a study examining the role spirituality plays in helping victims of human trafficking cope with the traumatic experiences they endured.

Pierrette Hondagneu-Sotelo received her Ph.D. in Sociology from the University of California, Berkeley and she is Professor of Sociology at USC, where she has taught for over twenty years. She is the author or editor of eight books and many articles, nearly all of which focus on the dramatic social issues that accompany immigration. Her book *Domestica: Immigrant Workers Cleaning and Caring in the Shadows of Affluence* (University of California Press 2001, 2007) won several awards, including the Max Weber Award and the C. Wright Mills Award. She is currently working on a book about social inequality, enchantment, and Los Angeles gardens.

Cynthia Hudley is Professor in the Gevirtz Graduate School of Education at the University of California, Santa Barbara. She received her Ph.D. at the

University of California, Los Angeles. Her research interests include Educational Psychology, Minority Education, Child Development, Social Development, Childhood Aggression, Youth Violence, Achievement Motivation, and Secondary Data Analysis.

Brett Johnson, who received his Ph.D. from the University of Colorado, is Assistant Professor of Sociology at Luther College in Decorah, IA. His current research projects include co-authoring *The Better World Handbook* (New Society 2007) and research on "lifestyle movements," especially the U.S. voluntary simplicity movement that advocates "simple living" as a tonic to personal, social, and environmental problems. He teaches courses on social conflict, social movements, environmental sociology, and research methods/statistics and is currently working with students to mobilize the Luther College community to invest in community development, such as microcredit, to reduce global poverty.

Ellis Jones is a Visiting Professor of Sociology at College of the Holy Cross in Worcester, MA. Since receiving his Ph.D. in Sociology from the University of Colorado, Boulder, he has focused his energies on bridging the gap between academics, activists, and the average citizen. His research interests include ethical consumerism, social responsibility, and global citizenship. He is the author of *The Better World Shopping Guide* (3rd edition, 2010), a comprehensive reference for socially responsible consumers, and co-author of *The Better World Handbook* (2nd edition, 2007), an exhaustive cataloguing of how individuals can impact global social problems. His next book, *The Social Responsibility Movement: Global Transformation in Everyday Life*, examines how growing numbers of people are using their private lives to affect large-scale social change.

Keith M. Kilty is Professor Emeritus of Social Work at Ohio State University. He was a co-founder of the *Journal of Poverty*, which he co-edited for 11 years. His research interests focused on poverty and inequality and their connections to race and ethnicity. After his retirement in June, 2007, he began working on a documentary about poverty. The film, titled "Ain't I a Person," was completed recently and is now available for screenings and classroom use.

Jonathan Kozol received a B.A. from Harvard University. His books include *Illiterate America, Death at an Early Age, Rachel and Her Children, Savage Inequalities, Amazing Grace, Ordinary Resurrections, The Shame of a Nation, On Being a Teacher,* and *Letters to a Young Teacher.*

Karyn R. Lacy received her Ph.D. in Sociology from Harvard University in 2000 and is currently Associate Professor of Sociology at the University of Michigan, is a Ford Fellow, and was a Visiting Scholar at the Russell Sage Foundation. Her areas of research include race and ethnicity, the sociology of culture, (sub)urban sociology, and stratification. Her book, *Blue-Chip Black: Race, Class and Status in the New Black Middle Class* (University of California Press 2007), examines how the black middle class defines itself in relation to whites, to the white middle class, and to blacks from other classes. Her current work explores the impact of an elite social organization on the construction and reproduction of class-based identities among middle-class blacks.

Liz Legerski is currently Assistant Professor of Sociology at the University of North Dakota. Her research interests include gender, social inequality, and health and social policy. She is particularly interested in the way employment opportunities and social policies shape the lives of working-class families. Her work has been published in *Social Forces, Women's Health Issues*, and *Gender & Society*. Her current research analyzes the welfare, work, and family factors that affect low-income women's access to health insurance over time.

Jonathan Leo is Professor of Neuroanatomy at Lincoln Memorial University DeBusk College of Osteopathic Medicine in Harrogate, TN. His work has appeared in such journals as the *Public Library of Science Medicine, Society, The Chronicle of Higher Education*, and *The Journal of Mind and Behavior*. He has served as a peer reviewer for the National Institute of Clinical Excellence (NICE) guidelines on ADHD. He and Sami Timimi are co-editors of the book *Rethinking ADHD* (2008).

Cynthia A. Lietz is Assistant Professor at the School of Social Work at Arizona State University. She conducts research that informs strengths-based practice with children and their families. She is interested in enhancing social work practice through research, education, and supervision. Prior to coming to ASU, she worked as a social work practitioner and clinical supervisor.

Michael Lovaglia is Professor in the Department of Sociology at the University of Iowa. His teaching applies the best recent research in social psychology to the important problems that students and professionals face in their daily lives. A new online course on the social psychology of leadership offered by the University of Iowa provides leadership techniques proven by research to a wide variety of students. His popular book, *Knowing People: The Personal Use of Social Psychology*, is available from Rowman and Littlefield.

Yvonne M. Luna is Associate Professor of Sociology at Northern Arizona University where she was awarded the College of Social and Behavioral Sciences' Teacher of the Year for 2008–2009. She earned her Ph.D. in Justice and Social Inquiry from Arizona State University in 2005. Her research and teaching is devoted to race and ethnic relations, poverty and welfare, domestic violence, and identity. She is also an avid BMW motorcycle rider.

William Marsiglio received his Ph.D. from The Ohio State University and is Professor of Sociology, University of Florida, and Fellow of the National Council on Family Relations. He primarily studies the social psychology of fathering, men's involvement with youth, and reproductive health. His books include *Men on a Mission: Valuing Youth Work in Our Communities; Situated Fathering: A Focus on*

Physical and Social Spaces; *Stepdads: Stories of Love, Hope, and Repair*; *Sex, Men, and Babies: Stories of Awareness and Responsibility*; *Procreative Man*; and *Fatherhood: Contemporary Theory, Research, and Social Policy*. He is currently completing a book with Kevin Roy for the ASA Rose Monograph Series which will be published with the Russell Sage Foundation. The book focuses on diverse fatherhood initiatives to promote nurturance.

Melissa A. Milkie is Professor of Sociology at the University of Maryland, College Park. Her research centers on gender, ethnicity, culture, social psychology, and family. She focuses on how social statuses and structures such as gender and ethnicity shape the self, well-being, and mental health. The way that people understand themselves and create meaning within oft-changing cultural boundaries is of great interest to her. She is author of many articles and two books, *Changing Rhythms of American Family Life* (with Suzanne M. Bianchi and John P. Robinson) and *Social Psychology: Sociological Perspectives* (in its 2nd edition, with David E. Rohall and Jeffrey W. Lucas) and is currently researching social aspects of children's, mothers', and fathers' mental health.

Ashley Mears is Assistant Professor of Sociology at Boston University. She completed her Ph.D. from New York University and was a Visiting Fellow at the Center for Gender Studies at the University of Cambridge. Her research focuses on the intersections of culture, markets, and inequalities. In her first book, *Pricing Beauty: The Making of a Fashion Model* (University of California Press 2011), she examined the backstage production of the "look" in New York and London fashion markets. She is currently researching international model scout networks as a way to understand global flows of culture.

C. Wright Mills received his Ph.D. in Sociology from the University of Wisconsin and was Professor of Sociology at Columbia University. His best-known works are *White Collar* (1951), *The Power Elite* (1956), and *The Sociological Imagination* (1959).

Adina Nack is Associate Professor of Sociology at California Lutheran University. She earned her Ph.D. in Sociology from the University of Colorado, Boulder and previously served on the faculty of the University of Maine. She has published a book, *Damaged Goods* (Temple University Press 2008), and articles in *Symbolic Interaction* and *Deviant Behavior*. Her research focuses on issues of deviance, sexuality, and gender in areas of medical sociology. Her recent projects focused on Latinos/as living with HIV/AIDS, and her current research collaboration examines traumatic childbirth narratives of women, their significant others, and health care providers.

Tamara K. Nopper earned her Ph.D. in Sociology at Temple University. She is currently a Lecturer in the Department of Sociology and the Asian American Studies Program at the University of Pennsylvania, where she teaches the courses Introduction to sociology and Race and Ethnic Relations and has developed new courses such as Immigrant Urban Labor and Ethnic Economies and Globalization. Her research on Asian immigrant entrepreneurship, colorblind racism and minority business policy, ethnic banking, racism and immigration enforcement, and Black-Asian conflict has been published in academic journals, edited anthologies, and popular print and online publications.

Devah Pager is Associate Professor of Sociology and Co-Director of the Joint Degree Program in Social Science and Social Policy at Princeton University. Her research focuses on institutions affecting racial stratification, including education, labor markets, and the criminal justice system. Her recent research has involved a series of field experiments studying discrimination against minorities and ex-offenders in the low-wage labor market. Her book, *Marked: Race, Crime, and Finding Work in an Era of Mass Incarceration* (University of Chicago 2007), investigates the racial and economic consequences of large-scale imprisonment for contemporary U.S. labor markets.

Lori Peek is Associate Professor of Sociology and Co-director of the Center for Disaster and Risk Analysis at Colorado State University. She has published widely on vulnerable populations in disaster and is Author of *Behind the Backlash: Muslim Americans after 9/11* and Co-editor of *Displaced: Life in the Katrina Diaspora*. In 2009 the American Sociological Association Section on Children and Youth honored her with the Early Career Award for Outstanding Scholarship.

Robert Perrucci is Professor of Sociology at Purdue University. His research has focused on the impact of the global economy on workers, communities, and structures of inequality. His work is reflected in two recent books with Carolyn Cummings Perrucci: an edited collection of readings entitled *The Transformation of Work in the New Economy* (Oxford University Press 2007) and *America at Risk: The Crisis of Hope, Trust, and Caring* (Rowman & Littlefield 2009). He has served as Editor of *The American Sociologist, Social Problems,* and *Contemporary Sociology,* and President of the Society for the Study of Social Problems.

Pamela Perry is Professor of Community Studies at the University of California, Santa Cruz. She is the author of *Shades of White: White Kids and Racial Identities in High School* (Duke University Press 2002).

Robert D. Putnam is the Peter and Isabel Malkin Professor of Public Policy at Harvard University. He is a member of the National Academy of Sciences, a Fellow of the British Academy, and past President of the American Political Science Association. Raised in a small town in the Midwest and educated at Swarthmore, Oxford, and Yale, he has written a dozen books, including *Bowling Alone* and *Better Together.* He is now studying strategies for building community in an increasingly diverse and time-strapped America.

Hernan Ramirez received his Ph.D., in Sociology from the University of Southern California and is Assistant Professor of Sociology at Florida State

University. His research interests include Latino immigrant entrepreneurship, gender and work, and the informal economy. He is currently working on a book that examines social mobility among self-employed Mexican immigrant gardeners and their US-born young adult children in Los Angeles.

Rashawn Ray is Assistant Professor of Sociology at the University of Maryland, College Park. He received a Ph.D. in Sociology from Indiana University, Bloomington in 2010. His research interests are social psychology, race and ethnic relations, and race-class-gender. His work addresses three key areas: the determinants and consequences of self-evaluated social class, men's treatment of women, and how racial stratification structures social life. He is Editor of *Race and Ethnic Relations in the 21st Century: History, Theory, Institutions, and Policy*. From 2010 to 2012, he was a Robert Wood Johnson Health Policy Research Scholar at the University of California, Berkeley/UCSF examining racial differences in barriers and incentives to physical activity. He has been awarded funding from the National Science Foundation, the National Institute of Mental Health, the American Sociological Association Minority Fellowship Program, the Society for the Study of Social Problems, and the Ford Foundation.

George Ritzer is best known for *The McDonaldization of Society*, but he has authored other works like it, including *The Globalization of Nothing, Enchanting a Disenchanting World*, as well as the forthcoming *The Outsourcing of Everything* (with Craig Lair). He is founding editor of *The Journal of Consumer Culture*. The vast majority of his work is in the area of social theory and metatheory. Sage (England) published two volumes of his collected works, one volume focusing on theory and the other on consumption. Among his edited works are *The Blackwell Companion to Globalization* and the 11-volume *Encyclopedia of Sociology*. He has been Distinguished University Professor at the University of Maryland since 2001.

Shelly Ronen received her B.A. from Stanford University in 2009 and is currently a Ph.D. student at New York University. Her interests include the intersection of gender and sexuality, public courtship behaviors, scripted interaction, and status. Having worked at a design consultancy, she hopes to forge a connection between the sociology of objects and the sociology of gender.

Jason A. Rosow received his graduate degree in sociology from Indiana University in 2005. As a graduate student, his primary interests were self and identity, peer cultures, and emotions.

Maura Ryan is a Lecturer in the Department of Sociology and affiliate faculty member of the Women's Studies Institute at Georgia State University. She specializes in the study of gender, sexualities, and social movements particularly in GLBTQ communities.

Howard Schuman is Professor of Sociology and Research Scientist (Emeritus) at the University of Michigan, but lives on the Maine coast, where he recently completed the book *Method and Meaning in Polls and Surveys* (Harvard University Press 2008) and has begun a new book, *Collective Memories*. He is also the senior author of *Questions and Answers in Attitude Surveys* (1981, 1996) and *Racial Attitudes in America* (1997). His writing has received awards from the American Political Science Association's Section on Public Opinion, the American Sociological Association's Methodology Section, and the American Association for Public Opinion Research.

David Schweingruber is Associate Professor of Sociology at Iowa State University. He received his B.A. from Bluffton College and his Ph.D. from the University of Illinois at Urbana-Champaign. His research, in the symbolic interactionist tradition, is concerned with the cultural and cognitive premises that guide social behavior in a variety of settings, including formal organizations, political demonstrations, and romantic events. He is currently researching Christian faculty at public universities.

Maynard Seider earned his Ph.D. in Sociology from the University of Wisconsin (Madison) in 1973 and has written *A Year in the Life of a Factory* (Singlejack 1984; reissued by Charles H. Kerr 1993). Now Professor Emeritus of Sociology at Massachusetts College of Liberal Arts, he periodically blogs for "Classism Exposed," a project of Class Action (www.classism.org). He has recently completed a seven-year stint as chapter president of the faculty union at Massachusetts College of Liberal Arts in North Adams, where he currently teaches. Besides continuing his research interests in social class and higher education, he is presently working on a film project focusing on globalization and change in a traditional New England mill town.

Jesse M. Smith is a doctoral candidate in the Department of Sociology at the University of Colorado, Boulder. He specializes and teaches in the sociology of religion, deviance, and identity. He is currently working toward completing his dissertation research project tentatively titled *Atheists in America: Investigating Identity, Meaning, and Movement in Contemporary Atheism*.

Joan Z. Spade is Professor Emeritus of Sociology at The College of Brockport. She is the author of several articles and books in the areas of gender, education, work, and family and is working on a study of the changing role of trustees in the governance of higher education.

Kaveri Subrahmanyam received her Ph.D. from UCLA and is currently Professor of Psychology at California State University, Los Angeles and Associate Director of the Children's Digital Media Center @ Los Angeles. Her research has examined the cognitive and social implications of interactive media use. In early work, she conducted one of the first training studies showing the effects of computer game use on spatial skills. Subsequently she studied the developmental implications of chat rooms, blogs, social networking sites, and virtual worlds, such as Second Life with a focus on the development of identity and intimacy.

She is now studying the role of interactive media in the transition to high school as well as the cognitive implications of multitasking. She has published several research articles on youth and digital media and has co-edited a special issue on social networking for the *Journal of Applied Developmental Psychology* (2008). She is the co-author (with David Smahel) of *Digital Youth: The Role of Media in Development* (Springer 2010).

Phillip Taylor is Professor and Director of Research in the School of Business at the University of Strathclyde, England. His research interests include call centers, offshoring of business services, occupational health and safety, trade union organizing, and information and consultation. He is the former co-editor of the journal, *Work, Employment, and Society*.

Jessica M. Vasquez received her Sociology Ph.D. from the University of California, Berkeley (2007) and is Assistant Professor of Sociology at the University of Kansas. She is author of *Mexican Americans Across Generations: Immigrant Families, Racial Realities* (New York University Press 2011). Her articles on race/ethnicity, Mexican Americans/Latinos, gender, and family have been published in *Ethnic and Racial Studies* (2009 article won the "Best Article Award" from the Latino/a Sociology Section of the American Sociological Association) and Sociological Perspectives. Her second book project is funded in part by the Russell Sage Foundation.

Maria Vidal de Haymes is the Lucien and Carol Matusak Professor of Social Work and Director of the Institute for Migration and International Social Work at Loyola University Chicago. Her research and teaching focuses on migration, inequality, and social policy. She is co-editor of the *Journal of Poverty*.

Monica T. Whitty is Professor of Contemporary Media in the Department of Media and Communication at the University of Leicester. Her current research is predominantly in social psychology with a focus on Internet relationships. She has conducted research on the following topic areas: Online dating, trust developed online, misrepresentation of the self online, possible selves, availability of social support online, cyberflirting and cybersex, Internet infidelity, cyberstalking, Internet privacy, security and surveillance issues in the workplace, online entertainment, cyber-ethics, and video games.

Earl Wysong is Professor of Sociology at Indiana University Kokomo. His current research interests include class analysis, organizations, intergenerational mobility, and worker access to family friendly workplace benefits. With Robert Perrucci, he is co-author of *The New Class Society: Goodbye American Dream?* (3rd edition, 2008). He is currently developing, with Robert Perrucci and David W. Wright, a substantially revised 4th edition of *The New Class Society* (forthcoming 2012).

PART I

The Sociological Vision

One of its early founders, August Comte, in developing his new approach to the study of human groups, dubbed sociology the "queen" of all the social sciences. By this he meant that sociology represented the most fully developed social scientific perspective, at the apex of all the others since, in many respects, it combines elements of the other disciplines. For instance, it builds on the understandings generated by *economics* into financial life and other aspects of commercial exchange; by *political science* into our government and official policies; by *psychology* into mental and cognitive processes; by *anthropology* into different cultures; by *history* into the large-scale trends of the past; and by *communication* into the ways we convey thoughts and information to each other, either face-to-face or through the mass media. Sociology has also served as the parent of several other "spin-off" disciplines, including *criminology* and *criminal justice*, the study of crime and our social institutions that deal with it (police, courts, jails); *gerontology*, the study of aging and the life course; *gender studies*, the examination of sex and gender; *ethnic studies*, the analysis of race and ethnicity; *urban studies*, the inquiry into life in cities; *social work*, an applied field that helps people navigate their way through social institutions; and several more that are still developing. Even *marketing*, a discipline in the field of business, is heavily influenced by sociology, by a need to know how people will act, their demographic traits, what they will like or dislike, what they will want and buy.

Positioned to both succeed its predecessors and lead its offspring, sociology's depiction, by Comte, as the "royalty" of the social sciences may have some truth. In its wide-ranging focus, one can imagine a sociology of almost anything, from the most minuscule worlds of how individuals pause to let others take a turn in a conversation, to middle-level scenes and subcultures such as hip-hop or auto thieves, to macro arenas such as the decline or evolution of the family, or the

1

global flow of markets and workers. In its substance, there can be a sociology of such diverse, idiosyncratic group behaviors as surfing, stamp collecting, religious fundamentalism, and dog breeding, as well as a sociology of large-scale trends relating to fertility rates, ethnic intermarriage, and international health care systems.

Yet, ironically, despite its omnipresence, sociology is in some ways still a "hidden" discipline. Quite possibly, you never even heard of it until you opened up your college catalogue and saw it listed. You've, no doubt, at least heard of political science through talk of voting polls, of psychology in conversations about neuroses and disorders, of communication in the news and the mass media, of anthropology in *National Geographic* or in films of foreign cultures, and of economics in the way the stock market fluctuates and the government regulates it. This lack of familiarity may be because sociology consists of an approach, a *perspective* for studying the world rather than a specific empirical focus within it. Sociology's subject matter focuses on the *trends and patterns* into which human behavior forms and the (sometimes invisible) *social forces* that mold it into these shapes. These social forces may be large-scale social institutions or policies, the *norms* (behavioral guidelines) and *values* (shared goals, desires, and beliefs) circulating within our groups, or specific, face-to-face encounters people have with others who influence them. Students of sociology, more than other disciplines, have the greatest opportunity to see their subject matter, including the sciences, the humanities, and the social sciences, in the everyday world. This sometimes leads people to think that sociology is not much more than the commonsensical, that which we experience in our daily lives. The readings presented in this book will show you that, quite to the contrary, sociology proves that things are not always what they appear to be.

In this book, we want to take you on a *sociological odyssey* through some of the fascinating and important revelations that this discipline has to offer. We want to show you what sociology has discovered about the everyday world and how you can use this perspective to keep discovering new insights into the patterns and connections of our world. Contemporary society is evolving and changing rapidly, and it is our goal to fill this book with fresh selections that keep you up-to-date with current developments that will have the most meaning and relevance to your lives.

In this first part of the book we introduce you to the "vision" of sociology, the unique perspective that differentiates it from your taken-for-granted realities. We begin by offering some writings that spell out the key elements of the sociological perspective. First, Randall Collins introduces us to the field of sociology by defining its domain of study. He shows how we can all recognize the subject matter and abstract forces of sociology in our routine, everyday experiences, and illustrates how these can inspire intellectual curiosity and satisfaction. Next, C. Wright Mills writes about the importance of recognizing the connection between concrete, everyday behavior and the larger, invisible forces of history and social structure. He suggests that a greater understanding and appreciation of the world can be gained by developing the "sociological imagination" and learning to recognize the invisible forces that surround us. Joel Charon then differentiates between a stereotype and a social scientific generalization, pointing out the dangers and flaws of the former and the careful construction of the latter. Keith M. Kilty and Maria Vidal de Haymes take us through some of the history of a sociological category to show how the concepts of race and ethnicity have changed and evolved. All of these selections are designed to alert you to this new way of sociological thinking that your professor is teaching you.

In the next section, our authors discuss the nature of theory and methods in sociology. Paul Colomy offers an explanation of the three main *theoretical perspectives* (functionalism, conflict theory, and symbolic interactionism) that form the foundation of sociology. These can be applied to help understand the larger functioning of society as a whole or the smaller workings of its subsections. Theories can help explain the "big picture" of how parts of society work separately or in interrelation, as well as the way people are guided to act within various social situations and contexts.

Macro-theories (functionalism and conflict) focus on the broad features of society. Sociologists who use this approach analyze such things as social institutions and social class to see how large entities are related to each other. They believe that to understand human behavior, we need to see how *social structure* both restricts and guides social behavior. Social structure is the framework of society that has already been laid out before you were born; it is the patterns of society, such as the relationships between men and women, education and religion, or ethnicity and politics. Critical to determining your behavior are such central sociological *demographic variables* as your ethnicity, religion, age, gender, occupational and marital status, or socioeconomic status. Most sociologists believe that the difference between people is not based on biological differences (nature) but on people's place in the social structure and what they learn there (nurture).

Micro-theories (symbolic interactionism) place the focus of study on social interaction. Here, we look at what people do when they come together and interact in a face-to-face environment. Micro-sociologists are more interested in forms of interaction, observing how people talk, gesture, and make sense of their everyday worlds. Rather than looking at the broad structures of society, they focus on the small details of social life, how people negotiate with each other, and how people's perceptions and interpretations of their world are influenced by each other.

Part I concludes with illustrations of three major methodological approaches in the social sciences (experimental design, survey research, and ethnography). Sociologists need to study the social world to learn about it, construct theories to explain how it works, and make social policies to help make it better. As social scientists, we are tied to many of the canons of science and the scientific method. However, ours is not a world of merely atoms and molecules. We study a distinctly social phenomenon: human beings. Studying the social world requires us to create methods that can capture the fullness of human behavior.

Human life is so complex, with so many variables to consider, that it is nearly impossible to study people with exact precision. Yet human behavior is also filled with patterns and regularities. We behave in fashions predictable enough for people to roughly anticipate what others will think and do. All of us are "raw" sociologists; every time we enter a scene, we take into account numerous variables about others, while at the same time we, ourselves, broadcast our own information. At a party, for instance, we make assessments about who looks interesting, who we want to talk to, or who we should stand or sit next to in a room. We classify and categorize people by their age, gender, race, class, fashion, appearance, actions, and expressions. We form expectations about the kind of people they are and how they might behave. If you are studying at a residential college, for instance, your first few days on campus were probably spent "checking out" various people to see who might likely become your friends. We also tend to act "normatively," following rules of behavior that operate under the surface of society, in such things as our seating patterns, traffic patterns, dress codes, and our rules of decorum and demeanor. In all of these respects, we think sociologically. It is our hope that with this book (and course), you may be able to hone these skills further.

As social scientists, sociologists have the problem of trying to study human behavior, even though humans are both unpredictable and highly patterned. Sociological methodology is the means by which social scientists solve this dilemma. The three methods highlighted in this book, as well as other methods (e.g., historical-comparative), offer us various systematic approaches to studying different aspects of society and its people. They offer us ways to be rigorous and controlled, broad and precise, or empathetic and deep, when studying people, so that we can invest our findings with some accuracy, validity, and reliability.

In the selections that follow, we present reflections on three major *methods* in the social sciences. Michael Lovaglia begins by discussing the logic and intent of one of the earliest methods used: the experiment. He carefully guides us through the logic of experimental design and its implementation, showing how a precise program of focus, matching, and control can help social scientists to isolate the influence of specific sociological variables on other factors. His chapter reviews some of the major experiments conducted in sociology and shows the type of knowledge that can be learned through this approach. Howard Schuman then lays out the logic of survey research in his review of that method. He traces the rise and refinement of surveys and points out the key dimensions of sampling, questionnaire design, and interviewing that need to be carefully addressed to bring them to their highest potential. Finally, in the selection on field research (also called ethnography or participant-observation), we describe the nature of fieldwork. Participant-observation offers researchers insight into people's deep, inner thoughts and feelings about what they do, shows how people and their situations change over time, and reveals the effects of multiple causal factors working together in the natural setting, where people actually live. We discuss the way this approach differs from the more objective, scientific methods and outline how subjectivity offers distinct methodological advantages. We offer precise prescriptions for how good ethnography should be done so that readers can understand how to evaluate the strength of this approach and assess the validity of ethnographic findings.

1

The Sociological Eye

RANDALL COLLINS

This short essay outlines the essence of sociology's core as a discipline. Collins, one of sociology's leading contemporary theorists, generates excitement about sociology by focusing attention on the lens, or eye, of sociology. He discusses the way it concentrates its particular attention on reality, from analyses of basic social interaction to larger social institutions and social structure, which can all be dynamically captured by the everyday life observer. Hopefully, after taking this course, you, too, will share in his excitement about the unique perspective that sociology offers. How is the sociological eye different from the commonsensical eye? What makes the sociological perspective unique?

Does sociology have a core? Yes, but it is not an eternal essence; not a set of texts or ideas, but an activity.

This is not the same as saying the discipline of sociology will always exist. Sociology became a self-conscious community only in the mid–1800s, about five generations ago, and has been an academic discipline for four generations or less. Disciplines go in and out of existence. The very concept of disciplinary specialization as we know it was created in the Napoleonic period at the time of reorganization of the French Academies, as Johan Heilbron has shown in *The Rise of Social Theory* (1995). There is no guarantee that any particular discipline will remain fixed. Biology, a discipline first recognized by Auguste Comte, has repeatedly shifted its boundaries, combining with physics and chemistry, or spinning off genetics and ecology, making up a shifting array of new fields. Discoveries do not respect administrative boundary lines. Major advances in research or theory tend to pull followers after them, who institutionalize themselves in turn for a while in some organizational form, if only until the next big round of discovery.

SOURCE: Randall Collins, "The Sociological Eye," *Contemporary Sociology* 27(1), 1998. Reprinted by permission of the American Sociological Association and the author.

In much the same way, sociologists keep forming hybrid communities on their borders, for example, with economics, literary theory, or computer science. In recent decades, hybrid disciplines have split off from, overlapped with, or encroached upon sociology as criminal justice, ethnic studies, gender studies, management, science and technology studies (i.e., what was once "sociology of science"), and no doubt more to come. There is nothing to lament in this. A glance at the history of long-term intellectual networks, and of academic organizations, shows that branching and recombining are central to what drives intellectual innovation. (The pattern of such long-term networks is documented in my book, *The Sociology of Philosophies* [1998].)

Sociology, like everything else, is a product of particular historical conditions. But I also believe we have hit upon a distinctive intellectual activity. Its appeal is strong enough to keep it alive, whatever its name will be in the future and whatever happens to the surrounding institutional forms. The lure of this activity is what drew many of us into sociology. One becomes hooked on being a sociologist. The activity is this: It is looking at the world around us, the immediate world you and I live in, through the sociological eye.

There is a sociology of everything. You can turn on your sociological eye no matter where you are or what you are doing. Stuck in a boring committee meeting (for that matter, a sociology department meeting), you can check the pattern of who is sitting next to whom, who gets the floor, who makes eye contact, and what is the rhythm of laughter (forced or spontaneous) or of pompous speechmaking. Walking down the street, or out for a run, you can scan the class and ethnic pattern of the neighborhood, look for lines of age segregation, or for little pockets of solidarity. Waiting for a medical appointment, you can read the professions and the bureaucracy instead of old copies of *National Geographic*. Caught in a traffic jam, you can study the correlation of car models with bumper stickers or with the types of music blaring from radios. There is literally nothing you can't see in a fresh way if you turn your sociological eye to it. Being a sociologist means never having to be bored.

But doesn't every discipline have its special angle on all of reality? Couldn't a physicist see the laws of motion everywhere, or an economist think of supply curves of whatever happens in everyday life? I still think sociology is uniquely appealing in this respect. What physicists or chemists can see in everyday life is no doubt rather banal for them, and most of their discoveries in recent centuries have been made by esoteric laboratory equipment. Fields like economics, it is true, could probably impose an application of some of their theories upon a great many things. But for virtually all disciplines, the immediate world is a sideshow. For sociologists, it is our arena of discovery, and the source at which we renew our energies and our enthusiasm....

All of us who are turned on by sociology, who love doing what we do, have the sociological eye. It is this that gives us new theoretical ideas and makes alive the theories that we carry from the past. The world a sociologist can see is not bounded by the immediate microsituation. Reading the newspaper, whether the business section or the personal ads, is for us like an astronomer training his or her telescope on the sky. Where the ordinary reader is pulled into the journalistic mode, reading the news through one or another political bias or schema of popular melodrama, the sociological eye sees suggestions of social movements mobilizing or winding down, indications of class domination or conflict, or perhaps the organizational process whereby just this kind of story ended up in print, defined as news. For us, novels depict the boundaries of status groups and the saga of social mobility, just as detective stories show us about backstages. Whatever we read with the sociological eye becomes a clue to the larger patterns of society, here or in the past. The same goes for the future: Today's sociologists are not just caught up in the fad of the Internet; they are already beginning to look at it as another frontier for sociological discovery.

I want to claim, in short, that all kinds of sociologists, microethnographers and statisticians, historical comparativists and theorists alike, have the sociological eye. I think that virtually all of the most productive sociologists among us do. We all went through a gestalt switch in our way of looking at the world, sometime early in our careers, that was the key moment in our initiation into sociology. Turn on the sociological eye and go look at something. Don't take someone else's word for what there is to see, or some common cliché (even a current trendy one), above all not a media-hype version of what is there; go and see it yourself. Make it observationally strange, as if you'd never seen it before. The energy comes back. In that way, I suspect, sociologists are probably more energized by their subject matter than practitioners of virtually any other discipline.

REFERENCES

Collins, Randall. 1998. *The Sociology of Philosophies*. Cambridge, MA: Harvard University Press.

Heilbron, Johan. 1995. *The Rise of Social Theory*. Minneapolis: University of Minnesota Press.

2

The Promise of Sociology

C. WRIGHT MILLS*

In a classic statement written in the 1950s, Mills points out the significance of the social context framing people and their actions. We sometimes overlook the role of larger historical and institutional factors in affecting our situations, failing to recognize the connection between the tangible, micro-level of how people act in everyday life and the invisible forces containing social trends and structures. Learning to recognize these forces is the challenge Mills poses to inquiring minds. He calls this an enriching yet frightening opportunity, for in discovering the role of the broader sweeps of history that guide and constrain us, we see ourselves as small parts of the larger microcosm. With this call to enlightenment, this essay, written by one of the premiere spokespersons of sociology, will help you develop your own "sociological imagination" and enter the intellectual domain of grand, exciting ideas. Hopefully, after taking this course, you will share his enthusiasm about the promise that sociology offers us. How is the sociological vision of society different from the commonsensical one? What makes it unique?

Nowadays men often feel that their private lives are a series of traps. They sense that within their everyday worlds, they cannot overcome their troubles, and in this feeling, they are often quite correct: What ordinary men are directly aware of and what they try to do are bounded by the private orbits in which they live; their visions and their powers are limited to the close-up scenes of job, family, neighborhood; in other milieux, they move vicariously and remain spectators. And the more aware they become, however vaguely, of ambitions and of threats which transcend their immediate locales, the more trapped they seem to feel.

---

*This introductory piece was written in 1959 before writers became sensitive to sexist language. Please excuse the use of the word "men" in this selection, where more appropriately the word "humans" should appear.

SOURCE: "The Promise," *The Sociological Imagination* by C. Wright Mills, ©1959, 2000 by Oxford University Press, Inc.

Underlying this sense of being trapped are seemingly impersonal changes in the very structure of continent-wide societies. The facts of contemporary history are also facts about the success and the failure of individual men and women. When a society is industrialized, a peasant becomes a worker; a feudal lord is liquidated or becomes a businessman. When classes rise or fall, a man is employed or unemployed; when the rate of investment goes up or down, a man takes new heart or goes broke. When wars happen, an insurance salesman becomes a rocket launcher; a store clerk, a radar man; a wife lives alone; a child grows up without a father. Neither the life of an individual nor the history of a society can be understood without understanding both.

Yet men do not usually define the troubles they endure in terms of historical change and institutional contradiction. The well-being they enjoy, they do not usually impute to the big ups and downs of the societies in which they live. Seldom aware of the intricate connection between the patterns of their own lives and the course of world history, ordinary men do not usually know what this connection means for the kinds of men they are becoming and for the kinds of history-making in which they might take part. They do not possess the quality of mind essential to grasp the interplay of man and society, or biography and history, of self and world. They cannot cope with their personal troubles in such ways as to control the structural transformations that usually lie behind them.

Surely it is no wonder. In what period have so many men been so totally exposed at so fast a pace to such earthquakes of change? That Americans have not known such catastrophic changes as have the men and women of other societies is due to historical facts that are now quickly becoming "merely history." The history that now affects every man is world history. Within this scene and this period, in the course of a single generation, one-sixth of mankind is transformed from all that is feudal and backward into all that is modern, advanced, and fearful. Political colonies are freed; new and less visible forms of imperialism installed. Revolutions occur; men feel the intimate grip of new kinds of authority. Totalitarian societies rise, and are smashed to bits—or succeed fabulously. After two centuries of ascendancy, capitalism is shown up as only one way to make society into an industrial apparatus. After two centuries of hope, even formal democracy is restricted to a quite small portion of mankind. Everywhere in the underdeveloped world, ancient ways of life are broken up and vague expectations become urgent demands. Everywhere in the overdeveloped world, the means of authority and of violence become total in scope and bureaucratic in form. Humanity itself now lies before us, the super-nation at either pole concentrating its most coordinated and massive efforts upon the preparation of World War Three.

The very shaping of history now outpaces the ability of men to orient themselves in accordance with cherished values. And which values? Even when they do not panic, men often sense that older ways of feeling and thinking have collapsed and that newer beginnings are ambiguous to the point of moral stasis. Is it any wonder that ordinary men feel they cannot cope with the larger worlds with which they are so suddenly confronted? That they cannot understand the meaning of their epoch for their own lives? That—in defense of selfhood—they become morally insensible, trying to remain altogether private men? Is it any wonder that they come to be possessed by a sense of the trap?

It is not only information that they need—in this Age of Fact, information often dominates their attention and overwhelms their capacities to assimilate it. It is not only the skills of reason that they need—although their struggles to acquire these often exhaust their limited moral energy.

What they need, and what they feel they need, is a quality of mind that will help them to use information and to develop reason in order to achieve lucid summations of what is going on in the world and of what may be happening within themselves. It is this quality, I am going to contend, that journalists and scholars, artists and publics, scientists and editors are coming to expect of what may be called the sociological imagination.

The sociological imagination enables its possessor to understand the larger historical scene in terms of its meaning for the inner life and the external career of a variety of individuals. It enables him to take into account how individuals, in the welter of their daily experience, often become falsely conscious of their social positions. Within that welter, the framework of modern society is sought, and within that framework the psychologies of a variety of men and women are formulated. By such means the personal uneasiness of individuals is focused upon explicit troubles and the indifference of publics is transformed into involvement with public issues.

The first fruit of this imagination—and the first lesson of the social science that embodies it—is the idea that the individual can understand his own experience and gauge his own fate only by locating himself within his period, that he can know his own chances in life by becoming aware of those of all individuals in his circumstances. In many ways it is a terrible lesson; in many ways a magnificent one. We do not know the limits of men's capacities for supreme effort or willing degradation, for agony or glee, for pleasurable brutality or the sweetness of reason. But in our time we have come to know that the limits of "human nature" are frighteningly broad. We have come to know that every individual lives, from one generation to the next, in some society; that he lives out a biography, and that he lives it out within some historical sequence. By the fact of his living he contributes, however minutely, to the shaping of this society and to the course of its history, even as he is made by society and by its historical push and shove.

The sociological imagination enables us to grasp history and biography and the relations between the two within society. That is its task and its promise. To recognize this task and this promise is the mark of the classic social analyst. It is characteristic of Herbert Spencer—turgid, polysyllabic, comprehensive; of E. A. Ross—graceful, muckraking, upright; of Auguste Comte and Emile Durkheim; of the intricate and subtle Karl Mannheim. It is the quality of all that is intellectually excellent in Karl Marx; it is the clue to Thorstein Veblen's brilliant and ironic insight, to Joseph Schumpeter's many-sided constructions of reality; it is the basis of the psychological sweep of W. E. H. Lecky no less than of the profundity and clarity of Max Weber. And it is the signal of what is best in contemporary studies of man and society.

No social study that does not come back to the problems of biography, of history, and of their intersections within a society has completed its intellectual journey. Whatever the specific problems of the classic social analysts, however limited or however broad the features of social reality they have examined, those who have been imaginatively aware of the promise of their work have consistently asked three sorts of questions:

1. What is the structure of this particular society as a whole? What are its essential components, and how are they related to one another? How does it differ from other varieties of social order? Within it, what is the meaning of any particular feature for its continuance and for its change?

2. Where does this society stand in human history? What are the mechanics by which it is changing? What is its place within and its meaning for the development of humanity as a whole? How does any particular feature we are examining affect, and how is it affected by, the historical period in which it moves? And this period—what are its essential features? How does it differ from other periods? What are its characteristic ways of history-marking?

3. What varieties of men and women now prevail in this society and in this period? And what varieties are coming to prevail? In what ways are they selected and formed, liberated and repressed, made sensitive and blunted? What kinds of "human nature" are revealed in the conduct and character we observe in this society in this period? And what is the meaning for "human nature" of each and every feature of the society we are examining?

Whether the point of interest is a great power state or a minor literary mood, a family, a prison, a creed—these are the kinds of questions the best social analysts have asked. They are the intellectual pivots of classic studies of man in society—and they are questions inevitably raised by any mind possessing the sociological imagination. For that imagination is the capacity to shift from one perspective to another—from the political to the psychological; from examination of a single family to comparative assessment of the national budgets of the world; from the theological school to the military establishment; from considerations of an oil industry to studies of contemporary poetry. It is the capacity to range from the most impersonal and remote transformations to the most intimate features of the human self—and to see the relations between the two. Back of its use there is always the urge to know the social and historical meaning of the individual in the society and in the period in which he has his quality and his being.

That, in brief, is why it is by means of the sociological imagination that men now hope to grasp what is going on in the world, and to understand what is happening in themselves as minute points of the intersections of biography and history within society. In large part, contemporary man's self-conscious view of himself as at least an outsider, if not a permanent stranger, rests upon an absorbed realization of social relativity and of the transformative power of history. The sociological imagination is the most fruitful form of this self-consciousness. By its use men whose mentalities have swept only a series of limited orbits often come to feel as if suddenly awakened in a house with which they had only supposed themselves to be familiar. Correctly or incorrectly, they often come to feel that they can now provide themselves with adequate summations, cohesive assessments, comprehensive orientations. Older decisions that once appeared sound now seem to them products of a mind unaccountably dense. Their capacity for astonishment is made lively again. They acquire a new way of thinking, they experience a transvaluation of values: In a word, by their reflection and by their sensibility, they realize the cultural meaning of the social sciences.

Perhaps the most fruitful distinction with which the sociological imagination works is between "the personal troubles of milieu" and "the public issues of social structure." This distinction is an essential tool of the sociological imagination and a feature of all classic work in social science.

Troubles occur within the character of the individual and within the range of his immediate relations with others; they have to do with his self and with those limited areas of social life of which he is directly and personally aware. Accordingly, the statement and the resolution of troubles properly lie within the individual as a biographical entity and within the scope of this immediate milieu—the social setting that is directly open to his personal experience and to some extent his willful activity. A trouble is a private matter: Values cherished by an individual are felt by him to be threatened.

Issues have to do with matters that transcend these local environments of the individual and the range of his inner life. They have to do with the organization of many such milieux into the institutions of an historical society as a whole, with the ways in which various milieux overlap and interpenetrate to form the larger structure of social and historical life. An issue is a public matter: Some value cherished by publics is felt to be threatened. Often there is a debate about what that value really is and about what it is that really threatens it. This debate is often without focus if only because it is the very nature of an issue, unlike even widespread trouble, that it cannot very well be defined in terms of the immediate and everyday environments of ordinary men. An issue, in fact, often involves a crisis in institutional arrangements, and often too it involves what Marxists call "contradictions" or "antagonisms."

In these terms, consider unemployment. When, in a city of 100,000, only one man is unemployed, that is his personal trouble, and for its relief we properly look to the character of the man, his skills, and his immediate opportunities. But when in a nation of 50 million employees, 15 million men are unemployed, that is an issue, and we may not hope to find its solution within the range of opportunities open to any one individual. The very structure of opportunities has collapsed. Both the correct statement of the problem and the range of possible solutions require us to consider the economic and political institutions of the society, and not merely the personal situation and character of a scatter of individuals.

Consider war. The personal problem of war, when it occurs, may be how to survive it or how to die in it with honor; how to make money out of it; how to climb into the higher safety of the military apparatus; or how to contribute to the war's termination. In short, according to one's values, to find a set of milieux and within it to survive the war or make one's death in it meaningful. But the structural issues of war have to do with its causes; with what types of men it throws up into command; with its effects upon economic and political, family and religious institutions, with the unorganized irresponsibility of a world of nation-states.

Consider marriage. Inside a marriage a man and a woman may experience personal troubles, but when the divorce rate during the first four years of marriage is 250 out of every 1,000 attempts, this is an indication of a structural issue having to do with the institutions of marriage and the family and other institutions that bear upon them.

Or consider the metropolis—the horrible, beautiful, ugly, magnificent sprawl of the great city. For many upper-class people, the personal solution to "the problem of the city" is to have an apartment with private garage under it in the heart of the city and, forty miles out, a house by Henry Hill, garden by Garrett Eckbo, on a hundred acres of private land. In these two controlled environments—with a small staff at each end and a private helicopter connection—most people could solve many of the problems of personal milieux caused by the facts of the city. But all this, however splendid, does not solve the public issues that the structural fact of the city poses. What should be done with this wonderful monstrosity? Break it up into scattered units, combining residence and work? Refurbish it as it stands? Or, after evacuation, dynamite it and build new cities according to new plans in new places? What should those plans be? And who is to decide and to accomplish whatever choice is made? These are structural issues; to confront them and to solve them requires us to consider political and economic issues that affect innumerable milieu.

Insofar as an economy is so arranged that slumps occur, the problem of unemployment becomes incapable of personal solution. Insofar as war is inherent in the nation-state system and in the uneven industrialization of the world, the ordinary individual in his restricted milieu will be powerless—with or without psychiatric aid—to solve the troubles this system or lack of system imposes upon him. Insofar as the family as an institution turns women into darling little slaves and men into their chief providers and unweaned dependents, the problem of a satisfactory marriage remains incapable of purely private solution. Insofar as the overdeveloped megalopolis and the overdeveloped automobile are built-in features of the overdeveloped society, the issues of urban living will not be solved by personal ingenuity and private wealth.

What we experience in various and specific milieux, I have noted, is often caused by structural changes. Accordingly, to understand the changes of many personal milieux we are required to look beyond them. And the number and variety of such structural changes increase as the institutions within which we live become more embracing and more intricately connected with one another. To be aware of the idea of social structure and to use it with sensibility is to be capable of tracing such linkages among a great variety of milieux. To be able to do that is to possess the sociological imagination....

3

Should We Generalize about People?

JOEL CHARON

Charon's description of sociology highlights its generalizing potential, discussing how categorizations and extrapolations are made accurately. Although generalization is an important aspect of everyday human life, people often mix stereotypes in with their empirical categorizations. Charon sensitively distinguishes between stereotyping and generalizing, showing how the inaccurate and potentially harmful characteristics of the former are replaced by measured and carefully constructed research and rigorous and systematic collection of data, which yield empirically rich, valid, and accurate information. In so doing, he offers you a guide for how to assess useful and damaging categories, and shows the careful work undergirding sociological analysis. Is generalizing, as a goal for sociology, good or bad? What is the difference between social scientists' generalizations and laypeople's stereotypes?

CATEGORIES AND GENERALIZATIONS

The Importance of Categories and Generalizations
to Human Beings

Sociology is a social science, and therefore it makes generalizations about people and their social life. "The top positions in the economic and political structures are far more likely to be filled by men than by women." "The wealthier the individual, the more likely he or she will vote Republican." "In the United States the likelihood

SOURCE: Joel Charon, *Ten Questions* 5/e, pp. 242–245, 251–256. © 2004 Cengage Learning/Wadsworth. Used by permission of the publisher.

of living in poverty is greater among the African-American population than among whites." "American society is segregated." "Like other industrial societies, American society has a class system in which more than three-fourths of the population end up in approximately the same social class as what they were at birth."

But such generalizations often give me a lot of trouble. I know that the sociologist must learn about people and generalize about them, but I ask myself: "Are such generalizations worthwhile? Shouldn't we simply study and treat people as individuals?" An English professor at my university was noted for explaining to his class that "you should not generalize about people—that's the same as stereotyping and everyone knows that educated people are not supposed to stereotype. Everyone is an individual." (Ironically, this is *itself* a generalization about people.)

However, the more I examine the situation, the more I realize that all human beings categorize and generalize. They do it every day in almost every situation they enter, and they almost always do it when it comes to other people. In fact, we have no choice in the matter. "Glass breaks and can be dangerous." We have learned what "glass" is, what "danger" means, and what "breaking" is. These are all categories we apply to situations when we enter so that we can understand how to act. We generalize from our past. "Human beings who have a cold are contagious, and, unless we want to catch a cold, we should not get close to them." We are here generalizing about "those with colds," "how people catch colds," and "how we should act around those with colds." In fact, every noun and verb we use is a generalization that acts as a guide for us. The reality is that we are unable to escape generalizing about our environment. That is one aspect of our essence as human beings. This is what language does to us. Sometimes our generalizations are fairly accurate; sometimes they are unfounded. However, we do in fact generalize: all of us, almost all the time! The question that introduces this chapter is a foolish one. *Should we generalize about people* is not a useful question simply because we have no choice. A much better question is: *How can we develop accurate generalizations about people?*

The whole purpose of social science is to achieve accurate categorizations and generalizations about human beings. Indeed, the purpose of almost all academic pursuits involves learning, understanding, and developing accurate categories and generalizations.

For a moment let us consider other animals. Most are prepared by instinct or simple conditioning to respond in a certain way to a certain stimulus in their environment. So, for example, when a minnow swims in the presence of a hungry fish, then that particular minnow is immediately responded to and eaten. The fish is able to distinguish that type of stimulus from other stimuli, and so whenever something identical to it or close to it appears, the fish responds. The minnow is a concrete object that can be immediately sensed (seen, smelled, heard, touched), so within a certain range the fish is able to easily include objects that look like minnows and to exclude those that do not. Of course, occasionally a lure with a hook is purposely used to fool the fish, and a slight mistake in perception ends the fish's life.

Human beings are different from the fish and other animals because we have *words for objects and events* in the environment, and this allows us to *understand* that environment and not just respond to it. With words we are able to make many more distinctions, and we are able to apply knowledge from one situation to the next far more easily. We are far less dependent on immediate physical stimuli. So, for example, we come to learn what fish, turtles, and whales are, as well as what minnows, worms, lures, and boats are. We read and learn what qualities all fish have, how fish differ from whales, and what differences fish have from one another. We learn how to catch fish, and we are able to apply what we learn to some fish but not other fish. We begin to understand the actions of all fish—walleyes, big walleyes, big female walleyes. Some of us decide to study pain, and we try to determine if all fish feel pain, if some do, or if all do not. Humans do not then simply respond to the environment, but they label that environment, study and understand that environment, develop categories and subcategories for objects in that environment, and constantly try to generalize from what they learn in specific situations about those categories. Through understanding a category we are able to see important and subtle similarities and distinctions that are not available to animals who do not categorize and generalize with words.

Generalizing allows us to walk into situations and apply knowledge learned elsewhere to understanding objects there. When we enter a classroom we know what a teacher is, and we label the person at the front of the room as a teacher. We know from past experience that teachers give grades, usually know more than we do about things we are about to learn in that classroom, have more formal education than we do, and usually resort to testing us to see if we learned something they regard as important. We might have also learned that teachers are usually kind (or mean), sensitive (or not sensitive), authoritarian (or democratic); or we might have had so many diverse experiences with teachers that whether a specific teacher is any of these things will depend on that specific individual. If we do finally decide that a given teacher is in fact authoritarian, then we will now see an "authoritarian teacher," and we will now apply what we know about such teachers from our past.

This is a remarkable ability. We are able to figure out how to act in situations we enter because we understand many of the objects we encounter there by applying relevant knowledge about them that we learned in the past. This allows us to intelligently act in a wide diversity of situations, some of which are not even close to what we have already experienced. If we are open-minded and reflective, we can even evaluate how good or how poor our generalizations are, and we can alter what we know as we move from situation to situation.

The problem for almost all of us, however, is that many of our generalizations are not carefully arrived at or accurate, and it is sometimes difficult for us to recognize this and change them. Too often our generalizations actually stand in the way of our understanding, especially when we generalize about human beings.

To better understand what human beings do and how that sometimes gets us into trouble, let us look more closely at what "categories" and "generalizations" are.

The Meaning of Categorization

Human beings categorize their environment; that is, we isolate a chunk out of our environment, distinguish that chunk from all other parts of the environment, give it a name, and associate certain ideas with it. Our chunks—or categories—arise in interaction; they are socially created. We discuss our environment, and we categorize it with the words we take on in our social life: "living things," "animals," "reptiles," "snakes," "poisonous snakes," "rattlers." A category is created, and once we understand it, we are able to compare objects in situations we encounter to that category. The number of distinctions we are able to make in our environment increases many fold. It is not only nouns that represent categories (*men, boys*) but also verbs (*run, walk, fall*), adverbs (*slow, fast*), and adjectives (*weak, strong, intelligent, married*). Much of our learning is simply aimed at understanding what various categories mean, and this involves understanding the qualities that make up those categories and the ideas associated with them....

The Meaning of Generalization

A category is an isolated part of our environment that we notice. We generalize about that category by observing specific instances of objects included in it and by isolating common qualities that seem to characterize those included in that category, including other yet unobserved members we might observe in the future. We watch birds build a nest, and we assume that all birds build nests out of sticks (including birds other than the robins and sparrows we observed). We continue to observe and note instances where birds use materials other than sticks, and then we learn that some birds do not build nests but dig them out. More often, our generalizations are a mixture of observation and learning from others: we learn that wealthy people often drive Mercedes, and that police officers usually carry guns. On the basis of generalizing about a category, we are able to predict future events where that category comes into play. When we see a wealthy person, we expect to see a Mercedes (or something that we learn is comparable); and when we see a police officer, we expect to see a gun. That is what a generalization is.

A generalization describes the category. It is a statement that characterizes objects within the category and defines similarities and differences with other categories. "This is what an educated person is!" (in contrast to an uneducated person). "This is what wealthy people do to help ensure that privilege is passed down to their children." "This is what U.S. presidents have in common." "This is what Catholic people believe in."...

THE STEREOTYPE

When it comes to people, generalization is very difficult to do well. The principal reason for this is that we are judgmental, and too often it is much easier for us to generalize for the purpose of evaluating (condemning or praising) others than for the purpose of understanding them. When we do this we fall into the practice of stereotyping.

A stereotype is a certain kind of categorization. It is a category and a set of generalizations characterized by the following qualities:

1. *A stereotype is judgmental.* It is not characterized by an attempt to understand, but by an attempt to condemn or praise the category. It makes a value

judgment, and it has a strong emotional flavor. Instead of simple description of differences, there is a moral evaluation of those differences. People are judged good or bad because of the category.

2. *A stereotype tends to be an absolute category.* That is, there is a sharp distinction made between those inside and those outside the category. There is little recognition that the category is merely a guide to understanding and that, in reality, there will be many individuals within a category who are exceptions to any generalization.

3. *The stereotype tends to be a category that overshadows all others in the mind of the observer.* All other categories to which the individual belongs tend to be ignored. A stereotype treats the human being as simple and unidimensional, belonging to only one category of consequence. In fact, we are all part of a large number of categories.

4. *A stereotype does not change with new evidence.* When one accepts a stereotype, the category and the ideas associated with it are rigidly accepted, and the individual who holds it is unwilling to alter it. The stereotype, once accepted, becomes a filter through which evidence is accepted or rejected.

5. *The stereotype is not created carefully in the first place.* It is either culturally and simply accepted by the individual or created through uncritical acceptance of a few concrete personal experiences.

6. *The stereotype does not encourage a search for understanding why human beings are different from one another.* Instead of seeking to understand the cause as to why a certain quality is more in evidence in a particular category of people, a stereotype aims at exaggerating and judging differences. There is often an underlying assumption that "this is the way these people are," it is part of their "essence," and there seems to be little reason to try to understand the cause of differences any further than this.

Stereotypes are highly oversimplified, exaggerated views of reality. They are especially attractive to people who are judgmental of others and who are quick to condemn people who are different from them. They have been used to justify ethnic discrimination, war, and systematic murder of whole categories of people. Far from arising out of careful and systematic analysis, stereotypes arise out of hearsay and culture, and instead of aiding our understanding of the human being, they always stand in the way of accurate understanding.

SOCIAL SCIENCE: A REACTION TO STEREOTYPES

Creating categories about people and generalizing intelligently is very difficult to do unless we work hard at it. A big part of a university education is to uncover and critically evaluate stereotypes in order to obtain a better understanding of reality. Each discipline in its own way attempts to teach the student to be more careful about categorizing and generalizing.

Because this book focuses on the perspective of sociology and social science, I would like to show how social science tries to rid us of stereotypes through the careful development of accurate categories and generalizations about human beings. Social science is a highly disciplined process of investigation whose purpose is to question many of our uncritically accepted stereotypes and generalizations. Social science does not always succeed. There are many instances of inaccuracies and even stereotyping that have resulted from poor science or from scientists simply not being sensitive to their own biases. It is important, however, to recognize that even though scientists make mistakes in their attempts to describe reality accurately, the whole thrust and spirit of social science is to control personal bias, to uncover unfounded assumptions about people, and to understand reality as objectively as possible. Here are some of the ways that social science (as it is supposed to work) aims at creating accurate categories and generalizations about human beings:

1. Social science tries hard not to be judgmental about categories of people. We recognize that generalizations and categories must not condemn or praise but must simply be guides to understanding. To stereotype is to emphasize qualities in others that we dislike or to emphasize qualities in others that are similar to our own that we like....

2. Categories and generalizations in social science are rarely—if ever—absolute. Social scientists begin with the assumption that it is difficult to generalize about people and that every time we do exceptions are likely, and often a large number. By definition, all atheists do not believe in God, but there is absolutely nothing else we can say about all atheists. However, we might contend that atheists tend to be more educated (but there are many exceptions to this), male rather than female (but there are many exceptions to this), and raised by atheist parents (but there are many exceptions to this). We can tease out generalizations about atheists from carefully studying them, but we will never find a quality that all of them have other than their belief that there is no God. This goes for every category of people we try to understand: those who commit suicide, those who abuse drugs, those who commit violent acts against children, serial killers, and students who do not finish college. We can generalize, but we must be careful, and we must assume exceptions within every category we create. The scientific generalization is treated as a probability rather than an absolute....

3. Categories in social science are not assumed to be all-important for understanding the individual. A stereotype is itself an assumption that a certain category necessarily dominates an individual's life. We might meet a young African-American single male artist. The role of each of these categories may or may not be important to the individual. For some individuals, being male or single or an artist will be most influential; for others it will be being African American. For those of us who stereotype by race, it will almost always be African American....

4. Social science tries to create categories and generalizations through carefully gathered evidence. Stereotypes tend to be cultural; that is, they are taught by people around us who have generalized based on what they have simply accepted from others or what they have learned through personal experience (which is usually extremely limited in scope, unsystematic, subject to personal and social biases, and uncritically observed). Science tries hard to encourage accurate generalizations through making explicit how generalizations must be arrived at.

5. Generalizations in social science are tentative and subject to change because evidence is constantly being examined. Stereotypes, on the other hand, are unconditionally held. Once held, a stereotype causes the individual to select out evidence that only reaffirms that stereotype. A stereotype resists change. When we believe that whites have superior abilities to nonwhites, we tend to notice only those individuals who support our stereotype. If we believe that politicians are selfish bureaucrats, we tend to forget all those political leaders who are unselfish and who get things done. (Note: the category "politician" gives away that one is stereotyping rather than simply generalizing, because politician has come to mean someone who is not worth our respect.) Because the purpose of a stereotype is to condemn or praise a category of people, it becomes difficult to evaluate evidence. The stereotype is embedded in the mind of the observer, it takes on an emotional flavor, and evidence that might contradict it is almost impossible to accept.

   A generalization in social science about a category of people is subject to change as soon as new evidence is discovered. The final truth about people is never assumed to have been found. The generalization is always taken as a tentative guide to understanding rather than a quality that is etched in stone....

6. Scientists do not categorize as an end in itself. Instead, scientists categorize because they seek a certain kind of generalization: they seek to understand cause. In social science that means we seek to know *why* a category of people tends to have a certain quality. We generalize about categories of people *to better understand what causes* the existence of qualities that belong to a given category. We seek to understand the cause of schizophrenia, but we can only do this after we understand what characterizes those people who are schizophrenic....

Real generalization in science therefore is to uncover why certain qualities make up a category, and why they are less in evidence in other categories. Why is there increasing individualism among Americans? Why do some people graduate college and not others? Why are women absent from the top political and economic positions in American life? Why is there an increase in the number of people who are experiencing downward social mobility in the United States? Why is there a rising suicide rate among young people? In every one of these cases we find a category, we describe those who make up that category, and we attempt to generalize as to why a certain quality exists in that category. To judge? No. True of everyone in the category? No. The only category of importance? No. A fixed category that clearly and absolutely distinguishes between one group and another? No. A generalization that we can regard as true without reservation? No....

Finally, we should examine once more the question that we started with: Should we generalize about people? This is not an easy question to answer.

We must begin our answer by admitting that we all take whatever we know about categories of people and apply it to situations we encounter. When we see that the individual is a child or an elderly person, male or female, single or married, a professor or a physician, wealthy or poor, kind or insensitive, that information guides us in our actions. If we are careful, we will recognize that our view of the other must be tentative, that the individual may in fact be an exception in our category, and that we must be ready to change whatever we think as we get to know that person as a unique individual. In fact the category we use may end up being unimportant for understanding this particular person.

In a society where we cry out for individual recognition, few of us will admit we want others to place us into categories and generalize. "Do not categorize me. I'm an individual!" Yet, if we are honest, we will recognize that those who do not know us will be forced to categorize us. It actually is not too bad if the category is a positive one. If we apply for a job we want the employer to categorize us as dependable, hardworking, knowledgeable, intelligent, and so on. We will even try to control how we present ourselves in situations so we can influence the other to place us in favorable categories: I'm cool, intelligent, sensitive, athletically talented, educated. The doctor may try to let people know "I am a physician" so that they will think highly of him or her as an individual. The individual who announces himself as a boxer is telling us that he is tough, the rock musician is telling us that she is talented, the minister that he or she is caring—in many such cases it does not seem so bad if we are being categorized. For almost all of us, however, it is the negative categorization that we wish to avoid. And this makes good sense: no one wants to be put into a category and negatively judged without having a chance to prove himself or herself as an individual.

But no matter how we might feel about others categorizing us and applying what they know to understanding us as a member of that category, the fact is that, except for those we know well, human beings can only be understood if we categorize and generalize. If we do this carefully, we can understand much about them, but if we are sloppy, we sacrifice understanding and end up making irrationally based value judgments about people before we have an opportunity to know them as individuals....

If we have to generalize, let's try to be careful. Stereotyping does not serve our own interests well because it blocks understanding; nor does it help those we stereotype.

4

What's in a Name?

KEITH M. KILTY AND MARIA VIDAL DE HAYMES

*Sociologists' ability to work with categories is affected by changes in social defini-
tions that occur over time and place. Terms and their definitions have changed not
only within the field of sociology but also in the larger culture. Kilty and Vidal de
Haymes examine the shifting meaning of racial and ethnic classifications as they
chart the evolution of the terms Hispanic and Latino in the U.S. Census over the
last several decades. In addition to questions about "sex," issues of "race" or
"color" have been addressed in each Census since 1790. Yet racial categorizations have changed from one time period to another. The way census forms have
been filled out with respect to race and ethnicity has also been transformed, from
interviewer-identified to self-identified. Many new categories have been added.
While these categorizations are presented as a result of innocuous demographic
processes, can you think of any other factors that might go into the changing
terminology and definitions of these categories? Can you think of any political
ramifications that might arise as a result of how these categories shift and evolve?
How do you imagine they might change in the future?*

The first national census was conducted in 1790, and has been repeated at ten
year intervals ever since. While census taking has been consistent, the way
individuals have been counted and categorized on the basis of race and ethnicity
has varied over time. This paper examines how the official census definition of
Latinos has changed over the twenty-two census periods. The modifications of
the official definition of this group are discussed in relation to changes in national
borders, variations in methodology used for census data gathering, and shifting
political contexts.

One of the most common scientific activities is classifying different objects
into categories. In fact, classification is not only a scientific activity but a common

SOURCE: Keith Kilty and Maria Vidal de Haymes, "What's in a Name?" *Ethnic Studies Review* 27(1),
2004. © National Association of Ethnic Studies. Reprinted by permission.

human activity in general. "Sorting things out," as Bowker and Star (1999) put it, can be quite informal and part of our everyday lives as well as more formalized and organized as part of the scientific enterprise. In fact, classification is fundamental to science. Observation and measurement require at a minimum the identification of categories to describe concepts, with the categories of a particular concept defined in such a way as to allow for non-ambiguous sorting of objects into categories. That is the nature of the most basic level of measurement, the nominal, where all objects must be classified into one (and only one) category, and all objects must be accounted for by a category....

Hispanics or Latinos are currently recognized as one of the most rapidly growing nonwhite groups in this country. But what exactly is a Hispanic or Latino? The Bureau of the Census identifies Hispanics as an "ethnic" group whose members can be of any "race." At one time, though, it also identified Mexicans as a racial group.... Yet what does it mean to be Hispanic? Who becomes labeled and why?

CLASSIFYING (AND DEFINING) RACE
AND ETHNICITY

For the most part, race is no longer identified by social scientists as a biological variable. In fact, some argue that race is no longer as meaningful a force in American life as it once was, and there is an ambivalence on the part of many people, particularly whites, to deal with race. Other social scientists have taken to using the term "ethnicity" in place of race. In a way this may be a broadening of the concept and a return to the nineteenth and early twentieth century when race itself was used in reference not only to skin color but also to such characteristics as nationality. Use of ethnicity as a substitute for race may also be a way for whites, whether social scientists or ordinary citizens, who are uncomfortable with race and racism to downplay the power of those forces in modern life. As Nobles (2000: 14) points out, "Scholars treat ethnic categorization as a benignly descriptive marker—albeit one sometimes used for politically objectionable ends—but race has always had political meanings and uses."

All the same, most members of this society still identify race (whether labeling it as "race" or as "ethnicity") as a social construct that has some continuing significance in people's lives. The question is the extent to which race is important. Most whites have accepted the idea that racism is largely a thing of the past. Yet, as Feagin, Vera, and Batur (2001: 13) note, "The substantial white consensus on the decline of racism is not based on empirical evidence." People in this society look upon and act toward each other based on their perceptions of other people in terms of racial and ethnic categories. Media images of certain groups as "criminal" or "welfare cheats" have affected social policy in terms of sentencing provisions for possession of particular illicit substances and changes in public assistance legislation. Racism and discrimination cannot exist without the acceptance of racial classification. Clearly, there is a broadly based consensus on specific racial categories in the U.S. and, for better or worse, an acceptance of these categories as having an "objective" reality, as being "natural."...

Since the first U.S. census of 1790, racial (and ethnic) categories have been important facets of the information collected by the Census Bureau. Yet these categories themselves have helped to define what race is in America. For the first six censuses, the primary categories were White and Black. Then Mulatto, the mix of White and Black, started to show up, as well as Native Americans and particular kinds of Asians (e.g., Chinese). By the twentieth century, a wide array of racial categories was included in the census, but, as noted, within the context of racial theory of the time. As Nobles (2000) shows, the census classification systems helped to create the concept of race and what it means in this society.

Census authorities deny what Nobles is arguing: that they are engaged in creating definitions of race. They assert that they are merely applying objective categories for statistical and programmatic purposes. Their statements reflect institutional racism, which refers to institutional forms or practices that systematically benefit one racial group over another. Institutional racism can be conscious and deliberate, but it can also be unconscious and indirect. Those in the dominant group often fail to see that their practices are discriminatory and may even deny that discrimination exists. Sentencing provisions, for example, for possession of cocaine vary substantially depending on the form of cocaine (crack vs. powder), which has led to racial bias in prison sentences between Blacks and Whites. Yet the U.S. Sentencing Commission has flatly denied racial bias, arguing instead that there is simply an appearance of bias.

THE MEANING OF HISPANIC AND LATINO

What it means to be white or black or red or yellow or brown in this society is at least partly constructed by the social scientists and bureaucrats and public officials who endorse and carry out the census. Part of the experience of a particular minority of color, then, is shaped by these "official" definitions of who we are and who we are not. In fact, part of that experience is reflected in whether one is even defined. The numbers of Hispanics or Latinos generally remained low in census counts until the 1960s. Is that, indeed, due to small numbers, or to the ways in which population groups have been identified, labeled, and counted?

Hispanics occupy a unique place in the racial and ethnic history of the United States. Unlike American Indians, they were not treated as a menace that needed to be removed. Unlike Africans, they were not formally enslaved. Unlike Asians, they were not legally excluded or denied citizenship. For the most part, they represent a group that was acquired through imperialism and empire-building, particularly the conquest of the American southwest. The Treaty of Guadalupe Hidalgo, which concluded the Mexican-American War in 1848, ceded the northern half of Mexico to the United States. That treaty granted all residents of the area U.S. citizenship and continued rights of ownership of their lands. While the terms of the treaty were never enforced, questions about formal citizenship never were raised, nor was there concern about passage back and forth across the U.S.-Mexican border until the Great Depression era. Future land acquisitions by the U.S. involving territories inhabited by Latinos were also generally resolved by granting the peoples of those occupied territories formal independence (e.g., Cuba and the Philippines) or U.S. citizenship (e.g., Puerto Rico). That certainly does not mean that Hispanics were thought of as the equals of Anglos. For the most part, they were treated as different—as outsiders and inferiors....

The experiences of any racial or ethnic group, then, are shaped by how they are defined. Racial categories not only vary from one society to another, but they can also vary from one era to another within the same society. Looking at the racial categories identified in U.S. censuses since 1790 clearly shows such change (Nobles 2000). Currently, there is a racialization process occurring regarding Hispanics, where racialization refers to the attachment of racial meaning to a group. While such meaning may be imposed upon a group, it may also be sought by members of the group. In response to demands from various advocacy groups, the U.S. Office of Budget and Management allowed a public review of the 1977 Statistical Directive No. 15 beginning in 1993. This led to a revision in the race and ethnicity categories in the 2000 census (Nobles 2000), with the addition of "other race" and "multiple-race" categories. Latinos were particularly responsive to these changes:

> In accordance with a well-established pattern, 42% of Latinos identified themselves as 'other race,' and 97% of all respondents who declared themselves 'other race' were Latino—a significant trend not emphasized in the press. In addition, 6% of Latinos took advantage of the new 'multiple race' option, compared with only 2% of the non-Hispanic population. In fact, of all the multirace combinations made possible by the new option, the most common was 'white' and 'some other race,' which census officials said was checked mainly by Hispanics. (Morales 2001)

To understand what it means to be Hispanic or Latino, one needs to understand how a group has been defined within the context of a particular society. One way that a group is defined is through some "official" process, such as how it is codified by a census bureau. That is the focus of this paper, how the U.S. Census Bureau has (or has not) defined "Hispanic" and/or "Latino" during the past two centuries. Categorizing members of a society and then counting (or not counting) them clearly affects their sense of group identity and citizenship....

CLASSIFYING LATINOS: LATINOS AND THE U.S. CENSUS

While Whites, Blacks, and Native Americans have been counted in one way or another, albeit imperfect and incomplete, in each census since its inception, and some Asians since 1870, Latinos have not been counted in a systematic and uninterrupted way. In 1930 "Mexicans" were counted. Enumerators were offered the following choices in response to the item "Color or race": White, Negro, Mexican, Indian, Chinese, Japanese, Filipino, Hindu, or Korean. Instructions to enumerators read "In order to obtain separate figures for Mexicans, it was decided that all persons born in Mexico or having parents born in Mexico who were not definitely White, Negro, Indian, Chinese, or Japanese, would be returned as Mexicans" (U.S. Department of Commerce 1979: 52). In the 1940 census "Mexican" was dropped from the list of responses, and enumerators were instructed to classify Mexicans as White unless they were "definitely Indian or some race other than White" (U.S. Department of Commerce 1979: 61).

In 1960 these instructions were expanded: "Puerto Ricans, Mexicans, or other persons of Latin descent should be classified as 'White' unless they were definitely Negro, Indian, or some other race" (U.S. Department of Commerce 1979: 70)....

Effective with the Census 2000 the Revised Standards for the Classification of Federal Data on Race and Ethnicity changed the aforementioned 1977 Statistical Directive No. 15 of the Office of Management and Budget (OMB), Office of Information and Regulatory Affairs in several significant ways. First, the revised standards detailed five categories for data on race: American Indian or Alaskan Native, Asian, Black or African American, Native Hawaiian or Other Pacific Islander, and White. The Bureau of the Census was also directed to allow for the reporting of more than one race. Furthermore, two categories for data on ethnicity were specified: "Hispanic or Latino," and "Not Hispanic or Latino." The OMB defined Hispanic or Latino as "a person of Cuban, Mexican, Puerto Rican, South or Central American, or other Spanish culture or origin, regardless of race." For the purposes of the 2000 Census and a few other federal data collection activities, the OMB approved the inclusion of a sixth racial category, "some other race," for respondents unable to identify with any of the five preceding categories (U.S. Department of Commerce 1993: 15)....

OFFICIAL CLASSIFICATION AND
LATINO IDENTITY

One of the most remarkable findings from the Census 2000 was the tremendous growth of the Latino population. National Census figures indicate that the Latino population grew by 58% since the previous census, reaching 35.3 million by the time of the 2000 census. Just two years later, the Latino population had grown by an additional 9.8%, reaching 38.8 million or 13.4% of the total U.S. population (Ramirez and de la Cruz 2002). This number marked the Latinos as the nation's largest minority community, a demographic shift that many are predicting will have broad implications for the political landscape of the U.S. True to Goldberg's (1995) predictions, political jockeying to gain the Latino vote has heightened. At issue here is how the "Latino" vote is seen by various groups, including the dominant "white" or "Anglo" group in this country or by other minorities who may attempt to develop coalitions.

What it means to be Hispanic or Latino is a complex issue. The numbers seem to have skyrocketed in the recent past. In 1930, 1.3 million Mexicans were reported in the U.S. census, while 2.3 million "persons of Spanish surname" were reported in the 1950 census. Then in 1970, 9.1 million were counted, followed by 14.6 million in 1980 and 22.4 million in 1990 (U.S. Department of Commerce 1993). The 2000 census reported 35.3 million Hispanics (U.S. Department of Commerce 2001).

One of the critical questions here, though, is the extent to which Hispanics have been counted accurately in the past. By not being specifically identified, it would be a simple matter for authorities (and most members of the dominant racial/ethnic group) to ignore them—and therefore not to count them at all. Throughout the first century of their presence in the U.S. (i.e., from 1850 to 1950), there seems to have been little formal identification by the Census Bureau. Yet as we showed earlier, Anglos clearly saw and related to Hispanics in the U.S. Southwest (formerly the northern half of Mexico) in racial terms, i.e., in a discriminatory manner. Not identifying makes it easy to ignore a group, potentially limiting group awareness and the emergence of group solidarity, while ensuring institutionalized biased treatment.

REFERENCES

Bowker, Goeffrey C. and Susan Leigh Star. 1999. *Sorting Things Out: Classification and Its Consequences.* Cambridge, MA: MIT Press.

Feagin, Joseph, Hernan Vera, and Pinar Batur. 2001. *White Racism.* (2nd ed.). New York: Routledge.

Goldberg, David T. 1995. "Made in the USA." pp. 237–56 in N. Zack (ed.), *America Mixed Race: The Culture of Microdiversity.* Lanham, MA: Rowman & Littlefield.

Morales, P. 2001. "Latinos and the 'Other Race' Option: Transforming U.S. Concepts of Race." *NACLA Report on the Americas,* XXXIV(6): 40–44.

Nobles, Melissa. 2000. *Shades of Citizenship: Race and the Census in Modern Politics.* Stanford, CA: Stanford University Press.

Ramirez, Roberto R. and Patricia de la Cruz. 2002. "The Hispanic Population in the United States: March 2002, Current Population Reports," pp. 20–545. Washington, DC: U.S. Census Bureau.

U.S. Department of Commerce. 1979. "Twenty Censuses: Population and Housing Questions 1790–1980." Washington, DC: U.S. Census Bureau, Economics and Statistics Administration, U.S. Department of Commerce.

_____. 1993. "We the American Hispanics." Washington, DC: U.S. Census Bureau, Economics and Statistics Administration, U.S. Department of Commerce.

_____. 2001. "The Hispanic Population." (Census 2000 Brief, C2KBR/01-2). Washington, DC: U.S. Census Bureau, Economics and Statistics Administration, U.S. Department of Commerce.

5

Three Sociological Perspectives

PAUL COLOMY

In this chapter, Colomy overviews the three perspectives most often associated with sociology: functionalism, conflict theory, and symbolic interactionism. Each of these three theories has contributed a great deal to our understanding of human behavior and group life. The first two, functionalism and conflict, are considered to be macro theories of society, examining how the social structure operates to determine people's behavior. The last theory, symbolic interactionism, is a micro theory of society, focusing on social interaction and how people act in face-to-face meetings with one another. Each theory has its strengths and weaknesses as an explanatory scheme, but taken together, you will see the power of sociological reasoning. From this article, can you differentiate the major foundational ideas that the theories represent? Do you have other ideas not represented by these three approaches that account for social behavior?

When conducting research, sociologists typically draw on one or more perspectives. Sociological perspectives provide very general ways of conceptualizing the social world and its basic elements. A perspective consists of a set of fairly abstract assumptions about the nature of human action and the character of social organization. Each perspective can be likened to a spotlight that brightly illuminates select aspects of behavior and social relations while leaving other areas shrouded in darkness. Because a single perspective supplies only a partial or one-sided view, a comprehensive understanding of social life requires becoming familiar with several different perspectives.

Sociology contains a large number of distinct perspectives, and they can be divided into two broad categories: micro and macro. In very general terms, micro perspectives are oriented toward small time and small space, while macro perspectives are oriented toward big time and big space (Collins 1981). That is, micro perspectives are usually concerned with the conduct of individuals and

SOURCE: Paul Colomy, "Three Sociological Perspectives." Reprinted by permission of the author.

small groups as it unfolds in relatively small spatial contexts and over short dura-
tions of time. Macro perspectives, on the other hand, focus on larger entities—
not individuals and small groups, but institutions, entire societies, and even the
global system—and on how these entities emerge, maintain themselves, and
change over decades, centuries, and millennia. The following section outlines
one micro perspective (symbolic interactionism) and two macro perspectives
(functionalism and the conflict approach).

SYMBOLIC INTERACTIONISM

Symbolic interactionism's intellectual roots reside in pragmatism, a philosophical
tradition developed by such prominent, early twentieth-century American thin-
kers as John Dewey, William James, George Herbert Mead, and Charles Peirce.
The sociological implications of pragmatism were articulated by several innovative sociologists, including Robert Park, W. I. Thomas, Herbert Blumer, Everett
Hughes, and Erving Goffman, who taught or studied at the University of Chicago
between 1910 and 1960. Because it originated at the University of Chicago, symbolic interactionism is sometimes referred to as the Chicago School.

Symbolic interactionism is based on five core ideas. First, it assumes that
human beings act in terms of the meanings they assign to objects in their environment. (Interactionists define the term object very broadly to include material
things, events, symbols, actions, and other people and groups.) Using slightly different terminology to make the same point, interactionists maintain that people's
conduct is powerfully influenced by their definition of the situation. This
assumption can be clarified by contrasting it to a rudimentary model of social
action advanced by a psychological perspective known as behaviorism. The
behaviorist approach characterizes conduct as a response to objective stimuli,
and suggests that human behavior resembles a series of stimulus-response chains:

$$\text{stimulus} \rightarrow \text{response}.$$

Rejecting the notion that individuals respond directly to an objective stimulus, interactionists insist that people interpret, or assign meanings to, the stimulus
before they act:

$$\text{stimulus} \rightarrow \text{interpretation} \rightarrow \text{response}.$$

Athletes' reactions to coaches' criticisms, for instance, depend largely on
whether they interpret that criticism as a constructive attempt to improve their
play or as a malicious attack on their character.

Even when a definition of the situation is demonstrably false, it can still exert
a powerful effect on behavior. As W. I. Thomas once said, "A situation defined
as real is real in its consequences." Many adults, for example, perceive Halloween
as filled with potential danger, and believe that their young children are vulnerable to sadistic strangers dispensing drug-tainted candy or apples laced with razor
blades. The belief that such acts of Halloween sadism are widespread is, in fact,
an urban legend with virtually no factual basis (Best and Horiuchi 1985).

Nevertheless, millions of parents are convinced that the threat is genuine and, acting in terms of their definition of the situation, continue to inspect their children's treats for signs of tampering.

Symbolic interactionism's second assumption asserts that social action typically involves making a series of adjustments and readjustments as an individual's interpretation of the situation changes. Interactionists reject the notion that behavior is the unmediated product of a variable or cause. Instead, they view action as something that is continually being built up, modified, re-directed, and transformed (Blumer 1969). People's initial definition of the situation is always subject to change, and as they redefine the situation their conduct changes accordingly. Effective teachers, for example, routinely interpret students' comments, facial expressions, and other gestures to determine whether the subject matter is being communicated clearly. They rely on this feedback to define and redefine the unfolding classroom situation and to make corresponding adjustments in their presentations. When students look confused, they may introduce a familiar example; if students' attention should wander, the instructors may call on them; and if students are visibly upset, they may ask to meet privately with them after class.

Third, interactionists assume that the meanings imputed to an object are socially constructed (Berger and Luckmann 1966). Meanings do not, in other words, simply reflect a quality or essence built into the very nature of an object. Other than its size and color, the cloth used to make handkerchiefs is virtually identical to that used to produce American flags. Though handkerchiefs and flags are sewn from the same physical material, the meanings attached to these two objects differ in dramatic ways. Rather than being intrinsic to an object, then, meanings are attributed to it by individuals, groups, and communities.

Elaborating this logic, some interactionists treat the self as an object whose meanings are socially constructed. In other words, the kind of person you assume yourself to be, and that others take you to be, mirror the meanings that individuals and groups have assigned to you. If, from a young age, family members, friends, and teachers have said you were "brilliant" and have acted toward you in a manner consistent with that characterization, then one assumption you are likely to make about yourself—one meaning that you are likely to assign to yourself—is that you are a highly intelligent person.

Fourth, symbolic interactionism holds that in modern, heterogeneous societies, different groups often assign divergent meanings to the "same" object. Contemporary societies contain a wide variety of groups (e.g., occupational, religious, age-based, racial and ethnic groups). Since group members interact and communicate frequently with one another, they tend to develop a common "universe of discourse" (Mead 1934) or shared meanings about the objects comprising their social world. Not surprisingly, discrepancies are likely to arise between the distinctive systems of meanings devised by different groups. Parents and adolescents, for instance, commonly attach opposing meanings to curfews, underage drinking, and body piercing.

Discrepant meanings can be a significant source of social conflict, with rival groups mobilizing to insure that their definition of the situation is officially acknowledged and enforced by the larger society. A classic interactionist study discovered that Prohibition was largely a battle over the meanings assigned to drinking (Gusfield 1963). In the early twentieth century, small-town, middle-class, WASPS (white, Anglo-Saxon, Protestants) regarded drinking as sinful, while the working-class and largely Catholic immigrants from Southeastern Europe who settled in the nation's largest cities viewed drinking as an integral part of everyday life. More adept at organizing and lobbying politicians, the small-town Protestants succeeded in inscribing their interpretation into law: the Eighteenth Amendment, approved by Congress in 1919, outlawed (at least for a time) the manufacture and sale of alcoholic liquors.

Fifth, established meanings are always subject to transformation, and interactionists maintain that the emergence and diffusion of novel definitions of reality are a critically important feature of social change. At any given time, the meanings attached to some objects and practices are so entrenched that they appear natural and beyond question. Behavior that deviates sharply from these prescribed meanings is regarded as threatening, immoral, and even a little crazy. Interactionists examine how social movements, broad cultural shifts, and/or deviant individuals and groups sometimes challenge long-standing meanings and replace them with alternative conceptions of reality. From an interactionist perspective, one of the most significant consequences of the feminist movement is its redefinition of what it means to be a woman. A generation or two ago, it was simply assumed, particularly by members of the middle class and by women as well as men, that a woman's "proper place" was in the home taking care of her household, husband, and children. In the late 1960s and throughout the 1970s, feminists questioned this assumption and ushered in a strikingly different conception of women, one that affirmed a woman's right to work outside the home and to be treated as an equal, in every respect, to her male colleagues. Today, feminism's once novel and radical definition of reality has been institutionalized, and the earlier view, which at the time was widely accepted and regarded as obvious and commonsensical, has now been redefined as an arbitrary infringement on women's freedom.

FUNCTIONALISM

Functionalism is a macro perspective that examines the creation, maintenance, and alteration of durable social practices, institutions, and entire societies. Emile Durkheim, a great French sociologist who published several provocative books between 1890 and 1915, is often regarded as the classic founder of functionalism. This approach was articulated most forcefully, however, during the twenty-five years between 1945 and 1970 by a group of American sociologists, most of whom were trained at Harvard or Columbia. Key figures in this group include Robert Bellah, Robert Merton, Wilbert Moore, Talcott Parsons, Neil Smelser, and Robin Williams.

Functionalism assumes, first, that societies can be likened to problem-solving entities. If a society is to persist, functionalists argue, it must address a large (but not infinite) number of problems in a reasonably satisfactory way. (Functionalists sometimes refer to these problems as requirements, functions, prerequisites, or functional prerequisites.) An enduring society must, for example, socialize its youngest members, distribute food and other goods and services, and devise mechanisms to control deviance and contain conflict. If a society does not satisfactorily address these (and other) problems, it will experience considerable strain, and if its failure to address these problems continues, it will collapse.

Extending the metaphor that likens societies to problem-solving entities, functionalism portrays persisting practices and institutions as answers or solutions to the kinds of problems mentioned above. Customary practices and institutions are established to meet the problems every society must confront. Families and schools, for instance, are institutions that arise to answer the problem of socializing and educating the young; free markets, on the other hand, are created to address the problem of producing and distributing goods and services, while police and prisons are mechanisms for controlling deviance and containing conflict.

Second, functionalists assume that during the course of human history societies have developed many different answers to basic needs. This assumption can be termed the principle of institutional alternatives. In traditional societies, for instance, the extended family had sole responsibility for supporting dependents, whether they be very young or very old. In many modern societies, however, insurance policies, pensions, social security, and welfare programs share responsibility for the problem of caring for dependents. From a functionalist perspective, these programs are institutional alternatives to the extended family.

The principle of institutional alternatives implies that any single functional prerequisite can be met in many different ways. Many functionalists argue, however, that in an attempt to address prerequisites more efficiently and more effectively, modern societies have increasingly replaced multi-functional institutions with more specialized ones (Parsons 1977). Two hundred years ago, the family was a multi-functional institution in that it assumed primary responsibility for many different tasks, such as economic production, procreation, socialization, care for the infirm elderly, and social control. Today, however, many of these problems have been delegated to specialized institutions. Economic production, for instance, is no longer addressed by the family but by business enterprises located (for the most part) outside the home, while the control of deviance is a problem for specialized social control agents like the police and criminal courts. The family, too, has become a more specialized institution, one whose primary tasks include procreation, socialization of the very young, and emotional support for family members.

Third, functionalism presumes that the particular practices and institutions that arise in response to one problem have crucial repercussions for the practices and institutions devised to address other problems. A society, in other words, can be viewed as a system of practices and institutions. The notion that persisting practices or institutions are part of a larger system has led some functionalists to develop a distinctive protocol for studying the inter-relations between different parts of social systems. Referred to as functional analysis (Merton 1968), this method examines the effects a practice or institution has on other institutions and on the larger society. These effects or consequences assume four principal forms. Manifest functions refer to the consequences or objectives an institution explicitly attempts to achieve. Universities, for instance, are designed to impart knowledge and skills that will enable students to become productive workers and thoughtful citizens. Latent functions, on the other hand, identify effects that typically go unnoticed by the general public and frequently appear unrelated or starkly incompatible with an institution's (or the larger society's) explicit objectives. For example, while many citizens routinely denounce crime, functionalists argue that it—or, more specifically, the condemnation crime provokes—has the important latent function of clearly defining and affirming a community's normative boundaries (i.e., its sense of right and wrong). Prisons, too, have a latent function: by serving as "schools of crime," they insure that many ex-prisoners will commit new crimes which, in turn, will elicit still more community outrage and additional affirmation of a society's moral code.

In addition to being either manifest or latent, the functions served by a practice or institution can be positive or negative. Positive functions are evident when an institution facilitates the operation of other institutions and/or contributes to the overall stability and effectiveness of the larger society. In this vein, Davis and Moore (1945) hypothesize that attaching unequal financial and social rewards to different occupations has the positive effect of attracting the most talented and qualified individuals to a society's most "functionally important" positions. Negative consequences, which are sometimes called dysfunctions, occur when a practice or institution impedes the operation of other institutions and/or produces instability. For example, the "soft money" donated by corporations and other large contributors to political campaigns fosters the perception that contemporary politics is corrupt and convinces many citizens that their votes "don't count." That perception, moreover, is partially responsible for shockingly low rates of voter turnout.

Fourth, functionalism suggests that in contemporary societies containing scores of specialized institutions and hundreds of heterogeneous sub-groups, societal integration is a recurring but manageable problem. In modern social orders, societal integration is achieved in two primary ways. First, specialized integrative institutions and processes—e.g., religious ceremonies, athletic contests, media events, and nationally celebrated holidays—heighten cohesion among people who otherwise share little in common. Second, consensus or agreement on such core values as individualism, freedom, achievement, and equal opportunity also serves to integrate complex societies. Incorporated into different institutions and internalized by individuals (during the course of their early socialization), shared values enable the diverse components of a large, differentiated society to co-exist and bond rather than dissolve into chronic chaos or a "war of all against all." According to one prominent functionalist, the relative stability of American society over the last two hundred years (the Civil War being a glaring exception) is largely attributable to the continuing consensus on the values of achievement and liberty (Lipset 1979).

Fifth, functionalism asserts that deviance and conflict arise from social strains, or contradictions within an institution or between institutions. That is, the primary source of contention and crime are inconsistencies inherent in the social system itself. In an influential essay, Robert Merton (1938) contends that in American society everyone, regardless of his or her station in life, is encouraged to pursue the American Dream. At the same time, however, the institutional means (e.g., a quality education and well-connected friends or acquaintances on the job market) for attaining success are not equally distributed: middle- and upper-class people are, in general, much more likely to have access to these institutional means than are working-class people. Confronted with this contradiction between a cultural goal (i.e., success) and the institutional means (e.g., a quality education) to achieve it, some individuals and groups will turn to crime (e.g., selling drugs). Note that in Merton's terms, crime often involves using "innovative," illegal means to realize a cultural goal prized both by criminals and law-abiding citizens. Under certain conditions, the same contradiction between cultural goals and institutional means can prompt widespread rebellion, with various groups replacing established cultural goals and the standard, institutional means with radically different values and means.

THE CONFLICT PERSPECTIVE

Like functionalism, the conflict perspective is a macrosociological approach that examines the emergence, persistence, and transformation of long-standing practices, institutions, and societies. Karl Marx, whose work first appeared in the mid-nineteenth century, is usually credited with crystallizing the key principles of this approach. Max Weber, an early twentieth-century German sociologist, is also recognized as a founding figure of conflict sociology. Leading contemporary conflict theorists include William Chambliss, Randall Collins, Ralf Dahrendorf, William Domhoff, and C. Wright Mills.

The conflict perspective rejects the functionalist notion that societies can be accurately portrayed as problem-solving entities. It also disputes the complementary idea that long-standing practices and institutions represent reasonably satisfactory answers to problems and as such contribute to a society's general welfare. Conflict sociologists embrace a very different orienting assumption: societies are arenas in which groups with fundamentally antagonistic interests struggle against one another. Different theorists within this tradition differ in terms of which particular groups and struggles they emphasize. Marx, for instance, highlights the conflicts between social classes, while Weber focuses on competing status groups (e.g., racial, ethnic, religious, age-based, etc.) and Dahrendorf and Collins draw attention to the battles between those who have authority and those subject to that authority. Despite disagreements about which groups and struggles are most important, all conflict sociologists believe that the interests that divide groups (whether classes, status groups, and so on) are built into the very fabric of a social order; these opposing interests are not readily negotiated, compromised, or resolved, nor can they be wished away or papered-over.

Second, conflicts among classes, status groups, and between those exercising authority and those subject to it supply the energy and the motivation for constructing and maintaining (as well as challenging and transforming) practices and institutions. Platt's (1977) well-known study of the origins of the juvenile court, for example, contends that this institution was created (in 1899) by social and economic elites and was employed to target and control the children of working-class immigrants residing in large cities. Conflict sociologists argue that today the nation's newly constructed maximum security prisons are, in practice, reserved predominantly for young, minority males raised in inner-city areas where good jobs are scarce (Chambliss 1999). On the other hand, white-collar, middle-class criminals, if they receive a prison sentence at all, are rarely housed in these types of facilities.

Third, the conflict perspective characterizes on-going practices and institutions as structures of domination that promote the interests of a relatively powerful, superordinate group while subverting the interests of relatively powerless, subordinate groups, even though the latter are usually much larger, numerically, than the powerful elites. Consequently, this perspective's orienting question is: which group's interests are served by a specific practice or institution? Kozol's (1991) investigation of how public schools are funded found that schools located in well-to-do suburban areas receive substantially more support than inner-city schools, which often lack textbooks, desks, and even serviceable plumbing. Far from enabling students from economically disadvantaged backgrounds to compete fairly on a level playing field, the current school system simply reflects and reproduces existing class inequalities.

Fourth, the conflict perspective reconceptualizes what functionalism terms values as ideologies. The primary purpose of an ideology is to protect and promote the distinctive interests of a particular class (or status or authority group). This legitimating purpose is best served when the ideology is presented in universal terms; when its ideas are stated as if they apply to everyone equally. According to the conflict approach, achievement and equal opportunity are most accurately viewed not as widely shared values but as a dominant ideology that operates to preserve (and reproduce) existing systems of inequality. In essence, the ideology of achievement and equal opportunity asserts that individuals and groups with great wealth, prestige, and power are rightfully entitled to these rewards because they have sacrificed, worked hard, and/or displayed exceptional talent. This ideology also explains why many people have few or none of these rewards: they are lazy, unwilling to make the sacrifices necessary for success, and/or lack the requisite talent. This ideology justifies the unequal distribution of social rewards by referring to individuals' character and moral virtues (or lack thereof). At the same time, it draws attention away from the structural inequities that largely explain why members of some groups are much more likely to "succeed" than are members of other groups.

Fifth, the conflict approach holds that significant social change usually reflects the efforts of groups mobilizing to advance their collective interests, often at the expense of other groups' interests. In this regard, proponents of conflict sociology question the functionalist claim that the substantial financial rewards enjoyed by physicians are due to the fact that medical doctors perform tasks that are, objectively, of great functional importance to society. Physicians' impressive incomes are more persuasively explained, conflict sociologists contend, with the observation that the American medical profession has established, in effect, a monopoly on the provision of health care (Starr 1982). Prior to 1850, this monopoly did not exist, and physicians were poorly paid and given little esteem. After the Civil War, however, doctors began to organize in earnest, and by the late nineteenth century they secured legislation prohibiting other groups and individuals from providing health care. The exclusion of competitors paved the way for a remarkable surge in doctors' income and a parallel rise in their prestige.

CONCLUSION

Each of the readings in this book employs one or more of the perspectives outlined above. The authors of these readings, it must be acknowledged, are not always fully explicit about which perspective (or combination of perspectives) they have used. Nevertheless, the perspectives they draw on, implicitly or explicitly, powerfully inform how they formulate the empirical problem under investigation and the type of explanation they devise to account for it. By keeping the basic assumptions of symbolic interactionism, functionalism, and the conflict perspective in mind, you will acquire a deeper and more complete understanding of the chapters that follow.

REFERENCES

Berger, Peter and Thomas Luckmann. 1966. *The Social Construction of Reality*. New York: Doubleday.

Best, Joel and Gerald Horiuchi. 1985. "The Razor Blade in the Apple: The Social Construction of Urban Legends." *Social Problems* 32: 488–499.

Blumer, Herbert. 1969. *Symbolic Interactionism*. Englewood Cliffs: Prentice-Hall.

Chambliss, William J. 1999. Power, Politics, and Crime. Boulder: Westview Press.

Collins, Randall. 1981. "On the Microfoundations of Macrosociology." *American Journal of Sociology* 86: 984–1014.

Davis, Kingsley and Wilbert E. Moore. 1945. "Some Principles of Stratification." *American Sociological Review* 10: 242–249.

Gusfield, Joseph R. 1963. *Symbolic Crusade*. Urbana: University of Illinois Press.

Kozol, Jonathan. 1991. *Savage Inequalities*. New York: HarperCollins.

Lipset, Seymour M. 1979. *The First New Nation*. New York: W. W. Norton.

Mead, George H. 1934. *Mind, Self and Society*. Edited by Charles W. Morris. Chicago: University of Chicago Press.

Merton, Robert K. 1938. "Social Structure and Anomie." *American Sociological Review* 3: 672–682.

_____. 1968. *Social Theory and Social Structure*. Enlarged edition. New York: Free Press.

Parsons, Talcott. 1977. *The Evolution of Societies*. Edited and with an Introduction by Jackson Toby. Englewood Cliffs: Prentice-Hall.

Platt, Anthony M. 1977. *The Child Savers*. 2nd edition. Chicago: University of Chicago Press.

Starr, Paul. 1982. *The Social Transformation of American Medicine*. Cambridge: Harvard University Press.

6

From Summer Camps to Glass Ceilings: The Power of Experiments

MICHAEL J. LOVAGLIA

This lively selection takes you through a visit to some of the most famous social science experiments ever performed: Milgram's electroshock investigations of conformity to authority, Zimbardo's prison experiment on brutalization and authority, Sherif's boys' summer camp experiment on in-group and out-group dynamics, and many more. These experiments may shed insights into some of the complex reasons for the prisoner abuse that occurred in Iraq in the early 2000s. The article then takes the laboratory experiment into the realm of real life and shows how it can impact public policy by looking at the way it was used to determine mandatory arrest policies for domestic violence offenders. Social science experiments on a few individuals can give rise to strategies for coping with social problems ranging from intergroup conflict to women's inequality in the workplace. Lovaglia describes the questions and assumptions driving this type of research, the nature of experimental design, and some of these studies' unexpected findings. We learn about the types of insights experimental research can yield and its potential for cause-effect generalization. What are some of the problems you see with research based on experimental design? What are some of the strengths associated with this scientific method? Given that the logic of this method is based on procedures developed in the physical sciences, do you think this method is easily transferable to the social sciences? Why or why not? How does research on such narrow groups contribute to broad social understanding and insight?

A man in torn clothes sprawls across an urban sidewalk. He moans softly. Pedestrians hurry by with no more than a worried glance. No one stops to help. Someone watching from afar might wonder at such uncaring behavior; surely some conscientious person would stop. Moreover, these pedestrians are all young adults wearing clerical garb, seminarians studying for the ministry. They are hurrying to the church to deliver sermons on the Good Samaritan. Why did they not stop? Researchers who staged this test found that seminary students did not stop because they worried about being late. Their personal obligation to keeping an appointment outweighed their general commitment to helping others.

Experiments such as this one startle us into new ways of understanding people. Although we tend to explain why people do what they do—or, in this case, not do—as an expression of personal character, experiments show that the context of events determines behavior to a significant extent. Experimental studies carry great weight in the social sciences, gaining acceptance in prestigious journals and, in a high-profile example, last year's Nobel Prize in Economics. Some experiment results also get exposure in popular media, generating prime-time news coverage and Hollywood films.

Many people who hear about these experiments—and some social scientists, too—wonder how experiments achieve their power to convince, especially when their results often defy common sense. Experiments usually feature contrived conditions and record the behavior of at most a few hundred participants, many of whom are college students. Yet the results can tell us a lot about society.

THE ROBBERS CAVE EXPERIMENT
AND SUMMER CAMP MOVIES

A sociological experiment in the 1950s demonstrated the effectiveness of a now common strategy in which competing corporations form joint ventures that would appear to prevent one firm from gaining advantage over the other (much like the United States and Russia cooperating on the space station). In 1954, Muzafer Sherif, an early proponent of social science experiments, set up a summer camp near Robbers Cave State Park in Oklahoma to test theories about group conflict and how to avoid it. He believed that individuals develop a group identity when they work together toward a common goal. Groups become more cohesive and rigid when faced with competition from another group. This competition creates frustration, triggering hostility and conflict between the groups. Sherif thought a solution to the conflict might be found in the same process by which groups form: working toward a common goal. If hostile groups have to work together, then members might learn to see each other as part of a combined larger group, which would reduce their conflict.

A group of 22 boys—all white, middle-class and close to their 12th birthdays—came to the Robbers Cave summer camp. Sherif and his colleagues divided them into two teams, the Eagles and the Rattlers. Each team completed projects requiring the cooperation of members, such as building a diving platform at a swimming hole. In the second phase of camp activities, the two teams competed against each other in various contests. The results are familiar. Rivalry between teams generated hostility and even a little mayhem (exaggerated in subsequent summer camp movies), and threatened to spin out of control. Hostility emerged during the first contest—a baseball game. Boys in each group cursed members of the opposing group and called them names. At dinner, Eagles refused to eat with Rattlers. Later, the Eagles tore down the Rattlers' flag and burned it. The Rattlers retaliated by vandalizing the Eagles' cabin. A food fight erupted in the mess hall.

The experiment showed that hostility between groups develops spontaneously when individuals within a group work together and then compete as a team against another group. The final phase of the experiment showed how to reduce conflict. On a hot summer day, researchers disabled the water supply and asked volunteers to find the problem. Boys from both groups stepped forward, located the problem and worked together to solve it. Afterward, they all shared the water in a friendly manner. Finding water was important enough that it neutralized the groups' mutual antipathy, fostering cooperation and the beginning of trust.

An overarching cooperative task that requires the contributions of both groups for success reduced intergroup conflict. This principle is widely applied today, in contexts as distant as international relations, even though the experiment had nothing directly to do with such serious settings.

DESCRIBING THE WORLD OR TESTING THEORIES

The logic of social experiments differs from that of other social research. Survey researchers, for example, try to describe a population of people by selecting a large, representative sample and then asking questions to determine respondents' attitudes and other characteristics. In contrast, experiments test theories rather than describe a population. That is, they test for evidence of a specific social process in a small sample of people, chosen to be as similar as possible. If a theory predicts a particular result under certain conditions, experimenters then set up only those conditions. In this way, researchers can tell whether the predicted differences in behavior are produced by the conditions of the experiment instead of by individual differences among the participants.

Psychologist Philip Zimbardo's prison experiment at Stanford University is another famous example. He tested the theory that the brutal behavior of guards in prison camps (such as those in Nazi Germany) was a result of their being guards, rather than a result of their being individuals psychologically prone to act brutally. Zimbardo predicted that normal, mentally healthy, American men would become brutal or be brutalized simply because they became either prison guards or prisoners.

In the early 1970s, Zimbardo created a "prison" in the basement of the psychology building at Stanford. He selected only male Stanford undergraduates to participate, ruling out those with any prior psychological problems. He then randomly assigned the participants to be either prisoners or prison guards. The procedure is like flipping a coin. Heads and the participant becomes a guard, tails and he gets arrested. Random assignment helped to ensure that the two groups in the experiment—guards and prisoners—would be similar in other ways. Within a day of the prisoners' arrival, guards began acting brutally and prisoners showed signs of anxiety. Conditions rapidly deteriorated until the experiment had to be stopped. (Because social experiments directly change people's lives, extraordinary care must be taken to avoid causing harm. Some social experiments have the potential to be as dangerous as a clinical trial testing a new drug. Today, universities' Institutional Review Boards review proposed social experiments as stringently as they do medical and other scientific studies on people.)

The Stanford prison experiment helped shift thinking away from blaming German culture for the Holocaust and toward the social conditions that promote brutal behavior. The study received much media attention and was made into a popular German movie, *Das Experiment*. Ironically, the film version concluded that the solution to brutality is for individuals to take personal responsibility for their actions. But a solution that follows more consistently from the study itself is to construct social situations that discourage brutality. (See "Making Sense of the Senseless: Understanding Genocide," *Contexts*, Spring 2003.)

Why was this experiment so influential? It said nothing directly about German behavior during the Holocaust. Rather, it tested a theoretical prediction that a coercive setting can induce brutal behavior. A good experiment subtly shifts the burden of scientific proof, challenging other researchers to show whether a social process demonstrated in the experiment operates differently in a complex, naturally occurring setting. Simple experiments are convincing in part because they demonstrate a difference in the behavior of people in contrasting situations. Simplicity helps build agreement; most people observing the results of the Stanford and Robbers Cave experiments would interpret their meanings similarly. Controlled conditions also allow other researchers to repeat the experiments to see if the same results occur, perhaps using slightly different procedures. Good experiments can in these ways extend theories and produce new knowledge.

Of course, no single study, theory or method, no matter how good, establishes a scientific fact. Instead, science synthesizes different kinds of research from a variety of researchers to reach its conclusions. An experiment such as Zimbardo's Stanford prison makes a simple yet forceful statement that builds on earlier and inspires later research pointing to a conclusion. Eventually, we better understand the social processes underlying a problem and can attempt a practical intervention. Experiments also can be used to directly assess the effectiveness of alternative social policies.

ARRESTING DOMESTIC VIOLENCE: EXPERIMENTING WITH SOCIAL POLICY

In 1981, police in Minneapolis changed the way they responded to reports of domestic violence. Before 1981, police officers had the discretion to arrest the person who committed the assault, order him (or her) to leave the home for a short period or provide on-site counseling. Advocates expressed concern that police were treating episodes of domestic violence too leniently, thereby failing to deter future assaults. Lawrence Sherman and Richard Berk designed an experiment to test whether making an arrest in a domestic violence case deterred future assaults better than the other two options of separating the couple and counseling.

The experiment had important implications for public policy, but it also addressed a long-standing dispute between two theoretical traditions in criminology. Deterrence theory holds that punishment discourages future criminal behavior. This school of thought maintains that suspects who are arrested will be less likely to commit another assault than those who are separated or counseled. A second theoretical tradition, known as labeling theory, suggests that when individuals are arrested, they become stigmatized as criminals by both society and in their own eyes. Their new self-image as a criminal then increases the likelihood of subsequent criminal behavior. (Labeling theory is the reason that names of juvenile offenders are kept out of the media except for serious offenses.) If labeling theory is valid, then those arrested for domestic violence actually would be more likely to commit another assault.

During the Sherman–Berk experiment, whenever Minneapolis police officers responded to a domestic violence call, they determined which procedure—arrest, separation or counseling—to follow by random assignment. Researchers tracked the behavior of suspects in the study for six months following the domestic violence incident. Results showed a deterrent effect for arrest and no evidence for labeling theory. That is, suspects who had been arrested were slightly less likely to commit another assault during the subsequent six months than were those who had been separated or counseled.

Although the deterrence effect of arrest was small, the experiment had a large effect on public policy. Arrest in domestic violence cases became the preferred procedure in many police departments and 15 states passed mandatory arrest laws. Meanwhile, debate over implications for social theory continued. During the next decade, other researchers repeated the experiment in several other police jurisdictions. The new results were more complicated. Arrest deterred suspects who were employed, perhaps because arrest is more serious for those who have a lot to lose. For unemployed suspects, arrest had the opposite effect, as predicted by labeling theory. They were more likely to commit a subsequent assault than the unemployed men who had been separated or counseled. The theoretical advance was exciting, but it left policy implications unresolved. In practice, police officers are still uncertain whether making an arrest will be beneficial in a domestic violence case. More systematic research could better equip police and judges to make such critical, sometimes life-and-death decisions....

WHY DO SOME GROUPS SCORE LOW
ON STANDARDIZED TESTS?

Low intelligence seems the obvious explanation for low scores on a mental ability test. But what if something besides intelligence determines test scores? In the

1990s, psychologist Claude Steele's experiments yielded the startling discovery that scores on standardized tests depend not only on students' ability to answer, but also on what they expect the consequences of their test scores to be. Students who are stereotyped as having low ability may underperform when they are apprehensive about getting a low score.

Steele and his colleagues conducted a simple experiment. They gave a difficult standardized test—like the college SAT but harder—to a group of Stanford students. Instructions for taking the test varied. Some students, selected at random, were told the test results could be used to compare their performance to that of other students. Some students were told the test was only to familiarize them with similar tests they would encounter at the university. When students were told the tests were just for familiarization, black students scored about the same as white students of similar academic attainment. But when students thought they were going to be compared, black students scored lower than did comparable white students—as is common on standardized tests.

My colleagues and I conducted subsequent experiments showing that Steele's theory was not limited to particular racial groups, but applied to any stigmatized group. We randomly assigned white university undergraduates to be treated as an advantaged "majority" or disadvantaged "minority," by telling some students that their left- or right-handedness made it unlikely that they would be able to contribute to a group project, and also that other group members might resent their inability to contribute. Then, we gave the students a standard test of mental ability, explaining that the results of the test would be used to assign them to group positions such as "supervisor," "analyst," or "menial" in the group project. We found that students' test scores were substantially lower if they were treated as a disadvantaged "minority" for as little as 20 minutes.

The line of research begun by Claude Steele now includes many studies by different researchers. They show that when black and white students take the same standardized test, different expectations for the consequences of the test—not differences in mental ability—determine whether white students have an advantage. That is, while the best mental ability tests do a fair job of determining differences in cognitive skills among otherwise similar individuals, differences in test scores between racial and ethnic groups are created by social conditions rather than by the groups' mental abilities.

Applied programs based on this research show promise for increasing the academic performance of disadvantaged students. One surprising detail is that the performances of the best black students suffer the most. The threat of fulfilling a negative stereotype is felt most keenly by black students with the potential to excel; it is they who worry most about the potential backlash from their competition with white students. This may explain why remedial programs to improve academic performance of weaker students have not closed the gap between blacks and whites generally. Honors programs that encourage black students to undertake accelerated studies may have more effect, because promising black students have more academic ability than their grades and test scores suggest. Claude Steele helped develop a successful program to improve the performance of incoming minority students at the University of Michigan that emphasizes high academic standards, affirming students' ability to achieve those standards, and building trust that successful minority students can be accepted in the academic community.

HOW CAN WOMEN ATTAIN STATUS EQUAL TO MEN AT WORK?

Social experiments can also suggest strategies individuals can use to improve their lives. Status Characteristics Theory explains how individuals attain influence in work groups: people who are expected to contribute more to the group gain more influence in the group and receive greater rewards from the group. That is, expected contributions often count more than actual contributions. Individuals expected to perform well are more often followed by the group and rewarded accordingly. For example, a woman may make a brilliant suggestion that guarantees a successful project, but her suggestion may be ignored until a respected male coworker endorses it. He then gets the credit.

Research using the theory confirms that people expect men to contribute more to group success than women and that men do have more influence in decision making. Men get more credit for the group's successes and less blame for the group's failures. And when group members are evaluated, men get higher performance ratings and bigger rewards. To achieve the same level of rewards, women must work harder and contribute more than men. Status Characteristics Theory can also explain the familiar strategies women have used to break through to positions of influence in the workplace. Traditionally, they have out-competed men, following a masculine model that includes demonstrating competence through hard work and aggressive, even ruthless, competition. Successful women sometimes feel that they have sacrificed too much of themselves by following "male" strategies.

In the early 1980s, Cecilia Ridgeway conducted experiments using this theory that produced remarkable results for professional women struggling for career advancement under a glass ceiling. Ridgeway realized that people value not only the ability of a person to contribute, but also whether that person is motivated by a desire to help the group; they would not expect a person who is competent but selfish to contribute much of value. Ridgeway proposed that, because of gender stereotypes, however, people expect that even selfishly motivated men will contribute to the group, but expect contributions from women only when women demonstrate that they care about the group.

Ridgeway conducted an experiment to test this theory. Four team members worked together to reach a decision. One of the team members—secretly collaborating with the experimenters—made comments that were either group-motivated ("It is important that we cooperate") or self-motivated ("I want to win points for myself"). As predicted, in the self-motivated condition, male collaborators had more influence over the groups' decisions than female collaborators. In the group-motivated conditions, however, women collaborators' influence increased while the men's stayed at about the same high level as when they appeared selfish. Put another way, group-motivated women had as much influence as equally competent men regardless of the males' motivations.

The results suggest a strategy to succeed at work that women could use as an alternative to the competitive male one. Demonstrated competence is primary. Assertiveness also helps, but the focus on ruthless competition may be unnecessary for women's success. Instead, emphasizing a concern for other group members and the importance of working together to accomplish group goals can help competent women achieve recognition for their contributions. Future research in actual workplaces will help refine an effective strategy.

FROM THEORY TO PRACTICE

The power of experiments flows from their use to test general theories. Sherif's Robbers Cave experiment tested a theory that explains how cooperation forms within groups and competition develops between them. Ridgeway tested her theory that influence in groups flows from the expectations people have about the ability and motivation of group members to contribute to group success.

Alone, a social experiment only demonstrates some phenomenon in one restricted context. But when experiments test theories, and their results lead to more tests in wider contexts, as well as other research with other methods, then we gain knowledge capable of transforming society. The experiments described have inspired lines of research with the potential to increase cooperation among competing organizations, decrease domestic violence, reduce the racial gap in academic success, and remove the glass ceiling limiting women in business. They successfully made the leap from small groups to helping us understand society at large.

7

Sense and Nonsense
about Surveys

HOWARD SCHUMAN

*Understanding surveys, the most widely used methodology in sociology today, is
critical to being an informed citizen, but the popular media often report surveys
without any guidance on how to interpret and evaluate the results. This selection
uses clear language to explain some of the history and issues critical to understanding the logic and problems of survey research: sampling, response rate, and
the language of the questionnaire. Schuman discusses how surveys use scientific
techniques that minimize distortion and enhance generalizability. This research
shows us how sociologists formulate trends and patterns connecting people's
demographic characteristics (age, race/ethnicity, gender, religion, region, income)
to their attitudes and behavior. Surveys use larger samples than experiments
and aim to foresee, even predict, the behavior of more widespread groups. As
Schuman notes, some basic guidelines can promote more sophisticated readings of
survey results and help teach us when we should trust or be wary of public opinion polls. How do you feel about the survey as a research tool? Are there topics
that might be too sensitive for surveys to measure accurately? How valid do you
think the answers are that you have given on surveys you have taken? How
would you evaluate surveys against the strengths and weaknesses of experiments?
Are there certain types of questions that are more suited for one of these methods
than the other?*

Surveys draw on two human propensities that have served us well from ancient times. One is to gather information by asking questions. The first use
of language around 100,000 years ago may have been to utter commands such as

"Come here!" or "Wait!" Questions must have followed soon after: "Why?" or "What for?" From that point, it would have been only a short step to the use of interrogatives to learn where a fellow hominid had seen potential food, a dangerous animal, or something else of importance. Asking questions continues to be an effective way of acquiring information of all kinds, assuming of course that the person answering is able and willing to respond accurately.

The other inclination, learning about one's environment by examining a small part of it, is the sampling aspect of surveys. A taste of something may or may not point to appetizing food. A first inquiry to a stranger, a first glance around a room, a first date—each is a sample of sorts, often used to decide whether it is wise to proceed further. As with questions, however, one must always be aware of the possibility that the sample may not prove adequate to the task.

SAMPLING: HOW GALLUP ACHIEVED FAME

Only within the past century—and especially in the 1930s and 1940s—were major improvements made in the sampling process that allowed the modern survey to develop and flourish. A crucial change involved recognition that the value of a sample comes not simply from its size but also from the way it is obtained. Every serious pursuit likes to have a morality tale that supports its basic beliefs: witness Eve and the apple in the Bible or Newton and his apple in legends about scientific discovery. Representative sampling has a marvelous morality tale also, with the additional advantage of its being true.

The story concerns the infamous *Literary Digest* poll prediction—based on 10 million questionnaires sent out and more than two million received back— that Roosevelt would lose decisively in the 1936 presidential election. At the same time, George Gallup, using many fewer cases but a much better method, made the more accurate prediction that FDR would win. Gallup used quotas in choosing respondents in order to represent different economic strata, whereas the *Literary Digest* had worked mainly from telephone and automobile ownership lists, which in 1936 were biased toward wealthy people apt to be opposed to Roosevelt. (There were other sources of bias as well.) As a result, the *Literary Digest* poll disappeared from the scene, and Gallup was on his way to becoming a household name.

Yet despite their intuitive grasp of the importance of representing the electorate accurately, Gallup and other commercial pollsters did not use the probability sampling methods that were being developed in the same decades and that are fundamental to social science surveys today. Probability sampling in its simplest form calls for each person in the population to have an equal chance of being selected. It can also be used in more complex applications where the chances are deliberately made to be unequal, for example, when oversampling a minority group in order to study it more closely; however, the chances of being selected must still be known so that they can later be equalized when considering the entire population.

INTUITIONS AND COUNTERINTUITIONS ABOUT SAMPLE SIZE

Probability sampling theory reveals a crucial but counterintuitive point about sample size: the size of a sample needed to accurately estimate a value for a population depends very little on the size of the population. For example, almost the same size sample is needed to estimate, with a given degree of precision, the proportion of left-handed people in the United States as is needed to make the same estimate for, say, Peoria, Illinois. In both cases a reasonably accurate estimate can be obtained with a sample size of around 1,000. (More cases are needed when extraordinary precision is called for, for example, in calculating unemployment rates, where even a tenth of a percent change may be regarded as important.)

The link between population size and sample size cuts both ways. Although huge samples are not needed for huge populations like those of the United States or China, a handful of cases is not sufficient simply because one's interest is limited to Peoria. This implication is often missed by those trying to save time and money when sampling a small community.

Moreover, all of these statements depend on restricting your interest to overall population values. If you are concerned about, say, left-handedness among African Americans, then African Americans become your population, and you need much the same sample size as for Peoria or the United States.

WHO IS MISSING?

A good sample depends on more than probability sampling theory. Surveys vary greatly in their quality of implementation, and this variation is not captured by the "margin of error" plus/minus percentage figures that accompany most media reports of polls. Such percentages reflect the size of the final sample, but they do not reveal the sampling method or the extent to which the targeted individuals or households were actually included in the final sample. These details are at least as important as the sample size.

When targeted members of a population are not interviewed or do not respond to particular questions, the omissions are a serious problem if they are numerous and if those missed differ from those who are interviewed on the matters being studied. The latter difference can seldom be known with great confidence, so it is usually desirable to keep omissions to a minimum. For example, sampling from telephone directories is undesirable because it leaves out those with unlisted telephones, as well as those with no telephones at all. Many survey reports are based on such poor sampling procedures that they may not deserve to be taken seriously. This is especially true of reports based on "focus groups," which offer lots of human interest but are subject to vast amounts of error. Internet surveys also cannot represent the general population adequately at present, though this is an area where some serious attempts are being made to compensate for the inherent difficulties.

The percentage of people who refuse to take part in a survey is particularly important. In some federal surveys, the percentage is small, within the range of 5 to 10 percent. For even the best non-government surveys, the refusal rate can reach 25 percent or more, and it can be far larger in the case of poorly executed surveys. Refusals have risen substantially from earlier days, becoming a major cause for concern among serious survey practitioners. Fortunately, in recent years research has shown that moderate amounts of nonresponse in an otherwise careful survey seem in most cases not to have a major effect on results. Indeed, even the *Literary Digest,* with its abysmal sampling and massive nonresponse rate, did well predicting elections before the dramatic realignment of the electorate in 1936. The problem is that one can never be certain as to the effects of refusals and other forms of nonresponse, so obtaining a high response rate remains an important goal.

QUESTIONS ABOUT QUESTIONS

Since survey questions resemble the questions we ask in ordinary social interaction, they may seem less problematic than the counterintuitive and technical aspects of sampling. Yet survey results are every bit as dependent on the form, wording and context of the questions asked as they are on the sample of people who answer them.

No classic morality tale like the *Literary Digest* fiasco highlights the question-answer process, but an example from the early days of surveys illustrates both the potential challenges of question writing and the practical solutions.

In 1940 Donald Rugg asked two slightly different questions to equivalent national samples about the general issue of freedom of speech:

- Do you think the United States should forbid public speeches against democracy?

- Do you think the United States should allow public speeches against democracy?

Taken literally, forbidding something and not allowing something have the same effect, but clearly the public did not view the questions as identical. Whereas 75 percent of the public would not allow such speeches, only 54 percent would forbid them, a difference of 21 percentage points. This finding was replicated several times in later years, not only in the United States but also (with appropriate translations) in Germany and the Netherlands. Such "survey-based experiments" call for administering different versions of a question to random subsamples of a larger sample. If the results between the subsamples differ by more than can be easily explained by chance, we infer that the difference is due to the variation in wording.

In addition, answers to survey questions always depend on the form in which a question is asked. If the interviewer presents a limited set of alternatives, most respondents will choose one, rather than offering a different alternative of their own. In one survey-based experiment, for example, we asked a national sample of Americans to name the most important problem facing the country. Then we asked a comparable sample a parallel question that provided a list of four problems from which to choose the most important; this list included none of the four problems mentioned most often by the first sample but instead provided four problems that had been mentioned by fewer than 3 percent of the earlier respondents. The list question also invited respondents to substitute a different problem if they wished.... Despite the invitation, the majority of respondents (60 percent) chose one of the rare problems offered, reflecting their reluctance to go outside the frame of reference provided by the question. The form of a question provides the "rules of the game" for respondents, and this must always be kept in mind when interpreting results.

Other difficulties occur with survey questions when issues are discussed quite generally, as though there is a single way of framing them and just two sides to the debate. For example, what is called "the abortion issue" really consists of different issues: the reasons for an abortion, the trimester involved and so forth. In a recent General Social Survey, nearly 80 percent of the national sample supported legal abortion in the case of "a serious defect in the baby," but only 44 percent supported it "if the family has a low income and cannot afford any more children." Often what is thought to be a conflict in findings between two surveys is actually a difference in the aspects of the general issue that they queried. In still other cases an inconsistency reflects a type of illogical wish fulfillment in the public itself, as when majorities favor both a decrease in taxes and an increase in government services if the questions are asked separately.

SOLUTIONS TO THE QUESTION
WORDING PROBLEM

All these and still other difficulties (including the order in which questions are asked) suggest that responses to single survey questions on complex issues should be viewed with considerable skepticism. What to do then, other than to reject all survey data as unusable for serious purposes? One answer can be found from the replications of the forbid/allow experiment above. Although there was a 21 percentage points difference based on question wording in 1940 and a slightly larger difference (24 percentage points) when the experiment was repeated some 35 years later, both the forbid and the allow wordings registered similar declines in Americans' intolerance of speeches against democracy.... No matter which question was used—as long as it was the same one at both times—the conclusion about the increase in civil libertarian sentiments was the same.

More generally, what has been called the "principle of form-resistant correlations" holds in most cases: if question wording (and meaning) is kept constant, differences over time, differences across educational levels, and most other careful comparisons are not seriously affected by specific question wording. Indeed, the distinction between results for single questions and results based on comparisons or associations holds even for simple factual inquiries. Consider, for example, a study of the number of rooms in American houses. No God-given rule states what to include when counting the rooms in a house (bathrooms? basements? hallways?); hence the average number reported for a particular place and time should not be treated as an absolute truth. What we can do, however, is try to apply the same definitions over time, across social divisions, even across nations. That way, we gain confidence in the comparisons we make—who has more rooms than whom, for example.

We still face the task of interpreting the meaning of questions and of associations among questions, but that is true in all types of research. Even an index constructed from a large number of questions on the basis of a sophisticated statistical calculation called factor analysis inevitably requires the investigator to interpret what it is that he or she has measured. There is no escaping this theoretical challenge, fundamental to all research, whether using surveys or other methods such as field observations.

Surveys remain our best tool for learning about large populations. One remarkable advantage surveys have over some other methods is the ability to identify their own limitations, as illustrated by the development of both probability theory in sampling and experiments in questioning. In the end, however, with surveys as with all research methods, there is no substitute for both care and intelligence in the way evidence is gathered and interpreted. What we learn about society is always mediated by the instruments we use, including our own eyes and ears. As Isaac Newton wrote long ago, error is not in the art but in the artificers.

8

The Promise and Pitfalls of Going into the Field

PATRICIA A. ADLER AND PETER ADLER

Ethnography differs considerably from experimental and survey research, enabling researchers to interact directly with subjects in the natural world and see, firsthand, what their lives are like. It combines direct observation, casual conversation, social interaction, and in-depth interviews with people being studied. Because of this, ethnography is much more subjective and less rigidly scientific than those other methods; it has to be, to adapt to the human nature of the subjects. The subjective dimension in ethnography, far from being a detriment, is an asset; it allows researchers to use their own "humanness" to subjectively grasp the way people in the world perceive and interpret the events around them in order to make decisions on how to act. The social world is intricate and complex, often hard to reduce to a single variable, as experiments do, or to a static time frame, as surveys do. Ethnography allows social scientists the opportunity to watch the world unfold over time and see how people change and adapt to the situations around them. But in a subjective methodology, questions arise about the characteristics and perceptions of the researchers and whether they have navigated the complexities of the field adequately. In this selection, we offer some guidelines for what distinguishes between stronger and weaker ethnography, giving readers some idea of what they should expect when they attempt this method or assess the findings of others using it. You may want to imagine yourself as a social scientist doing experimental, survey, or ethnography and think about which approach suits you better. Each method has strengths and weaknesses. What do you think these might be? What kinds of questions do you think each is best able to answer? Which would you

SOURCE: Patricia Adler and Peter Adler, "The Promise (and Pitfalls) of Going into the Field," *Contexts*, Vol. 2, No. 2: 41–47. © 2003, The American Sociological Association. Used by permission. All rights reserved.

rather do yourself? Which would you trust? If you had to pick a group with which to conduct ethnography, what would it be?

Barbara Ehrenreich, a white, divorced Ph.D. in her 50s, spent a year working low-wage jobs as a waitress in Florida, a housecleaner in Maine and as a Wal-Mart sales clerk in Minnesota. Her detailed ethnography, the best-selling *Nickel and Dimed,* reveals how physically demanding and personally demeaning these jobs are, and how workers are trapped in them. Ehrenreich's book was received as "piercing social criticism backed by first-rate reporting."

Some ethnographies are, however, more controversial. William Foote Whyte, then a young, Protestant, graduate student at Harvard, wrote a classic ethnography of Italian-American youths in the early 1940s, *Street Corner Society,* describing the "corner boys" who hung around the neighborhood and participated in illegal activities. He described them as a "gang." Yet Marianne Boelen, an Italian immigrant to America who years later revisited his setting and re-interviewed his subjects, asserted that Whyte had made methodological and substantive errors in his work. These boys were not a gang, she claimed, but rather followed a typical Italian pattern: women occupied indoor space and men claimed the outdoors. He might have realized this had he paid greater attention to gender, she claimed. His errors also resulted, she alleged, from relying too closely on one key informant, "Doc," whose role he exaggerated.

These different reactions raise important questions about ethnography. How can readers know if researchers have gotten the evidence and its interpretation right? What kinds of stories should we believe? We need to be able to assess the validity and value of ethnographic work, just as we do with other methods. Herbert Gans, in *The Urban Villagers* (1982, pp. 414–15) notes that "every social research method is a mixture of art and science," but that participant-observation is the best empirical research method available because it allows us to study, firsthand, what people do, think, and believe, in their own groups. While all methods may be subject to problems such as shaping findings to fit preconceptions, Gans continues, "ethnography is most successful when it becomes an all encompassing 14- to 16-hours a day experience, with at least a year's full-time fieldwork, and a good deal of additional time to analyze and think about the data."

Ethnography, as we defined it when we edited the *Journal of Contemporary Ethnography,* includes observing social activities as an outsider, observing while participating in the activities, and conducting intensive interviews. Considered the most accessible to readers of all the social scientific methods, ethnography draws on the language and perspective of everyday members of society, and is often written like investigative journalism. A successful ethnography captures readers' fancies, bringing them closer to others' lives, and, like a good movie or book, offers insight into individuals' ordinary worlds. Literally translated as a "portrait of the people," ethnography should describe and analyze the beliefs, motivations, and rationales of a people in a particular setting or subculture. It should make the familiar distant and the distant familiar.

Although ethnography resembles journalism, it differs by requiring systematic, long-term gathering of data and by engaging general theories of human behavior rather than simply reporting the news. Ethnography resembles literature as well, but differs in focusing on social trends and patterns rather than character development. Finally, ethnography differs from common sense interpretations by drawing on meticulous ethnography rather than popular stereotypes. But as the controversies we noted teach us, it is not always obvious which ethnographic reports are sufficiently systematic, sufficiently accurate, or sufficiently useful.

THE ETHNOGRAPHIC GENRE

Ethnography can be divided into three crucial stages: data gathering, data analysis and data presentation. One might be an exemplary field researcher, able to fit into myriad social settings and to elicit the insiders' view from a variety of people, but this is not enough. Ethnographers need to step back as well, to take a detached look at people's worlds so they can analyze underlying patterns of behavior. These careful observations and astute interpretations must be backed up by prose that brings readers into people's complex lives. We will see how exemplary researchers optimize the rewards of fieldwork while avoiding its pitfalls.

DATA GATHERING

Good ethnography takes time. The strength of ethnographers' data depends on the quality and depth of the relationships they forge and the rapport and trust they establish with the people they study. Superficial relationships yield superficial insights. Researchers sometimes spend up to several years in the field, as we did in our studies of drug dealers and smugglers in *Wheeling and Dealing* and elite college athletes in *Backboards & Blackboards*.

Ethnographers, in having to gain people's trust, require highly developed social skills. They must be able to get along with all sorts of people, from powerful managers to weak employees. For instance, in an outstanding ethnography of the homeless, *Down on Their Luck,* Leon Anderson and David Snow spent parts of two years under bridges, in Salvation Army shelters and plasma centers, at the city hospital and police department, and on the streets of Austin, Texas (see "Street People," *Contexts,* Winter 2003). Deeper yet, ethnography requires intimacy and commitment. For example, in studying drug traffickers, our long-term relationships with central figures were often tested by crises or suspicions of betrayal, and loyalty was expected on both sides during the six years of explicit research and for many years afterwards.

According to current thinking, ethnographers should get as near to the people they are studying as possible. Even studying one's self (auto-ethnography), as Carolyn Ellis did in *Final Negotiations,* where she documented the changing emotions she and her partner experienced as he was dying of emphysema, or as Carol Rambo Ronai did in her writings on incest, has become acceptable. Some ethnographers combine the intimacy of autobiography with the more general approach of talking to others who have gone through similar traumas or events. Best illustrated by David Karp in *Speaking of Sadness,* a study of manic-depressives, the author recounts his own bouts with depression as well as data gleaned from numerous observations and interviews with self-help groups for this illness. Karp's own experiences helped him gain participants' trust and gave him a deeper understanding of the emotional complexity of mood fluctuation. In evaluating ethnography, then, readers should pay attention to not only the length of time researchers spent in the field (a year or two tends to be the minimum depending upon the locale and topic of study), but also the depth of involvement they established with their subjects.

Sometimes problems arise when researchers are either too close or similar to their subjects or too distant or different from them. Researchers who are too close may "go native," uncritically accepting their subjects' perspectives. Researchers too distant may fail to penetrate beyond the fronts people design for public presentation. For example, Richard Mitchell, in *Dancing at Armageddon,* a study of survivalists, became involved with people whose behaviors evoked some repugnance. To forge the necessary rapport, he had to overcome his initial feelings of alienation, to spend time getting to know participants, and to establish friendship and trust on other planes. Readers who suspect ethnographers may have such problems should look for frank and personal methodological discussions that specifically address how they encountered and dealt with these issues.

Good ethnography is systematic, rigorous and scientific. One of the chief criticisms leveled at ethnography is that it is anecdotal, careless, and casual, depending too much on researchers' subjectivity. Poor ethnography may result when researchers are biased by their own opinions or history, or when they carry their preconceived attitudes, either personal or professional, into the field and cannot transcend them. Derek Freeman aimed this charge against Margaret Mead, claiming that in *Coming of Age in Samoa* she uncritically accepted the assertions of a few adolescent girls about their uninhibited sexuality to support her mentor's views that nurture trumped nature. Bias may also result from researchers' poor location or sponsorship in the field, where their access to the group is somehow impeded. And researchers can generate problems when they fail to gather multiple perspectives or prefer their own beliefs to others'. (Recently, some "postmodern" ethnographers have concluded that the process is so idiosyncratic that there should be no claims to describe the world, only to describe researchers' reactions to the world.)

To overcome these problems, ethnographers should include the voices of a full spectrum of participants, not just the ones they can easily reach. Generally, it is easier for researchers to "study down," looking at the downtrodden, the powerless, and the underclass, who, unlike the powerful, do not have the same ability to insulate themselves. Researchers may also more easily gather data from people like themselves, overlooking members of dissimilar groups. Part of Boelen's allegation against Whyte was that his perspective was skewed toward "Doc," his key informant, a man much like himself. Good ethnography gains the perspectives of all involved, so that the ultimate portrait is rounded and thorough. In Jack Douglas and Paul Rasmussen's study, *The Nude Beach,* the voices of the nudists, other beach-goers, residents, and police are all heard, providing this sort of completeness. An ethnography that only privileges some voices and perspectives to the exclusion of others may not be as representative.

An array of methodological tactics may help to generate the multiple perspectives required. Ethnographers may combine direct observation, participation, interviewing, and casual conversation to triangulate their findings. For instance, in researching drug traffickers, we cross-checked our observations against our own common sense and general knowledge of the scene, against a variety of reliable, independent sources, and against hard evidence such as newspaper and magazine reports, arrest records and material possessions. Similarly, Judith Rollins, in her study of domestics and their employers, *Between Women,* worked as a domestic for ten employers. In studying human-canine relationships for *Understanding Dogs,* Clinton Sanders not only drew on his own love of dogs and experiences as a dog owner, but also participated in the training of guide dogs and their owners, a "puppy kindergarten," observations of dogs and their owners in public settings, participant observation at a veterinary hospital for 14 months, and formal interviews with dog owners, veterinarians, and trainers.

To help readers assess what role the researchers' personal views played in their reports, an ethnographic report should have methodological reflections. Researchers use these "confessional tales" to explain problems, and then describe the ways they overcame them. Alan Peshkin confessed the problems he faced studying Bethany, a Christian fundamentalist community and school, for *God's Choice:* "I discovered, so to speak, that being Jewish would be the personal fact bearing most on my research…. They taught their children never to be close friends, marry, or to go into business with someone like me. What they were expected to do with someone like me was to proselytize…. To repeat, Bethany gored me." Yet, Peshkin was able to surmount his role as "odd man out" and to forge close research ties by living in the community for 18 months, attending all regular church and school activities, dressing and speaking as a member, and interviewing a significant portion of the school's teachers, students, and parents.

Ethical concerns are often raised about ethnography, since researchers interact so closely with their subjects and could potentially deceive or harm them. A maelstrom of controversy surrounded Laud Humphreys' *Tearoom Trade,* a study of impersonal homosexual encounters in public restrooms, partly because he was covert, observing without telling the men he watched that he was a researcher. Humphreys rejoined that he caused no harm to his subjects, and would not have been able to conduct the research under the strictures of "informed consent," rules that require the permission of those studied. New "Institutional Review Board" regulations at universities now require researchers to relinquish their data to the authorities, often raising conflicts between their loyalty to the people they studied and to the state. When faced with this dilemma, Rik Scarce went to jail for six months rather than turn over his field notes on environmental activists in the state of Washington to the police. Not everyone will take such drastic steps, though, and recent guidelines have been designed to safeguard subjects from their researchers by making sure people know that what they say *cannot* be protected.

Some feminist and "action" ethnographers believe researchers improve their ethical stance by eschewing the traditional "value neutral" position and openly aligning themselves with their subjects, "making the personal political" and working for social change. Others seek an ethical stance in "taking their findings back to the field," showing their writings to subjects and asking for feedback. At the same time, however, some ethnographers believe this leads researchers to censor themselves from writing things their subjects might interpret as too critical, pushing them toward "going native" in the field.

Perhaps most importantly, good ethnography conveys what it is like to "walk in the shoes" of the people being studied. No other method lets researchers adequately study hidden, secretive, and sensitive groups, since deviants, criminals, and others with something to hide are unlikely to talk to strangers. Jeffrey Ferrell's work on illegal graffiti artists and Jeffrey Sluka's investigation of violent political combatants in Ireland, for example, illustrate insightful ethnographic research into subterranean worlds. Readers should understand some of the participants' joys, feel their frustrations and sorrows, and know their problematic, complex and contradictory worlds. For instance, Karp's *Speaking of Sadness* delves deeply into the poignant fears and frustrations experienced by people who suffer from depression. One person Karp interviewed described the way depression stole away who she was and replaced her life with a black hole: "Depression is an insidious vacuum that crawls into your brain and pushes your mind out of the way. It is the complete absence of rational thought. It is freezing cold, with a dangerous, horrifying, terrifying fog wafting through whatever is left of your mind." In *Sidewalk,* Mitchell Duneier explains some of the practical problems that Greenwich Village African-American street vendors encounter in doing what we all take for granted: going to the bathroom. In the words of one of his informants: "I gotta get me a paper cup and I'm gonna be all right.... Now everybody out here gets a cup. You can't go to the bathroom in the stores and restaurants, because they don't want you in there if you ain't got no money to spend. So how you gonna piss? You gotta get a cup."

DATA ANALYSIS

Ethnographers begin forming their analyses early in their fieldwork, testing and refining them over time. Researchers usually remain near, or connected to their settings throughout the time they write up their data, to fill in holes they discover and to check their interpretations against participants'. Yet their observations about the specifics of a particular time and place must be joined by more far-reaching, general analyses. They want, for example, to speak about not just a poor neighborhood, but poor neighborhoods in general. One test of how well ethnographers have succeeded in capturing more general patterns comes when people in comparable settings recognize the descriptions they read. For instance, in our college athlete study, we were frequently satisfied when we gave lectures at universities and athletes in the audience came up afterwards to say that we "got it right."

Good ethnography generates, modifies, extends, or challenges existing understandings of social life. For instance, Pierrette Hondagneu-Sotelo's study of immigrant domestic workers in Los Angeles, *Domestica,* is powerful because it shows that American husbands' failures to share household duties and the influx of immigrant workers have combined to create a pattern in which housekeepers work in affluence but live in poverty.

DATA PRESENTATION

Ethnographers must write clearly and actively, and avoid jargon, highly technical terms or obscure phrases. Ethnography also should "give voice" to participants, enabling readers to get a sense of how people converse and what language they use. In *Code of the Street,* Elijah Anderson uses a voice from the neighborhood to explain why low-income African-American girls in vulnerable situations may become pregnant early: "I done see where four girls grow up under their mama. Mama working three to eleven o'clock at night. Can't nobody else tell 'em what to do. Hey, all of 'em pregnant by age sixteen. They can get they own baby, they get they own [welfare] check, they get they own apartment. They wanna get away from Mama."

What anthropologist Clifford Geertz called "thick description" is another hallmark of ethnography. Good ethnographies vividly present participants' stories, using colorful words, adjectives, or other literary devices to highlight the vibrancy of group culture. With sounds and action, Philippe Bourgois brings readers into the midst of the scene in this excerpt from *In Search of Respect,* his ethnography of Puerto Rican crack dealers in East Harlem: "But then when we stepped out of the room, she turns to me and whispers {snarling}, 'You motherfucker.' She like turns on me again. And then I went {burying his head in his hands}, 'Oh, my God.' And I got mad {making exaggerated whole-body wrestling motions}, and I grabbed her by the neck, and I threw her to the sofa. {pounding fist to palm} BOOM.... and I WHAAAAM, POOM {pounding again}, smacked her in the face with all my might." Even when the subject matter is disturbing, it should be easy, not hard, to read this type of social science.

Successful ethnography elicits the "uh-huh" effect in readers, presenting subjects' everyday behavior in ways that people can recognize. Lyn Lofland, an observer of public places, succinctly summarized behavior that we all do, but rarely acknowledge. She described how people get ready to enter a public space: they "check for readiness" (clothes, grooming, mirror glances), "take a personal reading" (pause, scan the area, check the layout), and "reach a position" (find a secure location or niche). These sorts of rich and resonating descriptions serve to authenticate ethnographic presentations.

THE CONTRIBUTIONS OF ETHNOGRAPHY

In making the familiar distant, researchers find new ways of looking at what we think we know and bring the unknown to light. Weak ethnography runs the risk of rediscovering the obvious. Poorly presented ethnography may stop at subjects' understandings of their worlds, or may analyze these in mundane, trivial or superficial ways. Gary Alan Fine's work is notable for introducing readers to the nuances of unusual subcultures, such as mushroom collecting, or taking familiar worlds, such as the Little League, and providing a framework for a much broader understanding of children's culture.

Good ethnography may also be socially influential. It may speak to social policy and public awareness as well as to scholarly knowledge and theoretical understanding. For example, Arlie Hochschild brought recognition to contemporary working women's dilemmas about housework with her research on *The Second Shift*. What seemed to be individuals' personal problems, she showed, emerged from social changes affecting many families. In the 1990s, government agencies implemented programs to distribute condoms, clean needles, and bleach after ethnographies of the drug world exposed the HIV dangers in the practices of street people. Whether or how ethnographic findings are used depends on the administration in power and the tenor of the times, however. The traditionally liberal leanings of sociologists have made their suggestions more appealing to Democratic politicians. Others believe, however, that ethnography should take theory-building, not political activism, as its goal. As famed ethnographer Erving Goffman put it: "I can only suggest that he who would combat false consciousness and awaken people to their true interests has much to do, because the sleep is very deep. And I do not intend to provide a lullaby, but merely to sneak in and watch the people snore."

Ethnography has the power to incite, infuriate, enthrall, and excite. Ethnographers need to be careful in their representation of others, scrupulous in how they relate to participants in order to obtain data, and true to their own integrity in not violating others' privacy. However, their stories are vital, allowing readers insight into worlds to which they will never be privy or to ones that they would otherwise never understand. The great ethnographies endure for decades because the evidence is accessible, the messages remain critical, and the stories of people's complex worlds continue to be fresh and insightful.

REFERENCES

Adler, Patricia A. 1993. *Wheeling and Dealing*, Second Edition. New York: Columbia University Press.

Adler, Patricia A., and Peter Adler. 1991. *Backboards and Blackboards*. New York: Columbia University Press.

Anderson, Elijah. 1999. *Code of the Street*. New York: Norton.

Bourgois, Philippe. 1995. *In Search of Respect*. New York: Cambridge University Press.

Douglas, Jack, and Paul Rasmussen. 1997. *The Nude Beach*. Beverly Hills, CA: Sage.

Duneier, Mitchell. 1999. *Sidewalk*. New York: Farrar, Straus, and Giroux.

Ehrenreich, Barbara. 2001. *Nickel and Dimed*. New York: Holt.

Ellis, Carolyn. 1995. *Final Negotiations*. Philadelphia: Temple University Press.

Gans, Herbert J. 1982. *The Urban Villagers*, Second Edition. New York: Free Press.

Hochschild, Arlie R. 1989. *The Second Shift*. New York: Viking.

Hondagneu-Sotelo, Pierrette. 2001. *Domestica*. Berkeley: University of California Press.

Humphreys, Laud. 1975. *Tearoom Trade*. New York: Aldine.

Karp, David. 1996. *Speaking of Sadness*. New York: Oxford University Press.

Mead, Margaret. 1928. *Coming of Age in Samoa*. New York: W. Morrow and Company.

Mitchell, Richard. 2002. *Dancing at Armageddon: Survivalism and Chaos in Modern Times*. Chicago: University of Chicago Press.

Peshkin, Alan. 1986. *God's Choice*. Chicago: University of Chicago Press.

———. 1984. "Odd Man Out: The Participant Observer in an Absolutist Setting." *Sociology of Education* 57: 254–64.

Rollins, Judith. 1985. *Between Women: Domestics and Their Employers*. Philadelphia: Temple University Press.

Sanders, Clinton. 1999. *Understanding Dogs*. Philadelphia: Temple University Press.

Snow, David, and Leon Anderson. 1993. *Down on Their Luck*. Berkeley: University of California Press.

Whyte, William Foote. 1943. *Street Corner Society*. Chicago: University of Chicago Press.

PART II

Culture and Socialization

Culture and socialization are foundational concepts. Culture alludes to the social context surrounding us, while socialization and interaction concern the behaviors that occur within this context. In Part II, we begin by considering some of the different types of cultures that exist in our society, from those characterizing elite groups to the subcultures of smaller groups.

Culture consists of the way people share a given space, their language, their relations to each other, the way they feel about each other, and their self-identities as part of the group. Its basic building blocks are norms and values—those behavioral "recipes" that tell us how to act (norms) and our shared beliefs about what is good (values). Values can also be thought of as the ends toward which we strive and norms as the standards or means that we legitimately use to work our way there. Each group has its own culture that makes it distinct and makes it similar to or different from others. In our contemporary society we all belong to more than one group, each having its own slightly different culture. In the United States most of us are, fundamentally, Americans and have some sense of the broad, overarching American culture into which we are socialized. We learn to cherish life, liberty, and the pursuit of happiness because this is how the early settlers established the credo of our country. We value individualism, freedom, materialism, loyalty, and creativity because they are important to our culture. We know what it means to be an American, although we often take it for granted because we are surrounded by it daily. We know that when we encounter Americans in other countries, we have a shared stock of knowledge and shared ways of communicating that bind us together.

Yet at the same time, we live in a highly diverse society, populated by many smaller groups whose standards, beliefs, and practices differ from each other. These things separate us and make groups distinct from each other. Each subgroup has its own *subculture* with its particular variety of norms and values. You may notice that you belong to several different subcultures that place their emphases on certain ways of thinking and acting. In addition to being an American, you may also be a member of your ethnic group, your religious group, your social class, your gender, and your age group. Each of these may give you some distinct ideas and a characteristic outlook. Beyond this, you may belong to subcultures relating to Greek life (a fraternity or sorority), musical tastes (rap, hip-hop, jam, rock, soul), theatrical performances (thespians, slam poets), clean-living punks (straight edgers), athletic endeavors (snow boarders, surfers, jocks), deviant groups (gangs, outlaw motorcyclists, homosexuals), video games (Warcraft, Ace Combat, John Madden), or drug use (hippies, ravers). Your membership in any of these groups shapes your beliefs, the way you present yourself, and the way you assess others. Subcultures have their own distinct language and jargon, norms and values, ideologies or beliefs, famous characters and stories about them, support systems, key information that is passed around, and system of status stratification through which people can assess the relative position of members in relation to each other. People's status among their peers is a key subcultural element, and their adherence to critical group standards is basic to determining this.

In the Culture section, we see some of the norms and values characterizing different groups in American society. We begin with a selection by Pamela Perry that focuses on one aspect of mainstream culture: Whiteness. Perry shows how Whites tend to view their group membership as the "null hypothesis," meaning that they have no color or race, since they are the default or dominant group. Yet going along with this mainstream position, Perry shows, is a hierarchical culture of entitlement that members of this racial/ethnic group take for granted and expect. We next present a selection on subculture, Ross Haenfler's depiction of the straight edge kids who go through their adolescence avoiding the use of alcohol and other drugs. Haenfler articulates and illustrates the core values of the straight edge movement, which grew out of the punk scene, embracing its fashion style and music, but rejecting its nihilism, drug abuse, and casual sex scene. Straight edgers reject consciousness-alteration, yet adhere to some more traditionally progressive ideals such as those involving environmentalism, feminism, and social activism. This "counterculture," which arose in opposition to both the mainstream and previous youth subcultures, offers them an alternative lifestyle and social crowd from the majority of high school youth. This study draws on sociological knowledge of the way smaller groups have their own codes, roles, statuses, terms, and behaviors through which they accord recognition and respect. Finally, Marie Sarita Gaytán breaks down the way culture can be commercially constructed in her exploration of how Mexican restaurants try to recreate Mexican culture and portray themselves as "authentic." She offers a continuum of authenticity, varied by these establishments' food, décor, and the language and ethnicity of their employees.

People are born into society with the basic features that nature has given them. Sociology contends that they are shaped and formed into individuals capable of functioning within our world by their socialization. *Socialization* involves individuals in a life-long process of learning the norms, values, roles, boundaries, and beliefs of a culture so that they become members of the group and develop a sense of self, or *identity* within it. The most central *agents of socialization* shaping and molding young people include family members, friends, teachers, and the media. The early part of our lives is characterized by an especially heavy dose of socialization, as we learn what it means to be a member of all our social categories and groups. This process, though, does not end after childhood, as we continue to develop and to change throughout our entire lives, taking on new roles and adding them to our repertoire. Thus, we may start out by learning what is involved in being a child, being a boy or girl, and following the rules of public behavior. Over time, we may augment this with socialization to the role of student, boyfriend or girlfriend, employee, and parent or grandparent. We never actually discard our old roles as we age out of them, but tuck them away somewhere and draw on them when it is appropriate. We even learn roles that we never enact, as this helps us to understand and to interact with people filling those roles. Thus, we know how to respond to teachers, doctors, ministers, and salespeople, even if we've never been one, because we know what their roles entail and what they need to do to fill them. This helps us to act competently in society because we learn how to anticipate other people's needs and behaviors to coordinate our actions with theirs.

George Herbert Mead said that the apex of socialization is learning the role and perspective of a social abstraction, "the generalized other" (people in general), to understand how numerous others, or society, will act or react to what we do. We become socialized to society's perspective so that we can anticipate how others will react to our behavior and thereby stay within social norms and values, be in synch with society's expectations and rhythms, and act as a successful person within the larger group. This ability becomes especially important as we increasingly interact, via electronic means, with individuals whom we never even meet face-to-face.

When we learn how to interact competently in society, it shapes our sense of self by enabling us to recognize our place in the world. Our identity consists of a set of perceptions about who and what kind of person we are. This is partly formed by our social position or status, and partly by the way we enact that status. It involves a dynamic tension between our developing awareness about a particular role (e.g., college athlete), the role-set members who interact with us in that role (coaches, teammates, other students, teachers, fans, boosters, the media), and the way each of us enacts that role individually (hot-dog, team player, star, regular guy, spirit-raiser, follower, role model, clown, leader, or rebel). There may be many people who learn a particular role, but each may enact it differently.

The selections in this section take us through a small portion of the way people become socialized and form their identities. We begin by looking at our own study of elementary school social cliques that focuses on the effects of peer socialization, particularly examining the power of popular clique leaders to influence other members of the group. Social group membership is one of the most critical features of young people's lives and their struggles to gain and retain inclusion in the popular clique are revealed in this study of preadolescent clique dynamics. You may relate to the way children learn the subtle vicissitudes of leadership and followership, in-group and out-group behavior, and how to twist and turn with the winds of shifting popularity and fads. While it may appear to portray children's behavior as excessively vicious, most people who read it find that it offers key insights into the micro-politics of everyday life throughout adulthood. Next, Melissa Milkie's article examines the way Black and White girls' self-images are differently shaped by media socialization. She shows how people's identities are affected by the way their subculture intervenes between them as individuals and the larger societal media messages they encounter. White and Black girls react in dissimilar ways to the potentially destructive images displayed of women's ideal bodies in teen magazines, largely because their racial subcultures accept or reject these images as relevant. Some groups develop a greater ability to resist the effects of social labeling and its identity consequences than others. Socialization to identities does not just occur in childhood, as people grow and change throughout their lives. Finally, Vichet Chhuon and Cynthia Hudley consider the influence of ethnic socialization, as they examine Cambodian-American high schoolers' identity presentations. Since they fall under the larger Asian rubric, they may incur the benefits of the broader Asian-American social status, receiving positive expectations and treatment from their teachers and peers rather than a more specifically Cambodian identity, which is stigmatized as lazy, disrespectful, and prone to trouble. Cambodian-American youth have to choose between these two very different self-presentations and incur both the benefits of their choice in some groups and resentment for their choice in others.

Social interaction, or the way people act individually and in groups in everyday life, forms the core of our interpersonal behavior. Socialization heavily influences interaction, as it teaches people how to behave and how to anticipate others' reactions to various behaviors. Part of learning how to interact in society is learning to read the response of others to your actions. Charles Cooley's concept of the "looking glass self" is particularly germane to helping us understand how we act. It is amazing, when we think about it, that we can simultaneously do something, think about how others are reacting to it, and think about how we might adjust our ongoing actions in light of our perception of their interpretations. That we can think separately and independently from our actions while we are engaged in them represents very sophisticated and skilled social competence.

Symbolic interactionists differ in the role they attribute to emotion in behavior. Some see our actions as involving a simple cost-benefit calculation, where we select the path that will net us the most gain. Others attribute a greater role to feelings, suggesting that we often choose paths that fall outside of these logical-rational economic models for emotional reasons. We may take a less advantageous path to further our self-image as good people, because of strong, driving feelings of jealousy, fear, or hatred, or just because we feel like it at the moment. Our behavior represents a fusion of these kinds of motivations.

Erving Goffman wrote extensively about the micro-interactional norms that govern behavior in public places. Although it may be clear why we want to look good in the eyes of significant others, Goffman pointed out that we also seem to care very much about the way we appear to complete strangers. He suggested that we are very conscious of our actions and the anticipated responses of all kinds of people, and that our lives represent acting out our private selves on the public stage of life. In making these self-presentations, we are often assisted by others in one of two ways: they may be insiders who help us to achieve our performances in front of outsiders or they may be members of our audience who help us to maintain a positive sense of self by not pointing out our public gaffes (our fly is unzipped, we have a speck of food in our teeth, we're calling someone by the wrong name). Goffman's ruminations about the micro-social rules of public interaction are a fascinating topic of study.

Applying some of Goffman's key concepts, we see Spencer Cahill and his students peering into a place we have all visited: public bathrooms. Their analyses of these "backstage" regions located within public space and the norms of demeanor within them will be sure to resonate with your experiences. The bathroom setting clearly illuminates the difference between the types of behavior that occur in relative privacy versus those that appear in the "frontstage" realms. How do social norms affect people's behaviors in even these private settings? The dramaturgical analogy is further carried out in the next chapter by David

Schweingruber, Sine Anahita, and Nancy Berns, who write about the dramaturgical aspects of the engagement proposal, a more private performance. Like asking someone on a first date or to an important dance, "popping the question" has developed a set of unspoken but powerful norms that guide actors' behavior. How people enact this symbolically important rite says a lot about their gender roles and values. This selection further shows how the behavior that publically occurs (and is recounted to others) is tightly scripted—underlying forms of communication belie the overt acts. It is amusing to see the strategies people employ to foster the best possible impressions of themselves for the public and private audiences.

Last, we look at the role of *crime and deviance* in shaping the contours and character of the sociological imprint. Although we often think of deviance as coming from some element intrinsic to the offense or offender, sociological research has shown that definitions of deviance are highly relative. Any act, no matter how heinous, can be defined as legitimate under certain circumstances. Even murder is deemed acceptable when it is committed by the State, in self-defense, in war, or, in "make my day" states, when someone threatens you on your property. Further, what one society, organization, or group deems deviant may also be considered offensive by another, yet something that is appropriate but still a different group. Thus, getting drunk and brawling may be commonplace in some bars, but not in church, and smoking pot may be normative among your friends, but not among your extended family members. Even cigarette smoking, a legal act, has become quite stigmatized in some communities so that those who engage in this activity are ostracized and segregated from others.

When we examine what lies at the root of social definitions of deviance, we discover the importance of social power. Definitions of deviance are forged through "morality campaigns," where some groups are successful in raising awareness, accumulating the testimonials of experts and the endorsements of influential opinion leaders, spreading this information, and recruiting various organizations to support their drives. They may vie against other politicized groups to determine whose opinions become legislated into morality and law, such as we see with the continuing struggle over abortion. In the final analysis, groups that have the power to mobilize these critical resources are able to enforce their opinions onto others, thus legitimating their behavior and criminalizing that of weaker groups. The legal status of the drugs preferred by the White middle class (alcohol, tobacco, marijuana, Prozac) compared to those preferred by inner-city, lower class, minority populations (crack, heroin) is testament to this situation, as the latter drugs are not necessarily inherently more harmful than the former, yet they are significantly more heavily criminalized and enforced.

Social norms and values exist in a state of relative flux, as they evolve and change over time. Some things that used to be considered deviant have been normalized, such as divorce, tattoos, and Ritalin, while others that were acceptable have now come to be redefined in a deviant light, such as sexual harassment, date rape, and smoking cigarettes. Social power plays a prominent factor in influencing definitions of deviance and the way these are applied. Devah Pager explores the affect of race on attributions of deviance through her clever experiment on job discrimination. Constructing matched sets of job candidates whose records differed primarily in their race and their past criminal record, Pager traced how these variables hindered or helped applicants to positions advertised in the newspaper. Her findings highlight the enduring presence of racial stereotypes and prejudice in our country today. Next, A. Ayres Boswell and Joan Z. Spade focus on the social context of two different fraternities, showing how one fosters a cultural environment that enables sexual violence against women to a greater degree than the other. They describe the settings, gender relations, and treatment of women that characterize these groups, tying these norms and values to their different behavioral outcomes.

Deviance serves many functions in society, some of which include laying the groundwork for social change and fostering full employment (for lawyers, judges, prison officials, treatment administrators, police, private investigators, and criminologists). Deviance fosters boundary maintenance by reinforcing normative behavior and sanctioning acts outside the acceptable limits. People are led to deviance by many factors, some lodged within the social structure, some culturally derived, and some learned through interaction with knowledgeable others. While Pager's chapter highlights structural factors outside the control of specific individuals such as racism, and Boswell and Spade examine the different cultural norms and values that shape fraternity rape behavior, Adina Nack provides an interactionist examination of sexually transmitted diseases (STDs), showing how women who contract these try to manage people's knowledge about their conditions to mitigate the effects of social stigma and developing a spoiled sexual identity. Do you think that the stigma of STDs has abated in our society or is it still as hidden as before? How conscientious do you think college students are about protecting themselves from contracting sexually transmitted infections? Where do you think most people learn the facts about these diseases and their spread? Where do you think they could effectively learn more about them?

9

White Universal Identity as a 'Sense of Group Position'

PAMELA PERRY

Being White has long been defined in the United States as the default race, a form of "colorless" and somewhat "flavorless" ethnicity against which all others are contrasted. White culture and identity have largely been taken for granted and overlooked by White Americans, regarded as mainstream and "normal." Historically, White people have enjoyed a position of power based on their dominance in social, financial, political, legal, and other arenas. Yet while Whites assume the perquisites of the majority group, they are often socialized to ignorance about the privileged nature of their position. Perry's research in two high schools of different racial composition, one predominantly White and the other multiracial with Whites representing a minority group, shows how White youth feel about their ethnic identity, heritage, and culture and how they react when their position of dominance is challenged. How have you been taught about your cultural heritage and how do you define yourself in relation to the mainstream? To what extent do you share the feeling of Perry's students, White or non-White, about your own racial/ethnic identity? What does the response of young Whites to the threat to their universal group position tell us?

N early fifty years ago, Herbert Blumer published "Race Prejudice as a Sense of Group Position" (1958). Much has transpired in race relations since then, and an entire field of study examining "whiteness" has come to flourish and reshape the paradigms by which studies of race, generally, are approached....

SOURCE: Pamela Perry, "White Universal Identity as a 'Sense of Group Position.'" *Symbolic Interaction* 30(3): 375–93, 2007. Reprinted by permissions of the University of California Press and the Society for the Study of Symbolic Interaction.

Blumer's group position model invites us to critically consider the role that dominant group identity plays in the reproduction of racial inequalities. I assert that identities seat both material and ideal concerns and "white" identities, in particular, provide a type of "ontological security" that white people will defensively protect. I make this argument with ethnographic research I conducted in 1994–96 in two demographically distinct high schools. Young whites in both schools expressed identities that positioned them as "universal," and they responded reactively, even prejudicially, when their universal group position was challenged.

BLUMER'S "SENSE OF GROUP POSITION"

In his classic and seminal piece, Blumer makes several assertions that offer a comprehensive explanatory framework for understanding the beliefs, behaviors, and feelings of white people that resist change to the racial status quo. He asserts that race "prejudice" is a matter of racial identification or a "sense of group position" that group members have of themselves and an outside group. I put "prejudice" in quotes because Blumer does not use the term in the conventional way, that is, to refer to individual and irrational antipathies people may have for those of a different race. He uses it to refer to *collective* and *rational* sentiments—negative feelings and discriminating behaviors resulting from a perceived threat to the sense of group position.

To make his argument, Blumer points out that the feelings commonly related to racial prejudice of dominant group members all speak to their positional arrangement. He lists these as (1) feelings of natural superiority; (2) feelings that the subordinate group is "fundamentally different stock" and must be socially excluded; (3) feelings of entitlement and proprietary claim to exclusive rights in both public and private areas of life; and (4) "fear or apprehension that the subordinate racial group is threatening, or will threaten," the dominant group's sense of superiority and entitlements (Blumer 1958:4). Blumer emphasizes that the first three feelings in this list may engender feelings of aversion or antipathy, but the last feeling—of threat—is essential for igniting discriminatory, exclusionary, and/or hostile behaviors of whites toward others....

WHITE "UNIVERSAL" IDENTITY

Among most white Americans today, white racial identity is elusive, at best, and is, indeed, experienced more as a "sense" of group position than as a clearly defined identity. This is largely because white culture and identity are unmarked and positioned as "normal" in the wider American mainstream. Whiteness studies emerged in large part to examine and disrupt the social construction of whiteness as an unmarked norm and the ways that whiteness sets a standard by which "difference" is measured. There are two interrelated ways of understanding the processes by which the intangibility and invisibility of white culture and identity are constituted: through hegemonic processes of "normalization" (Doane 2003:12), which for current purposes I call "universalization," and through the long career of colonial discourses that position whites as signifiers of *the* universal.

Since the civil rights era, de jure racism and explicit ideologies of white supremacy are no longer part of the explicit mainstream (white) American culture and identity. However, those shifts did not significantly change the racial state and white racial rule, which, as Omi and Winant (1986) argue, over time had come to be exercised less through direct discourse and action and more through processes of hegemony—through discursively embedding the political and economic interests and privileges of whites into a transparent "common sense" of the nation as a whole. As Doane (2003:12) writes, "The combination of existing domination with transparency enables 'whiteness' to be cast—but not named—as the larger society, the cultural mainstream, and the nation." The universalization of white culture and privilege has impacted not only the workings of the white racial state but the racial identities and subjectivities of whites themselves. This era is marked by the relative *absence* of public, collective racial definition among whites, and that, coupled with the unmarked, universal character of white culture, leave whites believing that being white has nothing to do with the ways they think and act in the world and with the entitlements they enjoy.

However, the normalization of the dominant group culture and identity is not particular to whites but, according to Rosaldo (1989), is true for every group that has numerical and political dominance over a nation. What *is* particular to whiteness and has shaped historical definitions of what being white means is the collective definition of whiteness as the model of the universal—the perfect and ultimate expression of humanity. As Kant exclaimed, "Humanity is at its greatest perfection in the race of the whites" (quoted in Eze 1997:63). This is the fundamental grounding on which whites have historically claimed a sense of natural superiority. Although it is difficult to pinpoint exactly when this sense of group position arose, it is clear that by the late sixteenth century, processes of collective definition of what constituted a human and, by extension, the white Western self were well underway. Over time, as the Enlightenment notion of universal humanity gained ground and white European colonial interests and practices needed justification, it became public business to define who and what a human was. As Balibar (1994:197) notes, "No definition of the human species, or simply the human—something that is so crucial for universalism … has ever been proposed which would not imply a latent hierarchy." That is, defining the perfect, universal model of humanity necessitated defining what *is* universal against what is *particular* and, by extension, defining who stands on the "superior" rung of the human scale and who not. In the minds of white, Western males, they, themselves, made a perfect template against which the particular and the subhuman could be measured, so their concepts of rationality, civilization, and aesthetics became measuring sticks for differentiating the human "species." As Hume wrote, "There never was a civilized nation of any other complexion than white…. No ingenious manufactures among [the negroes], no arts, no sciences" (quoted in Eze 1997:33). He implies, thus, that "civilization," the arts and sciences, distinguished the "superior" beings from the inferior. Human perfection had a moral and spiritual dimension as well, as was illustrated in the great chain of being, where white men were represented as the highest order of earthly beings, one step from the realm of angels and divine perfection.

Contemporary white culture and identity, though no longer wedded to blatant expressions of white racial superiority, are nonetheless still shaped by colonial discourses of self-other/universal-particular that invoke and underhandedly sustain notions of white supremacy. Frankenberg (1993) found this in her interviews with white women, and I found them in my discussions with white teenagers. Unlike Frankenberg, however, my work also explored the emotional investments whites attach to their sense of group position.

To summarize, white racial identity has three key components that may varyingly rise to salience depending on the circumstances and interrelational context. The first is, simply, the psychological and ontological need for narrative continuity of the self, which identities tend to provide; next is the universalized character of white culture and identity, which whites tend to experience as emptiness; and last, white identity *as* universal, the signifier of perfect human rationality and morality....

METHODS AND DATA

From January 1994 through June 1996, I conducted ethnographic research in two high schools, one predominantly white and the other multiracial and minority white, examining what differences, if any, the contrasting social contexts made on the racial consciousnesses and identities of white students. Initially this research was in large part motivated by personal questions. As a white woman who grew up in a predominantly white suburb in the 1960s, I was interested in examining the construction of my own racial consciousness and identity, which, indeed, seemed to be "empty." When, in my early twenties, I began to become engaged in social justice activism and considered myself antiracist, it came as something of a shock for me to realize that I might inadvertently reproduce racist norms and structures because of the way my mind and identities had been structured by white supremacy. This research was an opportunity to go back to my own "routes" with an insider-outsider eye, and my personal interest in the project framed the way I conducted the research and how I've interpreted it (and continue to interpret it).

I spent two and a half years in the field. The first year and a half I attended, fulltime, "Clavey High School," the multiracial school in my study, while visiting the other school once a week. Then I spent a full year, full time, at the predominantly white school, "Valley Groves," while returning to Clavey once a week or so. At the time of this research, Clavey High School resided in one of the most racially and culturally diverse cities in the nation. The high school population itself was 54 percent African American, 23 percent Asian, 12 percent white, 8 percent Latino, and the rest of mix of Filipino, Pacific Islander, and Native American. At

Valley Groves High School, "non-Hispanic" white students made up 83 percent of the school population, followed by Latinos (7 percent), Asians (5 percent), then Filipinos and African Americans (2 percent each). The two towns from which Valley Groves High drew students were 83 percent and 93 percent white....

DEFINITIONAL PROCESSES

... One of the main reasons I chose to conduct this research in schools is because they are widely understood to be primary sites for shaping and organizing individuals and in ways that reproduce the status quo, including white dominance. Dominant group definitions of race and other social relations pervade the practices and discourses of teachers, staff, and students. For my research, I closely examined students' experiences of race in the school context and how those experiences were explicitly or implicitly "defined" by teachers, staff, and students themselves. I believe that three main school experiences and the discourses surrounding them were operative in defining white identity as universal: racial isolation, racialized tracking, and multiculturalism.

Racial Isolation

Racial isolation is an effective means of universalizing white culture, identity, and racial dominance. At Valley Groves, the predominantly white school, white students had little or no experience of racialized differences; white people and culture were ubiquitous and unchallenged. As Lloyd (1991:70) argues, white ubiquity legitimizes domination, "since the capacity to be everywhere present becomes an historical manifestation of the white man's gradual approximation to the universality he everywhere represents." Along the same lines, racial isolation helps construct a white habitus that is everywhere confirmed without being named. When whiteness as a symbolic structure shapes social practice in a hegemonic way, it is taken for granted, and being white is understood as a natural, universal way of being. Moreover, at Valley Groves, "big events" (Blumer 1958:7) of racial significance, like the O. J. Simpson trial and the Million Man March, were either not discussed or discounted as racist *because* they focused on racial issues and tensions. White students heard no counter arguments or perspectives to challenge their assumptions about race and whiteness.

TRACKING

At Clavey, the multiracial school, whiteness could not be wholly constructed as a universal and hegemonic norm through processes of cultural ubiquity, because white students had considerable association with people of different cultures and values. White culture was put in relief in that setting and reflected upon.

Instead, a felt sense of universal group position was defined through active discursive processes, namely colonial-legacy discourses of self-other that positioned whites as more universal (more rational, intelligent, orderly, etc.) than students of color. Tracking was a key site for promulgating such discourses.

As in so many multiracial schools, students were tracked such that whites were overrepresented in the honors and advance placement classes, and blacks, Latinos, and immigrant Asians in the low-level and remedial classes. High-tracked classes at Clavey tended, more than low-tracked classes, to teach more culturally valuable and universalistic modes of thought and behavior, such as abstract creativity, liberal individualism, and entrepreneurship. Students in low-tracked classes tended to have more rote learning types of assignments and to be taught "discipline."

Moreover, racialized tracking at Clavey structured an "us-them" paradigm that fundamentally reproduced rationalist binaries of good/bad, rational/irrational, universal/particular—and positioned whites at the more valued end of each. Teachers, staff, and students alike participated in processes of interpreting and defining the tracking system and, by extension, racial group positions. Students in the accelerated classes were generally described as "well-behaved" and "good," and inside those classrooms they were highly controlled—listening attentively and taking notes, speaking only when called upon. In contrast, students in the regular classes were referred to as "bonehead," "rowdy," and "out of control." I overheard teachers in the lounge complaining vociferously about "those children," the "problem students." Jeannie Oakes (1994:87) asserts that racialized tracking reinforces and "institutionalizes racist conception of intellectual capacity." I suggest it also defines, reinforces, and institutionalizes white group position as the model of universal goodness and rationality.

Multiculturalism

At both schools, "multicultural" programs and discourses were school practices that defined whites as "not particular," "general," and not ethnic. This was most true at Clavey, however, because Valley Groves had only one multicultural event a year, whereas Clavey had two main, formal multicultural practices with events throughout the year: the "cultural assemblies" and "multicultural week." Once every other month or two an ethnic club—the African American student union, the Asian student union, the Latino student union, or the Intertribal student union—would put on a schoolwide assembly. Commonly, these were staged spectacles created by and for students of color that illustrated the history, arts, and folk cultures of their respective racial-ethnic group. There were no similar spectacles of white culture. When I asked teachers and students why not, they tended to say, "Because white culture is everywhere."

Multicultural week, which occurred once or twice a year, was another opportunity for students to show the styles, tastes, and practices of their cultural heritage. It consisted of lunchtime events each day for a week, in which students carried out ethnic fashion shows, performed dances, played music over loudspeakers, and offered various games. On rare occasions a white student would participate; once, for example, a girl of German ancestry paraded a traditional German costume at a fashion show. However, one white girl told me that a fellow student would not allow her to play some punk rock during multicultural week because it was "white people's music," drawing a distinction between those who have "culture" (people of color) and those who don't (whites).

The merits of these events were that they gave voice and visibility to the cultures and perspectives of people historically silenced by white domination. However, multicultural events also reproduced white supremacist, universal tenets of white identity. They consisted largely of public displays of the particularities of "ethnic" students before an audience in which white students were always spectators and never participants. White students tacitly understood that they had no cultural particularities of their own. Many white youth I spoke to appreciated these events for the opportunity to "learn about other cultures," but, as Rosaldo (1989:202) has observed, "as the Other becomes more culturally visible, the self becomes correspondingly less so." At the same time, I understood white students' appreciation for "learning other cultures" to support their "eclectic" identities through helping them develop diverse social skills. In short, multicultural events and the ways students interpreted them helped define and reinforce a sense of whiteness as not particular, as eclectic and everywhere—as universal.

THREATS TO GROUP POSITION

... Blumer's consistent use of the terms "feelings," "general feelings" and "sense" to describe group position has been particularly useful in analyzing my findings. Because white culture and identity are undefined, for the white students in my study, their sense of group position and threats to it were a *felt* sense. They spoke a lot in terms of feelings, and I interpreted many of their feelings to derive from a core sense of self. When I was talking with white students at Valley Groves and Clavey, there was rarely a time when I didn't hear them express—either in words or affect—deep, authentic feelings, both positive and negative, about issues concerning race, whiteness, and inequalities. At times I recognized feelings of threat that related to feelings of propriety or competitive threat, such as resistance to affirmative action on the basis of their entitlement to college entry and scholarships because of good grades ("If I have a 4.0 and they give my spot in college to a black guy with a 3.6 because of race, well I can't accept that"). In these instances, whites saw themselves in competition for scarce resources with people of color, a common defining experience of group position. Usually, however, I could not identify a reactive response that was due to feelings of competitive threat per se. These were fifteen- to seventeen-year-olds, themselves members of a subordinated group—childhood. And while some came from fairly privileged backgrounds, I did not get a sense that fear of losing economic livelihood or political or cultural power factored into what were often very strong feelings about race that sometimes sounded protective of the racial status quo. The threat that I read into the various discomforts and fears of whites had more to do with challenges to the foundations of their senses of selfhood.

Because white students experienced their group position differently between the two schools in my study, what they felt most threatened by differed as well. At Valley Groves, whites felt threatened by being racially marked and, hence, particularized—an upset to the sense of universal selfhood. They also responded defensively to threats to universalistic core values. At Clavey, white students were accustomed to being marked and particularized, and this, in itself, did not disturb them. However, more so than the whites at Valley Groves, Clavey whites were threatened by the challenge to maintain narrative cohesion as "good" while facing anger from students of color over white domination. Here are a few cases in point.

White students at Valley Groves fairly uniformly told me that they either felt or believed they would feel "uncomfortable" in the company of a lot of people of color. Some expressed that they wouldn't know how to act, but most mentioned something having to do with being marked and stereotyped. A young man named Howie said, "I'd have to say, if I am the only white person around, I feel uncomfortable.... But I mean, it's not that I'm afraid of them. It's just that I think they look at me and they say, 'He's a racist.'"

Another student told me of going to a track meet in which he was the only white male among African Americans running in an 800-meter race. He said that he saw the other sprinters talking together before the race, and he was sure they were talking about him. "I felt like I was being discounted as a factor in the race," he said. "Like I was not taken seriously."

Both of these examples illustrate the process Mead ([1934] 1962) described as "roletaking," seeing yourself through the other's eyes and interpreting their judgments of you. In these and other student's narratives, what they saw was the other *racializing* them, marking them as white and making judgments based on that. White students also found themselves in situations where they were asked to racialize themselves, and this was equally disconcerting. As part of the consent forms that I asked students to sign for our interviews, I included a request for students to check a box that best described their racial group membership. This frequently incited strong and animated comments by white students about the awkwardness of and their resistance to checking such boxes. They would complain about the seeming ubiquity of the racial identification boxes on test forms and applications and about how "stupid" the boxes were. They would say that they didn't know what to check. "I don't know, I'm a mutt"; "I guess I'm 'white' but I don't *feel* white!" As an expression of their resistance to such requests they would "mark everything" or "leave them blank."

A recent experience further illustrates the strong feelings that being particularized can raise among whites. In a class in which I teach race and whiteness, usually just after mid-quarter, white students in the class start erupting into anger at being racialized. One time, a student was giving a presentation on a rock band, pointing out that the band and its fans were white, but, she said, "race didn't matter." When I suggested that "race not mattering" was what *whiteness* was about, the class of over 120 students broke into pandemonium. White students began yelling, "That's not white!" "Race doesn't matter!" I replied, "To *whom* doesn't race matter?" and white students yelled back, "To youth!" or "To me!" Finally, one of the few African American students in the class raised her hand and said, "We've been in this class for six weeks talking about black culture, Latino culture, and all kinds of issues about race, but now when we are talking about *white* culture, you can't stand it. You're all upset because we're *naming* it!"

White students at Valley Groves became even more emotionally excited over issues that challenged their moral sense of the "rightness" of a universalistic orientation toward everyone—blacks, whites, Latinos, everyone. This response would especially emerge when the topic of affirmative action came up. For example, when Jonathan, a young conservative man, spoke the following, he punctuated his words with passionate inflections.

> [I] have no prejudices based on color. I mean it goes into affirmative action and stuff, you know? I feel everyone should be treated equally. And equally based on their color or race. Everyone should be treated as the same person. 'Cause everyone's really the same. Everyone's alike. You know? Color doesn't make a difference.

Comments such as Jonathan's have been quoted to illustrate "color-blind racism," the ways whites use color-blind and universalistic ideologies to dismiss the significance of race in perpetuating contemporary racial inequalities. While I believe that color blindness has emerged in recent years as a political ideology and discourse and is used to rally support behind antiaffirmative action measures and the like, I suggest that the appeal of the ideology is, at least in part, the resonance it has with deep-seated universalistic values and sense of self-hood among whites. It's not that universalistic values are Machiavellian in intent, they are two-faced—they consist of progressive beliefs in the natural equality of all human beings but are predicated on the belief that some are more human than others and, hence, more equal than others (Balibar 1994:197–98). "Color-blind" beliefs and sentiments are in part expressions of wider universalistic values of whites and embody the same contradictions as those values—in principle they are a good idea, but in practice they are reproductive of white dominance. White students at Valley Groves tended to deeply embrace universalistic values—even more so than at Clavey—but they stemmed more from a sense of the "goodness" of the principle and an obliviousness to the practice. Indeed, the majority of times that I engaged students in conversations about sustained inequalities because of ongoing forms of racial discrimination in education and hiring, they would begin to reconsider their position and take on a more positive outlook on affirmative action or other race-targeted reparative policies....

CONCLUSION

... More research is needed to examine whether the perceptions and feelings of white identity that I interpreted among youth persist into adulthood, and, if so, what effects white identity has on adult political behavior. The research of Kinder and Sears (1981; Kinder 1986; Sears 1988) may be indicative, with its finding of the relevance of values for influencing the political choices of whites. However, I see the possibility of attachments to white identity showing up in what Blumer (1965:335) called the "inner citadel" of the color line. He argued that even after subordinated groups have come to receive full civil rights and equal economic opportunity, an inclusion-exclusion axis pertaining to intimate and private areas would still need to be addressed and may be particularly resistant to change. This suggests to us that structural change is undoubtedly a necessary condition to end racism, but it may not be definitive. We need to work at dismantling all the ramparts of the color line. I have found in my own work as a teacher of young political activists—most of whom are white, savvy about white privilege, not mystified by ideologies of color blindness, and full of passionate outrage over social injustices—that the most difficult yet rewarding passages whites take in becoming effective antiracists is in developing the tools for critically reflecting on the assumptions they hold about what "white" means and how their taken-for-granted values and sense of selfhood reproduce the racial status quo.

REFERENCES

Balibar, Etienne. 1994. "Racism as Universalism." Pp. 191–204 in *Masses, Classes, Ideas: Studies on Politics and Philosophy before and after Marx*, translated by J. Swenson. New York: Routledge.

Blumer, Herbert. 1958. "Race Prejudice as a Sense of Group Position." *Pacific Sociological Review* 1(1): 3–7.

———. 1965. "The Future of the Color Line." Pp. 322–36 in *The South in Continuity and Change*, edited by J. McKinney. Durham, NC: Duke University Press.

Doane, Ashley W. 2003. "Rethinking Whiteness Studies." Pp. 3–20 in *White Out: The Continuing Significance of Racism*, edited by A. W. Doane and E. Bonilla-Silva. New York: Routledge.

Eze, Emmanuel Chukwudi. 1997. *Race and the Enlightenment: A Reader*. Oxford: Blackwell.

Frankenberg, Ruth. 1993. *White Women, Race Matters: The Social Construction of Whiteness*. Minneapolis: University of Minnesota Press.

Kinder, Donald. 1986. "The Continuing American Dilemma: White Resistance to Racial Change 40 Years after Myrdal." *Journal of Social Issues* 42: 151–71.

Kinder, Donald and David O. Sears. 1981. "Prejudice and Politics: Symbolic Racism versus Racial Threats to the Good Life." *Journal of Personality and Social Psychology* 40(3): 414–31.

Lloyd, David. 1991. "Race under Representation." *Oxford Literary Review* 13(1–2): 62–94.

Mead, George H. [1934] 1962. *Mind, Self, and Society*. Chicago: University of Chicago Press.

Oakes, Jeannie. 1994. "More Than a Misapplied Technology: A Normative and Political Response to Hallinan on Tracking." *Sociology of Education* 67(2): 84–88.

Omi, Michael and Howard Winant. 1986. *Racial Formation in the United States: From the 1960s to the 1980s.* New York: Routledge and Kegan Paul.

Rosaldo, Renalto. 1989. *Culture and Truth: The Remaking of Social Analysis.* Boston: Beacon.

Sears, David. 1988. "Symbolic Racism." Pp. 53–84 in *Eliminating Racism: Profiles in Controversy,* edited by P. A. Katz and D. Taylor. New York: Plenum.

10

Core Values of the Straight Edge Movement

ROSS HAENFLER

Straight edge (sXe) kids represent a subculture that stands in opposition to the dominant, mainstream culture. Members tend to be young and alienated from the customary norms of high school-aged students. Haenfler traces the development of oppositional youth subcultures from the hippies to the skinheads, punks, and straight edgers. He shows how each group evolved from the previous ones, casting themselves against the mainstream and against other youth countercultures. Straight edge youth reject the use of alcohol or other drugs, reject the casual sex of the "hooking up" scene, and may dabble slightly or more seriously in environmental, vegetarian, and feminist movements. How important do you think oppositional youth subcultures are for people growing up? To what extent do young people need to reject the lifestyle and values of their parents and community to help them forge their own identities? Do you think young people hold onto their high school and college values in becoming adults, or do they just stray from their parents' values temporarily and return to them after a brief period? In other words, does youthful idealism fade, or does it become incorporated into people's adult lives? To what other subcultures does sXe compare?

R esistance has been a core theme among both subcultural participants and the scholars who study them. Early subcultural theorists associated with Birmingham University's Centre for Contemporary Cultural Studies (CCCS) concentrated on the ways youth symbolically resisted mainstream or "hegemonic" society through style, including clothing, demeanor, and vernacular (Hebdige 1979). Subcultures emerged in resistance to dominant culture, reacting

SOURCE: Ross Haenfler, "Rethinking Subcultural Resistance: Core Values," *Journal of Contemporary Ethnography* 33, 4, pp. 407–436. Copyright © 2004 by SAGE Publications. Reprinted by permission of SAGE Publications, Inc.

against blocked economic opportunities, lack of social mobility, alienation, adult authority, and the "banality of suburban life" (Wooden and Blazak 2001, 20). Theorists found that young working-class white men joined deviant groups to resist conforming to what they saw as an oppressive society (Hebdige 1979; Hall and Jefferson 1976). Scholars have given a great deal of attention to whether these youth subcultures resist or reinforce dominant values and social structure. The CCCS emphasized that while subcultural style was a form of resistance to subordination, ultimately resistance merely reinforced class relations. Therefore, any such resistance was illusory; it gave subculture members a feeling of resistance while not significantly changing social or political relations. In fact, according to this view, subcultures often inadvertently reinforce rather than subvert mainstream values, recasting dominant relationships in a subversive style....

As a relatively unstudied movement, straight edge (sXe) provides an opportunity to rethink and expand notions of resistance. The straight edge[1] movement emerged on the East Coast of the United States from the punk subculture of the early 1980s. The movement arose primarily as a response to the punk scene's nihilistic tendencies, including drug and alcohol abuse, casual sex, violence, and self-destructive "live-for-the-moment" attitudes. Its founding members adopted a "clean-living" ideology, abstaining from alcohol, tobacco, illegal drugs, and promiscuous sex. Early sXe youth viewed punk's self-indulgent rebellion as no rebellion at all, suggesting that in many ways punks reinforced mainstream culture's intoxicated lifestyle in a mohawked, leather-jacketed guise.

Straight edge remains inseparable from the hardcore[2] (a punk genre) music scene. Straight edge bands serve as the primary shapers of the group's ideology and collective identity. Hardcore "shows" (small concerts) are an important place for sXers[3] to congregate, share ideas, and build solidarity. Since its beginnings, the movement has expanded around the globe, counting tens of thousands of young people among its members. In the United States, the typical sXer is a white, middle-class male, aged fifteen to twenty-five. Straight edgers clearly distinguish themselves from their peers by marking a large X, the movement's symbol, on each hand before attending punk concerts. While scholars have thoroughly researched other postwar youth subcultures such as hippies, punks, mods, skinheads, and rockers (e.g., Hall and Jefferson 1976; Hebdige 1979; Brake 1985), we know little about sXe, despite its twenty-year history.

The basic tenets of sXe are quite simple: members abstain, completely, from drug, alcohol, and tobacco use and usually reserve sexual activity for caring relationships, rejecting casual sex. These sXe "rules" are absolute; there are no exceptions, and a single lapse means an adherent loses any claim to the sXe identity. Members commit to a lifetime of clean living. They interpret their abstention in a variety of ways centered on resistance, self-realization, and social transformation. Clean living is symbolic of a deeper resistance to mainstream values, and abstinence fosters a broader ideology that shapes sXers' gender relationships, sense of self, involvement in social change, and sense of community....

PREVIOUS YOUTH SUBCULTURES

Studies of hippies, skinheads, and punks demonstrate both similarities and profound differences between these groups and the sXe movement. Hippies evolved in the mid-1960s from the old beatnik and folknik subcultures (Irwin 1977; Miller 1999). Their lifestyle was a reaction to the stifling homogeneity of the 1950s, emphasizing communalism over conformity and deliberate hedonism over reserve.... "If it feels good, then do it so long as it doesn't hurt anyone else" was the scene's credo. Hippie core values included peace, racial harmony, equality, liberated sexuality, love, and communal living.... They rejected compulsive consumerism, delayed gratification, and material success (Davis 1967). "Dope," however, was one of the group's most visible characteristics.... Dope differed from drugs; dope, such as LSD and marijuana, was good, while drugs, such as speed and downers, were bad. For hippies, dope expanded the mind, released inhibitions, boosted creativity, and was part of the revolution. It was the means to discovering a new ethic, heightening awareness, and "understanding and coping with the evils of American culture" (Miller 1991, 34). Like dope, sex, in its own way, was revolutionary. "Free love" rejected the responsibilities normally associated with sexual relationships: marriage, commitment, and children.... By practicing what most at the time would call promiscuous sex, the hippies deliberately threw their irreverence for middle-class values in the face of dominant society (Irwin 1977).

Skinheads received a great deal of attention during the 1990s, as reports of their growing membership in neo-Nazi groups infiltrated both popular media and scholarly work (Bjorgo and Wilte 1993; Moore 1994; Young and Craig 1997). Skinheads emerged in late-1960s Britain as an offshoot of the mod subculture (Cohen 1972; Hebdige 1979). While most of the fashion-conscious mods listened to soul music, frequented discotheques, and dressed in impeccably pressed trousers and jackets, the "hard mods," who eventually became the skinheads, favored ska and reggae, local pubs, and a working-class "uniform" of heavy boots, close-cropped hair, Levi jeans, plain shirts, and braces (suspenders) (Brake 1985). While the mods attempted to emulate the middle-class, hip 1960s style, the skins were ardently working class. Nearly everything about skinheads revolved around their working-class roots. Hard work and independence were among their core values; they abhorred people, such as some hippies, who they believed "live off the system." Skinheads were extremely nationalistic and patriotic, adorning themselves with tattoos, T-shirts, and patches of their country's flag. After a long day at work, they enjoyed drinking beer with their friends at the local pub. Although there were some women skins, males dominated the subculture and often reinforced traditional patriarchal ideals of masculinity.

The original skinheads borrowed heavily from the West Indian culture, adopting their music, mannerisms, and style, including among their number a variety of races. While they were not violently racist at the level of the current neo-Nazi groups, these skins, both black and white, engaged in violence against Pakistani immigrants ("Pakibashing") (Hebdige 1979, 56). Eventually, with reggae's turn to Rastafarianism and black pride, many white skinheads became increasingly racist. At the turn of the century, three main types of skinheads prevailed: neo-Nazis (racist), skinheads against racism (e.g., Skinheads Against Racial Prejudice), and nonpolitical skinheads, who took neither a racist nor an antiracist stand (Young and Craig 1997). Skinheads were quite visible at punk, ska, and Oi! music shows, though the nonpolitical and antiracist skins were more prevalent. Very rarely, a skinhead was also sXe.

In many ways, punk was a reaction to "hippie romanticism" and middle-class culture; punk celebrated decline and chaos (Brake 1985, 78; Fox 1987; O'Hara 1999). In mid-1970s Britain, youth faced a lack of job opportunities or, at best, the prospect of entering a mainstream world they found abhorrent (Henry 1989). They attempted to repulse dominant society by valuing anarchy, hedonism, and life in the moment. Early punks borrowed heavily from the styles of Lou Reed, David Bowie ("Ziggy Stardust"), and other glam-rock and new-wave artists. Adorned with safety pins, bondage gear, heavy bright make-up, torn clothing, flamboyant hairstyles, and spiked leather jackets, punks lived by their motto "No Future," celebrating rather than lamenting the world's decline. They embraced alienation, and their "nihilist aesthetic" included "polymorphous, often willfully perverse sexuality, obsessive individualism, a fragmented sense of self" (Hebdige 1979, 28).

Like the skinheads, punks disdained hippies; the preeminent punk band the Sex Pistols titled one of their live recordings "Kill the Hippies" (Heylin 1998, 117). Unlike the skins, and like the hippies, however, punks chose to reject society, conventional work, and patriotism. Many used dangerous drugs to symbolize "life in the moment" and their self-destructive, nihilistic attitude (Fox 1987). Straight edge emerged relatively early in the punk scene and has shared certain values and styles with punks, hippies, and skins ever since. While some punks today are sXe, the two scenes have become relatively distinct, and the sXe movement has replaced many of the original antisocial punk values with prosocial ideals.

METHOD

My first encounter with sXe occurred in 1989 at the age of fifteen through my involvement in a Midwest punk rock scene. As I attended punk shows and socialized with the members, I noticed that many kids scrawled large Xs on their hands with magic marker before they went to a concert. I eventually learned that the X symbolized the clean-living sXe lifestyle and that many punks in our scene had taken on a totally drug- and alcohol-free way of life. Having tried the alcohol-laden life of most of my peers, I quickly discovered it was not for me. I despised feeling I had to "prove" myself (and my manhood) again and again by drinking excessively. I could not understand why the "coolest," the most highly regarded men were often the ones who most degraded women. Furthermore, given my family's history of alcoholism, I wanted to avoid my relatives' destructive patterns. Finally, the local sXers' involvement in progressive politics and activist organizations connected with my interest in social justice and environmentalism. My association with sXers led me to adopt the sXe ideology as what I viewed, at the time, to be an alternative to peer pressure and a proactive avenue to social change. After a period of careful consideration (like many punks, I was suspicious of "rules"), I made known my commitment to avoid consuming alcohol, drugs, and tobacco, and the group accepted me as one of their own. Since then, I have attended more than 250 hardcore shows, maintained the lifestyle, and associated with many sXers on a fairly regular basis. The data I present result from more than fourteen years of observing the sXe movement in a variety of settings and roles and interviewing members of the scene.

During college, my involvement with sXe waned, and for several years I had little contact with the group. After completing my undergraduate career, I moved to "Clearweather," a metropolitan area in the western United States, to begin graduate training....

I gathered data primarily through longitudinal participant observation... with sXers from 1996 to 2001. The sXers I studied were mostly area high school or university students from middle-class backgrounds. My contacts grew to include approximately sixty sXers in the local area and another thirty sXe and non-sXe acquaintances associated with the larger metropolitan hardcore scene. My interaction with the group occurred primarily at hardcore shows and simply socializing at sXers' houses.

To supplement my participant observation, I conducted unstructured, in-depth interviews with seventeen sXe men and eleven women between the ages of seventeen and thirty. To learn from a variety of individuals, I selected sXers with differing levels of involvement in the scene, including new and old adherents, and individuals who had made the movement central or peripheral to their lives. I conducted in-depth interviews at sXers' homes or at public places free from disturbances, recording and later transcribing each session. Though I organized the sessions around particular themes, I left the interviews unstructured enough that individuals could share exactly what sXe meant to them. I sometimes asked for referrals in a snowball fashion, though I knew most participants well enough to approach them on my own. The variety of participants allowed me continually to cross-check reports and seek out evidence disconfirming my findings. Through participant observation, I was able to examine how participants' behaviors differed from their stated intentions. I consciously distanced myself from the setting to maintain a critical outlook by continually questioning my observations and consulting with colleagues to gain an outsider perspective....

STRAIGHT EDGE CORE VALUES

A core set of sXe values and ideals guided and gave meaning to members' behavior: positivity/clean living, reserving sex for caring relationships, self-realization, spreading the message, and involvement in progressive causes. Adherents maintained that sXe meant something different to each person assuming the identity, and as with any group, individual members' dedication to these ideals varied. However, while individuals were free to follow the philosophy in various ways, often adding their own interpretations, these fundamental values underlay the entire movement.

Positive, Clean Living

The foundation underlying the sXe identity was positive, clean living. It was, as Darrell Irwin (1999) suggested, fundamentally about subverting the drug scene and creating an alternative, drug-free environment. Clean living was the key precursor to a positive life. Many sXers shunned caffeine and medicinal drugs, and most members were committed vegetarians or vegans.[4] Positive living had broad meaning, including questioning and resisting society's norms, having a positive attitude, being an individual, treating people with respect and dignity, and taking action to make the world a better place. Straight edgers claimed that one could not fully question dominant society while under the influence of drugs, and once one questioned social convention, substance use, eating meat, and promiscuous sex were no longer appealing. Therefore, clean living and positivity were inseparable; they reinforced one another and constituted the foundation for all other sXe values....

Refusing drugs and alcohol had a variety of meanings for individual sXers, including purification, control, and breaking abusive family patterns. Purification literally meant being free from toxins that threatened one's health and potentially ruined lives. Refusing alcohol and drugs symbolized refusing the "popular" clique altogether as well as the perceived nihilism of punks, hippies, and skinheads.

The movement provided young people a way to feel more in control of their lives. Many youth felt peer pressure to drink alcohol, smoke cigarettes, or try illegal drugs. For some, this pressure created feelings of helplessness and lack of control; acceptance often hinged on substance use. Straight edgers reported that the group gave them a way to feel accepted without using and helped them maintain control over their personal situations. Many sXers celebrated the fact that they would never wake up after a night of binge drinking wondering what had happened the previous evening. Adherents reported that sXe allowed them to have a "clear" mind and be free to make choices without artificial influence. Walter, a reserved twenty-one-year-old university student, explained,

> I don't make any stupid decisions.... I like to have complete control of
> my mind, my body, my soul. I like to be the driver of my body, not
> some foreign substance that has a tendency to control other people. I
> get a sense of pride from telling other people, "I don't need that stuff. It
> might be for you but I don't need that stuff." And people are like,
> "Whoa! I respect that. That's cool."

In addition to the personalized meanings the identity held for adherents, sXers viewed their abstinence as a collective challenge. The group offered a visible means of separating oneself from most youth and taking a collective stand against youth culture and previous youth subcultures, including punks, skinheads, and hippies. Furthermore, for many positivity and refusing drugs and alcohol were symbolic of a larger resistance to other societal problems including racism, sexism, and greed....

Reserving Sex for Caring Relationships

Reserving sex for caring relationships was an extension of the positive, clean life-style. Straight edgers viewed casual sex as yet another downfall of dominant society, their counterparts in other youth subcultures, and their more mainstream peers. It carried the possibility of sexually transmitted diseases and feelings of degradation and shame. Whereas hippies viewed liberated sex as revolutionary, punks saw it as just another pleasure, and skinheads valued sex as a supreme expression of masculinity, sXers saw abstinence from "promiscuous" sex as a powerful form of resistance. Rejecting the casualness of many youth sexual encounters, they believed that sexual relationships entailed much more than physical pleasure. They were particularly critical of their image of the "predatory," insatiable male, searching for sex wherever he could get it. Kent, a twenty-one-year-old university student with several colorful tattoos, said, "My personal views have to do with self-respect, with knowing that I'm going to make love with someone I'm really into, not a piece of meat." Kyle, a twenty-three-year-old senior architecture major at Clearweather University, said, "For me personally, I won't sleep around with a bunch of people just for health's sake. A good positive influence. [Sex] doesn't mean anything if you don't care about a person." Walter, the university student, said,

> For me it's just choosing how I want to treat my body. It's not something I'm just going to throw around. I'm not going to smoke or use drugs. My body is something that I honor. It's something we should respect. I think sex, if you're gonna do it you should do it, but you shouldn't throw your body around and do it with as many people as you want. If you love your body so much as to not do those things to your body you should have enough respect to treat women and sex how they deserve to be treated.

Though sXe values regarding sexuality appeared conservative when compared to many other youth subcultures, sXers were neither antisex nor homophobic as a group. Premarital sex was not wrong or "dirty" in the sense of some traditional religious views, and numerous sXers and sXe bands took a strong stance against homophobia. Sex could be a positive element of a caring relationship. Believing that sex entailed power and emotional vulnerability, sXers strove to minimize potentially negative experiences by rejecting casual sex....

The movement's "rule" against promiscuous sex was more difficult for members to enforce, and thus there was greater variation in belief regarding sex than substance use. Several of my participants, both males and females aged twenty-one to twenty-three, had consciously decided to postpone sex because they had not found someone with whom they felt an intimate emotional attachment. Most of the young women believed not drinking reduced their risk of being sexually assaulted or otherwise put in a compromising situation....

Self-Realization

Like members of other subcultures, sXers sought to create and express a "true" or "authentic" identity amid a world that they felt encouraged conformity and mediocrity. Straight edgers claimed that resisting social standards and expectations allowed them to follow their own, more meaningful path in life toward greater self-realization. Like punks, they abhorred conformity and insisted on being "true to themselves." Similar to hippies, sXers believed that as children we have incredible potential that is "slowly crushed and destroyed by a standardized society and mechanical teaching" (Berger 1967, 19). Subcultures, like social movements, engage in conflict over cultural reproduction, social integration, and socialization; they are often especially concerned with quality of life, self-realization, and identity formation. Straight edgers believed toxins such as drugs and alcohol inhibited people from reaching their full potential. This view sharply contrasted with the hip version of self-realization through dope (Davis 1968). For sXers, drugs of any kind inhibited rather than enabled self-discovery; they believed people were less genuine and true to themselves while high. A clear, focused mind helped sXers achieve their highest goals. Kate, the activist, said, "If you have a clear mind you're more likely to be aware of who you are and what things around you really are rather than what somebody might want you to think they are. A little bit more of an honest life, being true to yourself."…

Straight edgers rarely spoke openly about self-realization, and they would likely scoff at anything that suggested mysticism or enlightenment (which they would connect to hippies and therefore drugs). Nevertheless, for many, underlying the ideology was an almost spiritual quest for a genuine self, a "truth." Some connected sXe to other identities: "queer edge," feminism, and activism, for example. For others, sXe offered a means of overcoming abusive family experiences. Mark, a quiet sixteen-year-old new to the scene, claimed sXe as a protest: "Straight edge to me, yeah, it's a commitment to myself, but to me it's also a protest. I don't want to give my kids the same life I had from my father."…

INVOLVEMENT IN SOCIAL CHANGE

Like members of the other subcultures, sXers often became involved in a variety of social causes. The sXe youth with whom I associated insisted that working for social change was not a prerequisite of sXe. Indeed, only a few belonged to the substantial activist community in our city. However, many viewed involvement in social change as a logical progression from clean living that led them to embrace progressive concerns and become directly involved at some level. Clean living and positivity led to clear thinking, which in turn created a desire to resist and self-realize. This entire process opened them up to the world's problems, and their concerns grew….

In the mid-1980s to late 1980s, sXe became increasingly concerned with animal rights and environmental causes. Influential leaders in bands called for an end to cruelty against animals and a general awareness of eco-destruction. At least three out of four sXers were vegetarian, and many adopted completely cruelty-free, or vegan, lifestyles. Among the approximately sixty sXers I associated with regularly, only fifteen ate meat. Several individuals had "vegan" tattooed on their bodies. Others led or actively participated in a campus animal defense organization. Essentially, the movement framed animal rights as a logical extension of the positivity frame underpinning the entire lifestyle, much like reserving sex for caring relationships and self-realization. Brian, an extremely positive and fun-loving twenty-one-year-old, explained vegetarianism's connection to sXe: "sXe kids open their minds a lot more. They're more conscious of what's around them. Some people think it's healthier and other people like me are more on the animal liberation thing." Elizabeth, the older veteran, said,

> If you are conscientious and care about the environment or the world, which perhaps more sXe people are than your average population, then [animal rights is] just going to be a factor. You're going to consider "How can I make the world a better place?" Well, being vegetarian is another place you can start. I'm glad it's usually a part of the sXe scene because it just goes along with awareness and choices. What kind of things are you doing to yourself and how is that impacting the world and the environment? The big corporate-owned beef lots and cutting down the rainforests … the most impactful thing you can do for the environment is to stop eating meat.

Many sXe women disdained more traditional female roles and appreciated the scene as a space in which they felt less pressure to live up to gender expectations, and the movement encouraged men to reject certain hypermasculine traits and challenge sexism on a personal level. A majority of bands wrote songs against sexism, and many young sXe men demonstrated an exceptional understanding of gender oppression given their ages and experiences. However, despite the movement's claims of community and inclusivity, some sXe women felt isolated and unwelcome in the scene. Men significantly outnumbered women, often creating a "boys club" mentality exemplified by the masculine call for "brotherhood." The almost complete lack of female musicians in bands, the hypermasculine dancing at shows, and the male cliques reinforced the movement's own unspoken gender assumptions that women were not as important to the scene as men and ensured that many women would never feel completely at home.

While some sXers joined animal rights, women's rights, environmental, and other groups, most strove to live out their values in everyday life rather than engage in more conventional "political" protest (e.g., picketing, civil disobedience, petitioning). Instead of challenging tobacco, beer, or beef companies directly, for example, a sXer refuses their products and might boycott Kraft (parent company of cigarette manufacturer Phillip Morris), adopt a vegetarian lifestyle, or wear a shirt to school reading "It's OK not to drink. Straight Edge" or "Go Vegan!" In sXe and other youth movements, the personal was political. Subcultures are themselves politically meaningful, and they often serve as a bridge to further political involvement.

Looking at resistance through the lens of meanings, sites, and methods forces us to reexamine the "success" of subcultural resistance. Analyzing sXe's core values shows that members' understandings of resistance are many layered and contextual. The issue of resistance goes beyond whether a subculture resists dominant culture to how members construct resistance in particular situations and contexts. Certainly, sXe, like other subcultures, has illusory tendencies; the movement's contradictions include its antisexist yet male-centered ideology. However, examining sXe with the framework I suggest shows that involvement has real consequences for the lives of its members, other peer groups, and possibly mainstream society. Personal realization and social transformation are not mutually exclusive. Although sXe has not created a revolution in either youth or mainstream culture, it has for more than twenty years, however, provided a haven for youth to contest these cultures and create alternatives.

NOTES

1. Straight edgers abbreviate straight edge as sXe. The *s* and the *e* stand for straight edge, and the *X* is the straight edge symbol.

2. Hardcore is a more aggressive, faster style of punk. Though punk and hardcore overlap, in the 1990s the two scenes increasingly became distinct. While present in both scenes, sXe is considerably more prevalent in the hardcore scene. The hardcore style is more clean-cut than punk. Punks and sXers draw a sharp distinction between "shows" and "concerts." Shows attract a much smaller crowd, are less expensive, feature underground bands, often showcase local bands, and are set up by local kids in the scene at little or no profit. Concerts are large, commercialized, for-profit ventures typically featuring more mainstream bands.

3. Straight edge individuals never refer to themselves as *straight edgers* and find the term quite funny. It likely comes from media portrayals of the group. Adherents call themselves sXe "kids," no matter their ages. I use *straight edger* in this article simply for ease of communication.

4. Veganism had become such a significant part of sXe by the late 1990s that many sXers gave it equal importance to living drug and alcohol free. Thus, many sXe vegans would self-identify as "vegan straight edge," and some bands identify as "vegan straight edge" rather than simply "straight edge." Veganism, while still widely practiced, had a declining presence after 2000.

REFERENCES

Berger, B. M. 1967. Hippie morality—More old than new. *Trans-Action* 5 (2): 19–26.

Bjorgo, T., and R. Wilte, 1993. *Racist violence in Europe*. New York: St. Martin's.

Brake, M. 1985. *Comparative youth culture: The sociology of youth culture and youth subcultures in America, Britain, and Canada*. London: Routledge Kegan Paul.

Cohen, S. 1972. *Folk devils and moral panics: The creation of the mods and the rockers*. Oxford, UK: Martin Robertson.

Davis, F. 1967. Focus on the flower children. Why all of us may be hippies someday. *Trans-Action* 5 (2): 10–18.

_____. 1968. Heads and freaks: Patterns and meanings of drug use among hippies. *Journal of Health and Social Behavior* 9 (2): 156–64.

Fox, K. J. 1987. Real punks and pretenders: The social organization of a counterculture. *Journal of Contemporary Ethnography* 16 (3): 344–70.

Hall, S., and T. Jefferson, 1976. *Resistance through rituals: Youth subcultures in post-war Britain*. London: Hutchinson.

Hebdige, D. 1979. *Subcultures: The meaning of style*. London: Methuen.

Henry, T. 1989. *Break all rules! Punk rock and the making of a style*. Ann Arbor: University of Michigan Research Press.

Heylin, C. 1998. *Never mind the bollocks, here's the Sex Pistols: The Sex Pistols*. New York: Schirmer Books.

Irwin, D. 1999. The straight edge subculture: Examining the youths' drug-free way. *Journal of Drug Issues* 29 (2): 365–80.

Irwin, J. 1977. *Scenes*. Beverly Hills, CA: Sage.

Miller, T. 1991. *The hippies and American values*. Knoxville: University of Tennessee Press.

_____ 1999. *The 60s communes: Hippies and beyond*. Syracuse, NY: Syracuse University Press.

Moore, D. 1994. *The lads in action: Social process in an urban youth subculture*. Aldershot, UK: Arena.

O'Hara, C. 1999. *The philosophy of punk: More than noise*. London: AK Press.

Wooden, W. S., and R. Blazak. 2001. *Renegade kids, suburban outlaws: From youth culture to delinquency*. 2nd ed. Belmont, CA: Wadsworth.

Young, K., and L. Craig. 1997. Beyond white pride: Identity, meaning and contradiction in the Canadian skinhead subculture. *Canadian Review of Sociology and Anthropology* 34 (2): 175–206.

11

Constructing Authenticity and Ethnicity in the Mexican Restaurant Industry

MARIE SARITA GAYTÁN

Just as Whites represent the dominant American culture and ethnicity, this country has long been characterized as a "melting pot" where people of different groups have immigrated and been absorbed. Once here, they want to find a way to resolve the juxtaposition of their former and new cultural experiences. Entrepreneurs have been quick to commercialize a vast array of cultural forms, in theme parks, lodging, clothing, and dining. In this article, Gaytán explores the commercial construction of "authentic" Mexican culture as it is presented along a spectrum of restaurant types. She shows the difficulty, usually overlooked by most diners, of smoothly integrating the tastes and desires of American consumers with the customs and preferences brought from the former country. You may find yourself surprised to learn about the role of the burrito in Mexican cooking. How might this compare to the role of pizza in Italian restaurants? As you read this article, stratify your favorite Mexican restaurants along this continuum to understand how the cultural symbols express commercialized authenticity to varying degrees. At the same time, think about extending this analysis beyond the restaurant setting to other milieus. What other kinds of non-American or subcultural expressions do you experience in your everyday life and how might they be affected by this pull between authenticity and commercialism?

E xamining how individuals negotiate the social world and make sense of their experiences reveals not only how people understand each other, but

SOURCE: From Sombreros to Sincronizados: Authenticity, Ethnicity, and the Mexican Restaurant Industry. *Journal of Contemporary Ethnography*. Reprinted by permission of Sage Publications.

illustrates how they view themselves in relation to those with whom they inter-act. Authenticity, as an index of these encounters, provides the opportunity to determine and communicate one's place as an individual in an increasingly global marketplace. While the search for authenticity highlights the desire for "genuine" forms of expression, it also calls attention to concerns regarding the interpretation and negotiation of contemporary consumer life. In a world where soft drinks are marketed as "the real thing" and restaurants are named "The Authentic Café," consumption serves as a means through which authenticity becomes a measure of the quality, efficacy, and legitimacy of one's experience in a particular setting.

Despite its fabricated character, performances of authenticity influence how people identify with and become involved in real world social, political, and cultural arrangements. Central to these arrangements is the ability to enact distinction and recognize difference. MacCannell (1973), for instance, noted that tourists consis-tently sought to enter the back regions of their destinations in an attempt to en-counter how people different from themselves authentically lived. Like tourists, customers of "ethnic" restaurants also desire backstage experiences and the "illusion of authenticity" when participating in different culinary traditions. These traditions are often linked to the meanings that people attach to their own ethnic identities and the ethnic identities of others. As a measure of symbolic ethnicity, authenticity contributes to the process, performance, and preservation of culture.

While scholars agree that performances of authenticity and ethnicity express social relations and demonstrate the socially constructed character of identity, we know little about how these interactions shape the politics of everyday life. By engaging in participant observation in Mexican restaurants, drawing on open-ended interviews with owners, managers, customers, and reviewers of Mexican restaurants, and considering the content of a range of available literature (i.e., restaurant industry reports, articles, and books) related to restaurant culture, I argue that the accomplishment of Mexican authenticity, whether maintained by Mexican owners or performed by large restaurant chains, is a social construc-tion. However, despite its socially created qualities, performances of authenticity and ethnicity affect not only how individuals understand each other, but illus-trate the challenges faced by different groups of people in the commercial pro-duction and consumption of identity. Focusing on the consequences and contradictions that are generated through different claims to authenticity, I examine the salience of ethnicity in day-to-day situations and explore the dynamics of cultural production....

EXPLORING THE CULTURE OF MEXICAN RESTAURANTS

Restaurants provide an empirically rich point of entry for observing interpersonal exchanges as part and parcel of the creation, enactment, and reproduction of symbolic meaning. As unique sites for observing "public ethnicity," restaurants bring together social actors as participants and spectators in performances of culture. From September 2000 through June 2001, I became a participant observer in Mexican restaurants in Amherst, Northampton; Springfield, Massachusetts; and Hartford, Connecticut. Additionally, I conducted interviews with customers, reviewers, owners, and managers of these restaurants. To gather a wide range of materials concerning the social relations of Mexican restaurants, I collected copies of flyers, menus, web pages, restaurant industry reports, restaurant reviews, and take-out containers. Thus, this project draws on participant observation, open-ended interviews, and content analysis to examine the culture of Mexican restaurants....

In total, I interviewed eight owners and managers of Mexican restaurants (two women and six men). These interviews were conducted face-to-face and included open-ended questions. They lasted from 1 to 2 hours, were tape recorded, and then transcribed. My conversations with customers of Mexican restaurants (12 women, 7 men) lasted from 15 minutes to an hour and were collected in restaurants and coffee shops in the towns of Amherst and Northampton. I approached people as they ordered, ate, and left the various establishments. Initially I was nervous about making contact with people in such an informal manner, but folks were quite eager to share their ideas about Mexican restaurants. Equally enthusiastic to talk to me were restaurant reviewers. My three interviews with restaurant reviewers (three women) took place after I contacted area newspapers that published restaurant reviews. In one case, a newspaper editor would not reveal the "true" name of its reviewer because he wanted to protect the reviewer's anonymity. Of the three interviews, one was conducted face-to-face and two were carried out over the telephone....

CATERING MEXICAN AUTHENTICITY

Sipping hot chocolate in the subterranean dining room of a local coffee shop, Adrian, a restaurant reviewer whose career in journalism spanned over 20 years, revealed "I like it authentic, real, from back home, the motherland." Although she had only recently acquired the position of senior food writer, Adrian was adamant about the importance of a reviewer's ability to accurately evaluate the authenticity of a restaurant's cuisine. When she was unable to gauge a restaurant's authenticity, Adrian brought along friends or colleagues to assist her. Such was the case when she reviewed an Argentinean restaurant. In her words, "I decided to take a friend from Chile with me to gauge how authentic it was. Everyone spoke Spanish, they were all from Argentina." Similarly, Catherine, a reviewer whom I talked to by telephone, explained how she intuitively knew when a restaurant was authentic by observing the rapport an owner established with customers. As she put it, "The owner is cooking, takes his time. The restaurant is full of warmth, of something that is common, that is family ... If you and I are eating here, then we're family."

Adrian and Catherine's observations suggest that authenticity is an important measure in determining the quality of a dining experience. However, their responses also show that evaluating a restaurant's authenticity requires more than just an appraisal of a restaurant's menu, recipes, or ingredients. Specifically, a restaurant's atmosphere is vital to an assessment of authenticity. Like reviewers, customers described the ambiance of a restaurant in terms of people and not décor. Frequently commenting on the presence or absence of "Hispanic" or "Mexican" employees, the ability to interact with Spanish-speaking employees was a central means by which many assessed their experiences. One customer recognized that a particular Mexican restaurant was authentic because she could "order in Spanish and the order was given back in Spanish," while another stated that when "all the cooks [were] Hispanic," he could more readily determine if some restaurants were more authentic than others. Customers and reviewers also looked for other cues. As a customer put it, "When you see Mexicans eating there it gives you the feeling that it's authentic and not just yuppies eating burritos and sipping margaritas," while a reviewer noted that she knew a restaurant was authentic because labor activist César Chavez dined there when he used to visit the area....

Although customers and reviewers readily acknowledged the importance of authenticity, they were uncritical of the cuisine or atmosphere offered at the local Mexican restaurants that they considered "inauthentic." In one case, a reviewer observed, "They make no pretense of serving traditional Mexican fare ... it's more whimsical and Americanized than loyal to any particular ethnic cuisine," while a customer remarked, "No, it's not authentic Mexican, but they don't claim to be." Even when a restaurant was unsuccessful at maintaining an authentic décor, reviewers still expressed confidence in making determinations about its authenticity. Such was the case with Jennifer, who explained that despite a restaurant's Formica tables and plastic chairs, she could still recognize "something real about the cuisine." Taken together, customers and reviewers' comments imply that there is more than one way to achieve authenticity and to evaluate those accomplishments. In addition, their responses suggest that there is something about *how* particular claims of authenticity are expressed that contributes to their assessment of their experiences. Turning to presentations of authenticity and ethnicity in Mexican restaurants, I examine the production of identity and the conditions of representation.

COMMEMORATIVE AUTHENTICITY: "IT'S GOT A MEXICAN SOUL ... AND THAT SAYS A LOT"

Maintaining loyalty to tradition was central to how Enrique organized the overall layout of his restaurant. For instance, he labeled menu entrées "*comida de la gente*" (food of the people), and adorned the windows of his restaurant with decals that read "*hecho en Mexico*" (made in Mexico) and "*viva la raza!*" (a rallying cry made famous during the Chicano Rights Movement of the 1960s). Describing what he called an "essence," Enrique elaborated,

> For me, when I think about Mexican food I think about what kind of food you find at Mexican kitchens at home, not so much at a restaurant. It's more like what I would find at a *quinceañera* (cotillion), what I would find at a *boda* (wedding).

According to Enrique, the "essence" of authentic Mexican food is linked to cultural celebrations such as *quinceañeras* and weddings. While unable to fully replicate these environments within the context of his restaurant, he is able to offer his customers Mexican "home-style" cooking that celebrates Mexican identity by both challenging and reinforcing people's perceptions of what and when "real" Mexican food is consumed.

Differentiating himself from his competition, he described how his restaurant, Los Cabos, stood out from the other local offerings:

> If you go to that restaurant, what it does is takes a variety of ethnic foods, wraps them into a large flour tortilla and calls it a burrito. So what you have is a stir-fried chicken burrito with roasted bell peppers, mushrooms, and yogurt, you know? [That] has nothing to do with what my *Tía* (aunt) Ana serves.

As he saw it, the use of unconventional burrito ingredients symbolized a departure from traditional food preparation.

Expressing his disappointment at the alterations restaurateurs commonly made to accommodate people's perceptions of Mexican food, Enrique detailed his refusal to compromise the traditional names of his menu items. In one case he challenged people's ideas about quesadillas:

> It would be easy for me to sell the quesadilla with your choice of flour or corn tortillas. Instead, what I say is let's call the flour tortillas with cheese inside for what they are, *sincronizadas*. Sin-chro-neez-ahh-dah [in an American English accent], well, that's made with flour tortillas. That's the difference so one day when they [his primarily Anglo clientele] go to Mexico, they're not shocked that their quesadilla isn't made with a flour tortilla, and they'll know what a *sincronizada* is.

Enrique's resistance against mainstream ideas about Mexican cuisine in general, and the local versions of Mexican food in particular, manifested not only in his menu, but also in the advertising and décor of his restaurant. When Los Cabos first opened, a college newspaper published a lukewarm review that praised the cuisine but commented extensively on its ambiance and "lack of Mexican charm." Recounting the incident, Enrique, clearly bothered by the remark explained:

> There's no *sombreros*, *zarapes*, or *piñatas*… it has a Mexican soul and you know that says a lot. It's not salt it isn't pepper, it's not nice tile or a beautiful chair. There's a spirit about it only those who are very proud of what Mexican is or how it should be represented can get it.

Symbols such as sombreros, *zarapes*, and piñatas, as Enrique put it, presented a limited version of Mexican identity, one that neglected to consider the "huge history of great art" produced and appreciated by Mexicans on both sides of the U.S.-Mexican border. In his estimation, stereotypical imagery could not contend with the "soul" and "spirit" fostered by presenting a commemorative approach to Mexican culture. With ads stating "*Evite el estereotipo!*" (avoid the stereotype), "Mexican food that Mexicans would want to eat," and "Mexican food like you've never had it in New England," Enrique actively engaged his customers in a political and cultural relationship that invoked authenticity as a significant and essential aspect of ethnic identity. Additionally, Enrique's comments suggest that even in commemorative versions of authenticity, consumers' preconceived ideas about how specific people create "charm" produce conflict when challenges are presented that extend the possibilities of available ethnic representations.

Loyalty to tradition was also a theme that emerged when I spoke with Ricardo, a restaurateur who recently moved to New England. Working collaboratively with his seven brothers, their first restaurant, Las Rosas, was named after their mother, and subsequent restaurants were named after other family members. This kind of environment, Ricardo explained, "emphasizes that we are family and that people can tell we are family and that people will feel like part of our family." Serving food typical of their home state of Jalisco, Mexico, they are accustomed to altering preparation styles and levels of spiciness to suit local palates. Even with such changes, Ricardo maintained:

> Our food is yes, really Mexican ... It's authentic in the way that if you go to Mexico, to our hometown ... well, it's a little different there, more spicy. Yes, it is authentic, but with a style from this side of the border.

According to this statement, while Ricardo asserts that his restaurant is "really Mexican," it is only authentic in comparison with Mexican food served within the United States.... However, this does not stop Ricardo from promoting an authentic "family-style" experience so "that people will feel like part of our family." While he concedes that the food he serves is not entirely authentic, he presents it as if it were. Ricardo reveals that Mexican food does not have to be authentic to *appear* authentic. Moreover, his comments suggest that performances of Mexican identity are inextricable from assumptions about class, gender, and family even in commemorative representations of ethnicity.

David, originally from New Mexico, opened his restaurant because "the area needed an authentic, traditional-type Mexican restaurant." Emphasizing the importance of preserving his family's recipes, David was interested in reaching out to the community "to hand-down a tradition or share a tradition to the people of New England from the Southwest." Seeing his restaurant as representing Mexican culture, David explained the differences between his menu and that of a large Mexican chain restaurant located nearby:

> ... a taco is a taco [but] it's what you put inside your taco [that] makes it different. Burritos are an American invention, there's no such thing as burritos in Mexico and the only reason that I serve them is because people love them.

Other alterations that David made to his menu included the addition of quesadillas and taquitos because "people kept requesting them." Décor was less relevant in David's opinion, especially when it came to displaying his heritage, "It's not important in a sense because really and truly I'm a Mexican–American, and the food I serve is traditional Mexican and New Mexican home-style cooking." In David's estimation, his Mexican–American background qualified him to represent authentic Mexican and New Mexican food. David's performance of authenticity, like that of Enrique and Ricardo's, is informed by claims that are unavailable to non-Mexican restaurateurs—in this case, tradition, family, and ethnic heritage rooted in Mexican culture.

As a social construction, authenticity generates a unique set of contradictions for those owners and managers who situate themselves as representatives of a specific collective identity. While they are able to display personal values associated with their presentation of ethnic heritage, they must also make concessions to fulfill certain customer expectations. In their research on the presentation of "ethnic" authenticity, Lu and Fine (1995) observed that the owners of Chinese restaurants often made changes to their cuisine to suite the "standards of the American palate" (p. 540). These alterations were based on aesthetic values linked to cultural discourses and the cost of particular food ingredients. Customers desired the "illusion of authenticity" regardless of modifications pertaining to the use of spices, methods of preparation, and styles of service (p. 541). Invoking innovative and adaptive measures, owners and chefs of Chinese restaurants cooked according to tradition even when adhering to customer preferences. Similarly, Ferraro (2002) found that the owners of Mexican restaurants in Los Angeles, California, used various names to describe menu items as a way to display their observance of authentic culinary styles and methods of preparation. According to this research, abiding by the guidelines of authenticity is an important and complicated endeavor that requires a combination of flexibility, foresight, and strategy....

In commemorative presentations of authenticity, ethnicity is depicted as both challenging and conforming to a dominant set of ideas that portray identity as fixed, stable, and recognizable. For example, when Enrique displays Chicano/Latino artwork throughout his restaurant, he sees this decision as a symbolic act that expands his customers' knowledge of Mexican art and Chicano heritage. Despite Enrique's desire to influence his customers' ideas about the diversity of Mexican culture, his strategies were unable to communicate a "different way of looking at Mexican food" because of expectations that he explained were put forth by "the standards of *lo gringo*" (that of American or non-Hispanic origin).

These types of struggles, as Gray (2004, p. xv) points out, call attention to the relationship of representations to "actual structural positions, social practices, cultural meanings, and the complex constellation of power" in which performances of authenticity take place. For members of historically marginalized communities, consumption is a key site where they are able to construct an identity that manages, contests, and extends common cultural categorizations that reproduce both oppositional *and* conventional representations. However, as these owners' responses suggest, despite attempts to affirmatively represent their traditions or present their own versions of authenticity, they are often unable to disrupt mainstream ideas of what is and what is not "authentic." When established sets of expectations are challenged, these efforts conflict with customer demand and desire—both shaped by long standing social forces that impose market-oriented notions of authenticity that cater to "unreal" events over "real" ones.

HYBRID INAUTHENTICITY: "THE SUPERBOWL AND SALSA ARE SYNONYMOUS"

Doug, an Anglo restaurateur, had the "burrito concept" in mind for several years before opening his restaurant. Living in a college town, he explained that a "burrito joint" was an appropriate addition to the area's other "ethnic" restaurants. He described his restaurant's food as follows:

> It's Mexican-American, well, more American than anything else, what we do as I see it. Sort of like pizza. We produce the Americanized version of something that's supposedly Mexican, and, in a way, authentic, but for the most part, it's an Americanized version of some foreign food like Chinese food that ends up appealing to the palate of people in this country, like fried rice, 'cause it's more American than anything else.

Remarking that the burrito is both "Mexican-American" and "more American than anything else," Doug is able to distinguish his restaurant's burritos from more "traditional" burritos. His observations are based on a distinction between different types of ingredients:

> Most of what's out there for a Mexican burrito has got a lot of fatty stuff in it. Like a lot of people perceive refried beans as made with lard, and they are, for the most part. We don't use refried beans, in fact, we don't use pinto beans. Predominantly we use black beans because they are more appealing to a broader market than pinto beans.

Ironically, part of Doug's success in the Mexican restaurant industry is a result of the perception that Mexican food is unhealthy. Capitalizing on what he calls the "burrito trend," he explains, "as an Americanized version of food, it [the burrito] fits into a nice classification system that's somehow there for whatever reason." Indeed, this "classification system" works to his benefit, even when it comes to the preparation of his restaurant's tacos:

> Our taco is really no more than a burrito than anything else. We call it a taco because it's smaller than the burrito and that was an arbitrary decision that seemed to make intuitive sense to our customers, and so that's what we called it.

Relying on an "arbitrary" decision to create tacos as similar, yet smaller versions of their burritos, reveals an approach to authenticity that adjusts to customer demand through a hybrid preparation and production of Mexican cuisine, one that in this case presents tacos as a different version of burritos. As he sees it, his customers are aware that they are not eating "truly authentic" Mexican food, but are instead consuming "healthy" and "funky" Mexican food. Even when it comes to the salsa, for Doug, "it's an illustration of something Americanized, and now, salsa's our thing. The Super Bowl and salsa are synonymous.

During our interview, I inquired about an El Salvadoran flag propped in a corner of the large indoor menu situated above the open kitchen. Explaining that the majority of his employees were El Salvadoran immigrants, he described how their input with regard to music and recipes was reflected throughout his restaurant. For example, one employee suggested the creation of an El Salvadoran burrito made with ingredients she cooked with in her own kitchen. Doug remarked, "I tried it, and liked it. It's a simple recipe and we adopted it and it was a sorta ethnic thing to have as an option … we've never really taken it off of the menu." As a "sorta ethnic" option, the inclusion of an El Salvadoran burrito reflects the hybrid inauthenticity through which Doug presents and markets his restaurant.

Sara, originally from California, described her style of cooking as "yuppified" and distinct from the food prepared by her Mexican mother. When referring to her menu, she stated, "I stay away from traditional Mexican food that I love because I've created this type of very quick, fresh food." Emphasizing that her menu does not reflect a traditional or authentic Mexican fare, she explained her interest in flavors and spices that parallel those found in Mexico:

> I've found this similarity with Asian food and Mexican food. There's a very big similarity with the freshness and the produce such as tomatoes, squashes, all the salsas. The cooking techniques are also very similar to say, Southeast Asian food, which I love, and it's very accessible to the common palate.

Discovering a link between cross-ethnic cooking styles, Sara builds on similarities which accentuate the seasonings used in different cuisines. For instance, in many of her dishes she adds peanut sauce for a "Thai flavor" and feta cheese for a "Mediterranean flavor." When describing the combination of ingredients, she elaborates on a hybrid culinary approach, "When I first opened, people weren't used to experiencing a Thai chicken burrito with a rice and bean burrito. I've been able to slowly bring in different things and mix them all up." Like Doug, Sara refrains from making any claims to traditional Mexican cooking.

Doug and Sara's customers, while identifying their restaurants as serving Mexican cuisine, take notice of their hybrid inauthentic approach. Able to distinguish the differences from more traditional fares, customers and reviewers shared a range of comments about their restaurants specifically:

> It's a gourmet burrito place; they've glamorized it into something not traditional.

> If you're looking for authentic Mexican food, don't come here. However, if you want Americanized Mexican food that's much better/ cheaper than Taco Bell, this place is great.

> I'm from California, so I know what authentic Mexican food is like…
> and it's not anything like La Fiesta restaurant.

These remarks show that customers and reviewers carefully evaluate the extent to which the cuisine they consume abides by a standard of authenticity that adheres to particular aesthetics, menu items, and ingredients.

According to Tuchman and Levine (1992), the inevitable "naturalness" of products related to specific ethnic traditions is often associated with the created character of ethnicity. Examining the internal logic of "ethnic" cultural invention, they show how "eating Chinese" became a New York Jewish custom that reflected articulations of self-identity. In some cases, Chinese food provided a "biographical continuity" for different generations of New York Jews through the linking of cultural themes and fond memories when dining in Chinese restaurants (p. 401). As they put it, socially constructed cultural differences are "… social facts, enduring and real in their consequences, yet malleable to reinterpretation by future generations" (p. 401). In the context of performances of authenticity, these social facts symbolize innovative cultural representations and reveal the potential for diverse meanings of "eating ethnic."

While Mexican culinary culture is central to Doug and Sara's advertising, décor, and menu, they do not make assertions of Mexican authenticity. In its place, they emphasize qualities such as "freshness" and "alternative" methods of food preparation. Through the use of hybrid inauthenticity, their restaurants highlight the vitality of different styles of cooking and observe the unique contributions (e.g., ingredients and seasonings) of a variety of cultures (e.g., El Salvadoran and Thai). More importantly, however, they respect the role and maintain the integrity of Mexican cultural forms. In other words, instead of appropriating small parts of Mexican culture in a piecemeal fashion, Doug and Sarah's restaurants honor Mexican ethnicity by drawing on it as the primary means to create something new (El Salvadoran burrito and Thai chicken burrito). Hence, implementing a hybrid inauthentic approach is a valuable method of honoring tradition even as it departs from tradition in its pristine form.

AMERICANIZED AUTHENTICITY: "NOW YOU CAN'T DO MEXICAN"

James, the manager of a popular international Mexican restaurant chain, began his career at El Torito, a California-based corporation. When I asked James why there were no El Torito restaurants in New England, he explained that New Englanders were not "ready for that type of Mexican food." Conversely, the chain he currently managed would not be well received in California, "because it's too Americanized; we haven't been brought up on Mexican food like people in California, not to mention produce availability year round; that kind of drives the menu." From James's perspective, because of their different degrees of exposure to Mexican food, customers in California and New England diverged in their Mexican food preferences. His restaurant is "too Americanized" for Californian palates, but sufficiently "Americanized" for New England palates. Comparing the restaurant chain he manages now to his experiences at El Torito, he explained:

> In recent years we've gotten a little more traditional than we used to be. Well, El Torito, it's more authentic. We took on some of El Torito's ways. Like we have a certain salsa, have gone to thicker corn tortillas, which are way better than our old ones used to be, way more authentic, you know.

In moving toward the "more authentic" ways of El Torito, James's menu offered foods such as "traditional nachos," "fajita pizza," and a "Buffalo chicken sandwich" served on a "Mexican-style bun."

James remarked that it was important for his restaurant to "try to become everything" to reach a wider customer base. As a result, the menu evolved to satisfy consumer trends:

> We have ribs, grilled burritos, so a bit of the Tex-Mex has melded into our menu as a necessity to compete with Chili's … Now you can't do Mexican. I think you have to offer other food to capture that one of four people that doesn't eat Mexican food.

Again and again, James emphasized the need to "capture" potential customers who do not normally eat Mexican food. To accomplish this, the restaurant must adhere to an "Americanized" menu. As he sees it, people who do not normally eat Mexican food "think it's hot, spicy; they think that's the only way you can get Mexican food. I compare it to Italian food, it's very flavorful, but not everything's hot." James' description of people's perceptions of Mexican food as spicy or hot, in combination with his comparison with Italian food, corresponds to those values put forth by large corporations, whereby Mexican food is marketed as "familiar" as opposed to distinct in and of itself.

Down the road from James's restaurant, Daniella, a manager of another chain restaurant, told me the story of the Mexican vaquero:

> The name [of the restaurant] is based on the Mexican vaquero, who is the Mexican cowboy. If you look at the front of the menu it gives you the story of the Mexican vaquero—so our mood is based on food that you cook on the range, the Mexican range.

The restaurant's mood, as she explained, was vital to its theme. With its Mexican vaquero logo, Daniella's restaurant is able to offer a wide range of both "Mexican cowboyish stuff," and "authentic cowboy stuff." For example, menu items include "campfire queso," "portabello fajitas," and a "border T-bone steak." To maintain an "authentic atmosphere," Daniella's restaurant must "readapt those philosophies [pertaining to traditional Mexican cuisine] to go along with the guests, the customers. It has been changed to where you can't go on without your theme, it [the restaurant] just won't survive."

The image of the Mexican cowboy worked well with the restaurant's theme. For instance, offering selections such as "smoked barbeque ribs" and "mesquite-grilled salmon," the restaurant's menu contributed to ideas about "traditional Mexican favorites" through the inclusion of nontraditional Mexican entrees. According to Daniella, creating a comfortable atmosphere is also a big part of the restaurant's success:

> Atmosphere has a lot to do with people's perceptions. They walk in here and they already have a preconceived notion [of how the food will taste] by how clean our restaurant is, by what they are looking at. If they walked in here [and saw] four walls painted white, with no Mexican music on, with 1970s music, or whatever the case may be, they aren't going to feel that homey comfort. I'm sitting by the campfire range eating Mexican food, it's huge.

Apparently Daniella's "guests" appreciate the atmosphere of the restaurant: "we get it on comment cards all the time; the atmosphere is awesome, it's warm, it's cozy, you feel like you're sitting in Mexico, you know, it's authentic."

Elaborately decorated with giant saddles, colorful blankets, and rustic lanterns, the dining area stood out from the cordoned-off bar area that was adorned with neon-lights, large screen televisions, and life-size beer advertisements. Daniella explained:

> Well, we try to keep it all the same flare, you know. We don't have that Mexican bar-type atmosphere because you can't generate business out of that. It has to be functional as well, so it's not as authentic in the bar area as it is out here.

According to Daniella, although customers enjoy eating in a Mexican atmosphere, they do not enjoy drinking in one. The "authentic" dining area and the "not as authentic" bar area fittingly capture a corporate approach to "ethnic" dining: ethnic, but not *too* ethnic, authentic, but not *too* authentic....
"Successful" authentic "ethnic" products:

> will respect that spectrum's boundaries, sticking close to traditional cuisines that reflect American tastes. Once a product goes past the line of comfort and accessibility ... it will not be seen as a real choice. It will be seen as something that is not convenient and not comfortable. And then it becomes foreign again. (Decker 2003, 113–114)

In her article "Eating the Other: Desire and Resistance," hooks (1992) explores the commodification of racial "otherness." Looking specifically at desire through the lens of race, she shows how discourses and practices successfully promote the enjoyment of racial difference as a "new delight, more intense, more satisfying than normal ways of doing and feeling" (p. 21). Concerned with cultural arrangements that celebrate ethnicity as "a spice, [a] seasoning that can liven up the dull dish that is mainstream white culture," hooks describes the unconscious as a site of contestation where capitalism, patriarchy, and colonialism continue to re-inscribe relational hierarchies (p. 21). As she sees it, this type of longing is deeply embedded in ideas of white superiority and the West's crisis with its own identity—both rooted in the learned libidinal aspiration to dominate. Shaping a large-scale restaurant chain according to the desires of an "American market" situates ethnicity not as a central expression of heritage or culture, but instead as "a spice" that accommodates the "desires" of an Anglo customer base whose consumption practices are considered hierarchically distinct.

Adhering to the logic of an "American market" systematically fails to recognize non-Anglo individuals as consumers of "ethnic" cuisine. With structures of exclusion built into the commercial organization of ethnic restaurants, it is no surprise that attributes such as "comfort" and "accessibility" trump qualities of cultural uniqueness. In such circumstances, people who are not classified as "American" remain unrecognized for their participation in and contribution to social, economic, and cultural expressions of public consumption. If consumption is one "place" where scholars insist the terms of citizenship are put into practice, then those consumers who are not "mainstream Northern and Western European" are excluded from particular privileges of citizenship. These processes illustrate not only the limited value of cultural agency in large-scale marketing practices, but also reveal the limitations that hierarchies of exclusion place on different groups of people. Through the promotion of an Americanized authenticity, the vibrancy of cultural forms that celebrate identity and heritage are collapsed into narratives that emphasize similarity by staying "close to traditional cuisines that reflect American tastes" (Decker 2003, 114). In such arrangements, "authenticity itself becomes a hot commodity," fostering conditions that commemorate similarity as opposed to difference (Halter 2000, 18)....

CONCLUSION

The contention that authenticity is a highly negotiated interaction that represents a romanticized representation of reality is by now a scholarly norm. Similarly, the notion that ethnicity is a socially constructed category that reflects a group's common origin and sense of unique collective solidarity, is also a widely accepted belief. Within the culture of the Mexican restaurant industry, the accomplishment of authenticity and ethnicity is a social construction, one that is produced by Mexicans and Mexican-Americans as much as it is scripted by marketing experts of large restaurant chains. However, in spite of its socially constructed character, practices of authenticity have real-world implications that illustrate the limits and constraints that less powerful populations face in the marketplace—especially when it comes to accessing the conditions that enable the production and consumption of their identities.

Mexican restaurants are important sites where ideas about Mexican culture are generated and maintained. While the restaurants in this study primarily catered to an Anglo customer base, future research should consider how authenticity is negotiated in Mexican restaurants that have a more diverse clientele.

REFERENCES

Decker, K. J. 2003. Expanding culinary horizons with dried mixes. *Food Product Design* (June): 106–130.

Ferraro, S. 2002. *Comida sin par*. Consumption of Mexican food in Los Angeles: "foodscapes" in a transnational consumer society. In *Food nations: selling taste in consumer societies*, ed. W. Belasco and P. Scranton, 194–219. New York and London: Routledge.

Gray, H. 2004. Introduction. In *Watching race: television and the struggle for blackness*, xiii–xxv. Minneapolis: University of Minnesota Press.

Halter, M. 2000. *Shopping for identity: the marketing of ethnicity*. New York: Schocken.

hooks, b. 1992. Eating the other: desire and resistance. In *Black looks: race and representation*, 21–39. Boston: South End.

Lu, S., and G. A. Fine. 1995. The presentation of ethnic authenticity: Chinese food as a social accomplishment. *Sociological Quarterly* 36(3): 535–53.

MacCannell, D. 1973. Staged authenticity: arrangements of social space in tourist settings. *American Journal of Sociology* 79(3): 389–603.

Tuchman G., and H. Levine. 1992. New York Jews and Chinese food: the social construction of an ethnic pattern. *Journal of Contemporary Ethnography* 22(3): 382–407.

12

Inclusion and Exclusion
in Preadolescent Cliques

PATRICIA A. ADLER AND PETER ADLER

This study of third to sixth graders takes an insider's look at children's social worlds, focusing on the cliques that stand atop their social hierarchy and dominate their interest and attention. In a study we conducted in the 1990s using our children and children's friends, we probe the question of where cliques and their leaders get so much power, enabling them to coerce followers to engage in outrageous and self-demeaning behavior while commanding from them high status and submission. We conclude that this power stems from the inclusionary/exclusionary dynamic found in these elite and highly restricted friendship circles. By tracing the parameters of how leaders include and exclude others (and induce their followers to follow suit), and by showing how nearly all children's social standing is, at one time or another, systematically undercut, we illustrate how children learn that the price of supporting others is social exclusion themselves, and that, in this dog-eat-dog world, they must sacrifice their friends to save themselves. These interactional dynamics are where children learn the essence of in-group and out-group behavior that carries into later adulthood, and forms the foundation of the micro-politics of everyday life. Do these forms look familiar to you? Were you a member of such a clique or an outcast from these powerful groups? What are the implications of these groups for adult life? Do adults engage in this type of behavior, or is this just indicative of the relations between young children? How powerful do you think the effect of peer socialization is on people's future lives?

SOURCE: Patricia A. Adler and Peter Adler, "Dynamics of Inclusion and Exclusion in Preadolescent Cliques," *Social Psychology Quarterly* Vol. 58, No. 3, pp. 148–162. © American Sociological Association. Reprinted by permission of the authors.

O ne of the dominant features of children's lives during the later elementary school years (fourth through sixth grades) is the popular-clique structure that organizes their social worlds. The fabric of their relationships with others, their levels and types of activity, their participation in friendships, and their feelings about themselves are tied to their involvement in, around, or outside the cliques organizing their social landscape. Cliques are basically friendship circles, encompassing a high likelihood that members will identify each other sociometrically as mutually connected (Hallinan 1979; Hubbell 1965; Peay 1974). Yet cliques are more than that: they have a hierarchical structure, being dominated by leaders, and are exclusive, so that not all individuals who desire membership are accepted. They function as bodies of power within grades, incorporating the most popular individuals, offering the most exciting social lives, and commanding the most interest and attention from classmates (Eder and Parker 1987). As such they represent a vibrant component of the childhood experience....

TECHNIQUES OF INCLUSION

Cliques maintained exclusivity through careful membership screening. Cliques are not static entities; they shifted irregularly and evolved their membership as individuals moved away or were ejected from the group and others took their place. In addition, cliques were characterized by frequent group activities designed to foster some individuals' inclusion while excluding others. Cliques embodied systems of dominance whereby individuals with more status and power exerted control over others' lives.

Recruitment

Potential new members could be brought to the group by established members who had met and liked them. The leaders then decided whether these individuals would be granted a probationary period of acceptance in which they could be informally evaluated. If the newcomers were liked, they were allowed to remain in the friendship circle; if they were rejected, they were forced to leave. Alexis, a popular, dominant seventh-grade girl, reflected on the boundary maintenance that she and her best friend, Hope, two clique leaders, had exercised in sixth grade:

Q: Who defines the boundaries of who's in or who's out?

ALEXIS: Probably the leader. If one person might like them they might introduce them, but if one or two people didn't like them, then they'd start to get everyone up. Like in sixth grade, there was Dawn Bolton and she was new. And the girls in her class that were in our clique liked her, but Hope and I didn't like her, so we kicked her out. So then she went to the other clique, the Margo clique.

Most individuals felt that an invitation to membership in the popular clique was irresistible. They asserted repeatedly that the popular group could get any people they wanted to join with them. One of the strategies used by the cliques was to try to select new desirables and seek them out. This usually entailed separating those people from their established friends. Melody, an unpopular fourth-grade girl, described her efforts to hold on to her best friend, who was being targeted for recruitment by the popular clique:

> She was saying that they were really nice and stuff. I was really worried. If she joined their group she would have to leave me. She was over there and she told me that they were making fun of me, and she kind of sat there and went along with it. So I kind of got mad at her for doing that. "Why didn't you stick up for me?" She said, "Because they wouldn't like me anymore."

Melody subsequently lost her friend to the clique.

When clique members wooed someone to join them, they usually showed only the better side of their behavior. The shifts in behavior associated with leaders' dominance and status stratification activities did not begin until the new person was firmly committed to the group. Julie recalled her inclusion in the popular clique, and its aftermath:

> In fifth grade I came into a new class and I knew nobody. None of my friends from the year before were in my class. So I get to school a week late, and Amy comes up to me and she was like, "Hi Julie, how are you? Where were you? You look so pretty." And I was like, wow, she's so nice. And she was being so nice for like, two weeks, kiss-ass major. And then she started pulling her bitch moves. Maybe it was for a month that she was nice. And so then she had clawed me into her clique and her group, and so she won me over that way, but then she was a bitch to me once I was inside it, and I couldn't get out because I had no other friends. 'Cause I'd gone in there and already been accepted into the popular clique, so everyone else in the class didn't like me, so I had nowhere else to go.

Eder (1985) also has noted that popular girls are often disliked by unpopular people because of their exclusive and elitist manner (befitting their status).

Realignment of Friendships

Status and power in a clique were related to stratification; those who remained more closely tied to the leaders were more popular. Individuals who wanted to be included in the inner circle often had to work regularly to maintain or improve their position.

This was sometimes accomplished by people striving on their own for upward mobility. Danny was brought into the clique by Tim, a longtime member who went out of his way to befriend him. Soon after joining the clique, however, Danny abandoned Tim when Jesse, the clique leader, took an interest in him. Tim discussed the feelings of hurt and abandonment caused by this experience:

> I felt really bad, because I made friends with him when nobody knew him and nobody liked him, and I put all my friends to the side for him,

and I brought him into the group, and then he dumped me. He was my friend first, but then Jesse wanted him.... He moved up and left me behind, like I wasn't good enough anymore.

Ingratiation

In addition to being wooed into the elite strata and breaking up friendships to consolidate or use power in the group, currying favor with people in the group was another dynamic of inclusion found in clique behavior. Group members curried favor with the leader to enhance their popularity and obtain greater respect from other group members. One way of doing this was by imitating the group leaders' style and interests. Marcus and Andy, two fifth-grade boys, described how borderline people fawned on their clique and its leader to try to gain inclusion:

MARCUS: Some people would just follow us around and say, "Oh yeah, whatever he says, yeah, whatever his favorite kind of music is, is my favorite kind of music."

ANDY: They're probably in a position where they want to be more in because if they like what we like, then they think more people will probably respect them. Because if some people in the clique think this person likes their favorite group, say it's REM or whatever, so it's, say, Bud's [the clique leader's], this person must know what we like in music and what's good and what's not, so let's tell him that he can come up and join us after school and do something.

Not only outsiders and peripherals fawned on more popular people. This was also common practice among regular clique members, even those with high standing. Melanie, the second-tier fourth-grade girl mentioned earlier, described how, in fear, she used to follow the clique leader and parrot her opinions:

I was never mean to the people in my grade because I thought Denise might like them and then I'd be screwed. Because there were some people that I hated that she liked and I acted like I loved them, and so I would just be mean to the younger kids, and if she would even say, "Oh she's nice," I'd say, "Oh yeah, she's really nice!"

Clique members, then, had to stay abreast of the leader's shifting tastes and whims if they were to maintain status and position in the group. Part of their membership work involved a constant awareness of the leader's fads and fashions, so that they could align their actions and opinions accurately with the current trends, in a timely manner (also see Eder and Sanford 1986).

Another technique involved acting in different ways toward different people. Bill, a sixth-grade boy, recalled how the clique leader in fifth grade used this strategy to maintain his position of centrality:

Mark would always say that Trevor is so annoying. "He is such an idiot, a stupid baby," and everyone would say, "Yeah, he is so annoying. We don't like him." So they would all be mean to him. And then later in the day, Mark would go over and play with Trevor and say that everyone else didn't like him, but that he did. That's how Mark maintained control over Trevor.

Mark employed similar techniques of manipulation to ensure that all the members of his clique were similarly tied to him. Like many leaders, he shifted his primary attention among the different clique members, so that everyone enjoyed the power and status associated with his favor. Then, when his followers were out of favor, they felt relatively deprived and strove to regain their privileged status. This process ensured their loyalty and compliance.

To a lesser degree, clique members curried friendship with outsiders. Although they did not accept them into the group, they sometimes included them in activities and tried to influence their opinions. While the leaders had their in-group followers, lower-status clique members could look to outsiders for respect, admiration, and imitation if they cultivated them carefully. This attitude and this behavior were not universal, however, because some popular cliques were so disdainful and so unkind to outsiders that nonmembers hated them. Diane, Jennifer, and Alyssa, three popular junior high school girls who had gone to two different elementary schools, described how the grade school cliques to which they had belonged displayed different relationships with individuals of lesser status:

DIANE: We hated it if the dorks didn't like us and want us to be with them. 'Cause then we weren't the populatest ones, 'cause we always had to have them look up to us, and when they wouldn't look up to us we would be nice to them.

JENNIFER: The medium people always hated us.

ALYSSA: They hated us royally and we hated them back whenever they started.

Thus, despite notable exceptions (as described by Eder 1985), many popular-clique members strove from time to time to ingratiate themselves with people less popular than themselves, to ensure that their dominance and adulation extended beyond their own boundaries, throughout the grade.

TECHNIQUES OF EXCLUSION

Individuals enhanced their own and others' status by maneuvering into more central and more powerful positions and/or recruiting others into such positions. These inclusionary techniques reinforced their popularity and prestige while maintaining the group's exclusivity and stratification. Yet the inclusionary dynamics failed to contribute to other, essential clique features such as cohesion and integration, the management of in-group and out-group relationships, and submission to the clique's leaders. These features are rooted, along with other sources of domination and power, in the exclusionary dynamics of cliques. Exclusionary techniques illuminate how clique leaders enhanced their elite positions by disdaining and deriding others lower in the prestige hierarchy both inside and outside their cliques, thus supporting their power and authority on the foundation of others' subservience. These very techniques fostered clique solidarity, however, because members developed internal cohesion through their collective domination over others, and were tied to the leaders by their fear of derision and exclusion by the leader-dominated group.

Subjugation of the Out-Group

When clique members were not being nice to outsiders to try to keep them from straying too far outside their influence, they largely subjected them to exclusion and rejection. Insiders were entertained by picking on these lower-status individuals. As one clique follower remarked, "One of the main things is to keep picking on unpopular kids because it's just fun to do." Eder (1991) observed that this kind of ridicule, in which the targets are excluded and are not encouraged to join in the laughter, contrasts with teasing, in which friends make fun of each other in a more lighthearted manner but permit the targets to remain in the group by also jokingly making fun of themselves. Hilary, a fourth-grade clique leader, described how she acted toward outsiders:

> Me and my friends would be mean to the people outside of our clique.
> Like, Eleanor Dawson, she would always try to be friends with us, and
> we would be like, "Get away, ugly."

Interactionally sophisticated clique members not only treated outsiders badly, but managed to turn others in the clique against them. Parker and Gottman (1989) observed that gossip is one way of doing this. Hilary recalled how she turned all the members of her class, boys as well as girls, against an outsider:

> I was always mean to people outside my group like Crystal, and Emily
> Fiore; they both moved schools.... I had this gummy bear necklace,
> with pearls around it and gummy bears. She came up to me one day
> and pulled my necklace off.... It was my favorite necklace, and I got all
> of my friends, and all the guys even in the class, to revolt against her.
> No one liked her. That's why she moved schools, because she tore my
> gummy bear necklace off and everyone hated her. They were like,
> "That was mean. She didn't deserve that. We hate you."

Turning people against an outsider solidified the group and asserted the power of the strong over the vulnerability of the weak. Other classmates tended to side with the dominant people over the subordinates, not only because they admired their prestige but also because they respected and feared the power of the strong.

In the ultimate manipulation in leading the group to pick on outsiders, insiders instigated the bullying and caused others to take the blame. Robert, a fourth-grade clique follower, described with some mystification and awe the skilled maneuvering of Scott, his clique leader:

ROBERT: He'd start a fight and then he would get everyone in it, 'cause
everyone followed him, and then he would get out of it so
he wouldn't get in trouble.

Q: How'd he do that?

ROBERT: One time he went up to this kid Hunter Farr, who nobody liked, and said, "Come on Farr, you want to talk about it?" and started kicking him, and then everyone else started doing it. Scott stopped and started watching, and then some para professional came over and said "What's going on here?" And then everyone got in trouble except for him.

Q: Why did he pick on Hunter Farr?

ROBERT: 'Cause he [Farr] couldn't do anything about it, 'cause he was a nerd.

Subjugation within the In-Group

A second form of domination occurred through picking on people within the clique. More central clique members commonly harassed and were cruel to those with lesser standing. Many of the same factors that prompted the ill-treatment of outsiders motivated high-level insiders to pick on less powerful insiders. Craig, a sixth-grade clique follower, articulated the systematic organization of downward harassment:

Basically the people who are the most popular, their life outside in the playground is picking on other people who aren't as popular, but are in the group. But the people just want to be more popular so they stay in the group. They just kind of stick with it, get made fun of, take it…. They come back every day, you do more ridicule, more ridicule, more ridicule, and they just keep taking it because they want to be more popular, and they actually like you but you don't like them. That goes on a lot, that's the main thing in the group. You make fun of someone, you get more popular, because insults is what they like. They like insults.

The moving finger of ridicule was capricious, and could stop at any individual but the leader. It might turn toward a person because he or she did something deserving insult; it might be directed toward someone who the clique leader felt had become an interpersonal threat; or it might fall on someone for no apparent reason (see Eder 1991). Melanie, the second-tier fourth-grade girl discussed earlier, described the ridicule she encountered and told of her feelings of mortification when the clique leader derided her hair.

Like I remember, she embarrassed me so bad one day. Oh my God, I wanted to kill her! We were in music class and we were standing there and she goes. "Ew! What's all that shit in your hair?" in front of the whole class. I was so embarrassed 'cause I guess I had dandruff or something.

Derision against insiders often followed a pattern: leaders started a trend and everyone followed it. This multiple force intensified the sting of the mockery. Jeff, a fifth-grade boy, compared the behavior of people in cliques to the links on a chain:

Like it's a chain reaction. You get in a fight with the main person, then the person right under him will not like you, and the person under him won't like you, and etcetera, and the whole group will take turns against you. A few people will still like you because they will do their own thing, but most people will do what the person in front of them says to do, so it would be like a chain reaction. It's like a chain. One chain turns and the other chain has to turn with them, or else it will tangle.

Compliance

When leaders or other high-status clique members initiated such negative and wounding power dynamics, others followed, participating either actively or passively in the derision. Active participation occurred when instigators persuaded other clique members to become involved in picking on their friends. This often happened in telephone prank calling, when leaders conceived the idea of making trick calls and convinced their followers to do the dirty work. They might start the call and then place followers on the line to finish it, or they might pressure others to make the entire call, thus keeping one step away from becoming implicated if the victim's parents should complain.

In passive participation, followers went along when leaders were mean and manipulative, as when Ryan acquiesced in Brad's scheme to convince Larry that Rick had stolen his money. Ryan knew that Brad was hiding the money, but he watched while Brad whipped Larry into a frenzy, pressing him to deride Rick, destroy Rick's room and possessions, and threaten to expose Rick's alleged theft to others. Only when Rick's mother came home, interrupting the bedlam, was the money revealed and Larry's onslaught stopped. The following day at school, Brad and Ryan could scarcely contain their glee. Rick was demolished by the incident and was cast out by the clique. Ryan was elevated to the status of Brad's best friend by his conspiracy in the scheme.

Many clique members relished the opportunity to go along with such exclusive activities, welcoming the feelings of privilege, power, and inclusion. Others appreciated the absence of ridicule towards themselves. This situation sometimes was valued by new members, who often feel unsure about their standing in a group (Sanford and Eder 1984). Two fifth-grade clique followers expressed their different feelings about such participation:

Q: What was it like when someone in your group got picked on?

GARY: If it was someone I didn't like or who had picked on me before, then I liked it. It made me feel good.

NICK: I didn't really enjoy it. It made me feel better if they weren't picking on me. But you can't do too much about it, so you sort of get used to it.

Like outsiders, clique members knew that complaining to persons in authority did them no good. Quite the reverse: such tactics made their situation worse. So did showing their vulnerabilities to the aggressors. Kara, a popular fifth-grade girl, explained why such declarations had the opposite of the intended effect:

Because we knew [it] bugged them, so we could use [it] against them. And we just did it to pester 'em, aggravate 'em, make us feel better about ourselves. Just to be shitty.

When people saw their friends in tenuous situations, they often reacted passively. Popular people who got into fights with other popular people might be able to count on some of their followers for support, but most people could not command such loyalty. Jeff, the fifth-grade boy discussed earlier, explained why people went along with hurtful behavior.

> It's a real risk if you want to try to stick up for someone because you could get rejected from the group or whatever. Some people do and nothing happens because they're so high up that other people listen to them. But most people would just find themselves in the same boat. And we've all been there before, so we know what that's like.

Clique members thus cooperated in picking on their friends, even though they knew it hurt, because they were afraid (also see Best 1983). They became accustomed to living in a social world where the power dynamics could be hurtful, and accepted it.

Stigmatization

Beyond individual incidents of derision, clique insiders often were made the focus of stigmatization for longer periods. Unlike outsiders, who commanded less enduring interest, clique members were much more deeply involved in picking on their friends, whose discomfort held their attention more readily. Jeff described how negative attention could focus on a person for longer than a single incident:

> Usually at certain times, it's just a certain person you will pick on all the time, if they do something wrong. I've been picked on for a month at a time, or a week, or a day, or just a couple of minutes, and then they will just come to respect you again.

When people became the focus of stigmatization, as happened to Rick above, they were rejected by all their friends. The entire clique rejoiced in celebrating their disempowerment. They were made to feel alone whenever possible. Their former friends might join hands and walk past them at recess, physically demonstrating their union and the discarded person's aloneness.

Worse than being ignored was being taunted. Anyone who could create a taunt was favored with attention and imitated by everyone else (also see Fine 1981). Taunting included verbal insults, put-downs, and sing-song chants. Even outsiders, who normally were not privileged to pick on a clique member, could elevate themselves by joining such taunting (also see Sanford and Eder 1984).

The ultimate degradation was physical. Although girls generally confined themselves to verbal humiliation, the culture of masculinity allowed boys to injure each other (Eder and Parker 1987; Oswald et al. 1987; Thorne 1993). Fights occasionally broke out in which boys were punched in the ribs or stomach, kicked, or given black eyes. When this happened at school, adults were quick to intervene. After hours or on the school bus, however, boys could be hurt. Physical abuse was also heaped on people's homes or possessions. People spat on each other or on others' books or toys, threw eggs at their families' cars, and smashed pumpkins in front of their houses.

Expulsion

Most people returned to a state of acceptance after a period of severe derision (see Sluckin 1981 for strategies used by children to help attain this end), but this was not always the case. Some people were excommunicated permanently from the clique. Others could be cast out directly without undergoing a transitional phase of relative exclusion. This could happen to clique members from any stratum of the group, although it was more likely among people with lower status. Jason, a sixth-grade boy, described how expulsion could occur as a natural result of the hierarchical ranking, in which a person at the bottom rung of the popularity ladder was pushed off. He described the ordinary dynamics of clique behavior.

Q: How do they decide who they are going to insult that day?

JASON: It's just basically everyone making fun of everyone. The small people making fun of smaller people, the big people making fun of the small people. Nobody is really making fun of people bigger than them because they can get rejected.... then they can say, "Oh yes, he did this and that, this and that, and we shouldn't like him anymore." And everybody else says, "Yeah, yeah, yeah," 'cause all the lower people like him, but all the higher people don't. So the lower-case people just follow the higher-case people. If one person is doing something wrong, then they will say "Oh yeah, get out, good-bye."

Being cast out could result either from a severely irritating infraction or from individuals' standing up for their rights against the dominant leaders.

Sometimes expulsion occurred as a result of breakups between friends or realignments in friendship leading to membership challenges (described earlier), in which higher-status people carried the group with them and turned their former friends into outcasts. Adam was able to undercut Kevin's rising popularity and to co-opt his friendship by cutting off the support of Kevin's longtime friend Nick. Adam pretended to be best friends with both Kevin and Nick, but secretly encouraged each one to say bad things about the other so that he could carry each one's nasty remarks back to the other. By escalating these hurtful remarks, Adam finally wounded Nick severely. Nick retaliated against Adam in a way that Adam used to turn both Kevin and the whole clique against Nick, and to expel him from the group.

On much rarer occasions, high-status clique members or even leaders could be cast out of the group (see Best 1983). One sixth-grade clique leader, Tiffany, was deposed by her former lieutenants for continued petulance and self-indulgent manipulations. She recounted the moment of her expulsion:

TIFFANY: Do you want to know why I turned dweeb? Because they kicked me out of the clique.

Q: Who kicked you out?

TIFFANY: Robin and Tanya. They accepted Heidi into their clique and they got rid of me. They were friends with her. I remember it happened in one blowup in the cafeteria. I asked for pizza and I thought I wasn't getting enough attention anymore, so I was pissed and in a bitchy mood all the time and stuff, and so I asked them for some, so she said like "Wait, hold on, Heidi is taking a bite," or something, and I got so mad I said "Give the whole fuckin' thing to Heidi" and something like that, and they got so sick of me right then, and they said, like, "Fuck you."

When clique members are kicked out of the group, they leave an established circle of friends and often seek to make new ones. Some people find it relatively easy to make what Davies (1982) called "contingency friends" (temporary replacements for their more popular friends), and were described by a fifth-grade teacher as "hot items" for the unpopular crowd. James, a sixth-grade clique follower, explained why people expelled from a popular clique might be in demand with nonclique members:

Because they want more people, who are bigger, who have more connections, because if you get kicked out of the group, usually you still have a friend who is still going to be in the group, so then they can say, "Oh yeah, we'll be more popular even though this person isn't respected anymore. At least there is one person who still respects them in the group, so he'll get a little higher up or more popular, or we just should give him a chance."

Many cast-outs, however, found new friendships harder to establish. They went through a period when they kept to themselves, feeling rejected, stigmatized, and cut off from their former social circle and status. Because of their previous behavior and their relations with other classmates, they had trouble being accepted by unpopular children. Others had developed minimum acceptability thresholds for friends when they were in the popular crowd, and had difficulty stooping to befriend unpopular people. Todd, a fifth-grade boy who was ejected from his clique, explained why he was unsuccessful in making friends with the unpopular people:

Because there was nobody out there I liked. I just didn't like anybody. And I think they didn't like me because when I was in the popular group we'd make fun of everyone, I guess, so they didn't want to be around me, because I had been too mean to them in the past.

Rejects from the popular clique occasionally had trouble making friends among the remainder of the class because of interference by their former friends. If clique members became angry at one of their friends and cast that person out, they might want to make sure that nobody else befriended him or her. By soliciting friendship with people outside the clique, they could influence outsiders' behavior, causing their outcast to fall beyond the middle crowd to the status of pariah or loner. Melanie, a fourth-grade popular girl, explained why and how people performed such manipulations:

Q: Have you ever seen anyone cast out?

MELANIE: Sure, like, you just make fun of them. If they don't get accepted to the medium group, if they see you, like "Fuck, she's such a dork," and like you really don't want them to have any friends, so you go to the medium group, and you're like "Why are you hanging out with that loser, she's such a dork, we hate her," and then you be nice to them so they'll get rid of her so she'll be such a dork. I've done that just so she'll be such a nerd that no one will like her. You're just getting back at them. And then they will get rid of her just 'cause you said to, so then you've done your way with them. If you want something, you'll get it.

People who were cast out of their group often kept to themselves, staying in from the playground at recess and going home alone after school. They took the bus to school, went to class, and did what they had to do, but they didn't have friends. Their feelings about themselves changed; this was often reflected in the way they dressed and carried themselves. Being ejected from the clique thus represented the ultimate form of exclusion, carrying severe consequences for individuals' social lives, appearance, and identity.

REFERENCES

Best, Raphaela. 1983. *We've All Got Scars*. Bloomington: Indiana University Press.

Davies, Bronwyn. 1982. *Life in the Classroom and Playground: The Accounts of Primary School Children*. London: Routledge & Kegan Paul.

Eder, Donna. 1985. "The Cycle of Popularity: Interpersonal Relations Among Female Adolescents." *Sociology of Education* 58: 154–65.

———. 1991. "The Role of Teasing in Adolescent Peer Group Culture," pp. 181–97 in *Sociological Studies of Child Development*, Volume 4, edited by Spencer Cahill. Greenwich, CT: JAI.

Eder, Donna and Stephen Parker. 1987. "The Cultural Production and Reproduction of Gender: The Effect of Extracurricular Activities on Peer-Group Culture." *Sociology of Education* 60(3): 200–13.

Eder, Donna and Stephanie Sanford. 1986. "The Development and Maintenance of Interactional Norms Among Early Adolescents." Pp. 283–300 in *Sociological Studies of Child Development*, Volume 1, edited by Patricia A. Adler and Peter Adler. Greenwich, CT: JAI.

Fine, Gary Alan. 1981. "Friends, Impression Management, and Preadolescent Behavior." Pp. 29–52 in *The Development of Children's Friendships*, edited by Steven Asher and John Gottman. Cambridge: Cambridge University Press.

Hallinan, Maureen. 1979. "Structural Effects on Children's Friendships and Cliques." *Social Psychology Quarterly* 42: 43–54.

Hubbell, C. H. 1965. "An Input-Output Approach to Clique Identification." *Sociometry* 28: 377–99.

Oswald, Hans, Lothar, Krappmann, Irene, Chowdhuri, and Maria von Salisch. 1987. "Gaps and Bridges: Interactions Between Girls and Boys in Elementary School." Pp. 205–23 in *Sociological Studies of Child Development*, Volume 2, edited by Patricia A. Adler and Peter Adler. Greenwich, CT. JAI.

Parker, Jeffrey G. and John M. Gottman. 1989. "Social and Emotional Development in a Relational Context: Friendship Interaction from Early Childhood to Adolescence." Pp. 95–132 in *Peer Relationships in Child Development*, edited by Thomas Berndt and Gary Ladd. New York: Wiley.

Peay, Edmund R. 1974. "Hierarchical Clique Structures." *Sociometry* 37: 54–65.

Sanford, Stephanie and Donna Eder. 1984. "Adolescent Humor During Peer Interaction." *Social Psychology Quarterly* 47(3): 235–43.

Sluckin, Andy. 1981. *Growing Up in the Playground*. London: Routledge & Kegan Paul.

Thorne, Barrie. 1993. *Gender Play*. New Brunswick, NJ: Rutgers University Press.

13

The Impact of Pervasive Beauty Images on Black and White Girls' Self-Concepts

MELISSA A. MILKIE

Not only does our mainstream culture posit health, exercise, and body norms and values but also a stream of beauty images is carried through the media that conveys images of how young women are expected to look. We largely assume that these media images influence people significantly, but we may not understand how this occurs. Milkie uses a combination of qualitative interviews and quantitative measures from a larger study to show that media influences on identity are filtered through others, rather than direct. She offers us a study design comparing the self-concepts of White and Black ninth- and tenth-grade girls to examine this process. This selection shows us how the White girls are negatively affected by the media-transmitted beauty images, a fundamental precursor to obsessive exercise and dieting behaviors as well as eating disorders, whereas the Black girls manage to deflect the internalization and self-application of these images and thereby avoid the consequent negative self-conceptions. Why do you think there are differences by racial group? What does this say about the cultural expectations of these groups? If you are a young woman, how have media images affected your sense of self? If you are a young man, do you see similar issues occurring as for the girls in this study?

SOURCE: Melissa A. Milkie, Social Comparisons, Reflected Appraisals, and Mass Media: The Impact of Pervasive Beauty Images on Black and White Girls' Self-Concepts. Social Psychology Quarterly, Vol. 62, © 1999 American Sociological Association. Reprinted by permission.

Questions of whether and how media influence self-concept—both self-identities and self-evaluations—as well as their impact on beliefs, values and behaviors underlie much media research. First, researchers have shown how the social context of media use is crucial, particularly in that significant others are relevant to the way people interpret and are affected by media. Second, they have focused on people's power to select and be critical of media content, and thus to discount media messages. Although recent qualitative work has increased our understanding of media processes by revealing the complexities of people's understandings of images, it has done so to the detriment of assessing how media *influence* people. Interpretive researchers view people as powerful in relation to media content, but do not examine this assumption directly.

In this study, I provide a way to bridge perspectives which argue either that media content is powerful or that people are powerful in interpreting media. I do so by analyzing the extent to which people's power to make critical assessments of content (for example, believing that stereotyped portrayals of one's group are unrealistic or unimportant) may prevent that content from negatively affecting the self. I take the case of feminine beauty images in media, and assess the relative power of critical interpretations in countering harmful effects on the self-concept, specifically on self-esteem. To clarify how media can affect people indirectly, I draw on basic principles in social psychology which point to the key role of *others* in self processes, and discuss these in terms of some unique properties of mass media....

I examine the case of pervasive beauty ideals disseminated through mass media, which many suggest are harmful to young women. These images, particularly in regard to body shape, are extremely unlike "real" American women. The gap between the image and the reality has grown in recent years, as the media images have become slimmer and Americans have grown heavier (Wiseman et al. 1992). The in-depth interviews focus on a tangible, explicit embodiment of idealized femininity—girls' magazines—which saturate their target audience (Evans 1990). The images presented therein are also pervasive in other media such as movies and television. I address these broad research questions: How do girls interpret the female image in media, how do they critique it, and how do they perceive its influence? How do girls view peers' interpretations of these images? How important are critical views of the imagery in protecting girls' self-esteem?

RESULTS

Ethnic status sharply differentiated whether girls identified with the images, supposedly intended for and about all adolescent girls. This status created an important filter for social comparisons and reflected appraisal processes, and thus influenced the effectiveness of critical interpretations for shielding harm to self-concept. First I discuss how the magazines and the images they contain were a part of white girls' culture at both the rural and the urban school, but how black girls generally rejected the images as part of their reference group even though they occasionally read popular girls' magazines. Both white and black girls interpreted the images as largely unrealistic; many wanted more normal or more "real" girls in the images.

Reference Groups, Media Interpretation and Criticism of Images

White girls: Peer culture and media images as role models for the reference group. Explicit in cultural products labeled *Seventeen* or *Teen* is the notion that such products provide images and information relevant to particular groups of people....

Girls' magazines, like other media, were part of the white girls' peer culture in both schools. They helped these girls to assess how well they fit into, or were similar to, their reference group (also see Currie 1997). The magazines gave advice on, and were perceived to help with, girls' concerns about "fitting in" and being accepted by others. In the interviews, for example, many girls stated that any hypothetical girl who does *not* read girls' magazines does not care about others' opinions or is very independent. This comment implies that the information contained in these media pertains to conforming to the "norm" of adolescent femininity.

The respondents considered reading the magazines an enjoyable leisure activity: 95 percent of the white girls surveyed read them occasionally or more often; more than half read them "always." Magazine reading as a part of peer culture, and the relative amount of interaction centered around the cultural products themselves, differed somewhat in the two locales. At the rural school, where girls made slightly higher use of the imagery and evaluated it more positively (data not shown), cliques regularly discussed content during school hours and after school over the telephone. They read the magazines in the lunchroom, the hallways, and the school library, and even during class. Subscribers often shared their magazines with friends, reading them either together or to each other, and passing on copies to those who did not subscribe. Indeed, for rural white girls, a great deal of peer interaction surrounded these magazines. This is not surprising because in rural areas, media may be an important means of understanding the larger world and the variety of people in it, with which the rural dwellers have much less contact (Johnstone 1974; Morgan and Rothchild 1983).

Urban white girls also said they discussed the magazines or particular items in the magazines with friends, but they reported this experience less often. Perhaps because more varied activities are available to the urban girls, magazine reading is less salient. Yet in a quantitative analysis examining how often white girls read the magazines alone or with friends, I found no differences between schools. Indeed, more white girls at the urban school than at the rural school subscribed to at least one girls' magazine (64 percent versus 50 percent).

Black girls: "Maybe if there were more of us in there." Black respondents less often read mainstream girls' magazines, both individually and as a collective activity: 86 percent of the black girls surveyed read them at least occasionally, but only 11 percent always read them. Even though, in recent years, black models have appeared regularly in the four magazines with the greatest circulation, the magazines are perceived as largely for white girls. Most of the black girls read *Ebony* or *Essence*, aimed at black adults and black women respectively, magazines about music directed toward black youth, and hairstyle publications. Thus, in contrast to white girls, these respondents largely regard mainstream girls' magazines as something they do not want to or should not orient themselves toward because they view the magazines as for and about white girls. They define the images as irrelevant to their reference group for this social aspect of the self. This finding is not surprising because, as Collins (1991) and others have observed, mainstream media show an especially distorted image of black females—very thin, and with "whitened" hair and features. Tanya[1], in responding to how people would understand "girls" if they had only girls' magazines to look at, said:

> I think this is mainly toward ... white females.... you really wouldn't see too many black people in here—so if this is all you saw, you'd be kinda scared when you saw one like me or something. (May 9, 1994; urban black girl)

Minority girls were quite critical about the realism of the images. Part of this critique was that normative adolescent femininity was portrayed as white femininity. Although ethnicity differentiated the respondents' use of these mainstream magazines, both African-American and white girls seemed to hold common perceptions about the unreality of the images.

Interpretations and critical interpretations of media images of females.   When asked to describe the magazines to a girl who had never seen them, most of the respondents interpreted them as conveying very traditional aspects of femininity, such as appearance and romance. They mentioned fashion, makeup, styles—all related to appearance—and relationships with males. Previous analyses of these magazines (Evans et al. 1991; Peirce 1990) indicate that these reports represent accurate assessments of the content of girls' magazines. Barb's explanation is similar to how most girls described the magazines:

> They're about how girls can do their hair, what's in fashion. They give advice on boys; sometimes they give you advice on your body and stuff like that—how to get in shape. They've got how to do your makeup right, hair—I think I already said that—what's the right jewelry. They talk a lot about stars and stuff like that, they also talk a lot about boys. (May 9, 1994; urban Asian girl)

Secondarily, the respondents reported that the magazines were about girls' "problems." This view is closely related to the above observation. The information presented about appearance and relationships with boys was interpreted by the girls as advice about problems of traditional femininity which they were experiencing or which were common to adolescent females. A minority of the girls described the magazines more broadly as about "teenagers' lives" or "everything." Only two of 60 respondents described the magazines in what might be considered feminist terms, as about girls' "being independent," although these two girls also discussed appearance as an important component of the magazines.

The great majority of the respondents, even those who seldom read the magazines, liked them as a whole or liked certain parts. The girls stated that they read them because they were interesting, entertaining, and informative. An important feature of the girls' enjoyment and interest was learning about themselves and assessing their lives and their problems in relation to their peers. Linda, a grade 9 student, explained this:

> The girls will write in, and you kind of realize they have the same problems as you do.... you know they [other girls] kinda make you feel like you're not the only one. (May 19, 1994; urban white girl)

Researchers have suggested that one reason why people are critical of media is that the media distort reality and reflect groups in distorted ways. Most of the respondents were critical in that they said media images of girls were not realistic at all, and they made negative comments about the lack of "normal" girls. In general, the respondents indicated that the feminine images in the magazines presented an unrealistic appearance, both in the styles of clothing and in the perfection of their faces, hair, and bodies in comparison with the largely imperfect local girls. A few respondents said that the girls in the magazines were somewhat realistic; sometimes they referred to the pages that focused on "real" girls' problems or compared the images with the most popular or most beautiful girls in the school. The black girls were quite critical of the magazine models' physical appearance in general and tended to be critical about the lack of ethnic diversity or representation.

In discussing how the models looked, the respondents were likely to comment that they were *too* perfect, especially in body shape, weight, hair, facial features, and complexion. Sandra, a grade 10 student, discussed the message sent by the magazines:

> They mainly focus on models.... they make them look perfect, which nobody is. Makes everyone's expectations really high of their self, and they don't need it. I don't think they show the true girl. You know, nobody is perfect, and they all have their mistakes, and some of these people look like they never make a mistake. (April 21, 1994; rural white girl)

In fact, many viewed the images not merely as unrealistic, but as artificial. A girl who had recently lost a good deal of weight remarked that some models shown in the magazines have altered their "true" selves:

> I think some of them might be fake. Like get contacts to change their eye color, cake on their makeup, starve themselves. Like they're really not that skinny, but they just starve themselves. (May 25, 1994; urban white girl)

Generally, the respondents disliked the fact that these pervasive media images deviated so much from reality. They remarked, as noted above by Sandra, that the media created an uneasy gap between image and reality. Barb, while looking at the title of a girls' magazine article in front of her, observed that even the so-called "problem" bodies shown in the magazines are perfect:

> Oh, if I read that "Four Weeks to a Better Body," I'd probably ... these magazines are trying to tell you "Do this and do that." Sometimes they have ... swimsuits and stuff, and what you can do if you have a problem body. If you got a big butt, big chest ... what to do. And these girls that they are showing don't have that problem. I mean you can tell they don't, and that makes me mad.... They say if you got a stick figure, wear a one-piece and ... colorful and I'm looking at the girl and she doesn't have a stick figure. If you got big hips, if you got a big stomach—she doesn't have it—you can never understand that. (May 9, 1994; urban Asian girl)

In response to open-ended questions about whether they would change anything about the magazines, particularly anything that was emphasized too much or was not included, more than one-third of the respondents specified that the magazines should change the feminine image to be more realistic or "normal." Amy believes that "normal" people are missing from the images:

> One thing I guess would be just more normal people.... not like the models, but just average. Other people that haven't really had modeling experience.... (May 25, 1994; urban white girl)

In sum, most of the respondents regarded media images of females, particularly those which are common in ads or fashion pages, as unrealistic. Many disliked the images for this reason, considered them harmful to themselves or to others and advocated that media producers should alter their products to include more "real," ordinary, or "normal" girls....

Emulation of media images of females versus distance from them.   Both the white and the African-American respondents, but especially the white girls, liked the magazines, even though they criticized the lack of realism of the girls pictured therein. The white respondents used the images and ideas in the magazines to assess themselves. They said frequently that they "felt better" or more normal when reading about the problems and experiences of other girls their age. This feeling came from the numerous articles and advice columns that dealt with problems of relationships with boyfriends and family members, peer group pressures, and health, beauty, and fashion issues. The respondents particularly liked to assess themselves in relation to their reference group by taking quizzes that evaluated them on topics such as relationships (e.g., "How Good a Friend Are You?"). These quizzes provide scores that categorize the reader as a certain type of person and explain how she tends to act in situations in comparison with others. Jackie explains why she reads the magazines:

> I guess I like to see what ... the clothes, like what people are wearing. And like questions-answers, like what people are ... curious about, and see if I'm the same, I guess.... (May 23, 1994; urban white girl)

The minority respondents, in sharp contrast, did not emulate these images nor compare themselves as negatively with the models. Even though most of the black girls occasionally read the mainstream publications, they considered the images less relevant, belonging to "white girls'" culture and not part of a reference group toward which they oriented themselves. Strikingly, 10 of the 11 minority girls (nine black and one Asian-American) said unequivocally that they did not want to be like these girls; one mixed-race respondent (African-American and white) said that she "sometimes" did. The black girls indicated that they did not relate to the images and did not wish to emulate the rigid white beauty ideal.[2] Tamika described why she generally did not read mainstream girls' magazines:

> Well, I don't see a lot of black girls.... don't see a lot of us.... maybe if they had more, maybe I could relate to that. I don't know. 'Cause obviously we can't wear the same makeup or get our hair the same way.... things like that. So maybe if they had more. (May 11, 1994; urban black girl)

In sum, for the great majority of white girls in both locales, national media images and information about the reference group served as an additional social comparison introduced into the local context. The white girls evaluated their own behavior, problems, emotions, and importantly, physical appearance in comparison with these media others. Even though they knew that the images were unrealistic, the white girls saw themselves as part of the reference group being portrayed, and compared their "problems" with adolescent females' problems. They reported that they often (reluctantly) made social comparisons with the perfect physical appearance of media images because they knew that these images were what "everybody" wants. The minority respondents and a very few of the white respondents did not emulate these feminine images in media, did not bring them meaningfully into peer groups, and seemingly did not make social comparisons unfavorable to themselves.

Criticism, social comparisons and self-evaluations. Although the white girls liked the magazines a good deal and enjoyed finding out that they were "normal" on the basis of other girls' behaviors and problems, many said that they personally or that "girls" in general felt abnormal and inferior in relation to the idealized feminine image. A key influence of the magazines, then, is that the great majority of white respondents said they wanted to look like the girls pictured therein, *even though most saw the images as unrealistic and unattainable.* These girls necessarily experienced relative deprivation because they could not attain the valued image promoted by the pervasive display of this unique part of the reference group. Although they generally understood that the images were unrealistic, the girls perceived that other girls in the school, and especially males, valued such an appearance. Thus it was difficult for critical appraisal of media images to become meaningful in local interaction....

The white respondents made negative social comparisons even while they recognized the media distortion. They indicated that the comparisons were difficult to opt out of and made them or "girls" feel worse about themselves because the girls inevitably looked worse than the glamorized, exceptional females in the media. In quantitative data from the larger group of girls surveyed ($N = 210$) the white girls felt significantly worse about themselves compared with the images than the minority girls....

The black girls' criticisms of media imagery, in contrast to the white girls', may be effective in reducing the impact of media in this case, because the black subculture as a whole is more critical of mainstream beauty ideals. The black girls in this study, although as concerned about appearance as the white respondents, perceived themselves as better-looking and were more satisfied with their appearance than were the white girls, and their self-esteem was higher. Though the black girls objectively are farther from the mainstream ideals of beauty in skin color, hair style, and weight (see Dawson 1988), they compared themselves more favorably with mainstream media images than did the white girls. Evidence from interviews also indicates that black girls perceived the white ideal as narrow or as less applicable to them. Eliza discussed how minority girls may strive less often to be like the images of girls shown in these magazines:

> This is kind of a stereotype, but more of my white friends than my black friends are into [trying to be like feminine images in magazines]. I mean a lot of them are going on a diet or "I want that body so bad"— I don't know how anybody can be like that.... (May 27, 1994; urban black girl)

Thus, because media images were a part of the white girls' peer culture, and because these girls perceived that significant others—other girls and especially boys in their local networks—evaluated them on the basis of media ideals that were nearly unattainable, they were influenced regardless of how strongly they criticized the imagery. Especially important were body shape "norms" in the media, which tended to warp average-weight and thin girls' perceptions of their weight and attractiveness, or at least made them over concerned about weight at objectively normal, healthy weights. Even girls who articulated the distorted nature of peers' views of attractiveness seemed to feel compelled to abide by the shifted "norm" of body shape.

The wider range of physical appearances and body shapes that the black girls seemed to accept as good-looking in themselves and others was related to a more inclusive beauty ideal promoted in the "black" media. In addition, the black girls were more tentative about suggesting that males evaluated them on the basis of mainstream (white) media images. Most girls indicated that some males might do so but that others would not. This belief that males (often specified as black) rejected the "whitened" image was important in reducing black girls' negative self-evaluations especially related to body size.

NOTES

1. All names are fictitious.
2. Many of the black girls mentioned alternative publications focusing on hairstyles for black females. These publications did not contain articles, however—only hairstyle photos and information. (See Parker et al. 1995 for an assessment of African-American girls' culture and appearance.)

REFERENCES

Collins, Patricia Hill. 1991. *Black Feminist Thought: Knowledge, Consciousness, and the Politics of Empowerment.* New York: Routledge.

Currie, Dawn H. 1997. "Decoding Femininity: Advertisements and Their Teenage Readers." *Gender & Society* 11: 453–77.

Dawson, Deborah A. 1988. "Ethnic Differences in Female Overweight: Data from the 1985 National Health Interview Survey." *American Journal of Public Health* 78: 326–29.

Evans, Ellis D. 1990. "Adolescent Females' Utilization and Perception of Contemporary Teen Magazines." Presented at the biennial meetings of the Society for Research on Adolescence. Atlanta.

Evans, Ellis D., Judith Rutberg, Carmela Sather, and Chari Turner. 1991. "Content Analysis on Contemporary Teen Magazines for Adolescent Females." *Youth & Society* 23: 99–120.

Johnstone, John W. C. 1974. "Social Integration and Mass Media Use among Adolescents: A Case Study." pp. 35–48 in *The Uses of Mass Communications*, J. G. Blumler and E. Katz. Beverly Hills, CA: Sage.

Morgan, Michael, and Nancy Rothchild. 1983. "Impact of the New Television Technology: Cable TV, Peers, and Sex-Role Cultivation in the Electronic Environment." *Youth & Society* 15: 33–50.

Parker, Sheila, Mimi Nichter, Mark Nichter, Nancy Vickovic, Colette Sims, and Cheryl Ritenbaugh. 1995. "Body Image and Weight Concerns among African-American and White Adolescent Females: Differences and Making a Difference." *Human Organization* 54: 103–14.

Peirce, Kate. 1990. "A Feminist Theoretical Perspective on the Socialization of Teenage Girls through *Seventeen* Magazine." *Sex Roles* 23: 491–500.

Wiseman, Claire V., James J. Gray, James E. Mosimann, and Anthony H. Ahrens. 1992. "Cultural Expectations of Thinness in Women: An Update." *International Journal of Eating Disorders* 11: 85–89.

14

Asian American Ethnic Options: How Cambodian High School Students Negotiate Ethnic Identities

VICHET CHHUON AND CYNTHIA HUDLEY

Many factors contribute to the way we see ourselves, and sociologists have noted that we have a variety of different facets to our social selves. Growing up often entails negotiating the identities that we will construct from among those we are given and those we actively choose. Chhuon and Hudley's research on Cambodian high school students offers an interesting contrast to Perry's study of their White same-age peers. Here, we again see the influence of ethnicity on identity, and how people's attitudes, sense of self, and worldly success are affected by ethnic culture and position. Asians have often been compared to Whites through their depiction as the "model minority," but a closer examination of the various Asian sub-groups reveals some variations within this ethnic stereotype. How do these Cambodians compare to immigrants from other Asian countries? What do you think accounts for these differences? As you read it, how might you react if you found yourself in the position of these Cambodian high schoolers? How do their dilemmas and choices compare to those of Perry's White students? How do the studies of these groups of high school students affect your thoughts on other factors that may facilitate or hinder upward mobility?

SOURCE: Reproduced by permission of the American Anthropological Association from *Anthropology & Education Quarterly*, Volume 41, Issue 4, pp. 341–359. November 2010. Not for sale or further reproduction.

Cambodian students in the United States are often viewed within an Asian American collective that assumes all Asian-descent students are academically successful, as propagated by the model minority stereotype of Asian American students (Lee 1996; Ng et al. 2007). As a result, Cambodian students are critically underserved in U.S. schools and underresearched in the literature. At the same time, those few studies that have been carried out with Cambodian students as a disaggregated ethnic group have presented conflicting images in the research literature and in the schools. Research suggests that although Cambodian students tend to be perceived by the larger society through the lens of the model minority stereotype, they often endure low expectations from teachers and counselors in their local high schools (Chhuon et al. 2010; Ngo and Lee 2007; Reyes 2007; Um 2003). These local stereotypes cast Cambodian youth as low academic achievers, delinquents, and dropouts (Chhuon and Hudley in press; Chhuon et al. 2010; Um 2003). The extant literature, however, has paid scant attention to these complex and contradictory representations of Cambodian students. In this article, we examine how these conflicting perceptions might influence Cambodian students' decisions about their ethnic identity in school and the consequences of these decisions for their school experience.

Previous research with academically successful Cambodian students documented that students' awareness of the negative local perceptions of their ethnic group influenced them to distance themselves from their Cambodian ethnic identity in both their high school and their community (Chhuon 2009; Chhuon and Hudley 2008). Ethnic disidentification for some Cambodian students may be related to their understanding of the U.S. racial hierarchy and the model minority stereotype. For these students, conforming to the model minority label can come at the cost of their own ethnic identity, and may result in ethnic identity conflict. Internalizing an essentializing, albeit positive, view of Asian Americans may encourage some Cambodian students to adopt negative feelings and misconceptions toward their own ethnic group. Additionally, Asian American students who embrace this stereotype might feel unrealistic pressures to achieve that undermine their schooling. Hence, how Cambodian students resolve this conflict between societal stereotype and ethnic identification has implications for their academic achievement and well being in school.

Research on the school performance of Cambodian students consistently shows that Cambodian youth perform less well academically than students from other ethnic groups. Recent analyses from the 2005 American Community Survey revealed that less than half (40.1 percent) of all Cambodians 25 years and older residing in the United States reported attainment of a high school education. Moreover, only 11.1 percent of Cambodians in the United States have earned a four-year college degree, far below the national average for Asian Americans (44 percent) and the total population (17.2 percent). Given these data, Cambodian children are described as being at risk of assimilating into the lowest social and economic segments of U.S. society. However, a critical limitation of this literature is the lack of attention to how Cambodian students experience ethnic stereotypes at their school, and how these stereotypes relate to identity and school achievement. Although these studies have contributed to our understanding of Cambodian students as a disaggregated subgroup, our research extends this literature by examining the experiences of high and low achievers through an alternative framework that focuses on how Cambodian students understand and reconcile the conflicting images of their ethnic group in school. Specifically, we argue that Cambodian students adopted, rejected, and affirmed certain ethnic identities in relation to perceived advantages associated with their school contexts. Our analyses revealed two distinct groups of Cambodian students—those who identified panethnically as Asian American and benefited from the model minority stereotype, and those who embraced a Cambodian ethnic identity. Within the latter group of Cambodian identifiers, some students sought to disrupt the negative perceptions of Cambodian youth while others responded by generally upholding the problematic stereotypes of their ethnic group. This study adds to a small but growing body of work that complicates the view of Southeast Asian students in U.S. schools by zooming in on the process through which adolescent high school youth chose to ethnically identify, as well as the consequences of those choices....

COMMUNITY AND SCHOOL CONTEXT

The research presented in this article is part of a larger study on Cambodian high school students' academic achievement and identity conducted during the 2007–08 school year. The setting for this study was one high school in a large southern California school district. This site is significant because this district serves the largest Cambodian community in the United States. In contrast to other Asian American enclaves (e.g., Little Saigon in Orange County, California), this Cambodian enclave is an impoverished community, with limitations on educational resources and college attendance rates that are typical of poor communities everywhere. Although white European American residents comprise approximately 50 percent of the population in this city, the inner-city community in which this study takes place is largely made up of Cambodian, African American, and Latino families. Comprehensive High School (CHS) is a racially and economically diverse school. During the 2006–07 school year, the reported student enrollment was approximately 4,700, and 60 percent of students were eligible for free or reduced price lunch. The student body comprised 27 percent African American, 27 percent Asian (incl. Cambodian), 26 percent Latino, 12 percent white, 6 percent Filipino, and 2 percent Pacific Islander. In the school's surrounding neighborhood, streets are lined with Cambodian businesses including auto repair shops, jewelry stores, ethnic grocery stores, and restaurants.

CHS is the oldest high school in the district and is well known for its highly selective college preparation programs and its current school-within-schools system called "academies." These academies are curricular themed, small learning communities, and all students participate in one academy. Observations at the school and conversations with current and former students and counselors suggest that the boundaries between the different academies are clearly marked, and the academic hierarchy of these programs is well understood by students, teachers, and administrators. Students who are not admitted into the school's three most selective "magnet" academies (Intensive College Prep, Math–Science, World Commerce) must participate in one of the school's three less prestigious and less selective academies (Humanities, Business, Communication). Staff and students often referred to the school's magnet academies as the "higher" academies, and many Asian American students participated in these three programs.

Data Collection

Participant-observation was carried out by the first author for the majority of the 2007–08 academic year. Here, particular attention was paid to the ways students discussed and presented their ethnic and panethnic identities across different settings.... As well, Cambodian students ($n = 52$) drawn from across the school's various academic programs were recruited to be interviewed for the study. Our sample is balanced by both gender (28 female and 24 male) and academy assignment (23 from magnet academies and 29 from nonmagnet academies). Time spent in the school by the first author–researcher was advantageous for recruiting participants and familiarizing him with potential informants. In general, purposeful sampling techniques were used to recruit the appropriate range of participants.... Each interview, lasting approximately 75 minutes, was designed to more deeply understand, from various perspectives, factors that influenced Cambodian students' academic achievement and well being in school. In addition to student interviews, individual interviews were carried out with 15 teachers, 5 counselors, 4 administrators, 2 school psychologists, 1 librarian, and 2 teachers' aides. We aimed for breadth and depth in the selection of student and adult informants....

FINDINGS

The Lay of the Land: Cambodian Students at Comprehensive High

Although Cambodians were a major student group at Comprehensive High, their presence on campus was not reflected in the school and the district's data on student ethnic backgrounds. At the beginning of fieldwork, the researcher (Chhuon) asked for the number of Cambodian students enrolled at the school, but the administrator in charge of enrollment explained that the district classified Cambodians as "Asian." In this way, *Asian* represented an institutionally imposed label ascribed to many Asian-descent youth in this district despite the community's ethnic heterogeneity. Filipino students, interestingly, were not aggregated within this panethnic label in the district. This was likely because of the different sociohistorical positions that larger Asian American groups, such as Filipinos, occupy in U.S. society (particularly in California) as influenced by differences in immigration, population size, and political power. In contrast, the considerably fewer number of Cambodians in the United States often makes them a more statistically negligible community. The researcher's assessment of Cambodian student enrollment at CHS relied on identifying those students whose home language is Khmer. At the same time, some CHS staff hinted at the evolving demographic complexities experienced at the school. Early on, one administrator explained about CHS: "Actually, we're heavy in Cambodians; quite a few Cambodians here. Before at this school, when you talked about Asians, you think about Japanese and Chinese but it's not like that anymore." Early observations confirmed these comments. For instance, as Chhuon walked around the school's main quad area during lunch, he noticed many more young Southeast Asian faces than he had seen at any other high school he had visited including nearby schools. Although he could not say for certain which students were ethnically Cambodian, CHS was located in a large Cambodian community and many Cambodian children attended the high school.

Throughout the school year, CHS staff and students routinely described Asian American students as "motivated," "serious," "hardworking," and "bright." In short, Asian-background students were seen as the smartest students and were expected to be in the most selective academies. The following quote from one teacher is indicative of this perception:

> How do I think Asian American kids are seen at this school? (Pause) You know, I actually went to school not too far from here in the old days. Old days to you I'm sure (laughter). Anyways, I remember the Asians always kicking our butts in school, to be blunt. They always were more serious I guess than most of the kids. I had lots of Asian friends since I was in the [high track] at [names a local high school]. The truth is I don't think things have changed that, or too much. The Asians at this school are still the most serious students. I mean it's no wonder really if you look at how they're raised, right? Just go look at the [magnet academies] classes. Maybe I'm not supposed say this, but you'll see what I'm talking about.

In addition to reminiscing about how Asian American students have been doing well since he was a local high schooler in the 1980s, this teacher, a white male in his forties, hinted at a belief that this success stems largely from Asian family values, as consistent with the model minority stereotype. Unfortunately, this stereotype of Asian Americans usually came at the expense of Cambodian students' ethnic identities. As well, Cambodian youth often spoke about the successful image of Asian American students in contrast to less positive views of Cambodian students. For example, Brenda, a Cambodian student in a magnet academy, described how Cambodians were perceived:

> Like, the other Asian ethnicities, they're more advanced. We're use to living in the ghetto. Well most of the Cambodians here in [city of research site], and most of us, like especially guys and stuff they're like poor and into crime and stuff like that. But then it's different from other Asians, like Koreans, they're more into school and more concentrated on it.

Whereas Asian students from other ethnic backgrounds were associated with high achievement and academic motivation, Cambodians were assumed to be poor, low achieving, and involved in crime. And although Brenda brings up Koreans in particular, Cambodian students' reference to "other Asians" usually meant those students from East Asian backgrounds, including Chinese American, Japanese American, and Korean American peers. At CHS, many of these "other Asians" were enrolled in the elite magnet academies.

Although Cambodian students were well represented in the less selective Business, Humanities, and Communication academies, few Cambodian students were enrolled in the three magnet academies. Unlike Cambodian students in magnet classes whose peer groups often comprised white and other Asian-background students, Cambodian students from the "lower academies" tended to associate with other Cambodian students from similar academies. Gina, a Business academy student, shared that in her classes "there are a lot of Cambodians, the same, difference you know? Well you know, the classes I have, I think there are way more African Americans and Cambodians in the classes. I think they are one of the two main ethnicities that we have." In her interview, Sharon lamented the ethnic stratification she observed across CHS. When asked about her academy, she chose to comment about nonmagnet academies in general:

> That's where they put all of the minorities [nonmagnet academies].
> Well, what I know of because like everyone that knows each other,
> everyone that hangs out with each other and CHS like is always like
> that … there's a majority of minorities and Asians, like Cambodians, and
> Blacks, and Hispanics. I barely see any Whites. That's not right.

When describing "Asians" enrolled in nonmagnet academies, students usually clarified to mean Cambodian students. As Sharon keenly discerned, the CHS academy system represented an academic hierarchy where certain ethnic groups, including Cambodians, were largely absent near the top. To be sure, lower expectations and fewer resources found in the nonmagnet academies were experienced by students from other ethnic backgrounds as well as Cambodian students in the "lower" academies. However, Cambodian students in these academies were in an especially precarious position because they often had to straddle a double perception of their racial and ethnic group by staff and peers in school. Cambodian students were faced with ethnic identity decisions that shaped how they might be seen and treated as well as how they perceived themselves. Cambodian students' sense of social and academic self was profoundly influenced by whether others viewed them through a stigmatized ethnic label such as Cambodian, or through a positive distortion represented by the model minority stereotype and associated with a panethnic label.

Negotiating Ethnic and Panethnic Identities

Cambodian students at CHS were very aware of the varying levels of status that certain ethnic and panethnic identities held, both in school and in larger society, and they often negotiated their identities in response to their understanding of this hierarchy. Although Cambodian students all viewed themselves as both Cambodian and Asian American, students attached different meanings and usages to these ethnic identity labels. Ethnic and panethnic identities were not just descriptive labels but represented differing expectations and behaviors. We discuss how Cambodian students negotiated their ethnic identities by focusing on two groups: those that identified panethnically and those that embraced a Cambodian ethnic identity.

Panethnic Identifiers

Panethnic identifiers tended to be Cambodian students enrolled in magnet programs. Seventy-eight percent (18 of 23) of our magnet informants explicitly described a preference for panethnicity. Although these students identified as Cambodian, they were generally uncomfortable with solely a Cambodian ethnic identity. For example, Vanna explained:

> Well people always ask me, "What are you?" I say I'm Asian and
> they say, "No, but what are you?" I think it's funny to go on and on.
> I identify myself as being Asian just because it's easier to say that I'm part
> of this group. Because obviously I'm Asian. I look that way. But like
> within the Asian community they ask me what are you? And I say I'm
> Chinese because I guess I think it like just looks better sometimes.
> I mean, I'm half anyways.

Vanna's explanation illuminates students' ambivalence with being perceived as Cambodian in school. Although Vanna stated that she typically responds to questions about her background with "I'm Asian," she admits this should be obvious, while noting that she stresses her mixed background identity (Chinese Cambodian) because it "just looks better sometimes." Cambodian students like Vanna explained that they often emphasized their "other half" because of the negative images associated with the Cambodian ethnicity. More often, they pre-ferred the panethnic label at school. A number of Cambodian students described themselves as "less ghetto" and did not want to be associated with Cambodian peers whom they perceived as trouble makers or those that did not care about school. In these cases, the students they were generally referring to were Cambodians enrolled in the "lower" academies.

Although students identified as ethnically Cambodian generally, Cambodian magnet students usually considered the Asian American panethnic label to be more salient at school. For example, Steven, a senior, said that he identified panethnically "to give a broader sense for probably someone who has not met me before. Like 'oh yeah I'm Asian.' [My friends] would say they're Asian too." A major reason cited for this identification was their participation in ethnically diverse peer groups formed through class. Different from nonmagnet classrooms, magnet academies were largely comprised of white American students and Asian American students of ethnic backgrounds other than Cambodian. Davy shared: "I have a Korean friend, a Vietnamese friend, and a Jewish friend, yeah, so most of them aren't [Cambodian] because there weren't that many Cambodian people. Like, actually the only one … there's only one other Cambodian in [Intensive College Prep] that I know of." Observations of a number of magnet classes confirmed that few other Cambodians students were indeed enrolled. Steven's preference for pan-ethnicity was attributed to a lack of opportunity for getting to know Cambodian peers. Interestingly, he explained that he preferred to identify himself to others panethnically although he would already be phenotypically identified as "Asian." On the one hand, for Steven, panethnicity was linked to an image of himself and high achieving other Asian friends. He added that he regretted not having made many Cambodian friends at CHS, despite the school having "probably way too many Cambodians for one place (laughter)."

On the other hand, Davy explained that even if more Cambodian students were in her classes, she still would participate in more ethnically diverse peer groups that did not include Cambodians. When asked about this preference, she responded this way: "A lot of Khmer kids hanging out in one place will usually get into trouble. Period." During the nutrition and lunch breaks, Davy typically hung out with white American students and one Korean American friend near where other magnet students congregated. Davy explained that she was better off socially and academically by avoiding other Cambodian students at school. When asked about how she came to this decision, Davy, a first-year student, stated that she was mostly following advice from friends and older siblings that attended CHS. Her impression of Cambodian students was shaped largely by the circulating stereotypes of Cambodian students in her school and community. For students like Davy, panethnicity in school was developed in response to the negative images associated with Cambodian youth.

Cambodian students also chose to embrace panethnicity in school because they perceived it as a path to a positive academic identity. For example, an Asian American label meant that teachers and students would view them as high achievers, rather than academic strugglers. Brenda, a magnet student, stated that "When I'm seen as Cambodian, for some reasons it lowers the view. That makes me feel proud to be seen as Asian. I want to be seen as Asian. I don't know why, but it's like when you're Asian you're good at math, and stuff like that." Not only were youth like Brenda considered good in math when they are perceived panethnically but also they were not associated with negative characteristics attached to stereotypes of Cambodian students such as poverty and low school achievement. Although these students did not deny their ethnic heritage when asked, they preferred panethnicity because of the flattering nature of this label. These Cambodian students well understood the model minority stereotype of Asian American students and used it to their advantage.

Although the majority of panethnic identifiers were enrolled in magnet academies, some nonmagnet Cambodian students also preferred a panethnic identity because of its perceived advantages. Different from magnet classes, nonmagnet courses enrolled many Cambodian students; however, students from other ethnic backgrounds (primarily Latino and African American) often perceived their Cambodian peers in model minority ways. Pich, a Humanities academy student, described the advantage in this way:

> Like the other Asians, like the East Asians for example. Like, they all have degrees. Their parents all live in pretty kind of houses, you know? They all have money and I don't know I feel like they want, sometimes I'd rather say just Asian than to say Cambodian because of that stereotype. Some people wouldn't even know and some are like "Oh okay" and then they have this assumption and like yeah let people assume because I rather have them assume good things about me than bad things.

As Pich explained, positive assumptions about Asian Americans afforded Cambodian students more positive academic and ethnic identities at school. How Cambodian students understood these identities and how they ethnically identified themselves combined with how they felt others saw them held important consequences for their school experiences. For Pich, if she were ascribed an Asian American label, she would likely be considered to be from East Asian descent and people would "assume good things" about her. Further, her quote suggested that she herself believed that all East Asians were educated and well off. Likewise, this stereotype is alluded to in Yesenia's admission: "I think people think, like, Asian people are smart.... I'm okay with it because it's good for me.... But it's not like I'm smarter than them ... but then I know they think I'm smarter." Even though Yesenia knew that she was not brighter than her classmates, she perceived a benefit from being thought of as "smarter" by her non–Asian American peers. In class, this perception led to Cambodian students being asked by their peers to join them in group work and for help on assignments. This was especially true for nonmagnet Cambodian students in math class, a subject that Asian Americans were stereotypically expected to excel in. Generally however, the panethnic label was linked to a feeling of raised expectations and more positive academic identities for Cambodian students across all academies....

CAMBODIAN IDENTIFIERS

In contrast to panethnic identifiers, Cambodian students from less rigorous academic environments often held less negative attitudes toward their Cambodian identity. Cambodian students from nonmagnet academies often expressed a more explicit Cambodian identity and they understood well the negative perceptions that some people held about Cambodians. Twenty-three of the 29 (79 percent) nonmagnet students interviewed were quite clear about their preference for a Cambodian identity. Many discussed pride in their ethnic background. For instance, when asked to explain how she viewed herself ethnically, Crystal, a student in the Humanities academy, stated: "Like what am I? I'm Cambodian. That's it. I guess I was brought up to be proud of who I am. So I've never really denied who I was. I've always said I was Khmer. I never felt that I shouldn't say I wasn't." A number of students similarly remarked that they "never really denied" their Cambodian identity to suggest that many of their peers did. Another student, Catherine, wanted to pursue higher education and set a good example for her younger brother:

> Smart people like [other] Asians are more likely to go to college than others but since Cambodians are like more of the dropouts out of the whole Asians, I want to prove that wrong. I want to go to college. I know some Cambodians that dropped out, but there are some other Cambodians that can make it into college and stuff, and I'll be like the first one in my family, I would be the first one to go to college. I have only a small little brother, he's 12 so I got to do what's best because he looks up to me.

Although acknowledging that Cambodians are often seen in negative ways at CHS, this Cambodian identifier expressed pride in her Cambodian background. Catherine's awareness of Cambodian students' aggregate low school achievement served as a motivational backdrop toward becoming the first member of her family to attend college. For some Cambodian identifiers, their academic drive was influenced by a desire to defy negative stereotypes of Cambodians at school and in the community. For instance, Thomas asserted that

> I know about the stuff people think about Cambodians around here. For example, I know that people look at me and don't think I'm gonna do anything. Probably partly cause I'm Khmer but also since my sister didn't finish [high school]. I do look at it as like trying to prove haters wrong. Like, I want them to come here in June to see me, and my friends too. We're all Cambodians but we care about school. Anyways, sorry … to answer your question, I'm just Khmer I guess.

As Thomas poignantly remarked, being Cambodian in his community carries with it important assumptions about school achievement. The "haters" Thomas referred to include those students, teachers, and people from his neighborhood that have negative stereotypes of him and other Cambodian youth. Thomas understood that being Cambodian was often a stigmatized identity but transformed this identity into a badge of honor that fueled his motivation to graduate from CHS. Some Cambodian students were aware of the perception of Cambodian students at the school but did not accept the simple views of Cambodians in their community, particularly those in nonmagnet academies.

Cambodian students from nonmagnet academies often provided a more positive interpretation of what it meant to be Cambodian. Unlike many of their magnet peers, these students more often discussed positive images associated with being Cambodian, including good food and cultural dances. These attitudes were often shaped by positive interaction with other Cambodian students at the school. For some youth, a Cambodian ethnic identity was viewed as advantageous in school. For example, Darius was a tenth grade Communication academy student who explained:

> In my opinion, this is a special school because you can actually speak Khmer with other Cambodians here. I know my Khmer isn't real good but it's nice to joke around and other people don't know what you're saying. Even in English too. When I talk about something my parents said last night my Cambodian friends get it.

Throughout the school year Cambodian students from nonmagnet programs were observed speaking Khmer to discuss matters that they knew only Cambodians could relate to including issues with parents, difficulties with teachers, and discrimination. Research has found that first- and second-generation youth who can speak their heritage language view their ethnic backgrounds more positively and have higher self-esteem (Lee and Suarez 2008; Phinney et al. 2001). Students acknowledged that even though they may not speak Khmer very well, they viewed their heritage language as an opportunity for bonding with other Cambodian students.

Whereas some ethnic Cambodian identifiers maintained a sense of determination in their schooling, others responded to the negative views of their group in ways that undermined school success. During lunch, students tended to be segregated by their program affiliation. One magnet student, Jenny, explained, "Cambodian kids from the lower academies are way different. Like the Cambodian kids over there [pointing to a group in front of the library], they're ghetto, gang bangers, that's just what they do." Because of the presence of Cambodian gangs in the community, Cambodian boys at CHS were often stereotyped as "gang bangers," particularly if they dressed in a certain way, as reflected by this encounter with one counselor:

> Before 5th period I ran into Mrs. Toder in the hallway who inquired into my research project. She is a counselor in [one of the nonmagnet academies]. She warmly welcomed me to the school and commended me for working toward my Ph.D. When I asked her about her experiences working with Cambodian students, she replied "the Cambodians here are really a mixed bag. Some of the brightest kids we have are Cambodian…. But you're going to see the boys here need a lot of support." When I asked what she meant specifically, Mrs. Toder mentioned gangs as "a real big thing getting in the way." As she walked away into her office she advised that "you can tell by them sagging their jeans!" [field note, December 14, 2007]

Most of the young men Chhuon interacted with, however, who fit this image, were not gang affiliated. Male students explained that their manner of dress (baggy pants, large T-shirts) was consistent with an urban youth culture that is glamorized in media outlets including MTV and Vibe Magazine. Thus, the ethnic identity issues for many Cambodian boys at CHS were further complicated by negative stereotypes about gang membership.

Again, although the majority of these students were not gang members, these youth often referred to themselves in negative terms consistent with the unfavorable stereotypes of Cambodians. For example, Chhuon often engaged in lunchtime conversations with Cambodian students hanging out in front of the library. These students (the majority of whom were male and participated in the school's three nonmagnet academies) expressed pride in their Cambodian ethnicity and were critical of how teachers and other school personnel treated them. During one lunch period, a tenth grade Humanities student named Arun discussed why his academy is often perceived as "ghetto." He explained: "If the people around us treat us that way [ghetto] then that's how we're going to be. Like, you know, those up on that stage [pointing to Cambodian magnet academy students across the quad], they get treated better by everybody here. That's why they do better. But they're White-washed." Accusations of being "White-washed" levied at Cambodians hanging out with white Americans and other Asians represented Arun's disdain for those who did not take sufficient pride in their Cambodian heritage as well as those enrolled in higher academic academies. At CHS, this often meant the same thing for Cambodian youth. Arun's statement "that's how we're going to be" indicates a kind of self-fulfilling prophecy effect that influenced the identity work of some Cambodian youth. Standing nearby, James overheard the discussion and added that, "Security guards don't send those Asians to [detention] when they're late to school! We're like the bad Asians I guess." Hence, the good Asians were largely students from East Asian backgrounds as well as Cambodian students who associated primarily with East Asian and white American peers. For James and this group of Cambodian identifiers, being genuinely Cambodian meant associating with primarily other Cambodian peers and belonging to less selective academies.

Some ethnic Cambodian identifiers developed a clearer adversarial stance toward school. When asked about what it was like to be Cambodian at school, Danny explained that "it depends who you asking. I think most people are gonna tell you bad things. Like, some of it is true and some of it ain't. You know? But if you real Cambodian like me and Chris, and Sovan, the ones from lunch, then you're down.... If you ask the teachers they won't get it.... They just want to talk to the rich kids." For Danny and the "real" Cambodians, being "down" meant that you understood what it was like to grow up poor and be outside of the CHS conception of a good student. Later on, Danny bragged about how he and his friends could manipulate their teachers by acting out in class: "Sometimes like when we get bored we'll act stupid on purpose just to like see how far we can go. Like, just to see what the teacher is gonna do. Sometimes they don't care and ignore us. But like sometimes they kick us out. That's okay. We don't like that class anyways. We ain't missing nothing." Many Cambodian students in this group belonged to an after school program in the community with which Chhuon was involved. One of these youth, Krazy (self-selected pseudonym), enthusiastically volunteered to be interviewed for this research. He said that it was important that "people know what's up with this place." The place Krazy ostensibly refers to is CHS and his eagerness to share his thoughts about identity and schooling reflected his experiences of invisibility as a Cambodian student. The following exchange reflected Krazy's sense of frustration and hopelessness:

CHHUON: How do you see yourself at [CHS]?

KRAZY: How do I see myself? Like race?

CHHUON: Yes.

KRAZY: I'm Khmer. You know? I'm not gonna deny it. For what? Know what I'm saying?

CHHUON: I think so. What do you think people think of Khmer kids at this school?

KRAZY: Ghetto. That's it really.

CHHUON: What do you think about that?

KRAZY: True probably. To be honest. I mean we ain't got money like the other Asians. No education really.

CHHUON: Do you think this is true for yourself?

KRAZY: I think so. I know you trying to help out and stuff, but it's too late I think. Like, I tried. I mean, like, what do you expect? Like if teachers don't care if we pass [their classes] then who's going to care? [Teacher's name] don't care. [Another teacher's name] don't care. The truth is I don't give a shit about school no more. One time I tried to ask for help and [another teacher's name] didn't even want to help.

CHHUON: Why not?

KRAZY: Probably cause I'm failing already (laughter).

Students like Danny and Krazy dealt with their marginalization in school by misbehaving and giving up. Danny's low achievement was balanced by a sense of empowerment he felt from feeling that he could control a classroom by acting out. Unlike the first group of ethnic Cambodian identifiers, Danny, Krazy, and other boys like them, did not discuss any positive aspects of being Cambodian. Similar to other ethnic minority youth, their identity negotiation underscored ethnic pride in a manner that opposed school achievement (Fordham and Ogbu 1986). For these students, their disengagement served to protect their sense of self-worth. Nevertheless, Krazy's excerpt provides a window into a number of important things about Cambodian youth at CHS. In addition to presenting himself as a tough guy when discussing his Khmer pride, Krazy's interview revealed feelings of invisibility and vulnerability. His anger about his teachers' unwillingness to care and help him was quickly turned into humor not because it was unimportant to him, but as a way to cope with a hopelessness that characterized his academic experience.

To summarize, the first group of ethnic Cambodian identifiers seemed fundamentally determined to do well in school despite the negative stigma attached to their ethnic group. These students' interpretation of their Cambodian identity centered on defying the negative image of their ethnic group by working hard and acting as role models. Unfortunately, a distinctly different pattern emerged for the second group of Cambodian identifiers that were primarily male. Although this latter group likewise understood the negative perception of their group in school, they responded to the low expectations of Cambodian students less so by seeking to violate those expectations, but, rather, by fulfilling them.

DISCUSSION

To extend the scholarship on Asian American ethnicity and panethnicity, our study delved into the individual decision processes through which ethnic identification occurred. Whereas other scholarship on panethnicity emphasized the instrumental incentives for Asian American subgroups to adopt a collective panethnic label, our findings revealed how panethnicity offered a psychological and educational advantage for some Cambodian youth. Cambodian students were actively involved in ethnic identity politics at their school. Marginalized Asian American students may actively adopt the model minority profile as a means for attaining a positive academic image in the classroom in pursuit of their larger academic goals. Cambodian students expressed considerable agency in their school experiences by adopting, emphasizing, and rejecting particular identities in differing contexts. Thus, the model minority label, although inherently hegemonic, was actively negotiated by Cambodian students. These youth negotiated their ethnic identity choices in ways that resist stigmatization while considering whether certain ethnic and panethnic categories were advantageous in particular settings. These findings resonate with other identity research where some second-generation immigrant youth of color believed that accepting their ethnic identity was an invitation to be mocked and negatively labeled. Consequently, some Cambodian students' embrace of panethnic identities and its associated model minority stereotype represented their acceptance of an overgeneralization that further marginalizes the variability in the experiences of Asian American youth....

REFERENCES

Chhuon, Vichet. 2009. How School Structures Influence Ethnic and Panethnic Identity in Cambodian High School Youth. Ph.D. dissertation, Department of Education, University of California, Santa Barbara.

Chhuon, Vichet, and C. Hudley. 2008. Factors Supporting Cambodian American Students' Successful Adjustment into the University. *Journal of College Student Development* 49(1): 15–30. Chhuon, Vichet, and C. Hudley. In press. Ethnic and Panethnic Asian American Identities: Contradictory Perceptions of Cambodian Students in Urban Schools. Urban Review.

Chhuon, Vichet, C. Hudley, M. E. Brenner, and R. Macias. 2010. The Multiple Worlds of Successful Cambodian American Students. *Urban Education* 45(1): 30–57.

Fordham, Signithia, and J. Ogbu. 1986 Black Students' School Success: Coping with the "Burden of 'Acting White.'" *Urban Review* 18(3): 176–206.

Lee, Jin Sook, and D. Suarez 2008. A Synthesis of the Roles of Heritage Languages in the Lives of Immigrant Children. *In* The Education of Language Minority Students in the United States. Terrence Wiley, Jin Sook Lee, and Russell Rumberger, eds. Pp. 136–170. Bristol: Multilingual atters.

Lee, Stacey 1996. Unraveling the "Model Minority" Stereotype: Listening to Asian American Youth. New York: Teachers College Press.

Ng, Jennifer, S. S. Lee, and Y. K. Pak. 2007. Contesting the Model Minority and Perpetual Foreigner Stereotypes: A Critical Review of Literature on Asian Americans in Education. *Review of Research in Education* 31: 95–130.

Ngo, Bic, and S. J. Lee. 2007. Complicating the Model Minority Stereotype: A Review of Southeast Asian American Education. *Review of Educational Research* 77(4): 415–453.

Phinney, Jane, I. Romero, M. Nava, and D. Huang. 2001. The Role of Language, Parents, and Peers in Ethnic Identity among Adolescents in Immigrant Families. *Journal of Youth and Adolescence* 30(2): 135–153.

Reyes, Angela. 2007. Language, Identity, and Stereotype among Southeast Asian American Youth: The Other Asian. Mahwah, NJ: Erlbaum.

Um, Khatharya. 2003. A Dream Denied: Educational Experiences of Southeast Asian American Youth: Issues and Recommendations. Washington, DC: SEARAC.

15

Meanwhile Backstage: Behavior in Public Bathrooms

SPENCER E. CAHILL, ET AL.

Erving Goffman, the founder of the dramaturgical perspective in sociology, first proposed that people behave in ways that they consciously manage to foster the most favorable impressions of themselves. They do this by scrupulously adhering to the micro-social norms of individual and interactional behavior, using backstage regions to prepare themselves for their frontstage, public displays. To investigate these norms, Cahill and his students made systematic observations in men's and women's public bathrooms, carefully recording people's behavior patterns. In a selection that is sure to generate both recognition and amusement, Cahill describes and analyzes the landscape of public bathrooms, the common rituals found there, and the way people engage in backstage behavior designed to support their appearance on subsequent re-emergence into the public domain. The norms upheld in this private yet public setting assert more fully how loyal members of society are to the behavioral guidelines we share and the meanings that people attribute to them. Do you recognize some of the behaviors that Cahill describes? Have you ever wondered why you do these things? How is the social order maintained by acting in these ways?

Years ago the anthropologist Horace Miner (1955) suggested, with tongue planted firmly in cheek, that many of the rituals that behaviorally express and sustain the central values of our culture occur in bathrooms. Whether Miner realized it or not, and one suspects that he did, there was more to this thesis than his humorous interpretation of bathroom rituals suggests. As Erving

SOURCE: Spencer Cahill, William Distler, Cynthia Lachowetz, Andrea Meaney, Robyn Tarallo, and Teena Willard, "Meanwhile Backstage: Public Bathrooms and the Interaction Order," *Journal of Contemporary Ethnography* (formerly *Urban Life*), 14(1), pp. 33–34, 38–49, 56. Copyright © 1985 by SAGE Publications. Reprinted by permission of SAGE Productions, Inc.

Goffman (1959: 112–113) once observed, the vital secrets of our public shows are often visible in those settings that serve as backstage regions relative to our public performances:

> it is here that illusions and impressions are openly constructed.... Here the performer can relax; he can drop his front, forgo speaking his lines, and step out of character.

Clearly, bathrooms or, as they are often revealingly called, restrooms, are such backstage regions. By implication, therefore, systematic study of bathroom behavior may yield valuable insights into the character and requirements of our routine public performances....

THE PERFORMANCE REGIONS OF PUBLIC BATHROOMS

Needless to say, one of the behaviors for which bathrooms are explicitly designed is defecation. In our society, as Goffman (1959: 121) observed, "defecation involves an individual in activity which is defined as inconsistent with the cleanliness and purity standards" that govern our public performances.

> Such activity also causes the individual to disarrange his clothing and to "go out of play," that is, to drop from his face the expressive mask that he employs in face-to-face interaction. At the same time it becomes difficult for him to reassemble his personal front should the need to enter into interaction suddenly occur. (Goffman, 1959: 121)

When engaged in the act of defecation, therefore, individuals seek to insulate themselves from potential audiences in order to avoid discrediting the expressive masks that they publicly employ. Indeed, over 60 percent of the 1000 respondents to a survey conducted in the early 1960s reported that they "interrupted or postponed" defecation if they did not have sufficient privacy (Kira, 1966: 58).

In an apparent attempt to provide such privacy, toilets in many public bathrooms are surrounded by partially walled cubicles with doors that can be secured against potential intrusions. In fact, public bathrooms that do not provide individuals this protection from potential audiences are seldom used for the purpose of defecation. In the course of our research, for example, we never observed an individual using an unenclosed toilet for this purpose. If a bathroom contained both enclosed and unenclosed toilets, moreover, individuals ignored the unenclosed toilets even when queues had formed outside of the enclosed toilets. In a sense, therefore, the cubicles that typically surround toilets in public bathrooms, commonly called stalls, physically divide such bathrooms into two distinct performance regions.

Indeed, Goffman (1971: 32) has used the term "stall" to refer to any "well-bounded space to which individuals lay temporary claim, possession being on an all-or-nothing basis." Clearly, a toilet stall is a member of this sociological family of ecological arrangements. Sociologically speaking, however, it is not physical boundaries, per se, that define a space as a stall but the behavioral regard given such boundaries. For example, individuals who open or attempt to open the door of an occupied toilet stall typically provide a remedy for this act, in most cases a brief apology such as "Whoops" or "Sorry." By offering such a remedy, the offending individual implicitly defines the attempted intrusion as a delict and, thereby, affirms his or her belief in a rule that prohibits such intrusions (Goffman, 1971: 113). In this sense, toilet stalls provide occupying individuals not only physical protection against potential audiences but normative protection as well.

In order to receive this protection, however, occupying individuals must clearly inform others of their claim to such a stall. Although individuals sometimes lean down and look under the doors of toilet stalls for feet, they typically expect occupying individuals to mark their claim to a toilet stall by securely closing the door.[1] On one occasion, for example, a middle-aged woman began to push open the unlocked door of a toilet stall. Upon discovering that the stall was occupied, she immediately said, "I'm sorry," and closed the door. When a young woman emerged from the stall a couple minutes later, the older woman apologized once again but pointed out that "the door was open." The young woman responded, "[I]t's okay," thereby minimizing the offense and perhaps acknowledging a degree of culpability on her part.

As is the case with many physical barriers to perception (Goffman, 1963: 152), the walls and doors of toilet stalls are also treated as if they cut off more communication than they actually do. Under most circumstances, for example, the walls and doors of toilet stalls are treated as if they were barriers to conversation. Although acquainted individuals may sometimes carry on a conversation through the walls of a toilet stall if they believe the bathroom is not otherwise occupied, they seldom do so if they are aware that others are present. Moreover, individuals often attempt to ignore offensive sounds and smells that emanate from occupied toilet stalls, even though the exercise of such "tactful blindness" (Goffman, 1955: 219) is sometimes a demanding task. In any case, the walls and doors of toilet stalls provide public actors with both physical and normative shields behind which they can perform potentially discrediting acts.

Toilet stalls in public bathrooms are, therefore, publicly accessible yet private backstage regions. Although same-sexed clients of a public establishment may lay claim to any unoccupied stall in the bathroom designated for use by persons of their sex, once such a claim is laid, once the door to the stall is closed, it is transformed into the occupying individual's private, albeit temporary, retreat from the demands of public life. While occupying the stall, that individual can engage in a variety of potentially discrediting acts with impunity.

When not concealed behind the protective cover of a toilet stall, however, occupants of public bathrooms may be observed by others. For the most part, as previously noted, same-sexed clients of a public establishment can enter and exit at will the bathroom designated for their use, and it may be simultaneously occupied by as many individuals as its physical dimensions allow. By implication, therefore, occupants of public bathrooms must either perform or be ready to perform for an audience. As a result, the behavior that routinely occurs in the "open region" of a public bathroom, that area that is not enclosed by toilet stalls, resembles, in many important respects, the behavior that routinely occurs in other public settings....

THE RITUALS OF PUBLIC BATHROOMS

As Goffman (1971) convincingly argued, much of this behavior can best be described as "interpersonal rituals." Emile Durkheim (1965), in his famous analysis of religion, defined a ritual as a perfunctory, conventionalized act which expresses respect and regard for some object of "ultimate value." In a different context, moreover, he observed that in modern, Western societies,

> the human personality is a sacred thing; one dare not violate it nor infringe its bounds, while at the same time the greatest good is in communion with others.... (Durkheim, 1974: 37)

According to Durkheim, negative rituals express respect and regard for objects of ultimate value by protecting them from profanation. By implication, according to Goffman (1971: 62), negative interpersonal rituals involve the behavioral honoring of the sacred individual's right to private "preserves" and "to be let alone." For example, individuals typically refrain from physically, conversationally, or visually intruding on an occupied toilet stall. In doing so, they implicitly honor the occupying individual's right to be let alone and in this respect perform a negative interpersonal ritual.

Similarly, the queues that typically form in public bathrooms when the demand for sinks, urinals, and toilet stalls exceeds the available supply are also products of individuals' mutual performance of negative interpersonal rituals. Individuals typically honor one another's right to the turn claimed by taking up a position in such a queue, even when "creature releases" (Goffman, 1963: 69) threaten to break through their self-control. Young children provide an occasional exception, sometimes ignoring the turn-order of such queues. Yet even then the child's caretaker typically requests, on the child's behalf, the permission of those waiting in the queue. Between performances at a music festival, for example, a preschool-age girl and her mother were observed rapidly walking toward the entrance to a women's bathroom out of which a queue extended for several yards down a nearby sidewalk. As they walked past those waiting in the queue, the mother repeatedly asked: "Do you mind? She really has to go."

The interpersonal rituals that routinely occur in the open region of public bathrooms are not limited, however, to negative ones. If individuals possess a small patrimony of sacredness, then, as Durkheim (1974: 37) noted, "the greatest good is in communion" with such sacred objects. When previously acquainted individuals come into contact with one another, therefore, they typically perform conventionalized acts, positive interpersonal rituals, that express respect and regard for their previous communion with one another. In a sense, moreover, negative and positive interpersonal rituals are two sides of the same expressive coin. Whereas negative interpersonal rituals symbolically protect individuals from profanation by others, positive interpersonal rituals symbolically cleanse communion between individuals of its potentially defiling implications.[2] Although a positive interpersonal ritual may consist of no more than a brief exchange of greetings, failure to at least acknowledge one's previous communion with another is, in effect, to express disregard for the relationship and, by implication, the other individual's small patrimony of sacredness (Goffman, 1971: 62–94).

Even when previously acquainted individuals come into contact with one another in a public bathroom, therefore, they typically acknowledge their prior relationship. In fact, the performance of such positive interpersonal rituals sometimes interfered with the conduct of our research. On one occasion, for example, a member of the research team was in the open region of an otherwise unoccupied men's bathroom. While he was writing some notes about an incident that had just occurred, an acquaintance entered.

A: Hey_____! (walks to a urinal and unzips his pants) Nothing like pissin.

O: Yup.

A: Wh' da hell ya doin? (walks over to a sink and washes hands)

O: Writing.

A: Heh, heh, yea. About people pissin... That's for you.

O: Yup.

A: Take care.

O: Mmm Huh.

As this incident illustrates, individuals must be prepared to perform positive interpersonal rituals when in the open region of public bathrooms, especially those in public establishments with a relatively stable clientele. Whereas some of these may consist of no more than a brief exchange of smiles, others may involve lengthy conversations that reaffirm the participants' shared biography.

In contrast, when unacquainted individuals come into contact with one another in the open regions of public bathrooms, they typically perform a brief, negative interpersonal ritual that Goffman (1963: 84) termed "civil inattention":

> [O]ne gives to another enough visual notice to demonstrate that one
> appreciates that the other is present ... while at the next moment
> withdrawing one's attention from him so as to express that he does
> not constitute a target of special curiosity or design.

Through this brief pattern of visual interaction, individuals both acknowledge one another's presence and, immediately thereafter, one another's right to be let alone.

A variation on civil inattention is also commonly performed in the open region of public bathrooms, most often by men using adjacent urinals. Although masculine clothing permits males to urinate without noticeably disturbing their clothed appearance, they must still partially expose their external genitalia in order to do so. Clearly, the standards of modesty that govern public behavior prohibit even such limited exposure of the external genitalia. Although the sides of some urinals and the urinating individual's back provide partial barriers to perception, they do not provide protection against the glances of someone occupying an adjacent urinal. In our society, however, "when bodies are naked, glances are clothed" (Goffman, 1971: 46). What men typically give one another when using adjacent urinals is not, therefore, civil inattention but "non-person treatment" (Goffman, 1963: 83–84); that is, they treat one another as if they were part of the setting's physical equipment, as "objects not worthy of a glance." When circumstances allow, of course, unacquainted males typically avoid occupying adjacent urinals and, thereby, this ritually delicate situation.

It is not uncommon, however, for previously acquainted males to engage in conversation while using adjacent urinals. For example, the following interaction was observed in the bathroom of a restaurant.

> A middle-aged man is standing at one of two urinals. Another middle-aged man enters the bathroom and, as he approaches the available urinal, greets the first man by name. The first man quickly casts a sidelong glance at the second and returns the greeting. He then asks the second man about his "new granddaughter," and they continue to talk about grandchildren until one of them zips up his pants and walks over to a sink. Throughout the conversation, neither man turned his head so as to look at the other.

As this example illustrates, urinal conversations are often characterized by a lack of visual interaction between the participants. Instead of looking at one another while listening, as is typical among white, middle-class Americans (see LaFrance and Mayo, 1976), participants in such conversations typically fix their gaze on the wall immediately in front of them, an intriguing combination of the constituent elements of positive and negative interpersonal rituals. Although ritually celebrating their prior communion with one another, they also visually honor one another's right to privacy.

Due to the particular profanations and threats of profanations that characterize public bathrooms, moreover, a number of variations on these general patterns also commonly occur. In our society, as Goffman (1971: 41) observed, bodily excreta are considered "agencies of defilement." Although supported by germ theory, this view involves somewhat more than a concern for hygiene. Once such substances as urine, fecal matter, menstrual discharge, and flatus leave individuals' bodies, they acquire the power to profane even though they may not have the power to infect. In any case, many of the activities in which individuals engage when in bathrooms are considered both self-profaning and potentially profaning to others.[3] As a result, a variety of ritually delicate situations often arise in public bathrooms.

After using urinals and toilets, for example, individuals' hands are considered contaminated and, consequently, a source of contamination to others. In order to demonstrate both self-respect and respect for those with whom they might come into contact, individuals are expected to and often do wash their hands after using urinals and toilets. Sinks for this purpose are typically located in the open region of public bathrooms, allowing others to witness the performance of this restorative ritual.[4] Sometimes, however, public bathrooms are not adequately equipped for this purpose. Most commonly, towel dispensers are empty or broken. Although individuals sometimes do not discover this situation until after they have already washed their hands, they often glance at towel dispensers as they walk from urinals and toilet stalls to sinks. If they discover that the towel dispensers are empty or broken, there is typically a moment of indecision. Although they sometimes proceed to wash their hands and then dry them on their clothes, many times they hesitate, facially display disgust, and audibly sigh. By performing these gestures-in-the-round, they express a desire to wash their hands; their hands may remain contaminated, but their regard for their own and others' sacredness is established.

Because the profaning power of odor operates over a distance and in all directions, moreover, individuals who defecate in public bathrooms not only temporarily profane themselves but also risk profaning the entire setting. If an individual is clearly responsible for the odor of feces or flatus that fills a bathroom, therefore, he or she must rely on others to identify sympathetically with his or her plight and, consequently, exercise tactful blindness. However, this is seldom left to chance. When other occupants of the bathroom are acquaintances, the offending individual may offer subtle, self-derogatory display as a defensive, face-saving measure (Goffman, 1955). Upon emerging from toilet stalls, for example, such persons sometimes look at acquaintances and facially display disgust. Self-effacing humor is also occasionally used in this way. On one occasion, for example, an acquaintance of a member of the research team emerged from a toilet stall after having filled the bathroom with a strong fecal odor. He walked over to a sink, smiled at the observer, and remarked: "Something died in there." Through such subtle self-derogation, offending individuals metaphorically split themselves into two parts: a sacred self that assigns blame and a blameworthy animal self. Because the offending individual assigns blame, moreover, there is no need for others to do so (Goffman, 1971: 113).

If other occupants of the bathroom are unfamiliar to the offending individual, however, a somewhat different defensive strategy is commonly employed. Upon emerging from a toilet stall, individuals who are clearly responsible for an offensive odor seldom engage in visual interaction with unacquainted others. In so doing, they avoid visually acknowledging not only the presence of others but others' acknowledgement of their own presence as well. In a sense, therefore, the offending individual temporarily suspends his or her claim to the status of sacred object, an object worthy of such visual regard. The assumption seems to be that by suspending one's claim to this status, others need not challenge it and are, consequently, more likely to exercise tactful blindness in regard to the offense.

Despite Miner's humorous misidentification and interpretation of bathroom rituals, therefore, there is something to recommend the view that many of the rituals that behaviorally express and sustain the central values of our culture occur in bathrooms. Although these "central values do but itch a little," as Goffman (1971: 185) noted, "everyone scratches." And, it must be added, they often scratch in public bathrooms. However, routine bathroom behavior consists of more than the interpersonal rituals that are found in other public settings or variations on their general theme....

MANAGING PERSONAL FRONTS

When in a public setting, as Goffman (1963: 24) pointed out, individuals are expected to have their "faculties in readiness for any face-to-face interaction that might come" their way. One of the most evident means by which individuals express such readiness is "through the disciplined management of personal appearance or 'personal front,' that is, the complex of clothing, make-up, hairdo, and other surface decorations" that they carry about on their person (Goffman, 1963: 25). Of course, keeping one's personal front in a state of good repair requires care and effort (Gross and Stone, 1964: 10). However, individuals who are inspecting or repairing their personal fronts in public encounter difficulties in maintaining the degree of interactional readiness often expected of them; their attention tends to be diverted from the social situations that surround them (Goffman, 1963: 66). For the most part, therefore, close scrutinization and major adjustments of personal fronts are confined to backstage regions such as public bathrooms.

Most public bathrooms are equipped for this purpose. Many offer coin-operated dispensers of a variety of "personal care products" (e.g., combs and sanitary napkins), and almost all have at least one mirror. The most obvious reason for the presence of mirrors in public bathrooms is that the act of defecation and, for females, urination, requires individuals to literally "drop" their personal fronts. In order to ensure that they have adequately reconstructed their personal front after engaging in such an act, individuals must and typically do perform what Lofland (1972) has termed a "readiness check." For example, the following was observed in the men's bathroom of a neighborhood bar:

> A young man emerges from a toilet stall and, as he passes the mirror, hesitates. He glances side-long at his reflection, gives a nod of approval and then walks out the door.

When such a readiness check reveals flaws in the individual's personal front, he or she typically makes the appropriate repairs: Shirts are often retucked into pants and skirts, skirts are rotated around the waist, and pants are tugged up and down.

Because bodily movement and exposure to the elements can also disturb a disciplined personal front, the post-defecation or urination readiness check sometimes reveals flaws in individuals' personal fronts that are the result of normal wear and tear. Upon emerging from toilet stalls and leaving urinals, therefore, individuals sometimes repair aspects of their personal fronts that are not normally disturbed in the course of defecating or urinating. For example, the following was observed in the women's bathroom of a student center on a college campus.

> A young woman emerges from a toilet stall, approaches a mirror, and inspects her reflection. She then removes a barrette from her hair, places the barrette in her mouth, takes a comb out of her coat pocket, and combs her hair while smoothing it down with her other hand. With the barrette still in her mouth, she stops combing her hair, gazes intently at the mirror and emits an audible "ick." She then places the barrette back in her hair, pinches her cheeks, takes a last look at her reflection and exits.

Interestingly, as both this example and the immediately preceding one illustrate, individuals sometimes offer visible or audible evaluations of their reflections when inspecting and repairing their personal front, a finding that should delight proponents of Meadian sociological psychology. Public bathrooms may protect individuals from the critical reviews of external audiences, but they do not protect them from those of their internal audience.

In any case, public bathrooms are as much "self-service" repair shops for personal fronts as they are socially approved shelters for physiological acts that are inconsistent with the cleanliness and purity standards that govern our public performances. In fact, individuals often enter public bathrooms with no apparent purpose other than the management of their personal front. For example, it is not uncommon for males to enter public bathrooms, walk directly to the nearest available mirror, comb their hair, rearrange their clothing, and then immediately exit. In our society, of course, females are often expected to present publicly a more extensively managed personal front than are males. Consequently, females often undertake extensive repairs in public bathrooms. For example, the following was observed in the women's bathroom of a student center on a college campus:

> Two young women enter, one goes to a toilet stall and the other immediately approaches a mirror. The second woman takes a brush out of her bookbag, throws her hair forward, brushes it, throws her hair back, and brushes it into place. She returns the brush to her bookbag, smooths down her eyebrows, and wipes underneath her eyes with her fingers. She then removes a tube of lipstick from her bookbag, applies it to her lips, and uses her finger to remove the lipstick that extends beyond the natural outline of her lips. As her friend emerges from the toilet stall, she puts the lipstick tube back into her bookbag, straightens her collar so that it stands up under her sweater and then exits with her friend.

Even though individuals routinely inspect and repair their personal fronts in the open regions of public bathrooms, they often do so furtively. When others enter the bathroom, individuals sometimes suspend inspecting or repairing their personal fronts until the new arrivals enter toilet stalls or approach urinals. In other cases, they hurriedly complete these activities before they can be witnessed.

For example, the following was observed from inside a toilet stall in a women's bathroom:

> A young woman walks to the end of the sinks where there is a full-length mirror. She turns sideways, inspects her reflection and reaches up to adjust her clothing. The outer door of the bathroom begins to open, and the young woman quickly walks over to the sink on which her purse is laying, picks it up and heads for the door.

Despite the furtiveness that sometimes characterizes individuals' inspection and repair of their personal fronts, however, the open region of a public bathroom is often the only available setting in which they can engage in these activities without clearly undermining their front stage performances. As Lofland (1972: 101) observed in a somewhat different context, "[I]t is apparently preferable to be witnessed by a few.... In a brief episode of backstage behavior than to be caught ... with one's presentation down" on the frontstage....

In short, the systematic study of routine bathroom behavior reveals just how loyal members of this society are to the central values and behavioral standards that hold our collective lives together. Whatever else they may do, users of public bathrooms continue to bear the "cross of personal character" (Goffman, 1971: 185), and, as long as they continue to carry this burden, remain self-regulating participants in the "interaction order" (Goffman, 1983).

NOTES

1. Yet a closed door is not always a reliable indicator that a toilet stall is occupied, as anyone who has cared for children is aware. The young sometimes exit toilet stalls by crawling under locked doors.

2. Along these lines, casual sex may make an individual "feel so cheap" in part because this intimate communion with another has not been adequately cleansed of its defiling implications by the performance of positive interpersonal rituals.

3. One further expression of these defiling implications is the fact that cleaning bathrooms is an almost universally despised activity in our society. Apparently, such close contact with objects that are used for the elimination and disposal of bodily excrete profanes individuals. Indeed, those who routinely clean bathrooms, janitors in particular, are often treated as if they had abdicated their claim to a small patrimony of sacredness.

4. Although we did not record the frequency of hand washing, it was our impression that individuals are more likely to wash their hands after using toilets or urinals if they think others would notice their failure to do so. When a bathroom was not otherwise occupied and we were observing from within a toilet stall, for example, it was not uncommon for individuals to neglect this practice.

REFERENCES

Durkheim, E. (1974). *Sociology and Philosophy*. (D. F. Pocock, trans.) New York: Free Press. (originally published in 1924)

———— (1965). *The Elementary Forms of the Religious Life*. (J. W. Swain, trans.) New York: Free Press. (originally published in 1915)

Goffman, E. (1983). "The interaction order." *Amer. Soc. Rev.* 48 (February): 1–17.

————. (1971). *Relations in Public: Microstudies in Public Order*. New York: Basic.

————. (1963). *Behavior in Public Places: Notes on the Social Organization of Gatherings*. New York: Free Press.

———— (1959). *The Presentation of Self in Everyday Life*. Garden City, NY: Doubleday.

————. (1955). "On face-work: An analysis of ritual elements in social interaction." *Psychiatry* 18 (August): 213–231.

Gross, E. and G. Stone. (1964). "Embarrassment and the analysis of role requirements." *Amer. J. of Sociology* 70 (July): 1–15.

Kira, A. (1966). *The Bathroom: Criteria for Design*. Ithaca, NY: Cornell University Center for Housing and Environmental Studies.

LaFrance, M. and C. Mayo. (1976). "Racial differences in gaze behavior during conversation: Two systematic observational studies." *J. of Personality and Soc. Psychology* 33 (May): 547–552.

Lofland, L. (1972). "Self-management in public settings: Part I." *Urban Life* 1 (April): 93–108.

Miner, H. (1955). "Body ritual among the Nacirema." *Amer. Anthropologist* 58 (June): 503–507.

16

The Engagement Proposal as Performance

DAVID SCHWEINGRUBER, SINE ANAHITA,
AND NANCY BERNS

Marriage proposals have witnessed a return to gender role traditionalism from a period in the 1960s and '70s where gender equalitarianism prevailed and these took a more mutual and informal tone. While men seem to have regained the privilege of instigating the move towards marriage, they have become responsible for new pressures to make this act creative, romantic, and individualistic. Yet despite the outward dramaturgy that these proposals are fully the work of the men who present them, Schweingruber, Anahita, and Berns' analysis of the work behind the scenes reveals that these carefully constructed interactions are scripted by both members of the couple. In reading this piece you may find similarities between these proposals and other invitations to dances and weekend events common in high school and college. Do you think that these depictions are accurate? What is the importance of the external audience for the inner parties of this interaction and why? Can you tie this return to traditionalism to behaviors and trends in other arenas of life? Why do you think the engagement proposal has evolved into this complex and highly scripted form? How might you compare the way people use props to convey their identity as they move from high school to college to the way they use proposals to symbolize aspects of their relationships and selves? How might you want to experience these kinds of acts, and how do these performances compare to idealized versions of love and romance contained in classic fairy tales?

SOURCE: David Schweingruber, Sine Anahita, and Nancy Berns, "The Engagement Proposal as Performance," *Sociological Focus* 37(2), 2004. Reprinted by permission of the North Central Sociological Association.

The American engagement proposal is one of the world's best known rituals. Not only do most Americans participate in this ritual sometime during their lifetime, but it is known to a worldwide audience through American movies and television shows. The very commonplaceness of the proposal may explain the dearth of sociological research on the topic despite both its significance in people's lives and the questions the ritual raises about gender relations and inequality.

In this paper we examine the engagement proposal by interviewing members of 20 heterosexual engaged couples about their proposals. For these couples, the "engagement proposal" is actually a misnomer because no new proposal is put forward. Instead, the couple ratifies a decision they have already made through a "surprise" performance they have mutually planned. The man appears to take the lead in initiating the engagement even though his girlfriend has given him instructions about his performance. We argue that the version of the engagement proposal described by these couples should be understood as a performance for two audiences, first the woman and then the couple's friends and families. During these performances, the couple uses standard conventions, such as revealing a diamond ring, to communicate to each other and to their friends and family that they are a properly engaged couple and ought to be married. Couples choose these elements not because of symbolic or ideological meanings attached to them but because excluding them might result in confusion from the secondary audience—friends and family—casting doubt on the relationship and the reality of the proposal. In particular, participants in the proposals downplay or dismiss interpretations of the proposal elements as being sexist. We argue that in the absence of any ideological challenge to this "traditional" proposal, dramaturgical considerations remain foremost in the minds of those planning the proposals.

THE ENGAGEMENT PROPOSAL

The engagement proposal is part of a chain of romantic rites of passage that includes high school proms, bridal showers, bachelor and bachelorette parties, choosing the wedding dress, the cutting of the wedding cake, and the rising phenomenon of vow-renewal rituals. The contemporary romantic ritual that has received the most scholarly attention is the wedding.... Most romantic rituals, such as the prom and the wedding, are sites where femininity is on display. In preparation for these occasions, women are expected to devote considerable time, money, and effort to making themselves normatively feminine. Men have fewer choices to make (e.g., in clothing) and may be less invested in these events. Other research has indicated that front stage presentations in marital relationships may obscure actual marital power dynamics....

Romantic rituals are shaped by gendered power dynamics. Previous research has suggested that power dynamics in romantic relationships are complicated. Older approaches to understanding power in marriages in terms of who makes decisions have been refined to include power that is covert or hidden. Potential conflicts that would make power overt may be avoided because, by accepting standard notions of masculinity and femininity, members of the couple do not perceive anything problematic about the man's greater power in the relationship.

THEORETICAL FRAMEWORK

We situated our analysis within two analytical frameworks. First, drawing on the work of Becker (1982) and Swidler (1986, 2001), we viewed individuals planning and carrying out proposals as drawing from cultural toolkits. Becker (1982) emphasized the usefulness of choosing tools that are intelligible to other individuals. People are able to communicate and coordinate their actions with one another because of shared conventional meanings that are embedded in social practices and artifacts. These shared meanings make conventional activities easier to carry out while simultaneously making unconventional ones more difficult. Swidler (1986, 2001) focused on how people use culture to make sense of their lives. Second, we drew upon Goffman's (1959, 1974) dramaturgical framework. According to Goffman, individuals use performances to send signals about themselves, their situations, and their performance team. He also emphasized how backstage preparations set the stage for frontstage performances....

METHODS

This study was situated within a larger project on weddings and drew from interviews conducted with members of engaged couples. Each member of the couple was interviewed separately. The couples were solicited with advertisements in two newspapers in a Midwestern university town and were paid $40 per couple ($15 to the first member interviewed and $25 to the second). Nearly all of the individuals who responded to the advertisements were interviewed. (The exceptions were four people who failed to show up for their scheduled appointment, eight who responded to the advertisement after we had obtained our sample, and two who did not fit the criteria for the sample.) This paper is based upon 38 interviews—36 interviews with both members of 18 couples and two interviews with just the female member. (Two males did not show up for their scheduled interviews after we interviewed their fiancées. Because this paper did not focus on male-female comparisons, we used data from the interviews with the two women whose fiancés did not participate.)

The respondents were all white Americans, ranging in age from 19 to 28 with a mean and median age of 22. Twenty-three respondents were undergraduates, and 15 were college graduates (with associate or bachelor degrees). Of the college graduates, three were graduate students and one was a veterinary student. The other college graduates were employed in a range of positions, including mechanical engineer, auditor, golf professional, landscape foreman, and maintenance worker. At the time of their interviews, couples had been dating an average of more than two and a half years (mean = 2.7 years), with two couples dating for less than one year, and one couple dating for 7 years. Forty-three percent of the couples were cohabiting at the time of their interviews, just over half were not cohabiting (52 percent), and in two couples, the members differed in their responses about cohabitation. The length of time couples had spent dating before the proposal occurred ranged from less than six months to five years, with a mean of 2.25 years. Fifty-seven percent of the couples interviewed had dated between two and three years before they became engaged. While this sample was certainly not representative of all American couples, the findings from this research will help to direct further research on larger samples that include more diverse populations....

PERFORMING THE SCRIPT

The Proposal as Performance

The engagement proposals enacted by couples in this study were performances put on by the couples to communicate to each other and to secondary audiences of family, friends, and any other interested parties (e.g., rival suitors, wedding product salespersons) that they intended to be married, that they were entering a new phase of their relationship called "engagement," and that they would, in fact, one day marry one another. The proposal finalized an earlier understanding the couple had that they would marry, and because the decision was settled and public, planning for the wedding could begin. In addition, the proposal was fun, a special event that the couples enjoyed performing.

In constructing this performance, the members of the couple chose from their cultural toolkits the scripts, props, and collaborators needed to get this work done. In putting these performances together, individuals were constrained by what tools were immediately available to them and also by their desire to communicate successfully with their audiences. By choosing symbols and scripts that were intelligible to potential audiences, performers stood the best chance of carrying off a successful performance. More unconventional performances may have resulted in audiences who had doubts about the reality of the engagement.

The result of the desire to communicate using intelligible signals was that all couples in this study adopted some version of the "standard" American engagement proposal, which included, minimally, the man asking the woman to marry him and presenting her with an engagement ring....

Secondary Performances

The initial performance of the man for the woman is followed by a number of subsidiary performances (or one continuous performance) as the couple shares the engagement proposal story with their friends and family and takes on the role of an engaged couple. Two pieces of evidence are presented to this secondary audience: (1) the story of the proposal and (2) the engagement ring. One woman explained the importance of the proposal in this way:

> Nobody would take us too seriously if we started planning a wedding and there hadn't been some official engagement process. [We needed] a story and a ring.... Because ... if I would have said, "Mom, I'm getting married on this day," she would have been like, "Well, where's the ring? Where's the proposal? ... I'm not gonna start any of this until we have that."

The story and ring are offered as proof that the couple considers themselves engaged and that they have the qualifications (e.g., mutual commitment) necessary to be a successfully engaged and married couple. The response of friends and family can be seen as the "critical response" to the engagement proposal and the engagement. A poor performance in the original proposal or in the retelling may result in the secondary audience doubting the reality or appropriateness of the engagement, even if the couple has no such doubts. In addition to the primary message that the couple will be getting married, the couple may also try to communicate about the quality of the couple's relationship and the nature of the man, such as whether he is cheap, traditional, creative, or romantic.

THE PROPOSAL SCRIPT

The majority of the proposals in this study contained five elements. Three were found in every proposal: (1) the man asking the woman to marry him, (2) the man presenting the woman a ring, and (3) the man orchestrating the proposal as a "surprise." Another element, the man getting down on one knee, was found in 17 of the 20 proposals. The final element, the man asking permission of the woman's father, was found in 12 of the 20 proposals.

Man as Proposer

A universal element in the proposals of heterosexual couples we interviewed was that the proposer was the male member of the couple and the audience was the female member. Breaking this rule, as only the female member can do, did not result in an official engagement. In fact, switching the roles may turn the proposal into a parody that produces amusement, instead of the usual romantic feelings. As one woman explained:

> I actually, as a joke, the week before, on our one year anniversary, I put in the card, I'm like, "I love you so much. Will you marry me?" And he just looked at me and he's, like [with intonation of amused scolding], "Oh, Julie."

Another woman twice asked her boyfriend to marry her and even gave him a ring. He wore the ring but did not take her proposals seriously. The couple did not consider themselves engaged until he asked her to marry him....

The Importance of the Ring

Although the man's verbal request that the woman marry him was important, words without an engagement ring did not make a complete proposal. The presentation of a ring made it clear that what was going on was an engagement proposal. This was a strong necessity not only for the first audience, the woman, but also for the secondary audience. Without a ring, the proposal was not yet appropriate as a public story and did not provide the official marking of their future as an engaged couple. The following two examples illustrate the change in the woman's emotions from hope to disappointment when she realized that her boyfriend was not presenting a ring and thus the proposal was not a "real" one:

> I actually kind of faked her out that night [shortly before the actual proposal]. [laugh] I acted like I was getting something on the couch, and I was on one knee, and I had my hands closed, and she was getting all excited then. And so like I go, "Molly, will you marry me?" And I open my hands up and there's nothing there. And ... [she] told me how mean that was and that the next time I do that it'd better be real. [laughs] (male)
>
> He said, "... will you marry me?" And I said, "Yes." And then he pulled out this ring box, and in it was a ring made out of pine needles, and I started to cry because I was expecting my real ring. And so I started to cry and he's like, "I promise I'll give you the real ring as soon as I can" and all this stuff. And so I was trying to be really nice about it, but in my head I was going, "Oh my gosh. I want my diamond."... So after doing that for about two minutes, then he pulled a box out of the other pocket, and it had my ring in it.... I still have the pine ring, but I like this one better. (female)

In the first case the proposal was transformed into a humorous rehearsal. In the second, the joke (until it was resolved) ruined the performance and created the impression that the proposer did not understand the script and, consequently, might not be serious about becoming engaged. Not all women shared the opinion that a proposal without a ring would be illegitimate. However, if there was no ring, some sort of explanation, such as financial exigency, was required....

The ring was just as important, or even more important, during the secondary performances. The presence of a diamond ring on the woman's ring finger acted as *prima facie* evidence to the couple's friends and family that they were indeed engaged. The secondary audience may in fact demand that the ring be produced. For instance:

> I think the first thing my mom said is, "Let's see your finger," because she knew he was going to ask so we just kind of broke [the news], you know, told everyone what happened and that we were engaged and let everyone know. (female)

Surprise

The proposal was supposed to be a "surprise" but only in a limited sense because an agreement to marry had already been reached by all of these subjects. The woman expected that a proposal would be made. What was supposed to surprise her was when and where the proposal would take place. Whether surprise was achieved varied within our sample. At one extreme, the woman knew that the proposal was imminent. She knew the ring had been purchased, and she suspected that she would be proposed to that day. The surprise was in the details. Her fiancé said, "We both knew it was coming, and we both knew it was going to happen that night, but she didn't know exactly how it was going to happen."

At the other extreme, the woman knew that the man would someday propose to her, but the time and place of the proposal caught her off guard. "He surprised me. That was his big thing 'cause we both knew that we'd be getting married. He was afraid that I was going to say I thought this was coming, and I didn't." Note that although this woman "had no idea" the proposal was coming, the couple had discussed the ring. She had also previously given her boyfriend instructions about what she expected from the proposal.

One of the purposes of the surprise was to allow the woman to respond spontaneously to the proposal. As one woman reported:

> It kind of swept me off my feet … If I know something's coming up, I can kind of plan for it and kind of prepare myself for it. But this, I just had none of that so maybe it was just like complete raw emotion when you just don't expect anything, you can't have any feeling preplanned or anything. You just kind of react and it seemed … more genuine that way. If you don't plan out what you're going to say before or after … if it just happens, maybe it's more true.

This outpouring of emotion was an expected part of the proposal and not just by the women. Respondents reported the outpouring of emotion that accompanied the proposal occurred in various ways. First, respondents reported weeping and not understanding what was going on. Second, they reported that they could not remember parts of the proposal because they were overwhelmed by emotion. Finally, men reported that performances were disrupted by emotion and some of their plans were abandoned….

Down on One Knee

Seventeen of the 20 men got down on one knee during the proposal. Another was down on both knees. The two exceptions were explained by the circumstances. One couple got engaged while looking at the stars as they sat in a car (it was too cold to go outside). The other proposal involved the man turning around to show his girlfriend the ring balanced on the end of a hockey stick. In many cases the woman had earlier requested that her boyfriend get down on one knee. This element of the proposal exemplified much of the entire ritual in that it clearly indicated that a proposal was taking place, but its origin or original meaning was unknown to the people who were participating. Specifically, the man getting down on one knee signaled that the proposal was beginning and the presentation of the ring and the verbal request to marry were imminent. When respondents were asked about the meaning of this element, they were able to come up with possible meanings but indicated that they did not know the "real" meaning or had not even given it much thought. For instance:

> I don't know. Maybe it's like going back to tradition, man trying to be submissive to his wife. Who knows really, but it just seems like the chivalrous thing to do, I guess, you know, because you're asking and when you get lower than the other person, you're supposed to be submissive. I don't know what it means really, but I like it. (male)

The lack of a tradition of men's submission to their wives suggests that this man's hypothesis was incorrect. The elements of the engagement proposal are important because they convey that a proposal is going on, not because of other symbolic meanings. Getting down on one knee is uniquely useful in communicating a proposal is taking place because it is rare during typical interactions.

Asking Permission

Twelve of the 20 men asked permission from their girlfriend's father before proposing. As with the proposal itself, asking the father was a performance with dual audiences. By asking his girlfriend's father's permission, the man acted out his respect for his future father-in-law, his hope to have his father-in-law's blessing, and his desire to join the family. His secondary audience was his girlfriend, who viewed asking permission as a romantic gesture and often specifically requested it. In fact, she might have been a more important audience than her father.

> I told them [my girlfriend's parents] that after careful consideration I would really like to marry Shari and wanted to know if that was all right with them, kind of knowing that it was, but they said that that was really all right with them and that was about it. Really it was kind of more of a formality for her than directly involved with us. (male)

As with other sex-specific requirements of the script, most couples did not make the connection between the script and its seemingly sexist symbolism, i.e., its suggestion that the woman is owned by her father, who is empowered to turn her over to her future husband. Couples liked it not only because it was "traditional" but also because women liked to involve their fathers in the event. As one woman explained:

> I thought it was kind of nice.... I just think it would make [my dad] feel special to be involved in that way and stuff, and I think that it really did make my dad feel good that he called and asked him. I mean my dad obviously ... loves Dan, and we both knew that he wouldn't say "no" or anything, but it's just kind of nice to have him involved that way.

One woman who thought the process "takes out the independence of the woman" still encouraged her fiancé to "notify" her father because he was "old-fashioned" and "I knew he'd get really excited about it and feel like he was part of the process.... It gives my dad a story to tell." Only one woman indicated a strong objection to her boyfriend asking her father, saying, "I'm my own person, and my dad has nothing to do with it." Another woman, though, said she thought that asking her father was more significant than asking her, because "then I knew that he was really serious." Only four of the 12 men who asked permission from their girlfriend's father also asked permission from her mother.

Making the Proposal Romantic

Even though there were standard elements, the proposal script was brought to life by the proposers, who carried it out with varying skill levels and projected various images of the couple. Some were virtuosos who produced exquisitely romantic performances. Others performed with less skill and imagination but were able to produce an acceptable effect merely by carrying out the elements of the script. Some women indicated that woodenly following the script was not enough. For instance:

> [In the movies] it's always this big thing and the guy's always down on one knee and so ... it's not like it wouldn't have meant as much but it's like it almost would have been less real if we'd just been sitting there and he'd been like, "Hey, why don't you marry me?" and pulled out a ring. It wouldn't be quite as [good] because it wouldn't have been quite as big of a production.

Although the hypothetical proposal she described included important elements such as revealing a ring, it would be unsatisfactory, even less "real," because the performance showed no creativity or sensitivity to the importance of the occasion.

This was one reason why the men were nervous about proposing even though the answer to the proposal was already known. Because the proposal story was used as evidence of the quality of the couple's relationship, it was central to the couple's impression management. The man, then, was expected to make the proposal "romantic," often by modifying the script to take into account the couple's unique history. The most common way this was done was by choosing a location that was meaningful to the couple, such as a favorite beach, a campus landmark, a river view the couple had often visited, a park where the couple fed ducks, and an ice rink where they met. Other men planned generically romantic activities, such as watching a sunset, dancing under the stars, or giving a dozen roses. The proposals were also made unique by including a favorite object, such as a stuffed animal, hockey stick, or plant. Other men incorporated their special skills, such as writing and singing a song.

Another common strategy was to deliver the ring in a creative way. For instance, one man inserted the ring into a chocolate egg inside a plastic egg covered with ice cream and had the restaurant staff deliver it to the table as a dessert. Another man cut the stem of a wine glass, slipped the ring onto the stem, and glued it back together. His fiancée had to break the glass to get the ring. Rings were also tied to the bow of a stuffed animal and inserted into a rose.

PLANNING THE SURPRISE

Each of the couples in the study had earlier reached a decision to marry. Thus, although the proposal was a surprise, it was a surprise anticipated by the audience and, in fact, a surprise that was mutually planned by the audience and the performer. Although many surprises are planned by one person, here the surpriser and the audience planned the event together while working with the understanding that the woman had only partial information so that she could be surprised.

There was a radical change in roles as the couple moved from backstage to frontstage. Backstage the woman acted as a co-director, who instructed, suggested, or hinted to her boyfriend how the performance should be conducted. The man, taking his girlfriend's advice into account, planned the performance in greater detail. The actual proposal was initiated and performed by the man while the woman acted as audience.

Reflecting the tension between making the proposal a surprise and making it a successful performance, the women in the study varied as to how much information they provided their boyfriends. Some women reported they had provided substantial instruction about their expectations of the proposal. For example:

> I just told him that I'm kind of old-fashioned and that it's just some-
> thing I want, like I want him to propose to me and get the ring. And I
> told him that he had to make it really special.... I told him he had to ask
> my dad.... And he had to get on his knee.

These expectations typically covered four major topics. First, women specified particular parts of the script the man should be sure to follow, such as getting down on one knee or asking her father. Second, women emphasized that the moment was expected to be "romantic," "special," or "a day I'll remember." For example:

> I do remember talking to him about it, like I was telling him how some
> guy proposed to some girl, and it was, like, this really romantic story
> and stuff. And he's like, "Man, I have all this pressure on me to, like,
> make this magical moment." And I'm like, "Seriously, as long as it's just
> a day that I'll remember, that's all that I need, you know. I don't need
> you to awe me and ooh me with this, like, wonderful planned out
> evening, you know, that you spent all this money on. I just want it to
> be a nice moment that I'm going to remember." That's really all that
> I said to him and I wanted out of it.

Third, the woman instructed the man on the proper time and/or place. Some couples planned the approximate date of the wedding (often based upon their graduation schedules). Then, the woman informed the man about how long it would take to plan the wedding. It was then his responsibility to propose within that time frame. Finally, the woman gave instructions about the ring.

Other women stressed that it was more important for the proposal to be surprising than for them to be involved in the planning. This strategy should not be construed as not giving any instruction but as giving instructions that pointed the man toward secretly planning a romantic proposal without divulging any of the details. For instance:

> She was sure she wanted it to be a surprise. She didn't just want to go to a jewelry store and pick out a ring and, here you go, type of thing…. She basically just said that she wanted it to be a surprise. There was no hinting about it. (male).

Selecting the Ring

Selecting the ring exemplifies the careful balance between surprise and planning. The goal of selecting a ring that suits the woman's taste compromises the goal of the ring being a complete surprise and vice versa. As one man explained:

> I think she would have liked it too if we could go back again, and I could have just picked it out by myself. But I wanted her there to kind of pick it out with me just because when we had been shopping, I could never quite see what she wanted just because I think it's hard for me to see distinct features. So, she'd be, like, "Which one do you like?" And I could never quite pick which one I think she would have liked, even though I'm definitely positive that she would have liked anything I would have got her, but it's still hard to get by that. So, no, it's something we picked out pretty much together.

For this couple, the surprise was compromised by the woman helping to select the ring, but a complete surprise might have resulted in a ring the future bride did not like. We can think of couples' choices as lying along a continuum. At one end, when the couple chooses surprise as their prime goal, the man picks out the ring himself without any input from the woman. Only two couples followed this pattern. At the other end of the continuum, the couple picks out the ring together, with the woman sometimes taking the leadership role. In only one case did the woman pick out the ring in the absence of the man and later show it to him. In some cases, the woman told the man to get something similar to a particular ring she liked, and he returned shortly to buy that particular ring.

Most couples we interviewed took a middle position, in which the bride provided some information to the groom about such matters as the diamond's cut and the style and color of the band. This information was usually imparted as the couple looked at rings together. The way couples discussed these shopping trips hints at the tension between the woman selecting the ring and being the recipient of a surprise. Many of the ring-shopping trips were ostensibly not planned as such but took place when the couple "happened" to be near a jewelry store, as these two examples illustrate:

> I think I got her interested in platinum.... It's kind of a funny time because I think she knew that I was going to propose to her at some time, but whenever we were in a mall or anything, she'd kind of pull me aside, pull me over to a jewelry store, and we'd kind of look around. And we'd looked at least three times that I know specifically together and just kind of pointed out what we liked, and I think it was the second time that we looked that we really ... knew that we wanted a platinum metal. (male)
>
> If we'd take her promise ring to get cleaned at the jewelers, we'd stand there and wait and just kind of look over stuff. And we'd joke around about wedding rings that had huge gaudy diamonds, kind of laugh, and I'd ask her what she kind of liked, and she'd show me a few things, you know, "that's kind of neat."... It was really casual talk. I didn't really ask for specifics—"What do you want?" The main thing she just always told me—"Whatever you get I'm going to like, it doesn't matter what it looks like, it's going to be fine." And so I could try to get her to tell me some things about what she liked, she really wouldn't say a whole lot, she just wanted me to pick something out. (male)

Each couple was at or near the jewelry story for another reason—visiting the mall or getting another ring cleaned—and looked over and discussed engagement rings together. The first story balances the two ring-selection goals. The couple chose to get a platinum band, and, subsequently, a marquis cut. The proposer then chose the rest of the setting and the baguettes. The second story emphasizes the surprise because the proposer claimed that she gave him no specifics. (He remembered later that she said she did not want an oblong cut.) In other cases, though, the woman simply gave the man instructions about what she wanted. For instance:

> She told me—these aren't a lot of discussions, more, "When you do this, you can do it?" Like, she wanted a round one, and she wanted it as big as I could go, but she understood that I don't have a whole lot of money to be wasting on items such as this. *[What words did she use to say that?]* Oh boy, I would say that she said something to the effect of, "I'd really like [this], but I know that if you got [this], the quality would be less than if you got something smaller so whatever you decide is all right." But there's kind of an understanding when she says that that I kind of need to listen to her, or I'll be in trouble. (male)

An important characteristic of the woman's role in selecting the ring—whether she was a partner in selecting a particular ring or had a less visible role—was that the selection was followed by a period when she was in the dark about the final purchase of the ring and then prepared to be surprised with its presentation. Some women we interviewed specifically indicated at what point their own participation had ended.

> I didn't want it to be something where we picked out the rings, and I knew that he was buying it. I didn't want to know.... I showed him a ring that I liked, and then I just said, "Now you know what I like. It's all up to you now. I mean, you can get your mom to help you, or whatever you want." I'm, like, "I don't want to be involved in it." (female)

Even if the woman had known what ring was going to be bought, she was typically not present when the purchase was made and thus did not know exactly when the purchase was made. (The one man who reported that his girlfriend was present when he purchased the ring said that he regretted it.)...

DISCUSSION

Ingraham (1999) suggested that the standard "white wedding" (referring both to the color of the wedding dress and the event's alleged connection to racial hierarchies) is a result of people being duped by a "wedding-industrial complex" that serves the needs of patriarchal heterosexuality and transnational capitalism. Using a different model of culture, we have argued that the couples in our study are not "cultural dopes" (Garfinkel 1967), but active creators who used the components of the engagement proposal as cultural tools to send messages to one another and to a secondary audience. Although participants in engagement proposals may incidentally benefit capitalists, for example, by buying diamond rings, their purpose in doing so is dramaturgical, not ideological. Discovering who benefits from social action is not the same as establishing causation. Social actions have many consequences that are not intended by the actors.

The engagement proposal is part of such an "old order" because neither feminism, the gay/lesbian civil rights movement, nor any other ideology has been successful in calling into question the traditional model or proposing an alternative that accomplishes the same task as the traditional ritual—communicating unambiguously that the couple is engaged....

REFERENCES

Becker, Howard S. 1982. *Art Worlds*. Berkeley: University of California Press.

Garfinkel, Harold. 1967. "Studies of the Routine Grounds of Everyday Activities," pp. 35–75 in *Studies in Ethnomethodology*. Englewood Cliffs, NJ: Prentice-Hall.

Goffman, Erving. 1959. *The Presentation of Self in Everyday Life*. New York: Doubleday-Anchor.

_____. 1974. *Frame Analysis: An Essay on the Organization of Experience*. New York: Harper & Row.

Ingraham, Chrys. 1999. *White Weddings: Romancing Heterosexuality in Popular Culture.* New York: Routledge.

Swidler, Ann. 1986. "Culture in Action: Symbols and Strategies." *American Sociological Review* 51: 273–286.

_____. 2001. *Talk of Love: How Culture Matters.* Chicago: University of Chicago.

17

The Mark of a Criminal Record

DEVAH PAGER

To measure concretely the interactive effects of race and a criminal record, Pager devised an experimental design where she constructed a fabricated pair of job applicants who were matched on all features except their criminal history. She then sent out these matched pairs, White pairs and Black pairs, to apply for real jobs in Milwaukee, and noted how far the candidates got in the interview process. Along the way she recorded the employer's likelihood of dismissing the applicants right away, checking their references, calling them back for further interviews, and offering them the job. In a real demonstration of the effects of race and criminal record on employment opportunities, Pager found that whereas Whites were offered more jobs than Blacks, and applicants with no criminal history were offered more jobs than those who had served time, even Whites with criminal pasts were more likely to be hired than Blacks who had led law-abiding lives. She found that employers were more likely to hold stereotypes suspecting Blacks, especially young Black men, of being prone to crime and of being unreliable employees. What type of influence do you think race has on hiring, promotion, and social interaction? Are affirmative action programs needed? Are they effective? Are they discriminatory? How does this study interface with Perry's study of White high school students? In what ways might the design of this experiment have been more useful?

While stratification researchers typically focus on schools, labor markets, and the family as primary institutions affecting inequality, a new institution has emerged as central to the sorting and stratifying of young and disadvantaged men: the criminal justice system. With over 2 million individuals currently incarcerated, and over half a million prisoners released each year, the large and growing numbers of men being processed through the criminal justice system raises important questions about the consequences of this massive institutional intervention.

SOURCE: Devah Pager, "The Mark of a Criminal Record," *American Journal of Sociology* 108(5), 2003, pp. 935–975. Reprinted by permission of the University of Chicago Press.

This article focuses on the consequences of incarceration for the employment outcomes of black and white men. While previous survey research has demonstrated a strong *association* between incarceration and employment, there remains little understanding of the mechanisms by which these outcomes are produced. In the present study, I adopt an experimental audit approach to formally test the degree to which a criminal record affects subsequent employment opportunities. By using matched pairs of individuals to apply for real entry-level jobs, it becomes possible to directly measure the extent to which a criminal record—in the absence of other disqualifying characteristics—serves as a barrier to employment among equally qualified applicants. Further, by varying the race of the tester pairs, we can assess the ways in which the effects of race and criminal record interact to produce new forms of labor market inequalities.

TRENDS IN INCARCERATION

Over the past three decades, the number of prison inmates in the United States has increased by more than 600%, leaving it the country with the highest incarceration rate in the world (Bureau of Justice Statistics 2002; Barclay, Tavares, and Siddique 2001). During this time, incarceration has changed from a punishment reserved primarily for the most heinous offenders to one extended to a much greater range of crimes and a much larger segment of the population. Recent trends in crime policy have led to the imposition of harsher sentences for a wider range of offenses, thus casting an ever-widening net of penal intervention.[1]

While the recent "tough on crime" policies may be effective in getting criminals off the streets, little provision has been made for when they get back out. Of the nearly 2 million individuals currently incarcerated, roughly 95% will be released, with more than half a million being released each year (Slevin 2000). According to one estimate, there are currently over 12 million ex-felons in the United States, representing roughly 8% of the working-age population (Uggen, Thompson, and Manza 2000). Of those recently released, nearly two-thirds will be charged with new crimes and over 40% will return to prison within three years (Bureau of Justice Statistics 2000). Certainly some of these outcomes are the result of desolate opportunities or deeply ingrained dispositions, grown out of broken families, poor neighborhoods, and little social control (Sampson and Laub 1993; Wilson 1997). But net of these contributing factors [sic], there is evidence that experience with the criminal justice system in itself has adverse consequences for subsequent opportunities. In particular, incarceration is associated with limited future employment opportunities and earnings potential (Freeman 1987; Western 2002), which themselves are among the strongest predictors of recidivism (Shover 1996; Sampson and Laub 1993; Uggen 2000).

The expansion of the prison population has been particularly consequential for blacks. The incarceration rate for young black men in the year 2000 was nearly 10%, compared to just over 1% for white men in the same age group (Bureau of Justice Statistics 2001). Young black men today have a 28% likelihood of incarceration during their lifetime (Bureau of Justice Statistics 1997), a figure that rises above 50% among young black high school dropouts (Pettit and Western 2001). These vast numbers of inmates translate into a large and increasing population of black ex-offenders returning to communities and searching for work. The barriers these men face in reaching economic self-sufficiency are compounded by the stigma of minority status and criminal record. The consequences of such trends for widening racial disparities are potentially profound (see Western and Pettit 1999; Freeman and Holzer 1986).

The objective of this study is to assess whether the effect of a criminal record differs for black and white applicants. Most research investigating the differential impact of incarceration on blacks has focused on the differential *rates* of incarceration and how those rates translate into widening racial disparities. In addition to disparities in the rate of incarceration, however, it is also important to consider possible racial differences in the *effects* of incarceration. Almost none of the existing literature to date has explored this issue, and the theoretical arguments remain divided as to what we might expect.

On one hand, there is reason to believe that the signal of a criminal record should be less consequential for blacks. Research on racial stereotypes tells us that Americans hold strong and persistent negative stereotypes about blacks, with one of the most readily invoked contemporary stereotypes relating to perceptions of violent and criminal dispositions (Smith 1991; Sniderman and Piazza 1993; Devine and Elliot 1995). If it is the case that employers view all blacks as potential criminals, they are likely to differentiate less among those with official criminal records and those without. Actual confirmation of criminal involvement then will provide only redundant information, while evidence against it will be discounted. In this case, the outcomes for all blacks should be worse, with less differentiation between those with criminal records and those without.

On the other hand, the effect of a criminal record may be worse for blacks if employers, already wary of black applicants, are more hesitant when it comes to taking risks on blacks with proven criminal tendencies. The literature on racial stereotypes also tells us that stereotypes are most likely to be activated and reinforced when a target matches on more than one dimension of the stereotype (Quillian and Pager 2002; Darley and Gross 1983; Fiske and Neuberg 1990). While employers may have learned to keep their racial attributions in check through years of heightened sensitivity around employment discrimination, when combined with knowledge of a criminal history, negative attributions are likely to intensify.

A third possibility, of course, is that a criminal record affects black and white applicants equally. The results of this audit study will help to adjudicate between these competing predictions.

STUDY DESIGN

The basic design of this study involves the use of four male auditors (also called testers), two blacks and two whites. The testers were 23-year-old college students from Milwaukee who were matched on the basis of physical appearance and general style of self-presentation. Objective characteristics that were not already identical between pairs—such as educational attainment and work experience—were made similar for the purpose of the applications. Within each team, one auditor was randomly assigned a "criminal record" for the first week; the pair then rotated which member presented himself as the ex-offender for each successive week of employment searches, such that each tester served in the criminal record condition for an equal number of cases. By varying which member of the pair presented himself as having a criminal record, unobserved differences within the pairs of applicants were effectively controlled. No significant differences were found for the outcomes of individual testers or by month of testing.

Job openings for entry-level positions (defined as jobs requiring no previous experience and no education greater than high school) were identified from the Sunday classified advertisement section of the *Milwaukee Journal Sentinel*.[2] In addition, a supplemental sample was drawn from *Jobnet,* a state-sponsored web site for employment listings, which was developed in connection with the W-2 Welfare-to-Work initiatives.[3]

The audit pairs were randomly assigned 15 job openings each week. The white pair and the black pair were assigned separate sets of jobs, with the same-race testers applying to the same jobs. One member of the pair applied first, with the second applying one day later (randomly varying whether the ex-offender was first or second). A total of 350 employers were audited during the course of this study: 150 by the white pair and 200 by the black pair. Additional tests were performed by the black pair because black testers received fewer callbacks on average, and there were thus fewer data points with which to draw comparisons. A larger sample size enabled me to calculate more precise estimates of the effects under investigation.

Immediately following the completion of each job application, testers filled out a six-page response form that coded relevant information from the test. Important variables included type of occupation, metropolitan status, wage, size of establishment, and race and sex of employer.[4] Additionally, testers wrote narratives describing the overall interaction and any comments made by employers (or included on applications) specifically related to race or criminal records.

TESTER PROFILES

In developing the tester profiles, emphasis was placed on adopting characteristics that were both numerically representative and substantively important. In the present study, the criminal record consisted of a felony drug conviction (possession with intent to distribute, cocaine) and 18 months of (served) prison time. A drug crime (as opposed to a violent or property crime) was chosen because of its prevalence, its policy salience, and its connection to racial disparities in incarceration.[5] It is important to acknowledge that the effects reported here may differ depending on the type of offense.

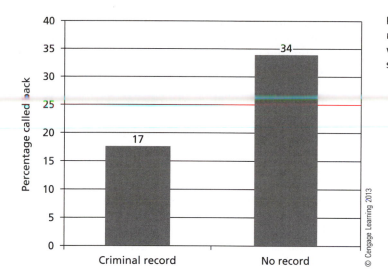

© Cengage Learning 2013

F I G U R E 17.1 The effect of a criminal record on employment opportunities for whites. The effect of a criminal record is statistically significant ($P < .01$).

THE EFFECT OF A CRIMINAL RECORD FOR WHITES

I begin with an analysis of the effect of a criminal record among whites. White noncriminals can serve as our baseline in the following comparisons, representing the presumptively nonstigmatized group relative to blacks and those with criminal records. Given that all testers presented roughly identical credentials, the differences experienced among groups of testers can be attributed fully to the effects of race or criminal status.

Figure 17.1 shows the percentage of applications submitted by white testers that elicited callbacks from employers, by criminal status. As illustrated below, there is a large and significant effect of a criminal record, with 34% of whites without criminal records receiving callbacks, relative to only 17% of whites with criminal records. A criminal record thereby reduces the likelihood of a callback by 50%.

There were some fairly obvious examples documented by testers that illustrate the strong reaction among employers to the signal of a criminal record. In one case, a white tester in the criminal record condition went to a trucking service to apply for a job as a dispatcher. The tester was given a long application, including a complex math test, which took nearly 45 minutes to fill out. During the course of this process, there were several details about the application and the job that needed clarification, some of which involved checking with the supervisor about how to proceed. No concerns were raised about his candidacy at this stage. When the tester turned the application in, the secretary brought it into a back office for the supervisor to look over, so that an interview could perhaps be conducted. When the secretary came back out, presumably after the supervisor had a chance to look over the application more thoroughly, he was told the position had already been filled. While, of course, isolated incidents like this are not conclusive, this was not an infrequent occurrence. Often testers reported seeing employers' levels of responsiveness change dramatically once they had glanced down at the criminal record question.

Clearly, the results here demonstrate that criminal records close doors in employment situations. Many employers seem to use the information as a screening mechanism, without attempting to probe deeper into the possible context or complexities of the situation. As we can see here, in 50% of cases, employers were unwilling to consider equally qualified applicants on the basis of their criminal record.

Of course, this trend is not true among all employers, in all situations. There were, in fact, some employers who seemed to prefer workers who had been recently released from prison. One owner told a white tester in the criminal record condition that he "like[d] hiring people who ha[d] just come out of prison because they tend to be more motivated, and are more likely to be hard workers [not wanting to return to prison]." Another employer for a cleaning company attempted to dissuade the white noncriminal tester from applying because the job involved "a great deal of dirty work." The tester with the criminal record, on the other hand, was offered the job on the spot. A criminal record is thus not an obstacle in all cases, but on average, as we see above, it reduces employment opportunities substantially.

THE EFFECT OF RACE

A second major focus of this study concerns the effect of race. African-Americans continue to suffer from lower rates of employment relative to whites, but there is tremendous disagreement over the source of these disparities. The idea that race itself—apart from other correlated characteristics—continues to play a major role in shaping employment opportunities has come under question in recent years (e.g., D'Souza 1995; Steele 1991). The audit methodology is uniquely suited to address this question. While the present study design does not provide the kind of cross-race matched-pair tests that earlier audit studies of racial discrimination have used, the between-group comparisons (white pair vs. black pair) can nevertheless offer an unbiased estimate of the effect of race on employment opportunities.

Figure 17.2 presents the percentage of callbacks received for both categories of black testers relative to those for whites. The effect of race in these findings is strikingly large. Among blacks without criminal records, only 14% received callbacks, relative to 34% of white noncriminals ($P < .01$). In fact, even whites *with* criminal records received more favorable treatment (17%) than blacks *without* criminal records (14%). The rank ordering of groups in this graph is painfully revealing of employer preferences: race continues to play a dominant role in shaping employment opportunities, equal to or greater than the impact of a criminal record.

The magnitude of the race effect found here corresponds closely to those found in previous audit studies directly measuring racial discrimination. Bendick et al. (1994), for example, found that blacks were 24 percentage points less likely

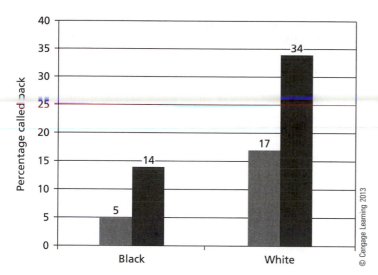

FIGURE 17.2 The effect of a criminal record for black and white job applicants. The main effects of race and criminal record are statistically significant (*P* < .01). The interaction between the two is not significant in the full sample. Grey bars represent criminal record; dark bars represent no criminal record.

© Cengage Learning 2013

to receive a job offer relative to their white counterparts, a finding very close to the 20 percentage point difference (between white and black nonoffenders) found here. Thus in the eight years since the last major employment audit of race was conducted, very little has changed in the reaction of employers to minority applicants. Despite the many rhetorical arguments used to suggest that direct racial discrimination is no longer a major barrier to opportunity (e.g., D'Souza 1995; Steele 1991), as we can see here, employers, at least in Milwaukee, continue to use race as a major factor in hiring decisions.

RACIAL DIFFERENCES IN THE EFFECTS OF A CRIMINAL RECORD

The final question this study sought to answer was the degree to which the effect of a criminal record differs depending on the race of the applicant. Based on the results presented in Figure 17.2, the effect of a criminal record appears more pronounced for blacks than it is for whites. While this interaction term is not statistically significant, the magnitude of the difference is nontrivial. While the ratio of callbacks for nonoffenders relative to ex-offenders for whites is 2:1, this same ratio for blacks is nearly 3:1. The effect of a criminal record is thus 40% larger for blacks than for whites.

This evidence is suggestive of the way in which associations between race and crime affect interpersonal evaluations. Employers, already reluctant to hire blacks, appear even more wary of blacks with proven criminal involvement. Despite the fact that these testers were bright articulate college students with effective styles of self-presentation, the cursory review of entry-level applicants leaves little room for these qualities to be noticed. Instead, the employment barriers of minority status and criminal record are compounded, intensifying the stigma toward this group.

The salience of employers' sensitivity toward criminal involvement among blacks was highlighted in several interactions documented by testers. On three separate occasions, for example, black testers were asked in person (before submitting their applications) whether they had a prior criminal history. None of the white testers were asked about their criminal histories up front.

DISCUSSION

There is serious disagreement among academics, policy makers, and practitioners over the extent to which contact with the criminal justice system—in itself—leads to harmful consequences for employment. The present study takes a strong stand in this debate by offering direct evidence of the causal relationship between a criminal record and employment outcomes. While survey research has produced noisy and indirect estimates of this effect, the current research design offers a direct measure of a criminal record as a mechanism producing employment disparities. Using matched pairs and an experimentally assigned criminal record, this estimate is unaffected by the problems of selection, which plague observational data. While certainly there are additional ways in which incarceration may affect employment outcomes, this finding provides conclusive evidence that mere contact with the criminal justice system, in the absence of any transformative or selective effects, severely limits subsequent employment opportunities. And while the audit study investigates employment barriers to ex-offenders from a microperspective, the implications are far-reaching. The finding that ex-offenders are only one-half to one-third as likely as nonoffenders to be considered by employers suggests that a criminal record indeed presents a major barrier to employment. With over 2 million people currently behind bars and over 12 million people with prior felony convictions, the consequences for labor market inequalities are potentially profound.

Second, the persistent effect of race on employment opportunities is painfully clear in these results. Blacks are less than half as likely to receive consideration by employers, relative to their white counterparts, and black nonoffenders fall behind even whites with prior felony convictions. The powerful effects of race thus continue to direct employment decisions in ways that contribute to persisting racial inequality. In light of these findings, current public opinion seems largely misinformed. According to a recent survey of residents in Los Angeles, Boston, Detroit, and Atlanta, researchers found that just over a quarter of whites believe there to be "a lot" of discrimination against blacks, compared to nearly two-thirds of black respondents (Kluegel and Bobo 2001). Over the past decade, affirmative action has come under attack across the country based on the argument that direct racial discrimination is no longer a major barrier to opportunity. According to this study, however, employers, at least in Milwaukee, continue to use race as a major factor in their hiring decisions. When we combine the effects of race and criminal record, the problem grows more intense. Not only are blacks much more likely to be incarcerated than whites; based on the findings presented here, they may also be more strongly affected by the impact of a criminal record. Previous estimates of the aggregate consequences of incarceration may therefore underestimate the impact on racial disparities.

Finally, in terms of policy implications, this research has troubling conclusions. In our frenzy of locking people up, our "crime control" policies may in fact exacerbate the very conditions that lead to crime in the first place. Research consistently shows that finding quality steady employment is one of the strongest predictors of desistance from crime (Shover 1996; Sampson and Laub 1993; Uggen 2000). The fact that a criminal record severely limits employment opportunities—particularly among blacks—suggests that these individuals are left with few viable alternatives.[6]

As more and more young men enter the labor force from prison, it becomes increasingly important to consider the impact of incarceration on the job prospects of those coming out. No longer a peripheral institution, the criminal justice system has become a dominant presence in the lives of young disadvantaged men, playing a key role in the sorting and stratifying of labor market opportunities. This article represents an initial attempt to specify one of the important mechanisms by which incarceration leads to poor employment outcomes. Future research is needed to expand this emphasis to other mechanisms (e.g., the transformative effects of prison on human and social capital), as well as to include other social domains affected by incarceration (e.g., housing, family formation, political participation, etc.); in this way, we can move toward a more complete understanding of the collateral consequences of incarceration for social inequality.

At this point in history, it is impossible to tell whether the massive presence of incarceration in today's stratification system represents a unique anomaly of the late 20th century, or part of a larger movement toward a system of stratification based on the official certification of individual character and competence. Whether this process of negative credentialing will continue to form the basis of emerging social cleavages remains to be seen.

NOTES

1. For example, the recent adoption of mandatory sentencing laws, most often used for drug offenses, removes discretion from the sentencing judge to consider the range of factors pertaining to the individual and the offense that would normally be taken into account. As a result, the chances of receiving a state prison term after being arrested for a drug offense rose by 547% between 1980 and 1992 (Bureau of Justice Statistics 1995).

2. Occupations with legal restrictions on ex-offenders were excluded from the sample. These include jobs in the health care industry, work with children and the elderly, jobs requiring the handling of firearms (i.e., security guards), and jobs in the public sector. An estimate of the collateral consequences of incarceration would also need to take account of the wide range of employment formally off-limits to individuals with prior felony convictions.

3.   Employment services like *Jobnet* have become a much more common method of finding employment in recent years, particularly for difficult-to-employ populations such as welfare recipients and ex-offenders. Likewise, a recent survey by Holzer and Stoll (2001) found that nearly half of Milwaukee employers (46%) use *Jobnet* to advertise vacancies in their companies.

4.   See Pager (2002) for a discussion of the variation across each of these dimensions.

5.   Over the past two decades, drug crimes were the fastest growing class of offenses. In 1980, roughly one out of every 16 state inmates was incarcerated for a drug crime; by 1999, this figure had jumped to one out of every five (Bureau of Justice Statistics 2000). In federal prisons, nearly three out of every five inmates are incarcerated for a drug crime (Bureau of Justice Statistics 2001). A significant portion of this increase can be attributed to changing policies concerning drug enforcement. By 2000, every state in the country had adopted some form of truth-in-sentencing laws, which impose mandatory sentencing minimums for a range of offenses. These laws have been applied most frequently to drug crimes, leading to more than a fivefold rise in the number of drug arrests that result in incarceration and a doubling of the average length of sentences for drug convictions (Mauer 1999; Blumstein and Beck 1999). While the steep rise in drug enforcement has been felt across the population, this "war on drugs" has had a disproportionate impact on African-Americans. Between 1990 and 1997, the number of black inmates serving time for drug offenses increased by 60%, compared to a 46% increase in the number of whites (Bureau of Justice Statistics 1995). In 1999, 26% of all black state inmates were incarcerated for drug offenses, relative to less than half that proportion of whites (Bureau of Justice Statistics 2001).

6.   There are two primary policy recommendations implied by these results. First and foremost, the widespread use of incarceration, particularly for nonviolent drug crimes, has serious, long-term consequences for the employment problems of young men. The substitution of alternatives to incarceration, therefore, such as drug treatment programs or community supervision, may serve to better promote the well-being of individual offenders as well as to improve public safety more generally through the potential reduction of recidivism. Second, additional thought should be given to the widespread availability of criminal background information. As criminal record databases become increasingly easy to access, this information may be more often used as the basis for rejecting otherwise qualified applicants. If instead criminal history information were suppressed—except in cases that were clearly relevant to a particular kind of job assignment—ex-offenders with appropriate credentials might be better able to secure legitimate employment. While there is some indication that the absence of official criminal background information may lead to a greater incidence of statistical discrimination against blacks (see Bushway 1997; Holzer et al. 2001), the net benefits of this policy change may in fact outweigh the potential drawbacks.

REFERENCES

Barclay, Gordon, Cynthia Tavares, and Arsalaan Siddique. 2001. "International Comparisons of Criminal Justice Statistics, 1999." London: U.K. Home Office for Statistical Research.

Bendick, Marc, Jr., Charles Jackson, and Victor Reinoso. 1994. "Measuring Employment Discrimination through Controlled Experiments." *Review of Black Political Economy* 23: 25–48.

Blumstein, Alfred, and Allen J. Beck. 1999. "Population Growth in U.S. Prisons, 1980–1996." pp. 17–62 in *Prisons: Crime and Justice: A Review of Research*, vol. 26. Edited by Michael Tonry and J. Petersilia. Chicago: University of Chicago Press.

Bureau of Justice Statistics. 1995. *Prisoners in 1994*, by Allen J. Beck and Darrell K. Gilliard. Special report. Washington, D.C.: Government Printing Office.

———. 1997. *Lifetime Likelihood of Going to State or Federal Prison,* by Thomas P. Bonczar and Allen J. Beck. Special report, March. Washington, D.C.

———. 2000. Bulletin. *Key Facts at a Glance: Number of Persons in Custody of State Correctional Authorities by Most Serious Offense 1980–99*. Washington, D.C.: Government Printing Office.

———. 2001. *Prisoners in 2000,* by Allen J. Beck and Paige M. Harrison. August. Bulletin. Washington, D.C.: NCJ 188207.

———. 2002. *Sourcebook of Criminal Justice Statistics.* Last accessed March 1, 2003. Available http://www.albany.edu/sourcebook/

Bushway, Shawn D. 1997. "Labor Market Effects of Permitting Employer Access to Criminal History Records." Working paper. New York: University of Maryland, Department of Criminology.

Darley, J. M., and P. H. Gross. 1983. "A Hypothesis-Confirming Bias in Labeling Effects." *Journal of Personality and Social Psychology* 44: 20–33.

Devine, P. G., and A. J. Elliot. 1995. "Are Racial Stereotypes Really Fading? The Princeton Trilogy Revisited." *Personality and Social Psychology Bulletin* 21(11): 1139–50.

D'Souza, Dinesh. 1995. *The End of Racism: Principles for a Multiracial Society*. New York: Free Press.

Fiske, Susan, and Steven Neuberg. 1990. "A Continuum of Impression Formation, from Category-Based to Individuating Processes." pp. 1–63 in *Advances in Experimental Social Psychology,* vol. 23. Edited by Mark Zanna. New York: Academic Press.

Freeman, Richard B. 1987. "The Relation of Criminal Activity to Black Youth Employment." *Review of Black Political Economy* 16(1–2): 99–107.

Freeman, Richard B., and Harry J. Holzer, 1986. *The Black Youth Employment Crisis*. Chicago: University of Chicago Press for National Bureau of Economic Research.

Holzer, Harry, Steven Raphael, and Michael Stoll. 2001. "Perceived Criminality, Criminal Background Checks and the Racial Hiring Practices of Employers." Discussion Paper no. 1254–02. San Francisco: University of Wisconsin-Madison, Institute for Research on Poverty.

Holzer, Harry, and Michael Stoll. 2001. *Employers and Welfare Recipients: The Effects of Welfare Reform in the Workplace*. San Francisco: Public Policy Institute of California.

Kluegel, James, and Lawrence Bobo. 2001. "Perceived Group Discrimination and Policy Attitudes: The Sources and Consequences of the Race and Gender Gaps." pp. 163–216 in *Urban Inequality: Evidence from Four Cities,* edited by Alice O'Connor, Chris Tilly, and Lawrence D. Bobo. New York: Russell Sage Foundation.

Mauer, Marc. 1999. *Race to Incarcerate*. New York: New Press.

Pager, Devah. 2002. "The Mark of a Criminal Record." Doctoral dissertation. Department of Sociology, Cambridge, Mass.: University of Wisconsin-Madison.

Pettit, Becky, and Bruce Western. 2001. "Inequality in Lifetime Risks of Imprisonment." Paper presented at the annual meetings of the American Sociological Association. Anaheim, August.

Quillian, Lincoln, and Devah Pager. 2002. "Black Neighbors, Higher Crime? The Role of Racial Stereotypes in Evaluations of Neighborhood Crime." *American Journal of Sociology* 107(3): 717–67.

Sampson, Robert J., and John H. Laub. 1993. *Crime in the Making: Pathways and Turning Points through Life*. Cambridge, Mass.: Harvard University Press.

Shover, Neil. 1996. *Great Pretenders: Pursuits and Careers of Persistent Thieves*. Boulder, Colo.: Westview.

Slevin, Peter. 2000. "Life after Prison: Lack of Services Has High Price." *Washington Post,* April 24.

Smith, Tom W. 1991. *What Americans Say about Jews*. New York: American Jewish Committee.

Sniderman, Paul M., and Thomas Piazza. 1993. *The Scar of Race*. Cambridge, Mass.: Harvard University Press.

Steele, Shelby. 1991. *The Content of Our Character: A New Vision of Race in America*. New York: Harper Perennial.

Uggen, Christopher. 2000. "Work as a Turning Point in the Life Course of Criminals: A Duration Model of Age, Employment, and Recidivism." *American Sociological Review* 65(4): 529–46.

Uggen, Christopher, Melissa Thompson, and Jeff Manza. 2000. "Crime, Class, and Reintegration: The Socioeconomic, Familial, and Civic Lives of Offenders." Paper presented at the American Society of Criminology meetings, San Francisco, November 18.

Western, Bruce. 2002. "The Impact of Incarceration on Wage Mobility and Inequality." *American Sociological Review* 67(4): 526–46.

Western, Bruce, and Becky Pettit. 1999. "Black-White Earnings Inequality, Employment Rates, and Incarceration." Working Paper no. 150. New York: Russell Sage Foundation.

Wilson, William Julius. 1997. *When Work Disappears: The World of the New Urban Poor*. New York: Vintage Books.

18

Fraternities and Collegiate Rape Culture

A. AYRES BOSWELL AND JOAN Z. SPADE

College fraternities represent a place where young men congregate to have fun, forge connections, and exclusively set themselves apart from nonmembers. These all-male groups have been identified as fostering a culture prone to nourishing sexist attitudes and behaviors. In particular, fraternities have been linked with the commodification and objectification of women that leads to rape. Not all frater- nities are the same, however, as some generate an environment that fosters the exploitation of women more than others. In this chapter, Boswell and Spade compare the climate at two different college fraternities at the same institution that stand in contrast to each other, noting clear distinctions between those fostering a high versus low risk of generating sexually assaultive behavior. Focusing on social interactions featuring men and women, they evaluate the cultural differences between high-risk and low-risk fraternity party and bar scenes, locating differences in their gender relations, treatment of women, and general attitudes toward rape. Although individual men are the ones who rape, Boswell and Spade found that the group norms at some settings promoted behaviors that reinforced a rape culture. In other words, there is a larger culture, besides the male college milieu, that might foster an atmosphere where these men can exploit women without reproach from their peers or even adult men. Could the behavior described in this study happen on your college campus? Why or why not? What is it about male culture that fosters these attitudes? Has society done anything in the last decade to prevent these kinds of milieu from developing? What are the implications for society of a privileged male culture that emphasizes these values?

SOURCE: A. Ayres Boswell and Joan Z. Spade, "Fraternities and Collegiate Rape Culture," *Gender & Society*, 10, 2, pp. 133–134, 136–138, 139–142, 143. Copyright © 1996 by Sociologists for Women in Society. Reprinted by permission of SAGE Publications, Inc.

Date rape and acquaintance rape on college campuses are topics of concern to both researchers and college administrators. Some estimate that 60 to 80 percent of rapes are date or acquaintance rape (Koss, Dinero, Seibel, and Cox 1988). Further, 1 out of 4 college women say they were raped or experienced an attempted rape, and 1 out of 12 college men say they forced a woman to have sexual intercourse against her will (Koss, Gidycz, and Wisniewski 1985).

Although considerable attention focuses on the incidence of rape, we know relatively little about the context or the *rape culture* surrounding date and acquaintance rape. Rape culture is a set of values and beliefs that provide an environment conducive to rape (Buchwald, Fletcher, and Roth 1993; Herman 1984). The term applies to a generic culture surrounding and promoting rape, not the specific settings in which rape is likely to occur. We believe that the specific settings also are important in defining relationships between men and women.

Some have argued that fraternities are places where rape is likely to occur on college campuses (Martin and Hummer 1989; O'Sullivan 1993; Sanday 1990) and that the students most likely to accept rape myths and be more sexually aggressive are more likely to live in fraternities and sororities, consume higher doses of alcohol and drugs, and place a higher value on social life at college (Gwartney-Gibbs and Stockard 1989; Kalof and Cargill 1991). Others suggest that sexual aggression is learned in settings such as fraternities and is not part of predispositions or pre-existing attitudes (Boeringer, Shehan, and Akers 1991). To prevent further incidences of rape on college campuses, we need to understand what it is about fraternities in particular and college life in general that may contribute to the maintenance of a rape culture on college campuses.

Our approach is to identify the social contexts that link fraternities to campus rape and promote a rape culture. Instead of assuming that all fraternities provide an environment conducive to rape, we compare the interactions of men and women at fraternities identified on campus as being especially *dangerous* places for women, where the likelihood of rape is high, to those seen as *safer* places, where the perceived probability of rape occurring is lower.

RESULTS

The Settings

Fraternity Parties   We observed several differences in the quality of the interaction of men and women at parties at high-risk fraternities compared to those at low-risk houses. A typical party at a low-risk house included an equal number of women and men. The social atmosphere was friendly, with considerable interaction between women and men. Men and women danced in groups and in couples, with many of the couples kissing and displaying affection toward each other. Brothers explained that, because many of the men in these houses had girlfriends, it was normal to see couples kissing on the dance floor. Coed groups engaged in conversations at many of these houses, with women and men engaging in friendly exchanges, giving the impression that they knew each other well. Almost no cursing and yelling was observed at parties in low-risk houses; when pushing occurred, the participants apologized. Respect for women extended to the women's bathrooms, which were clean and well supplied.

At high-risk houses, parties typically had skewed gender ratios, sometimes involving more men and other times involving more women. Gender segregation also was evident at these parties, with the men on one side of a room or in the bar drinking while women gathered in another area. Men treated women differently in the high-risk houses. The women's bathrooms in the high-risk houses were filthy, including clogged toilets and vomit in the sinks. When a brother was told of the mess in the bathroom at a high-risk house, he replied, "Good, maybe some of these beer wenches will leave so there will be more beer for us."

Men attending parties at high-risk houses treated women less respectfully, engaging in jokes, conversations, and behaviors that degraded women. Men made a display of assessing women's bodies and rated them with thumbs up or thumbs down for the other men in the sight of the women. One man attending a party at a high-risk fraternity said to another, "Did you know that this week is Women's Awareness Week? I guess that means we get to abuse them more this week." Men behaved more crudely at parties at high-risk houses. At one party, a brother dropped his pants, including his underwear, while dancing in front of several women. Another brother slid across the dance floor completely naked.

The atmosphere at parties in high-risk fraternities was less friendly overall. With the exception of greetings, men and women rarely smiled or laughed and spoke to each other less often than was the case at parties in low-risk houses. The few one-on-one conversations between women and men appeared to be strictly flirtatious (lots of eye contact, touching, and very close talking). It was rare to see a group of men and women together talking. Men were openly hostile, which made the high-risk parties seem almost threatening at times. For example, there was a lot of touching, pushing, profanity, and name calling, some done by women.

Students at parties at the high-risk houses seemed self-conscious and aware of the presence of members of the opposite sex, an awareness that was sexually charged. Dancing early in the evening was usually between women. Close to midnight, the sex ratio began to balance out with the arrival of more men or more women. Couples began to dance together but in a sexual way (close dancing with lots of pelvic thrusts). Men tried to pick up women using lines such as "Want to see my fish tank?" and "Let's go upstairs so that we can talk; I can't hear what you're saying in here."

Although many of the same people who attended high-risk parties also attended low-risk parties, their behavior changed as they moved from setting to setting. Group norms differed across contexts as well. At a party that was held jointly at a low-risk house with a high-risk fraternity, the ambience was that of a party at a high-risk fraternity with heavier drinking, less dancing, and fewer conversations between women and men. The men from both high- and low-risk fraternities were very aggressive; a fight broke out, and there was pushing and shoving on the dance floor and in general.

As others have found, fraternity brothers at high-risk houses on this campus told about routinely discussing their sexual exploits at breakfast the morning after parties and sometimes at house meetings (cf. Martin and Hummer 1989; O'Sullivan 1993; Sanday 1990). During these sessions, the brothers we interviewed said that men bragged about what they did the night before with stories of sexual conquests often told by the same men, usually sophomores. The women involved in these exploits were women they did not know or knew but did not respect, or *faceless victims*. Men usually treated girlfriends with respect and did not talk about them in these storytelling sessions. Men from low-risk houses, however, did not describe similar sessions in their houses....

Gender Relations

Relations between women and men are shaped by the contexts in which they meet and interact. As is the case on other college campuses, *hooking up* has replaced dating on this campus, and fraternities are places where many students hook up. Hooking up is a loosely applied term on college campuses that had different meanings for men and women on this campus.

Most men defined hooking up similarly. One man said it was something that happens

> when you are really drunk and meet up with a woman you sort of know, or possibly don't know at all and don't care about. You go home with her with the intention of getting as much sexual, physical pleasure as she'll give you, which can range anywhere from kissing to inter-course, without any strings attached.

The exception to this rule is when men hook up with women they admire. Men said they are less likely to press for sexual activity with someone they know and like because they want the relationship to continue and be based on respect.

Women's version of hooking up differed. Women said they hook up only with men they cared about and described hooking up as kissing and petting but not sexual intercourse. Many women said that hooking up was disappointing because they wanted longer-term relationships. First-year women students realized quickly that hook-ups were usually one-night stands with no strings attached, but many continued to hook up because they had few opportunities to develop relationships with men on campus. One first-year woman said that "70 percent of hook-ups never talk again and try to avoid one another; 26 percent may actually hear from them or talk to them again, and 4 percent may actually go on a date, which can lead to a relationship." Another first-year woman said, "It was fun in the beginning. You get a lot of attention and kiss a lot of boys and think this is what college is about, but it gets tiresome fast."

Whereas first-year women get tired of the hook-up scene early on, many men do not become bored with it until their junior or senior year. As one upperclassman said, "The whole game of hooking up became really meaningless and tiresome for me during my second semester of my sophomore year, but most of my friends didn't get bored with it until the following year."

In contrast to hooking up, students also described monogamous relationships with steady partners. Some type of commitment was expected, but most people did not anticipate marriage. The term *seeing each other* was applied when people were sexually involved but free to date other people. This type of relationship involved less commitment than did one of boyfriend/girlfriend but was not considered to be a hook-up.

The general consensus of women and men interviewed on this campus was that the Greek system, called "the hill," set the scene for gender relations. The predominance of Greek membership and subsequent living arrangements segregated men and women. During the week, little interaction occurred between women and men after their first year in college because students in fraternities or sororities live and dine in separate quarters. In addition, many non-Greek upper-class students move off campus into apartments. Therefore, students see each other in classes or in the library, but there is no place where students can just hang out together.

Both men and women said that fraternities dominate campus social life, a situation that everyone felt limited opportunities for meaningful interactions. One senior Greek man said,

> This environment is horrible and so unhealthy for good male and female relationships and interactions to occur. It is so segregated and male dominated....It is our party, with our rules and our beer. We are allowing these women and other men to come to our party. Men can feel superior in their domain.

Comments from a senior woman reinforced his views: "Men are dominant; they are the kings of the campus. It is their environment that they allow us to enter; therefore, we have to abide by their rules." A junior woman described fraternity parties as

> good for meeting acquaintances but almost impossible to really get to know anyone. The environment is so superficial, probably because there are so many social cliques due to the Greek system. Also, the music is too loud and the people are too drunk to attempt to have a real conversation, anyway.

Some students claim that fraternities even control the dating relationships of their members. One senior woman said, "Guys dictate how dating occurs on this campus, whether it's cool, who it's with, how much time can be spent with the girlfriend and with the brothers." Couples either left campus for an evening or hung out separately with their own same-gender friends at fraternity parties, finally getting together with each other at about 2 A.M. Couples rarely went together to fraternity parties. Some men felt that a girlfriend was just a replacement for a hook-up. According to one junior man, "Basically a girlfriend is someone you go to at 2 A.M. after you've hung out with the guys. She is the sexual outlet that the guys can't provide you with."

Some fraternity brothers pressure each other to limit their time with and commitment to their girlfriends. One senior man said, "The hill [fraternities] and girlfriends don't mix." A brother described a constant battle between girlfriends and brothers over who the guy is going out with for the night, with the brothers usually winning. Brothers teased men with girlfriends with remarks such as "whipped" or "where's the ball and chain?" A brother from a high-risk house said that few brothers at his house had girlfriends; some did, but it was uncommon. One man said that from the minute he was a pledge he knew he would probably never have a girlfriend on this campus because "it was just not the norm in my house. No one has girlfriends; the guys have too much fun with [each other]."

The pressure on men to limit their commitment to girlfriends, however, was not true of all fraternities or of all men on campus. Couples attended low-risk fraternity parties together, and men in the low-risk houses went out on dates more often. A man in one low-risk house said that about 70 percent of the members of his house were involved in relationships with women, including the pledges (who were sophomores).

Treatment of Women

Not all men held negative attitudes toward women that are typical of a rape culture, and not all social contexts promoted the negative treatment of women. When men were asked whether they treated the women on campus with respect, the most common response was "On an individual basis, yes, but when you have a group of men together, no." Men said that, when together in groups with other men, they sensed a pressure to be disrespectful toward women. A first-year man's perception of the treatment of women was that "they are treated with more respect to their faces, but behind closed doors, with a group of men present, respect for women is not an issue." One senior man stated, "In general, college-aged men don't treat women their age with respect because 90 percent of them think of women as merely a means to sex." Women reinforced this perception. A first-year woman stated, "Men here are more interested in hooking up and drinking beer than they are in getting to know women as real people." Another woman said, "Men here use and abuse women."

Characteristic of rape culture, a double standard of sexual behavior for men versus women was prevalent on this campus. As one Greek senior man stated, "Women who sleep around are sluts and get bad reputations; men who do are champions and get a pat on the back from their brothers." Women also supported a double standard for sexual behavior by criticizing sexually active women. A first-year woman spoke out against women who are sexually active: "I think some girls here make it difficult for the men to respect women as a whole."

One concrete example of demeaning sexually active women on this campus is the "walk of shame." Fraternity brothers come out on the porches of their houses the night after parties and heckle women walking by. It is assumed that these women spent the night at fraternity houses and that the men they were with did not care enough about them to drive them home. Although sororities now reside in former fraternity houses, this practice continues and sometimes the victims of hecklings are sorority women on their way to study in the library.

A junior man in a high-risk fraternity described another ritual of disrespect toward women called "chatter." When an unknown woman sleeps over at the house, the brothers yell degrading remarks out the window at her as she leaves the next morning such as "Fuck that bitch" and "Who is that slut?" He said that sometimes brothers harass the brothers whose girlfriends stay over instead of heckling those women.

Fraternity men most often mistreated women they did not know personally. Men and women alike reported incidents in which brothers observed other brothers having sex with unknown women or women they knew only casually. A sophomore woman's experience exemplifies this anonymous state: "I don't mind if 10 guys were watching or it was videotaped. That's expected on this campus. It's the fact that he didn't apologize or even offer to drive me home that really upset me." Descriptions of sexual encounters involved the satisfaction of men by nameless women. A brother in a high-risk fraternity, described a similar occurrence:

> A brother of mine was hooking up upstairs with an unattractive woman who had been pursuing him all night. He told some brothers to go outside the window and watch. Well, one thing led to another and they were almost completely naked when the woman noticed the brothers outside. She was then unwilling to go any further, so the brother went outside and yelled at the other brothers and then closed the shades. I don't know if he scored or not, because the woman was pretty upset. But he did win the award for hooking up with the ugliest chick that weekend....

DISCUSSION AND CONCLUSION

These findings describe the physical and normative aspects of one college campus as they relate to attitudes about and relations between men and women. Our findings suggest that an explanation emphasizing rape culture also must focus on those characteristics of the social setting that play a role in defining heterosexual relationships on college campuses (Kalof and Cargill 1991). The degradation of women as portrayed in rape culture was not found in all fraternities on this campus. Both group norms and individual behavior changed as students went from one place to another. Although individual men are the ones who rape, we found that some settings are more likely places for rape than are others. Our findings suggest that rape cannot be seen only as an isolated act and blamed on individual behavior and proclivities, whether it be alcohol consumption or attitudes. We also must consider characteristics of the settings that promote the behaviors that reinforce a rape culture.

REFERENCES

Boeringer S. B., C. L. Shehan, and R. L. Akers. 1991. "Social Contexts and Social Learning in Sexual Coercion and Aggression: Assessing the Contribution of Fraternity Membership." *Family Relations* 40: 56–64.

Buchwald, E., P. R. Fletcher, and M. Roth (eds). 1993. *Transforming a Rape Culture.* Minneapolis: Milkweed Editions.

Gwartney-Gibbs, P. and J. Stockard. 1989. "Courtship Aggression and Mixed-Age Groups." In *Violence in Dating Relationships,* edited by M. A. Pirog-Good and J. E. Stets. New York: Praeger.

Herman, D. 1984. "The Rape Culture." In *Women: A Feminist Perspective,* edited by J. Freeman. Mountain View, CA: Mayfield.

Kalof, L. and T. Cargill. 1991. "Fraternity and Sorority Membership and Gender Dominance Attitudes." *Sex Roles* 25: 417–23.

Koss, M. P., T. E. Dinero, C. A. Seibel, and S. L. Cox. 1988. "Stranger and Acquaintance Rape: Are There Differences in the Victim's Experience? *Psychology of Women Quarterly* 12: 1–24.

Koss, M. P., C. A. Gidycz, and N. Wisniewski. 1985. "The Scope of Rape: Incidence and Prevalence of Sexual Aggression in a National Sample of Higher Education Students." *Journal of Consulting and Clinical Psychology* 55: 162–70.

Martin, P. Y. and R. Hummer. 1989. "Fraternities and Rape on Campus." *Gender & Society* 3: 457–73.

O'Sullivan, C. 1993. "Fraternities and the Rape Culture." In *Transforming a Rape Culture,* edited by E. Buchwald, P. R. Fletcher, and M. Roth. Minneapolis: Milkweed Editions.

Sanday, P. R. 1990. *Fraternity Gang Rape.* New York: New York University Press.

19

Identity and Stigma of Women with STDs

ADINA NACK

In this chapter, Nack focuses her attention on a highly secretive group: college students with sexually transmitted diseases (STDs). She examines the way infected women deal with the deviance they have acquired. Employing a range of rich concepts, she discusses how women manage the stigma and identity associated with their sexually diseased condition. People initially go through a stage of nonacceptance, or denial, of their deviance, during which they try to distance themselves from the stigma by passing, covering, and/or deceiving others, all the while feeling guilty about their deceptions. They then deflect their diseased identities onto others by transferring responsibility and blame for becoming infected to those they imagine have given it to them. After a time, they begin to accept their deviant identities and make disclosures about their status to family members, friends, and sexual partners. We see the connection between stigma management behavior and self-concept as we follow the way the women's evolving strategies eventually lead them to change their vision of themselves, incorporating this deviant dimension into their core identity. Do you have any secret conditions that need to be "managed" as Nack describes? What is the nature of stigma in society? Why do some conditions or behaviors need to be hidden from others?

The HIV/AIDS epidemic has garnered the attention of researchers from a variety of academic disciplines. In contrast, the study of other sexually transmitted diseases (STDs) has attracted limited interest outside of epidemiology and public health. In the United States, an estimated three out of four sexually active adults have HPV infections (human papillomavirus—the virus that can cause genital

SOURCE: "Damaged Goods: Women Managing the Stigma of AIDS," by Adina Nack, *Deviant Behavior*, 21(2), 2000, pp. 95–115. Reprinted with permission of Taylor & Francis Ltd. http://www.tandf.co.uk/journals.

warts); one out of five have genital herpes infections (Ackerman 1998; CDC Server 1998).... This article focuses on how the sexual self-concept is transformed when the experience of living with a chronic STD casts a shadow of disease on the health and desirability of a woman's body, as well as on her perceived possibilities for future sexual experiences....

STIGMA AND THE SEXUAL SELF

For all but one of the 28 women, their STD diagnoses radically altered the way that they saw themselves as sexual beings. Facing both a daunting medical and social reality, the women employed different strategies to manage their new stigma. Each stigma management strategy had ramifications for the transformation of their sexual selves....

Stigma Nonacceptance

Goffman (1963) proposed that individuals at risk for a deviant stigma are either "the discredited" or "the discreditable." The discrediteds' stigma was known to others either because the individuals revealed the deviance or because the deviance was not concealable. In contrast, the discreditable were able to hide their deviant stigma. Goffman found that the majority of discreditables were "passing" as non-deviants by avoiding "stigma symbols," anything that would link them to their deviance, and by utilizing "disidentifiers," props or actions that would lead others to believe they had a non-deviant status. Goffman (1963) also noted that individuals bearing deviant stigma might eventually resort to "covering," one form of which he defined as telling deceptive stories. To remain discreditable in their everyday lives, nineteen of the women employed the individual stigma management strategies of passing and/or covering. In contrast, nine women revealed their health status to select friends and family members soon after receiving their diagnoses.

Passing.   The deviant stigma of women with STDs was essentially concealable, though revealed to the necessary inner circle of health care and health insurance providers. For the majority, passing was an effective means of hiding stigma from others, sometimes, even from themselves.

Hillary, a 22-year-old White senior in college, recalled the justifications she had used to distance herself from the reality of her HPV infection and facilitate passing strategies.

> At the time, I was in denial about it. I told myself that that wasn't what it was because my sister had had a similar thing happen, the dysplasia. So, I just kind of told myself that it was hereditary. That was kinda' funny because I asked the nurse that called if it could be hereditary, and she said "No, this is completely sexually transmitted." ... I really didn't accept it until a few months after my cryosurgery.

Similarly, Gloria, a Chicana graduate student and mother of four, was not concerned about a previous case of gonorrhea she had cured with antibiotics or her chronic HPV "because the warts went away." Out of sight, out of her sex life: "I never told anybody about them because I figured they had gone away, and they weren't coming back. Even after I had another outbreak, I was still very promiscuous. It still hadn't registered that I needed to always have the guy use a condom."

When the women had temporarily convinced themselves that they did not have a contagious infection, it was common to conceal the health risk with partners because the women, themselves, did not perceive the risk as real. Kayla, a lower-middle class, White college senior, felt justified in passing as healthy with partners who used condoms, even though she knew that condoms could break…. Tasha, a White graduate student, found out that she might have inadvertently passed as healthy when her partner was diagnosed with chlamydia. "I freaked out—I was like, 'Oh my God! I gave you chlamydia. I am so sorry! I am so sorry!' I felt really horrible, and I felt really awful." Sara, a Jewish, upper-middle class 24-year-old, expressed a similar fear of having passed as healthy and exposed a partner to HPV. "Evan called me after we'd been broken up and told me he had genital warts. And, I was with another guy at the time, doing the kinda-sorta-condom-use thing. It was like, 'Oh, my gosh, am I giving this person something?'" Even if the passing is done unintentionally, it still brings guilt to the passer.

The women also tried to disidentify themselves from sexual disease in their attempts to pass as being sexually healthy. Rather than actively using a verbal or symbolic prop or action that would distance them from the stigma, the women took a passive approach. Some gave nonverbal agreement to put downs of other women who were known to have STDs. For example, Hillary recalled such an interaction. "It's funny being around people that don't know that I have an STD and how they make a comment like, 'That girl, she's such a slut. She's a walking STD.' And how that makes me feel when I'm confronted with that, and having them have no idea that they could be talking about me." Others kept silent about their status and tried to maintain the social status of being sexually healthy and morally pure. Kayla admitted to her charade: "I guess I wanted to come across as like really innocent and everything just so people wouldn't think that I was promiscuous, just because inside I felt like they could see it even though they didn't know about the STD." Putting up the facade of sexual purity, these women distanced themselves from any suspicion of sexual disease.

Covering.   When passing became too difficult, some women resorted to covering to deflect family and friends from the truth. Cleo summed up the rationale by comparing her behavior to what she had learned growing up with an alcoholic father. "They would lie, and it was obvious that it was a lie. But, I learned that's what you do. Like you don't tell people those things that you consider shameful, and then, if confronted, you know, you lie."

Hillary talked to her parents about her HPV surgery, but never as treatment for an STD. She portrayed her moderate cervical dysplasia as a pre-cancerous scare, unrelated to sex. "We never actually talked about it being a STD, and she kind of thought that it was the same thing that my sister had which wasn't sexually transmitted." When Tasha's sister helped her get a prescription for pubic lice, she actually provided the cover story for her embarrassed younger sister. "She totally took control, and made a personal inquiry: 'So, how did you get this? From a toilet seat.' And, I was like, 'a toilet seat,' and she believed me." When I asked Tasha why she confirmed her sister's misconception, she replied, "because I didn't want her to know that I had had sex." For Anne, a 28-year-old lower-middle class graduate student, a painful herpes outbreak almost outed her on a walk with a friend. She was so physically uncomfortable that she was actually "waddling." Noticing the strange behavior, her friend asked what was wrong. Anne told her that it was a hemorrhoid: that was only a partial truth because herpes was the primary cause of her pain. As Anne put it, telling her about the hemorrhoid "was embarrassing enough!"

Deception and Guilt.   The women who chose to deny, pass as normal, and use disidentifiers or cover stories shared more than the shame of having an STD—they had also told lies. With lying came guilt. Anne, who had used the hemorrhoid cover story, eventually felt extremely guilty. Her desire to conceal the truth was in conflict with her commitment to being an honest person. "I generally don't lie to my friends. And I'm generally very truthful with people and I felt like a sham lying to her." Deborah, a 32-year-old, White professional from the Midwest, only disclosed to her first sexual partner after she had been diagnosed with HPV: she passed as healthy with all other partners. Deborah reflected, "I think my choices not to disclose have hurt my sense of integrity." However, her guilt was resolved during her last gynecological exam when the nurse practitioner confirmed that after years of "clean" pap smear results Deborah was not being "medically unethical" by not disclosing to her partners. In other words, her immune system had probably dealt with the HPV in such a way that she might never have another outbreak or transmit the infection to sexual partners.

When Cleo passed as healthy with a sexual partner, she started, "feeling a little guilty about not having told." However, the consequences of passing as healthy were very severe for Cleo:

> No. I never disclosed it to any future partner. Then, one day, I was
> having sex with Josh, my current husband, before we were married, and
> we had been together for a few months, maybe, and I'm like looking at
> his penis, and I said, "Oh, my goodness! You have a wart on your penis!
> Ahhh!" All of a sudden, it comes back to me.

Cleo's decision to pass left her with both the guilt of deceiving and infecting her husband.

Surprisingly, those women who had *unintentionally* passed as being sexually healthy (i.e., they had no knowledge of their STD-status at the time) expressed a similar level of guilt as those who had been purposefully deceitful. Violet, a middle class, White 36-year-old, had inadvertently passed as healthy with her current partner. Even after she had preventively disclosed to him, she still had to deal with the guilt over possibly infecting him.

> It hurt so bad that morning when he was basically furious at me think-
> ing I was the one he had gotten those red bumps from. It was the hour
> from hell! I felt really majorly dirty and stigmatized. I felt like "God,
> I've done the best I can: if this is really caused by the HPV I have, then
> I feel terrible."

When employing passing and covering techniques, the women strove to keep their stigma from tainting social interactions. They feared reactions that Lemert (1951) has labeled the *dynamics of exclusion:* rejection from their social circles of friends, family, and most importantly sexual partners. For most of the women, guilt surpassed fear and became the trigger to disclose. Those who had been deceitful in passing or covering had to assuage their guilt: their options were either to remain in nonacceptance, disclose, or transfer their guilt to somebody else.

Stigma Deflection

As the women struggled to manage their individual stigma of being sexually dis-eased, real and imaginary social interactions became the conduit for the conta-gious label of "damaged goods." Now that the unthinkable had happened to them, the women began to think of their past and present partners as infected, contagious, and potentially dangerous to themselves or other women. The com-bination of transferring stigma and assigning blame to others allowed the women to deflect the STD stigma away from themselves.

Stigma Transference.   I propose the concept of *stigma transference* to capture this element of stigma management that has not been addressed by other deviance theorists. Stigma transference is a specialized case of projection which … mani-fests as a clear expression of anger and fear. The women did not connect this strategy to a reduction in their levels of anxiety; in fact, several discussed it in relation to increased anxiety….

Transference of stigma to a partner became more powerful when the woman felt betrayed by her partner. When Hillary spoke of the "whole trust issue" with her ex-partner, she firmly believed he had lied to her about his sexual health status and he would lie to others. Even though she had neither told him about her diagnosis nor had proof of him being infected, she fully transferred her stigma to him.

> He's the type of person who has no remorse for anything. Even if I did
> tell him, he wouldn't tell the people that he was dating. So it really
> seemed pretty pointless to me to let him know because he's not
> responsible enough to deal with it, and it's too bad knowing that he's
> out there spreading this to God knows how many other people.

Kayla also transferred the stigma of sexual disease to an ex-partner, never confront-ing him about whether or not he had tested positive for STDs. The auxiliary trait of promiscuity colored her view of him: "I don't know how sexually promiscuous he was, but I'm sure he had had a lot of partners." Robin, a 21-year-old White undergraduate, went so far as to tell her ex-partner that he needed to see a doctor and "do something about it." He doubted her ability to pinpoint contracting genital warts from him and called her a "slut." Robin believed that *he* was the one with the reputation for promiscuity and decided to "trash" him by telling her two friends who hung out with him. Robin hoped to spoil his sexual reputation and scare off his future partners. In the transference of stigma, the women ascribed the same auxiliary traits onto others that others had previously ascribed to them....

In all cases, it was logical to assume that past and current sexual partners may have also been infected. However, the stigma of being sexually diseased had far-reaching consequences into the imaginations of the women. The traumatic impact on their sexual selves led most to infer that future, as yet unknown, partners were also sexually diseased. Kayla summed up this feeling: "After I was diagnosed, I was a lot more cautious and worried about giving it to other people or getting something else because somebody hadn't told me." They had already been damaged by at least one partner. Therefore, they expected that future partners, ones who had not yet come into their lives, held the threat of also being *damaged goods*.

For Hillary, romantic relationships held no appeal anymore. She had heard of others who also had STDs but stayed in nonacceptance and never changed their lifestyle of having casual, unprotected sex:

> I just didn't want to have anything to do with it. A lot of it was not trusting people. When we broke up, I decided that I was not having sex. Initially, it was because I wanted to get an HIV test. Then, I came to kind of a turning point in my life and realized that I didn't want to do the one-night-stand thing anymore. It just wasn't worth it. It wasn't fun....

Blame. The women's uses of stigma transference techniques were attempts to alleviate their emotional burdens. First, the finger of shame and guilt pointed inward, toward the women's core sexual selves. Their sexual selves became tainted, dirty, damaged. In turn, they directed the stigma outward to both real and fictional others. Blaming others was a way for all of the women to alleviate some of the internal pressure and turn the anger outward. This emotional component of the *damaged goods* stage externalized the pain of their stigma.

Francine recalled how she and her first husband dealt with the issue of genital warts. "We kind of both ended up blaming it on the whole fraternity situation. I just remember thinking that it was not so much that we weren't clean, but that he hadn't been at some point, but now he was." Francine's husband had likely contracted genital warts from his wild fraternity parties: "We really thought of it as, that woman who did the trains [serial sexual intercourse]. It was still a girl's fault kind-of-thing." By externalizing the blame to the promiscuous women at fraternity parties, Francine exonerated not only herself, but also her husband....

For Violet, it was impossible to neatly deflect the blame away from both herself and her partner. "I remember at the time just thinking, 'Oh man! He gave it to me!' While, he was thinking, 'God, [Violet]! You gave this to me!' So, we kind of just did a truce in our minds. Like, OK, we don't know who gave it—just as likely both ways. So, let's just get treated. We just kind of dropped it." Clearly the impulse to place blame was strong even when there was no easy target.

Often, the easiest targets were men who exhibited the auxiliary traits of promiscuity and deception. Tasha wasn't sure which ex-partner had transmitted the STD. However, she rationalized blaming a particular guy. "He turned out to be kind of a huge liar, lied to me a lot about different stuff. And, so I blamed him. All the other guys were, like, really nice people, really trustworthy." Likewise, when I asked Violet from whom she believed she had contracted chlamydia, she replied, "Dunno, it could've been from one guy, because that guy had slept with some unsavory women, so therefore he was unsavory." Later, Violet contracted HPV, and the issue of blame contained more anger: "I don't remember that discussion much other than, being mad over who I got it from: 'oh it must have been Jess because he had been with all those women.' I was mad that he probably never got tested. I was OK before him." The actual guilt or innocence of these blame targets was secondary. What mattered to the women was that they could hold someone else responsible.

Stigma Acceptance

Eventually, every woman in the study stopped denying and deflecting the truth of her sexual health status by disclosing to loved ones. The women disclosed for either preventive or therapeutic reasons. That is, they were either motivated to reveal their STD status to prevent harm to themselves or others, or to gain the emotional support of confidants.

Preventive and Therapeutic Disclosures. The decision to make a preventive disclosure was linked to whether or not the STD could be cured. Kayla explained, "Chlamydia went away, and I mean it was really bad to have that, but I mean it's not something that you have to tell people later 'cause you know, in case it comes back. Genital warts, you never know." Kayla knew that her parents would find out about the HPV infection because of insurance connections. Prior to her cryosurgery, Kayla decided to tell her mom about her condition. "I just told her what [the doctor] had diagnosed me with, and she knew my boyfriend and everything, so—it was kind of hard at first. But, she wasn't upset with me. Main thing, she was disappointed, but I think she blamed my boyfriend more than she blamed me." Sara's parents also reacted to her preventive disclosure by blaming her boyfriend: they were disappointed with their daughter, but angry with her boyfriend.

Preventive disclosures to sexual partners, past and present, were a more problematic situation. The women were choosing to put themselves in a position where they could face blame, disgust, and rejection. For those reasons, the women put off preventive disclosures to partners as long as possible. For example, Anne made it clear that she would not have disclosed her herpes to a female sexual partner had they not been, "about to have sex." After "agonizing weeks and weeks and weeks before trying to figure out how to tell," Diana, a 45-year-old African-American professional, finally shared her HPV and herpes status before her current relationship became sexual. Unfortunately, her boyfriend "had a negative reaction": "he certainly didn't want to touch me anywhere near my genitals." In Cleo's case, she told her partner about her HPV diagnosis because she wasn't going to be able to have sexual intercourse for a while after her cryosurgery. Violet described the thought process that lead up to her decision to disclose her HPV status to her current partner:

> That was really scary because once you have [HPV], you can't get rid of the virus. And then having to tell my new partner all this stuff. I just wanted to be totally up front with him: we could use condoms. Chances are he's probably totally clean. I'm like, "Oh my god, here I am tainted because I've been with, at this point, 50 guys, without condoms. Who knows what else I could have gotten (long pause, nervous laugh)?" So, that was tough....

Many of the therapeutic disclosures were done to family members. The women wanted the support of those who had known them the longest. Finally willing to risk criticism and shame, they hoped for positive outcomes: acceptance, empathy, sympathy—any form of nonjudgmental support. Tasha disclosed to her mother right after she was diagnosed with chlamydia. "My family died—'Guess what, mom, I got chlamydia.' She's like, 'Chlamydia? How did you find out you got chlamydia?' I'm like, 'Well, my boyfriend got an eye infection.' (laughter) 'How'd he get it in his eye?' (laughter) So, it was the biggest joke in the family for the longest time!" In contrast, Rebecca, a White professional in her mid-fifties, shared her thought process behind *not* disclosing to her adult children. "I wanted to tell my younger one ... I wanted very much for him to know that people could be asymptomatic carriers because I didn't want him to unjustly suspect somebody of cheating on him ... and I don't believe I ever managed to do it ... it's hard to bring something like that up."...

Consequences of Disclosure.   With both therapeutic and preventive disclosure, the women experienced some feelings of relief in being honest with loved ones. However, they still carried the intense shame of being sexually diseased women. The resulting emotion was anxiety over how their confidants would react: rejection, disgust, or betrayal. Francine was extremely anxious about disclosing to her husband. "That was really tough on us because I had to go home and tell Damon that I had this outbreak of herpes." When asked what sorts of feelings that brought up, she immediately answered. "Fear. You know I was really fearful—I didn't think that he would think I had recently had sex with somebody else ... but, I was still really afraid of what it would do to our relationship."...

Overall, disclosing intensified the anxiety of having their secret leaked to others in whom they would have never chosen to confide. In addition, each disclosure brought with it the possibility of rejection and ridicule from the people whose opinions they valued most. For Gloria, disclosing was the right thing to do but had painful consequences when her partner's condom slipped off in the middle of sexual intercourse.

> I told him it doesn't feel right. "You'd better check." And, so he checked, and he just jumped off me and screamed, "Oh fuck!" And, I just thought, oh no, here we go. He just freaked and went to the bathroom and washed his penis with soap. I just felt so dirty....

Disclosures were the interactional component of self-acceptance. The women became fully grounded in their new reality when they realized that the significant people in their lives were now viewing them through the discolored lenses of sexual disease.

CONCLUSION

The women with STDs went through an emotionally difficult process, testing out stigma management strategies, trying to control the impact of STDs on both their self-concepts and on their relationships with others. In keeping with Cooley's "looking glass self" (1902/1964), the women derived their sexual selves from the imagined and real reactions of others. Unable to immunize themselves from the physical wrath of disease, they focused on mediating the potentially harmful impacts of STDs on their sexual self-concepts and on their intimate relationships.

REFERENCES

Ackerman, S. J. 1998. "HPV: Who's Got It and Why They Don't Know." *HPV News* 8(2), Summer: 1, 5–6.

Centers for Disease Control and Prevention. 1998. "Genital Herpes." *National Center for HIV, STD, & TB Prevention*. Online. Netscape Communicator. 4 February.

Cooley, Charles H. [1902] 1964. *Human Nature and the Social Order*. New York: Schocken.

Goffman, Erving. 1963. *Stigma*. Englewood Cliffs, NJ: Prentice Hall.

Lemert, Edwin. 1951. *Social Pathology*. New York: McGraw-Hill.

Social Inequality

We turn from the culture and interactional socialization that shapes us in Part II to the social structure that frames our lives in Part III. Like culture, social structure is abstract and invisible, an idea that people on the street have a hard time grasping in their everyday explanations of how the world works. But the invisible forces of social structure affect our lives in profound ways. One of the key concerns of sociology is inequality, especially by race, class, and gender. We all recognize that some people have greater advantages in society than others, that not all people are treated the same way. No matter how much we subscribe to the American ethos of equal opportunity for all, we know that this is not quite true. Lines are drawn between privileged positions and those characterized by discrimination and prejudice. The strata, or hierarchical rungs of society, are not easily carved out, however, and are often complicated by the way race, class, and gender characteristics are combined, with some people having advantages here and disadvantages there, while others benefit or suffer in compounded ways.

We live in a heterogeneous society, filled with many types of people. The consequences of these differences range from apparent to subtle. Some are highly visible, and we can see them every day. Government policies favor some categories of people over others, social institutions privilege some groups of people over others, and individuals, in their daily functioning, treat some particular people better than others. Whereas we can see the discrimination being applied in certain cases, in others the patterns of differential treatment are so deeply embedded in the criteria by which institutional decisions are made that we cannot easily recognize them. This is called *institutional discrimination*. The guidelines by which people are treated differently are complex, and intersect according to several variables. They align people and the groups to which they belong into a *system of stratification*, with favored groups having greater access to the benefits of society such as money, privilege, prestige, and power. Thus, the ability to access wealth, to gain the respect of others, and to impose your will on others is unequally distributed in society.

The first way we see social inequality is in the system of *social class*. As Earl Wysong and Robert Perrucci note in our first selection, a very small percentage of the population owns most of the wealth in our society, and has for many years. During the 1980s, however, the gap between the rich and the poor, which had narrowed during the 1960s and 1970s as the result of Federal programs and the progressive tax system, began to widen steadily with government policies that favored the rich. America has traditionally been a country with three distinct social classes: upper, middle, and lower (although some might argue that there is a distinction between the working poor, or working class, and the poor who do not work in the legitimate economy, or underclass). This contrasts significantly with many Latin American countries that are composed more exclusively of a very rich upper class and a very large peasant class, with little in the way of middle class. Our largest class, throughout most of our history, has been our middle class, and we have conceived of America as a country dominated by that ideology and way of life. But the 1980s brought an increasing divide between the rich and poor, accompanied by a shrinking of the middle class. Most middle-class families responded to this challenge by coming increasingly to rely on the income of the wife, the second wage earner, and this has placed greater burdens on other social institutions such as education and the family. In the twenty-first century the growing schism between the rich and the poor widened, with economic policies that favored the *"uber"* rich such as taxation shifts away from the wealthy to the middle and working classes, easements on lobbying regulations and political contributions for big businesses, and restricted bankruptcy filings for the poor, the sick, the unemployed, and the divorced. Class inequality continues to be a growing problem in our society.

Next, Karyn R. Lacy follows with a fascinating analysis of the interplay between race and class. She examines the lives and residential strategies of middle-class Blacks, people who have moved occupationally into the mainstream White world, but who still desire to preserve their ethnic heritage and identity for their children. She outlines two different approaches taken, as one group pursues a more aggressive integrationist strategy by living in White neighborhoods and sending their children to White schools, while the other stays closer to their roots in Black communities and schools. How these families balance their straddle of mainstream routes to upward mobility with their ethnic identity is an ongoing negotiation. Finally, Yvonne M. Luna gives us a glimpse of the hardships and struggles single mother welfare recipients face in trying to navigate the system to get enough money for their family's survival. Beyond their subsistence difficulties, they have to try to hide their welfare identity as they purchase goods with the aid of welfare cards or face the considerable stigma and condemnation of people in more affluent sectors in society.

A second dimension along which people are unequally arrayed is *race* and *ethnicity*. *Ethnicity* refers to the cultural characteristics of a group of people that are often derived from their heritage and geographical origin. *Race* is a more difficult and increasingly problematic concept that at one time referred to membership in a group characterized by different physical features including skin color and appearance. But the intermixing of population has grown so pervasive that we can no longer speak of "pure" racial groups, and the concept has lost its original value. As we saw in Chapter 4, by Keith Kilty and Maria Vidal de Haymes, race now exists as a social construction that grounds its meaning in the way it is used in each society. In fact, even ethnicity is being eroded by inter-marriage. Many Americans begin their experience in this country with a clearly defined heritage and ethnicity in the first generation (the group who moves here) and then begin to assimilate in the second generation as young members move through the public schools, speak the language, and acquire American cultural values. People of different racial/ethnic groups are conceived and treated according to common stereotypes in this country, with some groups, notably Asians, benefiting from the "model minority" perception, and others, notably Blacks and Latina/os, still suffering profound prejudice that results in blocked upward mobility and social acceptance.

Americans have conceived of the United States as an immigrant-formed country, as most of our population derives its roots from elsewhere. Large waves of immigrants populated our land in the eighteenth, nineteenth, and early twentieth centuries. We considered ourselves a *melting pot* where people from all countries came together and formed into one, ethnic differences evaporating as new members were exposed to the dominant culture and mass education. But as time passed and distinctions remained, some groups still failing to attain parity, we began to recognize the relevance of a *cultural pluralist* model, our society consisting of many different groups living together. Recent increasing waves of immigration have raised issues of race and ethnicity to the fore once again, as our country struggles with its pluralistic composition. Troubling problems include segregation, racism, sexism, homophobia, anti-Semitism, the large influx of Asians and Latina/os, and the "browning" of America.

Our first selection, by Tamara K. Nopper, looks at some of the structural foundations Korean-Americans have forged to enhance their position in America. As a group, they have achieved a measure of success and upward mobility, starting with forming entrepreneurial businesses. They then reached out into more structural foundations of society to develop an ethnic banking industry to further support their business ventures. This chapter offers a more in-depth look at some of the mechanisms by which ethnic groups move up the economic hierarchy and enhance their accompanying social standing.

We then present Rashawn Ray and Jason A. Rosow's comparison of two different ethnic groups, Blacks and Whites, in their analysis of racially differentiated fraternities. Just as Boswell and Spade suggested in Chapter 18, there can be a huge variation between similar-aged male groups depending on their culture. In Chapter 24, Ray and Rosow look not only at cultural differences between Black and White fraternity members' views toward romance and sex but also at structural differences in the way their housing affects these, leading to their vastly disparate treatment of women.

Finally, Jessica M. Vasquez examines the way third-generation Latinos, who have become more assimilated, and often more racially mixed, navigate their self-presentations and receptions as they interact with Whites and other Latinos. Many of these people represent one of the fastest-growing racial/ethnic populations in our society: biracial and multiracial individuals. As such, her subjects may inhabit a variety of social scenes, modifying the way they speak, act, and are received as they navigate a complex and shifting American landscape.

Gendered inequality is our last focus in this Part, with men attaining a greater share of the world's valued resources than women. In many countries, the gap between men and women is wider than in the United States. India and China look to male children to support the family, while girls represent an economic drain and social loss, moving in with in-laws and having to be supported by a dowry. These countries struggle with difficult problems such as female infanticide, fetal (female) sex-screening and abortion, and the malnourishment and premature death of female children. Women in many parts of the Arab world remain secluded and veiled, symbolic properties of their male relatives, unable to drive, to hold jobs, or to travel freely in public. The United States, along with some of the socialist democracies in Western Europe, has the most liberated and powerful women in the world. But even in America we live in a *patriarchy*, a system of male dominance embedded in the culture and social structure. Unequal legal rights, opportunities, and treatment of women foster their reduced power economically, socially, politically, occupationally, and interpersonally.

Gender roles have been an area of struggle for decades, with people trying to break out of traditional constraints while more conservative forces hold them in place. The role of women underwent some profound changes during the women's movement in the 1970s, only to revert back during the anti-feminist backlash that followed. Avenues of structural opportunity for gender parity have opened to women that were closed before, but many blocks remain. At the same time, women have succumbed to the tyranny of cultural gender norms through the crippling focus on their appearance. It seems that no matter how well- or poorly-educated, how financially affluent or challenged, women find it hard to escape the judgment of themselves and others that resides in their appearance. Ashley Mears and William Finlay offer a poignant depiction of models, women prototypically trapped in their embodied self-concepts, to show the struggle these culturally defined "beauties" experience to adapt to media-induced gender ideals. They show the depths to which women will go to hold onto fading beauty and the toll this takes on their lives, their work, and their identities.

Men's lives have not been transformed as much over this period, as men have remained more trapped in traditional gender obligations and outlooks. But ironically, our next selection shows some of the progress evinced by certain groups of men in discarding traditional male stereotypes and phobias. Eric Anderson's portrayal of different groups of male cheerleaders illustrates some of the gendered hierarchies of masculinity, showing how some have higher status while others are more stigmatized. This chapter defies common impressions about men in female-dominated settings to show that not all men have to adhere to the most prestigious hypermasculine examples, and that some can successfully forge new types of masculinity.

Finally, Shelly Ronen looks at some of the intricacies of cross-gender inter-actions in her examination of how men and women literally navigate the dance of flirtation in a very public setting. Focusing on sexualized dancing, first described in Boswell and Spade's Chapter 18, she describes how college men and women engage in very intimate physical contact that may lead to romance, hooking up, a bad reputation, or all of the above. She deftly illustrates various dimensions of the approach, the acceptance, the break-off, and the culmination of these sexually charged interactions, and how they are socially managed by individuals with the occasional help of their friends.

20

The U.S. Class Structure

EARL WYSONG AND ROBERT PERRUCCI

For centuries the United States has been known as a predominantly middle-class society where people could break through class barriers, going from rags to riches as the result of their hard work and determination. Unlike England, where class structures were entrenched, we prided ourselves on a fluid class system that permitted upward mobility. Policies of earlier administrations such as Franklin D. Roosevelt's New Deal and Lyndon B. Johnson's Great Society fostered more equality, with a safety net operating to protect the poor and disabled. Yet at the same time, the American ethos has always embodied a rugged individualism where people were reluctant to give up too great a share of their earnings to ensure the common welfare. In their analysis of the U.S. class system, Wysong and Perrucci present us with the most current portrait of the composition and distribution of wealth in America. We can see that the American middle class has been replaced with a great squeeze: the distribution of wealth has shifted upwards, with a smaller group controlling earnings and assets, while the comfort class has shrunk, with many of its members moving downwards. Wysong and Perrucci explain the four main elements that are responsible for locating people in the new class system, and we can see that many of these are stable and enduring, passed from one generation to the next rather than attainable by individuals through dedication and hard work. What kinds of jobs do your family members hold and what types of assets do they own? Does this model accurately locate them? How do they stand in relation to other class groups in society? What are your economic goals and aspirations? How would you rate your chances of achieving these goals? How does this model of class in America fit the ideal that you think our country should embody? How does this model fit with the popular conception of the class

SOURCE: Earl Wysong and Robert Perrucci, Organizations, Resources, and Class Analysis: The Distributional Model and the US Class Structure, *Critical Sociology*, Volume 33, Numbers 1–2, 2007, pp. 211–246. Reprinted by permission of Brill. Table and figure reprinted by permission of Rowman & Littlefield Publishers.

structure? How do you think the American class structure may have been affected by the global recession that began in 2008, the nature of our system of tax breaks, the export of labor overseas, and the political movements against unions?

The last decade witnessed numerous mass media and scholarly reports on the growing polarization of wealth and income in the United States. Media accounts describe a variety of economic and social trends increasingly dividing the USA into two class-like groupings sometimes called the "haves" and "have-nots." During this same period, the relevance of class analysis in contemporary societies has been debated by sociologists with very different approaches to the topic. Some question the usefulness of class analysis. Others assert the continued importance of class analysis and call for substantial revisions in the ways in which class frameworks have been conceptualized and applied. As these developments indicate, class analysis has emerged as a topic with wide popular and scholarly appeal. More specifically, within sociology there is emerging interest in reevaluating existing approaches to class analysis and in exploring promising new theoretical and empirical avenues of inquiry.

This paper builds upon growing interest in new lines of sociological inquiry concerning class analysis and presents a theoretical approach that links the study of class with organizational resources and distributional processes....

THE DISTRIBUTIONAL MODEL AND
THE US CLASS STRUCTURE

We consider how occupations—a factor central to production models—plus organizational resources, structures, and distributional processes can be incorporated into what we call a *distributional model* of class analysis. We believe this model provides an effective theoretical approach for identifying the bases of class membership, conceptualizing a new map of the stratification system, and for considering the consequences of class inequalities. The model considers how the *distribution* of important resources, at unequal levels to individuals and families, occurs through linkages that connect people's occupations as well as other organizational affiliations (e.g., trustees, investors) to organizationally-based resources and distributional processes.

The concepts of *generative capital* and the *distributive order* are two central features of the distributional model relevant to our purposes in this paper. In our approach we define classes as collectivities of individuals and families who possess, *over time,* similar levels of the important economic and social resources we call generative capital. These resources are viewed as occurring in four forms: (1) consumption capital, (2) investment capital, (3) credentialed skill capital, and (4) social capital. We consider these forms to be *generative* because in some cases they can produce more of the same resource when invested, or in other cases they can contribute to the production of another resource (e.g., social capital can produce investment capital). The condition, *over time,* denotes that variations in the levels of resources distributed to and held by individuals and families tend to persist over time (i.e., within generations over periods of several years and over even longer periods as between generations) and that their impact on the life chances and experiences of people in different classes is substantial and enduring. We view the unequal and patterned distribution of these resources as the basis of class inequality in the USA....

THE US CLASS STRUCTURE TODAY: DESCRIPTION AND IMAGE

In our model, the top officers, managers, administrators, public officials, and upper-level staff members of large corporations and government bodies create and administer the laws, policies, and rules that direct the formal distributional processes component of the distributive order. These organizational leaders and staffers, though widely dispersed in terms of specific organizational affiliations, geography, and occupations are viewed as comprising an economically advantaged and affluent group we term the *Privileged Class*. This relatively large class consists of two distinct segments: (1) the top one percent of all US wealth and income holders which we term the *Superclass* and, (2) the next 19 percent which we term the *Credentialed Class*. The remaining 80 percent of Americans are members of what we term the *New Working Class*. This class is "new" in the sense that, based on recent economic and social developments, the individuals and families placed by our model in this class today—especially young adults— collectively face new and more daunting economic and social challenges than in the recent past. These challenges involve difficulties in accessing and maintaining economic resources at levels sufficient to sustain modest incomes and what may be called middle class lifestyles....

Table 20.1 lists the two major classes in our distributional model and the major segments of the current US class structure. It also provides summaries of "Class Characteristics" for each class segment and lists our estimates of the percentage of the US population associated with each segment. The summary entries describe the organizational affiliations and/or occupational groupings typically associated with persons in each segment. They also provide illustrations of the levels and types of generative capital forms—held over time—that persons located in each segment possess. The major classes and the class segments differ in terms of the extent to which persons located within them possess stable and secure access to varying levels of the four forms of capital resources over time. However, the *acquisition* of varying levels of capital and thus the process by which *placement* in the class structure occurs is determined by the operation of the distributive order.

Figure 20.1 illustrates how the distributional model envisions the current US class structure as a two-class, double-diamond form. The double-diamond image is scaled to visually represent the proportions of the population our model views as located in each major class: 20 percent in the privileged class and 80 percent in

TABLE 20.1   The U.S. Class Structure

Major Classes & Segments	Class Characteristics	Percentage of Population
Privileged Class (20%) **Superclass**	Owners and employers. Make a living from investments or business ownership; incomes at six- to seven-figure level, yielding sizable consumption and investment capital.	1–2%
Credentialed Class **Managers**	Mid- and upper-level managers and CEOs of corporations and public organizations. Incomes for upper-level CEOs in seven-figure range, others six-figures.	13–15%
Professionals	Possess credentialed skill in form of college and professional degrees. Use of social capital and organizational ties to advance interests. Wide range of incomes, from $100K to upper-six figures.	4–5%
New Working Class (80%) **Comfort class**	Nurses, teachers, civil servants, very small-business owners, and skilled and union carpenters, machinists, or electricians. Incomes in the $35–50K range but little investment capital.	10%
Contingent Class **Wage earners**	Work for wages in clerical and sales jobs, personal services, and transportation and as skilled craft workers, machine operators, and assemblers. Members of this group are often college graduates. Incomes at $30K and lower.	50%
Self-employed	Usually self-employed with no employees, or family workers. Very modest incomes, with high potential for failure.	3–4%
Excluded class	In and out of the labor force in a variety of unskilled, temporary jobs.	10–15%

SOURCE: Copyright, Rowman & Littlefield. Reprinted with permission.

the new working class. In our model, the four forms of capital are the basis of class membership, interests, and inequalities in the larger stratification system. Persons holding high levels of all forms of capital over time are viewed as possessing secure and substantial resources and are considered to be members of the privileged class. Persons who lack access to high levels of the four generative capital forms, over time, as well as those with only a tenuous hold on moderate levels of one or more of the four capital forms, are considered to be members of the new working class. At the macro level, our model views the unequal distribution of the scarce and valued forms of capital over time to individuals and families as the structural foundation of class inequality and the larger US stratification system.

It is important to keep in mind that one's location in the double-diamond class structure is related to one's occupation but not determined by it. Some lawyers are in the top diamond, and some in the bottom. Some engineers, scientists, and professors are in the privileged class, and some in the new working class. It is not occupation that determines placement in various classes (or class segments); rather, occupation *plus* other organizational affiliations and the influence of the other elements of the distributive order provide people—to a greater or lesser extent—with stable or unstable access to varying levels of the four forms of generative capital over time.

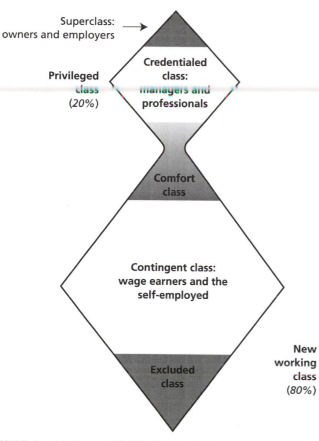

F I G U R E 20.1 Double-Diamond Diagram: The U.S. Class Structure and Class Segments.

CLASS INTERESTS

We maintain that most members of the privileged class (the top two segments) are bound together by shared class interests in developing, implementing, and legitimating corporate practices and public policies that promote and perpetuate a highly unequal, class-biased distribution of most class-relevant resources. Our view is that the privileged class dominates all levels of the distributive order, including the immediate organizational level; this means privileged class members, through their positions as organizational leaders, managers, and top staffers, ensure that the largest shares of economic and social resources are distributed to individuals and families who are members of this class. In this way, the affluence and advantages of privileged class members are perpetuated.

We see an important structural consequence occurring as a result of shared privileged class interests and the dominance of this class over the distributive order, viz: the class structure develops into the polarized shape shown by Figure 20.1.... We view this outcome, discussed at length elsewhere, as occurring as the result of organizationally-based changes (at multiple levels due to privileged class dominance) that have narrowed the distribution of resources to members of the new working class and expanded the distribution of resources to members of the privileged class....

Our model views the new working class as having shared class interests that are very different from those of the privileged class. Most obviously, members of the new working class have interests in shifting the current resource distribution pattern which favors the privileged class to a pattern that would produce more equitable resource outcomes across class lines. However, compared with those in the privileged class, we view members of the new working class as having far less influence over how the structures and processes of the distributive order operate to produce class-based advantages. Also, we posit that compared with the privileged class, members of the new working class are less likely to recognize and act on shared class interests. Due to these differences, we see the new working class, compared with the privileged class, as having less stable access to and possessing lower levels of all four capital forms and thus, substantially lower material resource levels.

GENERATIVE FORMS OF CAPITAL AND THE US CLASS STRUCTURE

Consumption Capital

This resource refers primarily to wages and salaries, but it also includes total annual incomes received by individuals and families from all sources. While we do not view income as the sole determinant of class, in our model total annual incomes of individuals and families provide useful and approximate empirical reference points for identifying social class membership. In the language of our model and as illustrated earlier in Table 20.1 and Figure 20.1, most people receiving high levels of consumption capital over time (e.g., individuals and families in the top one-fifth of all income groups) are likely to be members of the *privileged class* and those receiving extremely high levels of income (the top one percent) are likely to be members of the *superclass* (the top class segment within the privileged class). People with incomes that place them in the bottom 80 percent of all income groups are likely to be members of the *new working class*.

We believe income differences are relevant to class membership in part because empirical evidence suggests income in the USA is distributed over time to individuals and families in relatively stable patterns that approximate the two large classes in our model. For example, various studies have documented the receipt of high incomes by a relatively stable population made up of the same people over long periods of time—such as a decade or more. Also, recent research on intergenerational mobility suggests that in recent years as compared to a generation ago, a larger proportion of high income positions are being occupied by the adult children of parents who were themselves high income earners. While research suggests substantial movement occurs over time among individuals and families located in the bottom 80 percent of the income distribution, few individuals or families move up from the lower 80 percent into the top 20 percent and remain in the top quintile for long periods of time.

Investment Capital

Investment capital includes material resources owned or held (in full or in part) by individuals and families (as well as by organizations) that generate cash income or other forms of monetary value for the owner(s). If people have a surplus of consumption capital, it can be used to acquire investment capital in various forms such as stocks, bonds, and commercial real estate. These forms of investment capital generate income or monetary value through dividends, interest, rent, and capital gains. Investment capital forms a large portion of the assets held and used by corporations. Such firms produce income and distribute it to holders of investment capital according to the level of capital (e.g., stock shares) each person owns. The possession of substantial levels of investment capital typically yields far higher income levels than individuals need to meet basic needs. Thus, substantial levels of investment capital generate wealth for individuals and families—meaning higher levels of economic assets than the sum of all economic liabilities.

As with high levels of income and as shown in Table 20.1, the possession of high levels of investment capital and wealth by individuals and families is closely associated with membership in the *privileged class*. Those holding the highest levels of investment capital (and receiving the highest incomes) are likely to be members of the *superclass* segment of the privileged class. Persons holding modest or low levels of investment are likely to be members of the *new working class*.

Credentialed Skill Capital

This form of capital is primarily conceptualized as linking the prestige or eliteness of universities and the credentials they confer with the "organizational affiliations" (types and levels of organizational positions) graduates are likely to be afforded as employees or as holders of other positions within organizations. Our view of this resource differs from traditional notions of education and skills. Skill capital is generally discussed as the specialized knowledge that people accumulate through their work experience, training, or education. Conventional views of skill capital (or human capital) see it as being exchanged in the labor market in much the same way that investment capital is used in connection with the financial market. However, *credentialed skill capital* does not simply refer to the accumulation of years of education or work experience. The significance of this form of capital to the acquisition of high levels of other forms of capital and to placement in the class structure depends upon *where* the credentialed skill was obtained.

We maintain that elite universities are the most important sources of credential skill capital today in terms of generating and reinforcing inequalities in consumption and investment capital. The credentials provided by such institutions are increasingly necessary for advanced placement into upper level occupational positions with lengthy career ladders in resource-rich organizations. Such positions provide substantial income and opportunities to accumulate investment capital.

Social Capital

Following Lin, we view social capital as "assets in networks" (1999:28). The network of social ties that people have to family, friends, and associates is social capital that can provide emotional support, financial assistance, job information, and referrals to employer organizations. The basis of social ties can be found in kinship networks, family friendships, school affiliations, membership in social clubs and/or religious groups, political contacts or affiliations, and occupational links to other workers, professionals, and administrators. In our model, social capital is viewed as varying in terms of the *quantity* (number) of social ties that individuals have with others and with organizations and in terms of the *quality* and the duration of those ties. Quality refers to our view that the people or organizations individuals are linked to possess different levels of prestige or eliteness within social groups located in the economic, political, and cultural arenas. Duration refers to the length of time over which social capital ties are maintained.

An important qualitative feature of social capital is illustrated by people's involvement in social networks whose members have links to varying levels of organizational power, prestige, and investment capital. One's position in such networks, especially those with high levels of prestige, can provide access to information and opportunities that may be converted into important financial and social benefits. For example, individuals who hold paid or unpaid leadership positions in economic, political, or cultural organizations can use the social, economic, and political connections such positions provide to advance their personal careers and to create opportunities for friends or relatives. Also, positions in high-level corporate social networks (e.g., senior officers, directors, upper-level staffers) can facilitate access to high levels of investment capital. Positions in resource-rich kinship networks can lead to similar outcomes (e.g., the billionaire offspring of Wal-Mart Corporation founder, Sam Walton)....

CONCLUSION

As we stand at the cusp of the twenty-first century reviewing the past and anticipating the future, we think our approach illustrates how insights from production models can be fruitfully combined with emerging interest in organizational resources and distributional processes to better describe the US class structure and better understand how organizational structures, processes, and resources are crucial in creating, sustaining, and transforming the US class system.

REFERENCE

Lin, Nan. 1999. "Building a Network Theory of Social Capital," *Connections* 22(1): 28–51.

21

Black Middle-Class Suburbia

KARYN R. LACY

One group that has managed to attain movement into moderate-income brackets, without attaining the wealth about which Wysong and Perrucci spoke, is the Black middle class. Lacy's research on this group takes place outside of Washington D.C. and describes a population made up of professionals with incomes in the $60,000–70,000 range. Lacy sensitively describes the issues that confront this group as they move between White middle class occupational worlds and Black communities. How they balance these often divergent cultures and strive to help their children maintain the upward mobility they have achieved poses a challenge that families resolve differently. While they acknowledge that their children need to be able to navigate in White America to be occupationally successful, they do not want them to lose their roots in their Black culture. Some help their children sustain their "Blackness" by living in Black middle-class communities and going to Black schools, while others opt to integrate their children more fully, living in White neighborhoods and/or sending them to White schools. Lacy describes the steps these families take to ensure their children's acculturation into the language, values, and connections of their African-American heritage. Have any generations or groups in your family achieved upward mobility? Has this put you into a cultural clash with other family members? Do you, or people you know, move between social, ethnic, religious, or occupationally diverse worlds? To what extent do you hold onto the values and traditions of your ancestors? Of the two choices Lacy describes, which path would you take for yourself and your family?

Native-born blacks have been portrayed historically as "unassimilable," while white ethnic immigrants—and more recently, immigrants of color—are studied in terms of their attempts to assimilate over generations. Although

SOURCE: "Black Spaces, Black Places: Strategic Assimilation and Identity Construction in Middle-Class Suburbia," by Karyn Lacy, *Ethnic and Racial Studies*, 27(6), 2004, pp. 908–938. Reprinted by permission of Taylor & Francis Ltd., http://www.tandf.co.uk/journals.

researchers debate how best to characterize the assimilation patterns of the latter groups, there is little disagreement about the status of native-born blacks. According to the prevailing view, blacks have undergone cultural assimilation or acculturation, but racism impedes their structural assimilation, that is, integration into mainstream "social cliques, clubs and institutions ... at the primary group level" (Gordon 1964, p. 80), leading some theorists to the conclusion that the assimilation model is most useful for understanding the incorporation of voluntary immigrants, not native-born blacks who entered the US involuntarily and were selectively incorporated through enslavement, coercion and dehumanizing Jim Crow laws.... Simply put, the classic assimilation model assumes that a black racial identity is a liability in the US.

In contrast to the classic assimilation model, in which voluntary immigrants are said to naturally aspire to a single assimilation destination—the white, Anglo-Saxon mainstream—the theory of segmented assimilation advanced by Portes and Zhou (1993) posits that three paths of adaptation are possible for immigrants of color. Ostensibly, one destination is the white middle class, but because immigrants of color grapple with racial discrimination when they enter the US, they are more likely to either identify with the black underclass or carve out a third path by deliberately retaining the culture and values of their immigrant community (Portes and Zhou 1993). Along this third path, immigrants rely on their ethnic communities for social capital, employment leads, and relief from discrimination. In short, segmented assimilation theory characterizes an ethnic identity as an invaluable resource.

In this article, I propose that a variant of the third path of segmented assimilation applies to middle-class black Americans. How middle-class blacks think about their integration into the American mainstream is a topic that has been relatively neglected in the sociological literature. Indeed, a central criticism of segmented assimilation theory is that it, too, characterizes a black racial identity as a liability. The theory presents poor blacks (and their downward mobility trajectory) as representative of all black Americans, discounting middle-class blacks as a potential path of adaptation for immigrants (Neckerman et al. 1999). While Neckerman et al. (1999) outline an additional black middle-class path as a corrective, their primary purpose is to demonstrate that the black middle class is a realistic destination for immigrants of color, not to assess the effect of class position on the assimilation preferences of blacks.

I explore the relevance of an assimilation model for understanding the complexities of the black experience by examining middle class blacks' concerns about nurturing black racial identities. Specifically, I generate a theoretical framework to explain the processes by which middle class blacks from two suburban communities negotiate the racial dualism that accompanies having to shift incessantly from the black to the white world. Like immigrants associated with the third path of segmented assimilation, many middle class blacks with access to majority white colleges, workplaces, and neighborhoods continue to consciously retain their connections to the black world as well; through their interactions in these black spaces, middle class blacks construct and maintain black racial identities. I refer to this segmented form of incorporation as *strategic assimilation*.

To understand how strategic assimilation processes operate, I draw on data collected in 1997 and 1998 through in-depth interviews with thirty middle class black couples and ethnographic study in two distinct middle class suburban communities in the Washington, DC metropolitan area: Lakeview, a predominantly white suburb and Riverton, a majority black suburb.[1] I focus specifically on the symbolic distinctions drawn by respondents, as a means of marking off the black world and as a strategy for negotiating the transition from the black to the white world....

METHODS

The Suburban Sites

The data presented in this article were collected in 1997 and 1998 through in-depth interviews and participant observation in two middle-class suburban communities. Using the 1990 US census and literature on suburbanization trends, I selected Washington, DC, a metropolitan area with significant rates of black suburbanization and a sizeable black middle class. Comparing the racial composition, educational attainment, median income, and median mortgage payments characterizing DC census tracts, I identified a majority white and a majority black suburb. A census tract I dub Riverton met my criteria for the black site. In 1990, Riverton was 65 percent black, 21 percent of residents were professionals, and 23 percent were college graduates or above; median family income was $66,144, and the median mortgage payment was $1,212. A tract I dub Lakeview met my criteria for a majority white suburb. In 1990, Lakeview was 4 percent black, 31 percent of residents were professionals, 44 percent were college graduates or above; median family income was $78,907, and the median mortgage payment was $1,242. The mean individual income for my sample is about $72,000. I refer to respondents as "middle-class" rather than upper middle-class because this is how they define themselves.

Riverton is a mostly black lower to upper middle-class bedroom suburb. Prosperous blacks began to move there in the late 1970s, settling in subdivisions abandoned by whites. Over time, as older housing changed hands and inexpensive newly constructed housing was constructed, lower-middle-class blacks began to move in. As a result, living in Riverton means that residents come into close contact with a broad range of middle-class blacks. Blacks began to move into Lakeview, directly across the river from Riverton, in the late 1960s and 70s, but in much smaller numbers. Lakeview subdivisions are established, older communities and middle-class blacks are a marginal presence there. Consequently, blacks in Lakeview are less likely to encounter and interact with other blacks from their neighborhood....

The Desirability of Racial Integration

To properly understand the assimilation preferences of these middle-class blacks and the significance of black racial identities, I examine their perceptions of discrimination. Such perceptions reflect the broad racial backdrop against which identity construction processes are carried out. Studies designed to assess various indicators of racial progress lead to vastly different understandings of the persistence of racial hierarchy in the US. One set paints a bleak picture ... drawing our attention to ongoing disparities between blacks and whites, in wealth ... in access to housing ... and in the public sphere.... However, a more optimistic portrait, painted by a different group of social scientists, points to indicators such as improved black educational attainment ... and a narrowing of the black-white wage gap ... as measures of racial hierarchy deconstruction.

Widespread disagreement among scholars as to the nature and character of racial progress in the US underscores the need for greater attention to middle-class blacks' understanding of their integration into American society. Attitudinal studies have made considerable gains in this regard. As Bobo makes clear, "race relations involve more than just external and material ingredients like housing patterns, incomes, or educational attainments ... [t]here are also cultural, intra-personal, and interpersonal aspects that affect the subjective experience of Americans" (Schuman, Steeh and Bobo, 1985, p. 12). These researchers report the majority of blacks believe that racial discrimination has not diminished, but increased, and that the pace of racial progress is painfully slow. Moreover, this grim assessment is fairly pervasive among middle-class blacks. Wealthier blacks report more discrimination than poor blacks, are less optimistic about racial progress, and are less inclined to believe that whites positively type blacks.

To be sure, middle-class blacks' willingness to assimilate into the American mainstream is tempered by their belief that racial discrimination has not waned. How does this sentiment expressed among national samples of middle-class blacks resonate with Riverton and Lakeview blacks? Both populations of blacks realize that the primary locus of interaction for their children will be the white world where they will work, shop, and, possibly, live. Thus, both groups are concerned with preparing their children to function in predominantly white settings. Indeed both groups report support (sometimes lukewarm) for indicators of assimilation such as having their children marry interracially or join a white fraternity or sorority. However, this general support is undercut by their awareness of ongoing racial discrimination. Often relying on their own experiences as a gauge, respondents fear that their children will endure unnecessary hardship in these kinds of intimate interactions with whites. This perspective is typified by Philip, a Riverton resident, who explained how he would react if his young daughters were to marry a white man when they grew up.

> I'd talk to them about how interracial couples fare in this society, some of the drawbacks about being married to a white person, how you're not going to be able to relate to that family like you'd like to. Going to family functions is going to be real tough ... when you meet rednecks in your [husband's] family. And regardless of what their economic status is, you're still going to have that. And again, once all these things are made clear to them, if they still want to do it ... after I've painted that picture for them, then hey, it's on them.

Isabelle, a Lakeview resident, expressed similar concerns—and resignation. "I would not rule out that all three of [her children] at some point in their lives will date someone white," she lamented, "And I'm not gonna sit here and lie and say that that thrills me, because it doesn't."

Riverton and Lakeview blacks also are resistant to their children joining a white fraternity or sorority, although just as they would not intervene to prevent their child marrying a person of another race, they also would not preclude their children, once adult, from seeking membership in a white organization. Their concerns range from their own negative experience with fraternities to the ulterior motives of whites to the importance of black social spaces....

Though generally opposed to white fraternities and sororities, few respondents objected to their children attending a white college. Most attended white colleges themselves. Laughing, Terry, a Riverton resident who attended a white college, exclaimed, "I went to one! What's the difference?" Michael joked that his child could go to a white college "if he got the money!" [he laughs]. Then turning serious, Michael favors attending a white college because these institutions provide access to influential social networks.

> My wife went to [a white college], she says she missed some of the black experience.... I went to [a black college], I probably had too much of the black experience! ... I don't see it negative, I think they're equal. You go to an all-white school, you're rubbin' ... elbows with some of the next day's leaders ... not that black people can't be leaders, but, [my wife] was telling me ... she was talking to people who had maids and servants, and stuff. Now, I didn't know anybody like that.

Richard believes attending a black college is a good thing to do before going on to a white graduate school.

> I think at an undergraduate level, it's good to experience that black college university environment. I think it's nothing like it. Best four years of my life. Now when you think about going to graduate school, then I think that might be the time to get to the best school. I kind of took that path myself. I went to graduate school at [a white university]. And undergraduate, I went to [a black college].

And Michelle insisted that she would compel her children to attend a white college because that is the "real" world. "I expect my kids to go to white colleges. Because you need to know what real life is like. I don't think black colleges give you that experience. I went to a predominantly white college." When her ten-year-old daughter who is sitting in on the interview asks, "Are you saying that if I wanted to go to Spelman you wouldn't let me?" Michelle responded, "Not gonna happen."

Riverton and Lakeview blacks share a "cultural orientation" with respect to the degree of social integration they desire with whites (Barth 1969). Both groups report that a current of racial discrimination undergirds blacks' interactions with whites. Consequently, respondents are "not thrilled" about the possibility of their children engaging in interracial marriages. They want their children to marry blacks and to socialize with blacks. Using these measures of social integration, residents draw sharp boundaries around the black world. Yet, respondents are open to their children attending white colleges. Both groups perceive this form of integration as an important step on the path to economic success. Moreover, attendance at a white college or university may facilitate blacks' ability to move back and forth from the black to the white world in adulthood....

Growing up around Blacks: The Desire for
Racially-Distinct Spaces

Marked changes in the modern period such as growth in the percentage of the black population that is middle class and the introduction of Civil Rights legislation in 1964 and 1965 have led to a shift in the social meaning of black spaces.... Before desegregation, middle-class blacks maintained their own exclusive, black spaces because they had been denied full access to white institutions. While the *legal* barriers to most white spaces have now been dismantled, many *informal* barriers remain. Under these conditions of marginality, a middle-class black affinity for black spaces is still essential—to black identity formation. Charlotte, a Lakeview resident, summed up nicely the modern dilemma of identity negotiation: "[W]e live in this world, and it's kind of like being black, you know you have two faces. So you know how to present yourself in the white world, and you present yourself in the black world as yourself." This contemporary enactment of DuBois's ([1903], 1961) *double consciousness* construct reinforces the idea that black identity is bound by external *and* internal determinants.[2] Theorists who insist that black identity is entirely externally derived inappropriately give more weight to the way in which blacks are perceived by others than to how blacks perceive each other.... In this section, I demonstrate the importance of black spaces as crucial "construction sites" (Cornell and Hartmann 1998) for the formation of black racial identity. I begin with Riverton, a geographically black space.

Riverton

Michael, a corporate manager and 10-year resident of Riverton, reveals,
I can tell black people that didn't grow up around other black people, 'cause they act different.... I haven't been able to put my finger on it. It's either the expression, the way they give five, I mean it's just *something*. They missed something. Well, I shouldn't say *missing,* but they are *lacking* something. And that's not positive or negative, they just don't have an ingredient. Not that they aren't black, but ... they're just missing something.

Ironically, the view that blacks who did "grow up around other black people" hold a more salient racial identity than those who did not is shared by blacks from both Riverton and Lakeview. Both groups of blacks believe that racial identity is sutured primarily through social interaction in the black world, and that blacks who miss this experience fail to interpret correctly the cultural cues group members use to draw boundaries around the black world; these blacks are reduced to the status of "outsiders" (Merton 1972). The key distinction is that respondents disagree about what growing up around other black people means. Riverton blacks mean that their *neighborhood* serves as the "construction site" (Cornell and Hartmann 1998) for black identity. Their children develop an insider's sense of what it means to be black as they learn cultural cues through interaction with black neighbors. Because they are immersed in a mostly black environment, Riverton residents tend not to discuss racial identity outright with their children, under the assumption, as Greg explains, that "they learn on their own."…

> It's really hard sometimes for them to get a conscious understanding of what black is … It's not so much, "we doing this, we doing that," but they know that [their neighbor] is a judge, and they can see the different types of interactions and the people that come to the house. So … I feel that they will just grow up being who they are … it's not so much they're wearing blackness on their chest … walking around talkin' 'bout "I'm black and I'm proud," … if we lived in an all-white neighbor-hood, we'd probably have to do that, you know. But we don't because they just see it.

Riverton blacks ontologize the black experience (Anderson 1995), con-structing blackness as just "being who [you] are." But, this standpoint differs from the essentialist platform used by gender theorists to describe the motivation for gender-based social movements such as feminism (Fuss 1989). Rather than arguing that the essence of blackness is the socially constructed biological com-ponent, Riverton blacks argue that race is a given when they are ensconced in their heavily black suburb and that it can therefore be set aside. What remains is the identity they or their children choose to put forth in neighborhood interac-tions. John complained,

> You know, I hear people talking about, "that little girl don't know she's black." To me, that's the most ridiculous thing you could ever say! How am I, I mean, should I teach [his daughter] about being a woman? Should I teach her about being an earthling? No. I don't see any point in that. Now, I know a lot of people would criticize me for that. She is going to be black. There is nothing that I can teach her that's going to prevent her from being black. Or help her to be black.

Riverton blacks find that blackness is actualized as they socialize in a black environment. During their immersion in a black community, Riverton children prepare for their transition into the white world. "I want my children to be able to leave my house strong enough to make it in the [real] world," Lydia, a Riverton resident, told me. Lydia and her husband's residence in a predominantly black suburb is a significant part of their children's training.

To "make it" in the real world, Riverton blacks realize that their children need to be prepared to travel back and forth from the black to the white world.

Yet, the term "black world" does not refer solely to a *spatial* community of black residents (Wellman 1999). The black world is also an "interpretive community" (Gilroy 1993) with boundaries defined by a cultural repertoire that middle-class blacks rely on to define the black experience. It is this interpretive community that black middle-class residents of Lake view access in the construction of a racial identity.[3]

Lakeview.   At first glance, it would appear that Lakeview blacks have sacrificed their ties to the black community in order to take up residence in a white suburb, but this view is inaccurate. "When I moved here, it wasn't one of those things that I tried to move away from blacks." William, a social worker and five-year Lakeview resident informed me, "It was one of those things where I tried to move to a place where the kids could get the best of what the county had to offer." In other words, William chose the suburb where he believed his children would have access to the greatest amount of resources. Since they live in a highly mobile society, Lakeview blacks reason that they need not live near blacks in order to socialize with them. "I wasn't as concerned about how many blacks were gonna be living out there [in Lakeview]," Alana, a corporate attorney and Lakeview resident for twenty-two years explained, "because I used to say that would be my job, to supplement my kids with exposure to other blacks. So that was not a factor, and is still not, because there are a lot of other ways that they can interact with blacks."

What does concern middle-class blacks in Lakeview is the same thing that troubled Riverton blacks. Blacks in both suburbs struggle with how to prepare their kids for the "real world." Riverton blacks focus on how they see themselves in relation to their (majority) black neighbors. Lakeview blacks, whose children spend more time negotiating interactions with whites, stress identification by others, whites in particular....

> Some kids have a difficult time because they are not as comfortable in all types of settings. And they have to get used to that, because that's the *real* world. When you go on a job, if you go to Harvard, clearly you are not going to be in the majority, and a lot of kids, not even just kids, a lot of *people* can't deal with it. They've got to be around their own element.

But when they are not at work or at school, Lakeview blacks socialize in the black world. They maintain contact with this segment of their lives by participating in black religious and social organizations. Lakeview residents are more likely than Riverton residents to attend a black church, hold membership in a black sorority, fraternity, or social group, or to participate in the exclusive, black social organization *Jack and Jill*....

Blacks in Riverton and Lakeview seek a construction site where black identity is nurtured, a community where they can socialize and reconnect with other blacks after spending the bulk of their day in the white world. The key difference in Riverton and Lakeview blacks' affinity for black spaces is their access to these settings. Riverton parents are immersed in a black residential community, and can therefore assume that their children will have ample opportunities to interact with other blacks. Lakeview blacks live in a suburban community where blacks are rare. They have to take additional steps to expose their children to black spaces. Note, the socialization concerns of Riverton and Lakeview blacks differ considerably from those of poor and lower middle-class black parents in predominantly black neighborhoods (see Anderson 1999; Pattillo-McCoy 1999). This preference for socializing in the black world and academic achievement in the white constitutes a pattern of strategic assimilation, in which middle-class blacks make the transition back and forth from the black to the white world, prioritizing black culture and social ties in the black world so as to reconnect with other blacks and to benefit from the resources and social networks exclusive to their group....

NOTES

1. Names of respondents and suburban sites have been changed to maintain anonymity.

2. DuBois intended the term "double consciousness" to reflect the tension he believed black Americans experience as they negotiate two "unreconciled" identities: their status as Americans and their status as blacks. The structure of race relations in American society, he argued, created in blacks a unique "second-sight ... this sense of always looking at one's self through the eyes of others" (DuBois [1903] 1961, pp. 16–17).

3. Gilroy's (1993) "interpretive community" parallels Anderson's (1991) notion of an "imagined community" in which members feel a strong connection to one another on the basis of nationality despite the fact that they may never actually speak to or have contact with most members of their community. Race and ethnicity theorists have appropriated Anderson's concept in order to examine membership loyalties in racial communities.

REFERENCES

Anderson, Benedict. 1991. *Imagined Communities: Reflections on the Origin and Spread of Nationalism*, London: Verso.

Anderson, Elijah. 1999. *Code of the Street: Decency, Violence, and the Moral Life of the Inner City*, New York: W.W. Norton & Company.

Anderson, Victor. 1995. *Beyond Ontological Blackness*, New York: Continuum Publishing Company.

Barth, Fredrik. 1969. *Ethnic Groups and Boundaries*, Boston, MA: Little Brown.

Cornell, Stephen and Hartmann, Douglas. 1998. *Ethnicity and Race: Making Identities in a Changing World*, Thousand Oaks, CA: Pine Forge Press.

DuBois, W. E. B. [1903]. 1961. *The Souls of Black Folk*, Greenwich, CT: Fawcett.

Fuss, Diana. 1989. *Essentially Speaking: Feminism, Nature, and Difference*, New York: Routledge.

Gilroy, Paul. 1993. *Small Acts: Thoughts on the Politics of Black Cultures*, London: Serpent's Tail.

Gordon, Milton. 1964. *Assimilation in American Life: The Role of Race, Religion, and National Origins*, New York: Oxford University Press.

Merton, Robert. 1972. "Insiders and outsiders: a chapter in the sociology of knowledge," *American Journal of Sociology*, vol. 78, no. 1, pp. 9–47.

Neckerman, Kathryn, Carter, Prudence, and Lee, Jennifer. 1999. "Segmented assimilation and minority cultures of mobility," *Ethnic and Racial Studies*, vol. 22, no. 6, pp. 946–65.

Pattillo-McCoy, Mary. 1999. *Black Picket Fences: Privilege and Peril among the Black Middle Class*, Chicago, IL: University of Chicago Press.

Portes, Alejandro and Zhou, Min. 1993. "The new second generation: segmented assimilation and its variants," *Annals, American Academy of Political and Social Science*, vol. 530, pp. 74–96.

Schuman, Howard, Steeh, Charlotte and Bobo, Lawrence. 1985. *Racial Attitudes in America*, Cambridge, MA: Harvard University Press.

Wellman, Barry. 1999. *Networks in the Global Village: Life in Contemporary Communities*, Boulder, CO: Westview Press.

22

Single Welfare Mothers' Resistance

YVONNE M. LUNA

For a long time, our country has stigmatized welfare recipients. In particular, we attach a negative connotation to unmarried women who collect welfare, imagining them as having babies indiscriminately to boost their monthly stipends, often leaving the care of these children to others. We tend, also, to blame the poor for their poverty, rooting its cause in their personal decisions and choices. In this study, Luna looks at the way single mothers who collect welfare manage their stigma and the bureaucracies of the welfare system. How has this chapter challenged or reinforced your prior conceptions (or maybe a little of both)? How do you feel about the way the welfare system and its caseworkers treat these recipients? How do you feel about the way the women attempt to navigate the system and legitimize their manipulations? How does this portrayal of the poor contrast with the middle-class Blacks as described in the previous chapter?

The notion that poverty is caused by behavioral dysfunction may be traced to early American history. Since its inception, public assistance policy has been guided by this ideology of free market individualism that limits the role of government, ensuring equal opportunity for individuals' success through their own efforts. But for single mothers, free market individualism ignores the social value of mothering. Historically, public support for single mothers has been embedded in ambivalent and unstable views of unmarried mothers. Late nineteenth-century Mothers' Pensions defined legitimate recipients as mothers who were widows. When assistance extended to other single mothers, it was accompanied by evaluations of individual worthiness. Women of color were not viewed as deserving and were often excluded. The scrutiny of morality and worth applied to single mothers on welfare is unmatched by similar evaluations of other aid recipients.

Popular conceptions of welfare recipients are congruous with policy that ideologically attributes the responsibility for poverty to irresponsible decisions and lack of effort; there is an acceptance that poverty is self-made. This rationale is related to Oscar Lewis's *culture of poverty*, which provides that norms and values of the poor are dysfunctional, embedded cultural traits (Lewis, 1966). Other culture-of-poverty theorists argue that the poor fail to adopt middle-class values. Most often associated with Black single mothers, the portrayal of poverty as an individual problem often focuses on welfare recipients who are lazy, shun work, and pass on these values to their children. Public opinion surveys confirm this pejorative view of poverty and consequent negative views of welfare. Because of the negative social constructions associated with welfare, to many, the punitive nature of the policy seems justified. As Soss (2005) points out, people who seek Aid to Families with Dependent Children contend with many unpleasant procedures, including waiting for hours for required in-person appointments where welfare caseworkers sit behind protective glass barriers, dealing with intrusive questioning and continual threats of benefit terminations.

The focus of this research is on public assistance recipients who contend with negative stereotypes, "people who believe their welfare receipt makes them objects of public scorn" (Soss, 2005, p. 314). It investigates ways that welfare mothers resist these pejorative constructions. To do this, it is necessary to conceptualize and define *resistance,* then demonstrate how single mothers on welfare use both overt and covert methods of it (Hollander & Einwohner, 2004) to redefine themselves. "Avery Gordon suggests that we loosen the grip of conventional categories of analysis and consider *complex personhood*: both the visible, overt forms of power as well as the invisible, covert forms that shape experience" (Gordon, 1996, as cited in Ferraro, 2006, p. 5). Many of the 64 single welfare mothers who participated in this study engaged in resistance to shape an identity that transcends the stereotype. Welfare mothers should not be viewed as total victims or undeserving slackers but as complex persons with complex lives (Gordon, 1996).

METHOD

Data for this article are derived from focus groups with 64 single welfare mothers living in the central area of a metropolitan southwestern city and personal interviews of a smaller sample. I had no relationship with the participants prior to this study. To recruit them, I met with institutional leaders at several central city community organizations who provided meeting places for the focus groups and allowed flyers to be posted in their establishments. The flyers briefly described the study, asked for participation on specific dates, and explained that participants would be compensated. Ten groups were conducted from 1998–2004, three of which were held in Spanish and English. Personal interviews in 2004 comprised a meaningful sample of eight women to verify the voices of participants in selected focus groups....

All of the participants were raising children, the majority their own (95.3%), while a small percentage (4.7%) were raising grandchildren or other relatives. All were single (i.e., widowed, divorced, separated, never married) and receiving Temporary Assistance for Needy Families (TANF cash assistance), Food Stamps, and/or Medicaid. Thirty-four percent of the sample had been terminated from at least one type of public assistance. They were from diverse backgrounds and ages: 53% Latina, 28% Black, 8% White, 6% American Indian, and 5% mixed race; from late teens to early 60s, although 65% were between the ages of 29 and 39....

DEFINING RESISTANCE

A universal definition of *resistance* does not exist. Hollander and Einwohner (2004) provide an examination of the literature on resistance and demonstrate how its characterizations diverge on two issues—recognition and intent. Their typology provides seven categories of resistance that consider whether the act in question is *intended* as resistance by the actor (in this research, it is the single welfare mother) and whether the act is *recognized* by the target (e.g., the welfare system). Two major categories of resistance in Hollander and Einwohner's (2004) synthesis are relevant to this research—overt and covert resistance. Overt resistance is intended as resistance by actors and is recognized as resistance by targets and observers. Covert resistance is intended as resistance by actors, is not recognized as resistance by targets, but is recognized as resistance by observers. During this investigation, three overt and three covert resistance strategies emerged. Resistance strategies are methods the participants in this study used to minimize the pejorative constructions associated with a welfare identity. The strategies' labels are my creations for analytical clarity.

OVERT RESISTANCE

Overt resistance is not only intended by actors, but is recognized by targets and observers, and in fact, is intended to be recognized as resistance (Hollander & Einwohner, 2004). When people engage in overt resistance, they assume the risk of punishment. The types of overt resistance that emerged in the data were *oppositional culture*—a rejection of dominant views (Foucault, 1981; Scott, 1990), or what I call *fighting the system*—complaining to welfare caseworkers and supervisors and appealing terminations and/or sanctions of benefits, and *maintaining control in a McDonaldized welfare culture*. McDonaldization refers to a process where the principles of the fast-food restaurant are dominating more and more segments of American social structure (Ritzer, 1994). Building on Weber's work, Ritzer (1994) suggests the fast-food restaurant, McDonald's, serves as an organizational model that extends the process of rationalization, whereby efficiency and formalized social control are paramount. In a McDonaldized system, four main dimensions exist: efficiency, calculability, predictability, and control. Efficiency disallows for individuality by streamlining processes that may include the use of technology rather than human interaction. Calculability occurs when quantity over quality is stressed. Predictability is when production processes are organized to generate standardized outcomes. And finally, in a McDonaldized system, employers maintain control over their employees by assigning tasks that limit discretion (Ritzer, 1994).

Oppositional Culture

Employing oppositional culture offers an opportunity for marginalized groups to express dissatisfaction toward the dominant group. The women in this study did this by disregarding outsiders' views and/or embracing the welfare identity rather than avoiding, withdrawing, or dissociating from it. Close to what Cohen (1955) discusses as *reaction formation*, this involves rejection and a reversal of societal expectations.

The welfare system is set up to inhibit people from applying for benefits by reinforcing ideologies of individualism and self-reliance. Welfare recipients are, in essence, punished for their dependence. However, some of the women realized that no matter how much they subscribed to ideals of work and independence, they were unable to achieve self-sufficiency due to institutional barriers. With this realization came a rejection of others' views of them. Patricia, an American Indian, age 23, developed this response over time:

> Now, I guess because too because [sic] I got older and I matured a little more, so now I could care less what people say about me or the circumstance or the situation that I'm living right now. And they're living their own lives, so what they say does not bother me. They can make all the comments they want. I don't care because I'm doing this for my children and for myself and I'm getting by....

Rejecting others' judgments was expressed by a number of respondents. More than indifference to others' views, some expressed pride for those who receive public assistance. Essence, a 34-year-old Black woman, said, "Food Stamps. That's the proudest time of my life. Girl I don't got no problem with it." In the same discussion, when Lolly, a 38-year-old Latina, said she was on welfare, Valerie, a Black woman, age 26, responded with "Bless you honey. You go girl!" Being proud of one's welfare use is unique, as research has documented the public's animosity toward welfare, which may result in its recipients using strategies to protect themselves from the stigma. Nonetheless, some recipients expressed abhorrence toward those with negative views of them:

Q: How about the welfare workers?

VALERIE AND ESSENCE: They suck.

ESSENCE: They act like it's coming out of their pocket. They act like when they get their check, they are going to each one of their clients' houses and say okay, "here's your 50, here's your 50," and they ain't giving me a dime.

When stigmatized groups engage in oppositional culture or reject negative social constructions of themselves, they express their dissatisfaction with dominant ideals. Another way that the welfare identity is embraced is when recipients form an oppositional subculture of survival.

Goffman (1963) defines *the own* as "the first set of sympathetic others is of course those who share the stigma" (p. 20). Some of the participants felt an affinity with other welfare recipients; a few viewed other welfare recipients like their own families. Yvette, a 37-year-old Black woman said, "Some of my, I call 'em all my brothers and sisters...." Later, she shared:

> So when you move from one area to the next, and it's sad to say, but you're comfortable with your own kind. So it's like everybody that was on welfare has a race of their own. If that makes any sense? They have their own identity. But don't go across the street, 'cause if you go across the street, they gonna badger you ... we [welfare recipients] band together. We pull together. We hustle together, you know ... That's how a family works....

Some welfare mothers develop a connection and will disclose their identities to one another—*the own*. In doing this, the women oppose dominant culture by refusing to internalize the stigma and shame that accompanies public assistance. Others, however, will fight the system when they feel they are being treated unjustly.

FIGHTING THE SYSTEM

What I refer to as *fighting the system* is the clearest example of overt resistance. Although most of the study sample accepted, ignored, or adapted to negative treatment by agents of the system, a few fought back. They complained and appealed terminations and/or sanctions of benefits. These women intended for their actions to be recognized by agents and supervisors of the welfare system as oppositional, and they assumed substantial risk of punishment. Cathy, Shonna, a 25-year-old American Indian, and Janet, a 50-year-old Black woman, share:

Q: Do you have any advice for women in situations similar to yours?

JANET: Speak your mind. Speak your mind. When something's not right let them know.

Q: Does that help?

JANET: Sometimes you don't want to speak your mind because you're scared you might lose this or lose that.

CATHY: Be willing to stand up and fight back instead of letting them push you around like your nothing.

SHONNA: Don't be afraid of, to voice your opinion.

Like Janet, Cathy, and Shonna, Nancy, a 39-year-old White woman, believes in the power of this type of resistance:

> So a lot of them out there, you just have to call, you have to complain, that's the problem. Sometimes we don't know who to call, what agencies to call, even if you have to call legal services. Call legal services. The squeaky wheel gets oiled.

In a similar vein, Chris, 19-year-old Latina, describes the need to "go over the head" of the caseworker and talk to a supervisor when benefits have been terminated:

> They [welfare caseworkers] tell you, "you're going to be cut off by this if, and if you don't have this within a certain day, you're going to be cut off," but they don't do that, they just cut you off and then you get this letter in the mail. You've been cut off for no reasons and then you go over their head and talk to a supervisor....

Although fighting the system offers marginalized groups an opportunity to oppose dominant structures, speaking up may pose substantial risk, as Essence later shared:

Q: Have your benefits been cut?

ESSENCE: I was cut off, dismissed, see ya, peace.

Q: Why?

ESSENCE: You want the truth?

Q: Yes.

ESSENCE: Cause I'm Black, and I speak my mind.

Risking termination of benefits by confronting welfare system agents is an unquestionable act of resistance with substantial risk. Nonetheless, fighting injustices meant complaining to caseworkers and supervisors or appealing sanctions and/or terminations of benefits. Many women also employed resistance tactics that allowed them to maintain control.

Maintaining Control in a McDonaldized Welfare Culture

The concept of McDonaldization has been applied to several institutions, like education (Hartley, 1995), the criminal justice system (Bohm, 2006), and even exotic dance clubs. This same phenomenon is applicable to the welfare experience. The 1996 reform of welfare creates rules and regulations that *McDonaldized* the welfare system. Although states have flexibility in implementation of those rules, everyone must follow them. Application processes and management of welfare caseloads are efficient, calculable, and predictable. There is little room for flexibility, and welfare caseworkers have little discretion. Many welfare mothers engage in tactics that allow them to maintain control over their lives and keep their benefits. These strategies include untruthfulness and refusing to comply with formal and informal rules....

Edin and Lein (1997) argue that the rules of welfare policy necessitate untruthfulness. Yvette's comment puts this into perspective:

> They [welfare agents] could care less, you know…. They add that pressure to you. Then that's when it starts breaking you down. And that's when you get to the point where I'm just going to stay on public assistance if I got to lie, con, steal, cheat, whatever it's gonna take, I'm gonna stay on it.

One way dishonesty is forced is when welfare recipients are required to report all of their income. If they do so, however, they run the risk of their benefits being reduced or terminated, as Nancy, 39-year-old White woman, describes:

> Another thing that they do too is if you do get any help from your family, if they buy your children shoes, or the family … sends your children clothes, welfare wants to know about that! … You can't even get gifts from your family?! Your family can't help you out at all whatsoever?! But they will ask you that and like Lesa said, sometimes you got to be dishonest. You don't want to be dishonest, but you have to.

Since the policy requires that recipients work, Vanessa, a 38-year-old Hispanic woman, asked her supervisor to lie for her about the number of hours she was working:

> I was working part-time, like you say, and I was getting AFDC but I was only getting like a hundred and something, and I was only getting paid every 2 weeks. They told me I could work part-time and get AFDC. Well I was working part-time and then they were saying I had to work more than 36 hours a week. If I didn't, they were going to cut it off. So what I did, I had to get a note from my supervisor saying I was working more hours….

These women's experiences reflect how the system *creates* the need to be untruthful. As Dodson and Schmalzbauer (2005) note in their study of poor mothers' habits of hiding, many women will tell welfare workers what they want to hear. Essence says that welfare forces them to find ways to "beat the system," but when too many do this, additional punitive policies are put in place:

> You know what they make you do? They make you find ways to beat the system. And I don't mean fraud or nothing like that, but ways to beat the system and, but then when you find out how to beat the system and you tell somebody else how to beat the system and they find out that somebody done found out how to beat the system, then you in trouble. Then they adjudicate a law covering this.

Essence's comment is representative of the predicament of welfare mothers. In order to keep their welfare benefits, these women must engage in deceitful behavior.

COVERT RESISTANCE

James C. Scott (1985) argues that *everyday resistance* by powerless people who do not have the resources to resist, do so by denying or mitigating claims made by appropriating groups. Covert resistance is intended as resistance by the actors (single welfare mothers). However, it is often unnoticed and unpunished by the targets (agents of the welfare system), but may be recognized as resistance by culturally aware observers (Hollander & Einwohner, 2004). When discussing their experiences as welfare recipients, the women in this study revealed several covert resistance strategies, including avoidance, withdrawal, and dissociation.

... Agents of the welfare system may recognize these covert acts, but may not define them as resistance nor have authority to wield punishment. This is different than overt resistance strategies that risk punishment.

Avoidance

Avoidance occurs when individuals with negative or stigmatizing identities avoid situations where their identity may be problematic.... Seccombe (1999) shows how they engage in several *stigma management* strategies including grocery shopping at off hours. Although many of the participants had similar stories, Amy, a 28-year-old Black woman, said "I wouldn't go to the grocery store there [neighborhood where she worked] and use my card, because they look at you funny, or they're like, 'oh, are you ... using your EBT card?"...

Shopping at grocery stores seems to be the primary occasion at which a welfare identity may be made known; visible locations tend to engender the most feelings of stigma. Yvette discusses how her welfare identity is obscured until she uses her EBT card:

> People don't know, majority of the time that I'm on welfare, until they see me bring out my EBT card.... My kids used to go to the grocery store with me. They don't do it no more. That hurts, you know.

... Monica is conscious of paying with Food Stamps. She explains, "Most of the time when I get in line, Yvonne, I look around to see who's getting in the line behind me. Or I'll wait and let other people go ahead of me."...

Withdrawal

Strategies of withdrawal occur when individuals with stigmatized identities disengage from their identities by disclosing them only to select groups. As opposed to forming an oppositional subculture of survival, some women shared how they only feel comfortable disclosing to families and close friends—what Goffman (1963) refers to as *the wise,* "... namely, persons who are normal but whose special situation has made them intimately privy to the secret life of the stigmatized individual and sympathetic with it ..." (p. 28). Some participants said they feel a sense of understanding by family and friends. Patricia expresses this idea:

> I have my parents. I have my sisters and a brother. They all get together and help me. Like last night, my Dad and my sister came by ... I was frying my girls' egg sandwiches because I didn't have anything ... and my sister was like, "What are you making your kids?" and she saw what I was cooking ... She went and bought all my kids hamburgers and she bought me a hamburger too.... They are always there for me, my family and when they come and see that I'm in need, they'll help.

Like Patricia, a majority of the women in this study depend on their families and friends for financial and emotional support. This type of kinship network is documented in earlier literature. Stack (1974) noted the importance of these sharing patterns among Black families that allowed them to survive unpredictable external forces....

DISSOCIATION

Dissociation is the act of separating oneself from the group to which one belongs. It is an attempt to "pursue alternative identities that do not evoke the damaging, painful discourses associated with stigma ..." (Williams, 2008, p. 250). Two patterns of dissociation emerged from the data—what I refer to as *origin* and *continuity stories*. These ideas dovetail with Lyman and Scott (1970), who discuss the sociology of talk—*accounts* are talk that is subject to valuative inquiry. Two types of accounts—excuses and justifications—are invoked when someone may be accused of doing something wrong. In this study, *origin stories* are used to explain why the person became a welfare recipient, whereas, *continuity stories* focus on why the individual continues to receive welfare. In line with Lyman and Scott (1970), origin stories relate to what they refer to as *excuses*, where people admit their actions are unseemly in some way, but deny responsibility. Continuity stories are close to what Lyman and Scott call *justifications* where obligations are seen as reasons for continuation of negative actions (here, the receipt of welfare).

Origin Stories Origin stories help us understand how the women make the decision to apply for welfare. By doing this, it allows them to dissociate from the stigma associated with the welfare identity. Seccombe, James and Walters (1998) documented how welfare mothers' views of other welfare recipients are congruous with dominant constructions that tend toward an individualist perspective. When explaining their own reasons for turning to public assistance, however, they tend to focus on structural or fatalist accounts, such as lack of employment, transportation or childcare, or falling victim to domestic violence. Many women in this study used structural and fatalistic factors to explain why they applied for welfare. They used accounts of children and adverse situations with men, while others shared that work alone could not meet their needs. Some also discussed pregnancy as a catalyst. Rita's (a 23-year-old Latina) statement reflects her negative view of welfare recipients, but believes her circumstance is different:

> Well at first I would look down on people like when I was younger, you know before I had my child, and I would see people take advantage of it [welfare] and I'm like, no, people can work. And then when, you know, the tables turned, then I saw it as, you know, it was a way that I was able to be with my baby all the time.

… Other women exempted their behavior by explaining how male partners and/or fathers of their children created untenable situations. Sokila says, "My husband left me and he wouldn't help me, so I had no other alternative but to go to welfare." Stacy, a 37-year-old White woman, and Silvia, a 38-year-old Latina, shared how they were self-sufficient until the wrong men entered their lives….

While many of those who dissociated focused on fatalist accounts, Yvette and Nancy cited lack of jobs and high cost of living as reasons for turning to welfare:

Q: Why do you think you are in the situation that you're in?

Yvette: Lack of stable employment and no, what's that thing for like the house go up and the money stay down?

Nancy: Oh yeah, the cost of living.

Q: Anybody else?

Nancy: Lack of a job and also lack of my children's father.

Because of the negative connotations associated with welfare receipt, many single mothers use origin stories to mitigate societal stereotypes. For those who remain on welfare, dissociation also provides justification for continued involvement in the welfare system.

Continuity Stories   Negative actions may be initially excused but continuation often requires justification. Continuity stories are close to what Lyman and Scott (1970) call *justifications*, where obligations to others are perceived as reasons for continuation of negative actions. Although reliance on public assistance comes with a stigmatized, devalued identity, welfare mothers often engage in *good mothering* to resist the stereotype of the *bad welfare mother*. Shonna, a 25-year-old American Indian, Cathy, and Maria discuss mothering responsibilities and further dissociate by claiming they are not like other welfare recipients:

Shonna: But see the help is so limited. They got benefit caps, you know, and I can understand that a lot of people they do have babies just so they can have a fatter check. However, it took me 5 years just to have my second child. And I don't even get AFDC for her. I just get AFDC for one child.

CATHY: I can understand that there's people out there that are just sitting on welfare. But people like myself, I'm waiting for Social Security to go though … so that's my only way to support my kids…. And I've been classified totally disabled the last 10 years.

MARIA: The last time I was on it, I took my own self off and I went and got a job. But you know now I'm getting to where I'm getting bigger [pregnant] now and it's hard for me.

Similarly, Strata, a 22-year-old Latina, justifies her reliance on assistance from the government. She clarifies that, unlike the stereotype of lifelong dependency, public assistance is a temporary plan, "I have a baby to support and getting cash assistance is not something I plan on doing for the rest of my life. And I do plan to go back to school and study." Fawna, a 42-year-old American Indian, like Strata, says her dependence on public assistance is just until she is able to get back on her feet. She dissociates by proclaiming that she is "not gonna abuse it."

Others resist the stereotype of laziness by proclaiming that they would rather work. Some of the participants, particularly immigrants, discussed employment woes, but proclaim a good work ethic by saying work is preferable to public assistance:

BRISA (LATINA, AGE 24):   I don't like it [welfare], but I don't have a choice because in my case, I can't work because of my kids. I need to ask for help. What else can I do? I don't get help from nobody [sic].

MARIBEL (LATINA, AGE 29):   I do feel bad because they think that the money goes to us and not our kids. And then they want to know how we spend our money. I would rather be working.

CECY (LATINA, AGE 32):   My daughter is 8 years old and it's been 1 year that I've been asking for help [receiving assistance]. I didn't want to ask for help. If we had the opportunity to work and make our own money that would be better…. We do feel very bad because people look down on us, but we ask for help because we need it.

The participants who engaged in dissociation as a resistance strategy tended to do so by saying that their decisions were due to situations out of their control….

IMPLICATIONS

… Even though the resistance strategies revealed in this research are specific to welfare mothers, the process of resistance can be seen as trans-situational and temporally durable. The covert and overt resistance strategies discussed here transcend the welfare identity and are tools for managing stigma for other marginalized groups.

Resistance strategies for managing stigma tend toward individualized efforts. In light of pervasive, free-market individualism where individual merit is viewed as the key to independence, we cannot expect marginalized groups to effect change within systems. These women's lives are complex, and the welfare system should serve their unique needs. This type of change is possible with the combined efforts of scholars, practitioners, and policymakers where the lived experience of welfare recipients is central to the discourse.

REFERENCES

Bohm, R. M. (2006). McJustice: On the McDonaldization of criminal justice. *JQ: Justice Quarterly, 23*(1), 127–146.

Cohen, A. C. (1955). *Delinquent boys: The culture of the gang.* Glencoe, IL: Free Press.

Dodson, L., & Schmalzbauer, L. (2005). Poor mothers and habits of hiding: Participatory methods in poverty research. *Journal of Marriage and Family, 67,* 949–959.

Edin, K., & Lein, L. (1997). *Making ends meet: How single mothers survive welfare and low wage work.* New York: Russell Sage Foundation.

Ferraro, K. (2006). *Neither angels nor demons: Women, crime, and victimization.* Boston: Northeastern University Press.

Foucault, M. (1981). *The history of sexuality vol 1: An introduction.* New York: Vintage.

Goffman, E. (1963). *Stigma: Notes on the management of spoiled identity.* Englewood Cliffs, NJ: Prentice-Hall.

Gordon, A. (1996). *Ghostly matters: Haunting and the sociological imagination.* Minneapolis, MN: University of Minnesota Press.

Hartley, D. (1995). The *McDonaldization* of higher education: Food for thought? *Oxford Review of Education, 21*(4), 409–423.

Hollander, J. A., & Einwohner, R. L. (2004). Conceptualizing resistance. *Sociological Forum, 19*(4), 533–554.

Lewis, O. (1966). *The culture of poverty.* San Francisco: W. H. Freeman.

Lyman, S. M., & Scott, M. B. (1970). *A sociology of the absurd.* New York: Appleton-Century-Crofts.

Ritzer, G. (1994). *The McDonaldization of society: An investigation into the changing character of contemporary social life.* Thousand Oaks, CA: Pine Forge Press.

Scott, J. C. (1985). *Weapons of the weak: Everyday forms of peasant resistance.* New Haven, CT: Yale University Press.

Seccombe, K. (1999). *"So you think I drive a Cadillac?" Welfare recipients' perspectives on the system and its reform.* Boston: Allyn & Bacon.

Seccombe, K., James, D., & Walters, K. B. (1998). "They think you ain't much of nothing": The social construction of welfare mother. *The Journal of Marriage and the Family, 60*(4), 849–865.

Soss, J. (2005). Making clients and citizens: Welfare policy as a source of status, belief, and action. In A. L. Schneider & H. M. Ingram (Eds.), *Deserving and entitled: Social constructions and public policy* (pp. 291–328). Albany, NY: State University of New York Press.

Stack, C. (1974). *All our kin: Strategies for survival in a Black community*. New York: Harper & Row.

Williams, C. C. (2008). Insight, stigma, and post-diagnosis identities in schizophrenia. *Psychiatry, 71*(3), 246–256.

23

Korean Banking and Korean Immigrant Entrepreneurship

TAMARA K. NOPPER

Nopper paints a very different portrait of Asian immigrants from Chhuon and Hudley's research on Cambodians in this study of Korean-American banking. Although they may not have the higher status of Japanese- and Chinese-Americans, Korean-Americans have been very successful economically in the United States. In fact, Korean immigrants establish their own small businesses at a rate exceeding that of immigrants from any other nation, with more than one-third of all Korean immigrant adults involved in small businesses. Residents of urban areas may be particularly familiar with Korean-American–owned corner grocery store/fruit stands, a niche they have captured and dominated. In using the entrepreneurship available to them through the American capitalist system, they have transformed the Korean cultural identity from that of a lesser Asian group into a more highly respectable one characterized by stability and security. What Nopper's chapter focuses on is the banking industry that structurally underlies and supports Korean-Americans' entrepreneurial opportunities. After reading this chapter you may have a greater understanding of even more elements that go into facilitating upward mobility for various groups in America. What do you think of the collaboration between the government and the banking industry that this chapter discusses? Does people's status in their countries of origin translate into assets in their new countries? How do Koreans stand in relation to other Asian-Americans and why?

Scholars have studied Korean immigrant entrepreneurship for the last thirty years and for good reason. By 1980, Koreans were one of the most concentrated in small business of all ethnic groups, a pattern that persists. Although they

SOURCE: Permission granted by Jean Ait Belkhir, on behalf of Race, Gender & Class, *"The Globalization of Korean Banking and Korean Immigrant Entrepreneurship in the United States,"* by Tamara K. Nopper in Race, Class & Gender, 2009, 16(3–4): 248–270.

make up only 2.7 percent of all immigrants, Korean immigrants' business owner-ship rate is 22.6 percent, second only to those from Greece. Consistent with the national growth in business ownership, the number of Korean firms increased about 16 percent between 1997 and 2002. They currently own one percent of all U.S. firms, fourteen percent of Asian-owned firms, and four percent of businesses owned by people of color. Despite these patterns, few studies have empirically documented how Korean immigrant entrepreneurship is influenced by Korean banks and the globalization of banking. This article addresses this void. Drawing from interviews with 28 Korean bank employees, as well as a representative of the Federal Deposit Insurance Corporation (FDIC), my study explores how banks attempt to replicate Korean cultural practices in terms of service delivery and simultaneously serve as conduits of financial literacy among new immigrants, sometimes in collaboration with U.S. government institutions such as the FDIC. Also examined is how globalization informs banks' transnational approaches to dealing with the limited credit and business histories of recent Korean immigrants. Finally, the intersection between the minority banking industry in the U.S. with the globalization of banking is explored.

DATA AND METHODS

Between 2004 and 2006, semi-structured interviews were conducted with employees of Korean banks and the FDIC located in Los Angeles and New York. Although many cities are home to Korean bank branches and FDIC offices, this study focuses on activities in these two locations for two reasons. First, with the exception of one, all of the Korean banks have their headquarters or holding company offices in both cities; and second, the greater areas of Los Angeles and New York are home to the largest number of Korean immigrants and Korean owned businesses (U.S. Bureau of the Census, 2006). The bulk of data came from Korean bank employees. Hereafter known collectively as "Korean banks," this cat-egory includes both Korean and Korean American banks. Korean banks chartered in South Korea have subsidiaries or branches in the United States. Korean American banks are U.S.–chartered banks located in the United States that were founded and are run by Korean Americans. Regardless of the charter, Korean banks share the common features of having Koreans or Korean Americans comprising at least half of their board members, serving in major senior management positions, and promoting themselves as serving Korean communities. During the period of data collection, there were 14 Korean banks in the United States. Out of these, eleven banks are part of the sample, with three being South Korean. The sample size totals 28 employees. Because of limited funding, only 13 interviews were conducted in person; the rest were conducted over the telephone. With the exception of a few in managerial positions, staff members included commercial and business lenders, international trade departments, credit analysts, Small Business Administration (SBA) lenders, and Community Reinvestment Act (CRA) and compliance officers.

In the body of this article, respondents' quotes are provided regardless of English language grammatical errors. In other words, no quotes were changed but rather reflect what was actually said in the interview.

FINDINGS

This section examines how Korean banks replicate Korean norms of service delivery in the U.S. context, increase the financial literacy of immigrants, and engage in transnational forms of documentation to mediate the limited credit and business histories of their clientele. Before beginning, a brief bit of background information about the Korean banks is in order.

The number of full-time staff members and branches of banks range from 64 to 559 and 1 to 22, respectively. Most of the branches are concentrated in the cities or states in which the headquarters are based. However, many of the banks whose main branch base is in Los Angeles have branches in New York and vice versa. Almost all of the banks have SBA offices, loan centers, or loan production offices that do not provide full branch services. These offices allow banks to expand their lending bases by accepting loan applications in cities that have a relatively large or growing Korean American community but do not necessarily have a concentration of Korean banks. For example, there are non-branch offices in Chicago, Dallas, and Seattle.

THE CULTURE OF KOREAN BANKING
IN THE UNITED STATES

Korean banks, like Chinese American banks in southern California (Li et al., 2002), were established to facilitate Korean American economic development in the United States. Although many employees reported that their banks are actively pursuing non-Korean clientele in order to stay competitive, the institutions nevertheless share a narrative of initially being established in the United States to support Korean immigrant communities. For example, one bank recounts on its website how it "was founded specifically to serve the needs of Korean-Americans, one of the fastest growing segments of the Asian ethnic group over the past decade." Another describes its history as "focused on providing much needed financial services to the then under-served ethnic Korean community in the New York metropolitan area." During our interviews, Korean bank employees basically repeated what the websites said. When asked about his bank's history, a seniorvice president of a branch of a foreign bank remarked, "I think it is to, you know, be supportive of the Korean community. The major issue is support the Korean community, or an Asian community, or immigrant."

As an international trade representative based in New York put it, "Many people come here and have money but they can't deal with American banks."

According to several respondents, the main reason why Korean immigrants need banks to support them is language access. An employee at one of the largest Korean banks based in Los Angeles explained, "In the last 20, 25 years … there was a large immigration boom … of Koreans. And many of them maybe did not speak English." An SBA loan officer at another bank in the same city echoed this point: "Naturally, they have entrepreneurial minded. Not to mention, they don't have choice, they don't speak English much and they can't assimilate into corporate America." Another banker gave a similar conclusion: "A lot of new immigrants who doesn't speak English, they feel more comfortable coming to a Korean-owned bank. Have a lot of employees can speak, bilingual."

Given the targeting of the Korean community, Koreans, not surprisingly, comprise the majority of the clients at all banks, with staff members estimating that Koreans comprised about 80 to 90 percent of their clients. In some cases, the portion of Korean clients was estimated to be as low as 50 to 60 percent, but these estimates were given in only two interviews. The majority of Korean clients are immigrants; bankers described second-generation Koreans as more likely to use larger U.S. banks rather than smaller Korean ones. However, many of the immigrant clients have been in the country for an ample amount of time and in many cases, long enough to be naturalized.

Koreans also make up the bulk of the staff, with most speaking Korean. The majority of respondents were born in South Korea and had spent a significant part of their lives there. Ostensibly, English was not their first language. Some Korean immigrant staff worked in banking in South Korea. In some cases, interview subjects were second or "1.5" generation Koreans who were bilingual in Korean and English. Some of the second-generation employees had less banking experience than their first-generation peers and had obtained American degrees in finance or business.

In addition to Korean speaking staff members, most banks recruit Korean clients by having materials such as websites and brochures available in the Korean language that often feature Koreans or other East Asians. The websites allow viewers to engage in a variety of transactions on-line and have all of the information regarding business activities provided in Korean.

Along with providing spaces that are culturally amenable to Korean immigrants, the banks also go out of their way to demonstrate to clients that they understand the culture of "Korean banking." For example, a compliance officer at a Los Angeles bank explained how her company provides "personal attention" in a manner consistent with Korean banking customs:

> We still do the business the Korean way where the branch manager will come out and say, 'Let me help you' rather than just a sales agent. You know how it is in Korea where you need to know the right people to get things done. And maybe it isn't the way it is here but they still perceive it as such so when we, when a branch manager goes out to visit a customer or the president of the bank goes out to visit the customer, they like that and they would rather bank with us rather than go to Bank of America where they have to go and talk to a loan officer,

they have to go and ask about the loan product rather than have us go
to the customer.

As this comment indicates, some Korean banks will purposefully create an
environment that not only serves Koreans in their language but also reminds
them of how banks operate in South Korea. The compliance officer concluded
that the environment of the bank and ways in which it delivers services are more
important factors in attracting Korean clients than financial ones such as the
calculations of credit. As she put it: "I truly believe that really it is the language
and the culture rather than the credit with us than anything else." Thus, Korean
clients walk into environments that are relatively more accessible in terms of language and culture. They are able to learn about bank services and loan packages,
conduct transactions, and engage staff in the Korean language. In some cases,
their presumed cultural needs and traditional methods of banking are replicated.

KOREAN BANKS, FINANCIAL LITERACY, AND THE FDIC KOREAN-LANGUAGE MONEY SMART PROGRAM

Along with efforts to create culturally amenable environs, Korean banks try to
educate clients about "American" norms of banking and business in order
to receive better-prepared loan applications. In many cases, applicants simply do
not have enough documentation of financial transactions, an issue common
among those unfamiliar with how business is "done" in a new country. An
SBA loan officer addressed the difficulty of working with immigrants who lack
what financial institutions term "financial literacy":

> They have to learn how the financial system in this country works.
> A lot of people like to buy in cash, cash, cash. That doesn't work well
> with a banking institution like us. After 9-11, Fed gov to track cash
> transactions so we are having extra scrutiny of "black money." I'm not
> talking about drag money—cash payments cannot always be verified.

Along with lacking the appropriate documentation of business transactions,
many recent immigrants lack credit histories. Shown in the following exchange
with the same SBA loan officer, this may adversely impact loan approvals:

> We have, you know, requirements in terms of business experience or
> credit history. You know, there are areas where we can, because of the
> SBA guarantee, we can be a little bit more, I would say, aggressive than
> a conventional bank would to lend to start-up businesses or people with
> short credit history. But they are always mitigated by other items such as
> additional collateral ... well, the first-generation Koreans, generally it's
> not always the persons that are coming in immediately that we're giving
> loans to. It's the persons that have lived here and done business or have
> done business previously.

As this lender suggests, he may privilege Koreans with longer residencies because they are better able to demonstrate business experiences and credit histories developed in the United States. Another SBA loan officer at a different bank explained that his office tries to lend mainly to existing businesses as opposed to start-ups. Most of his Korean clients are immigrants who have lived in the United States for between ten to twenty years. A loan officer at another bank reiterated this point. In the case of larger loans made to Koreans, clients are usually immigrants who have established credit histories and already own a business. According to the loan officer, there is "less risk involved."

Like all banks, Korean banks must consider the issue of risk, which structures the banking industry. The process of calculating risk generally involves dimensions of transactional lending. Peek (2007) explains:

> much bank lending is transactional … For example, many large banks, and some smaller banks, lend to small firms using credit scoring models, basing their loan decisions on information that can be quantified … credit scoring applies statistical methods to 'hard' data, summarizing borrower characteristics to produce a 'score' that can be used to evaluate the likelihood of the loan applicant repaying the loan. (p. 4)

Although the banking industry is currently driven by transactional lending, more banks and U.S. government agencies are rediscovering the relevance of "relationship lending" for financing "riskier" loan applications (Li et al., 2001). Whereas transactional lending is characterized by a reliance on credit scoring models, relationship lending is determined by long-term relationships with firms and a reliance on "soft data," which includes "more qualitative information acquired through direct interaction with the firm" (Peek, 2007: 4). Specifically, "Relationship banking is the conduct of banking business on the basis of personalized bank customer relations, broad-based information exchanges between borrowers and lenders, and in some cases banks' roles as informal advisors" (Li et al., 2001: 1935). For those who have limited or negative business and credit histories, relationship lending may be more advantageous than transactional lending.

Like many other small banks and consistent with Chinese American banks (Li et al., 2001; 2002), Korean banks are sites of relationship lending. As described by an SBA loan officer:

> Mainstream banks such as Wells Fargo and Bank of America are good banks but I would say are catered more towards consumer lending and they're more credit-scoring based because of their sheer size of the bank. Whereas our bank, although our bank, we've gone big in recent years, we cater on a case by case the need of each individual and we I think, because of our experience, we recognize the businesses and we know how to finance those specific businesses, the retail businesses better than the mainstream lenders out there.

Relationship lending takes several forms at Korean banks. Many loan officers mentioned that they worked with Korean business owners "case by case." Some claimed that, due to their immigrant niche market, institutions were more lenient than banks that may strictly follow transactional lending guidelines. For example, some lenders were not as stringent regarding business records. Given the economic diversity of their clientele, relationship lending did not always result in approved loans. Similar to Chinese American banks (Li et al., 2002), lenders try to educate clients that were risky on paper, a process made all the easier given that everyone involved may converse in Korean. Lenders recounted how they encourage financial literacy so that clients may write better business plans. An SBA loan officer pointed out that SBA literature is not available in Korean and that he often receives "bad business plans." He mentioned that when this happens, "I try to educate them." These gestures demonstrate that Korean banks are not simply sources of economic capital but are also, in the case of recent Korean immigrants who are "risky," valuable sources of business information and financial literacy.

Although not given attention in the studies conducted on Chinese American banking (Li et al., 2001; 2002), one method for increasing the financial literacy among its clientele is by collaborating with U.S. government organizations. Banks, often in partnership with government agencies, sponsor Korean-language workshops for potential new business owners. In other cases, lenders refer clients to business development programs and services sponsored by U.S. government organizations such as the SBA. And, as described here, some Korean banks participate in the FDICs Korean-language Money Smart program.

Sponsored by the FDIC, Money Smart is an "adult education program" initiated to develop financial literacy among those "with little or no banking experience" (FDIC, 2008b). According to the FDIC website, the purpose of Money Smart is to promote financial stability among communities:

> Financial education fosters financial stability for individuals, families, and entire communities. The more people know about credit and banking services, the more likely they are to increase savings, buy homes, and improve their financial health and well being. The Money Smart curriculum helps individuals build financial knowledge, develop financial confidence, and use banking services effectively. (FDIC, 2008b)

Money Smart places a great deal of responsibility on banks for providing training and workshops to clients. In doing so, the FDIC is able to perform its general function of encouraging confidence in banking (FDIC 2008a). Participating banks benefit by developing clients' financial literacy. This symbiosis is alluded to in the FDICs explanation of its emphasis on collaborations with banks:

> The collaboration between the FDIC and banks as illustrated by Money Smart Collaboration is important to the success of any education effort. The FDIC encourages banks to work with others in their communities to deliver financial education and appropriate financial services to individuals who may be unfamiliar with the benefits of having a relationship with an insured depository institution. (FDIC, 2008b)

Ultimately, Money Smart serves the Korean banking industry by helping it grow a more financially sophisticated clientele. Greater financial literacy presumably helps clients broaden the range of financial activities in which they may participate. Therefore, Korean-language Money Smart ultimately helps Korean banks educate its clientele in the norms of U.S. banking and in turn, sustains the Korean banking industry. Additionally, increasing Korean immigrants' financial literacy may contribute to entrepreneurship.

GLOBALIZATION AND THE MEDIATION OF RISK

Generally at banks, loan applicants put up property as collateral in order to secure a loan. If the applicant defaults on the loan, the bank may sell the property to pay off the debt. Koreans with limited financial histories in the United States may have difficulty finding sources of collateral that are accepted at most banks. Thus, ethnic banks take on greater significance in these cases because opportunities to address limited financial histories in the United States are facilitated by the globalization of banking. As Li et al. (2001) point out, "transnational information and network advantages have been critical to these banks' ability [sic] aggressively to pursue loan and development opportunities" (p. 1938).

In some cases, and consistent with Chinese American banks (Li et al., 2001), Korean banks offer unique opportunities in terms of collateral that their non-Korean peers may not in that Korean immigrants may use property they own in South Korea as collateral. However, this is only an option for clients who use Korean banks that are affiliated with foreign-owned banks or overseas branches that may verify property ownership. Thus, American banks that do not have branches in South Korea cannot offer this option. As such, Korean foreign-owned banks may be more advantageous for recent Korean immigrants as long as clients own property in South Korea. When asked about the process, a lender explained:

> In Korea, if the person who has the property in the Korea, we appraise the property in Korea and then we collateral in Korea and then we made the decision for the value and then we lend the money here. So that's kind of the program for the appraisal base, yeah.

Do you have a lot of situations where you have to do that?

> Not really. Because, uh, that person, the borrower, need to have property under his, his or her name. I mean, yeah, in Korea, so but other country, we don't do that. Only Korea we do that … because the (bank) headquartered in Korea. It's called (name of bank). Yes, so they manage the property or can appraise the property. They make the decisions, so we do that in Korea.

Additionally, Korean banks affiliated with foreign-owned Korean banks or overseas branches are able to look into an applicant's credit or business history in South Korea. As a vice president of a Korean bank in New York put it, "We uh, check up to Seoul Korea credit history." A loan officer at the same bank elaborated on this point:

> I mean as long as we could actually um, um, search his, his business history in Korea. Or, you know, if, just like, just like, people who has been here and starting their, starting a new business, we could go through their business projection and see if it makes sense and we check their personal credit history or their financial credit history. And if it, if it makes sense, then we could, we could proceed with it.

As these comments demonstrate, Korean banks are sites of globalization that negotiate banking regulations of the United States with approaches made possible by the globalization of Korean banking. An aspect of this globalization involves the expansion of branches in other countries. More than one foreign bank reported having hundreds of branches in South Korea, which means that Korean immigrants can access a network of banks that may look into property ownership and pre-migrant business and credit histories. And as reported by some loan officers, Korean American banks are increasingly opening branches in South Korea. Thus, the number of Korean banks providing opportunities for overseas collateralization and credit and business checks is simultaneously growing in South Korea and the United States.

THE GLOBALIZATION OF KOREAN BANKING, THE MINORITY BANKING INDUSTRY, AND INEQUALITY

Although Korean banks are able to provide unique opportunities to Korean immigrants, the number of Korean banks in the United States is relatively small compared to the number of corporate "American" banks. Findings show that clients are reportedly attracted to Korean banks because of cultural amenability and the emphasis on relationship lending. There is debate, however, regarding the role of cultural sensitivity in determining bank clientele. A market research report published by Packaged Facts (2005) found that "unbanked" respondents, many of whom were lower- or middle-income African Americans and Latinos, tended to forgo relationships with banks not so much because they felt "unwelcomed" or could not speak English well, but because they lacked the required identification and had poor credit histories (p. 3). Nevertheless, racial discrimination at "mainstream" banks is not obsolete. As suggested by this study, ethnic banks, such as those catering to Koreans, may serve as useful alternatives to "mainstream" banks that are considered "unwelcoming."

Yet it is unclear if the ethnic banking industry can successfully mediate the discrimination racial minorities may experience at mainstream banks. According to the FDIC (2008c), as of June 30, 2008 there were, including those that have minority boards, 95 Asian American and Pacific Islander (AAPI) banks, 45 black banks, 52 Latino banks, and 21 Native American banks. The combined total assets of AAPI banks, black banks, Latino banks, and Native American banks amount to 58 billion, 7 billion, 136 billion, and 2 billion, respectively. While they should give us pause, these figures do not account for all minority banks since they are not required to register with the FDIC. For example, among FDIC minority institutions, there are only four Korean banks listed. However, these four banks have combined total assets of 3 billion, which, respectively, is about 40 percent and 150 percent of the amount of the combined total assets of the 45 black banks and 21 Native American banks registered with the FDIC Information about more Korean banks is provided by the Federal Financial Institutions Examination Council (2007). Including the four Korean banks registered with the FDIC, there are nine Korean banks with combined total assets of 14.6 billion dollars. Thus, nine Korean banks have twice as much in total assets compared to those of 45 black banks and seven times as much as 21 Native American banks. While U.S. government institutions try to build symbiotic partnerships with minority banks and in turn, connect with a variety of minority communities, such disparities among ethnic banks will largely shape the extent of such collaborations.

The significance of ethnic banks in the United States is under-scored by opportunities made possible by globalization. Whereas African American and Native American banks comprise the "first-generation" of ethnobanks, the "second-generationethnobanks"—of which Asian and Asian American banks are a part—serve "ethnic communities actively engaged in cross-border population and money flows" (Li et al., 2001: 1926). Thus, it is important for those looking to Asian ethnic groups as examples of minority business success to keep in mind that not all ethnic banks catering to racial minorities are part of a global network. But even if they were, not all clients have the ties to other countries that facilitate transnational economic exchanges. In terms of people of color, this is a salient issue for African Americans and Native Americans. Not only do they have smaller numbers of banks and total assets in them, they also have, compared to other immigrant-based groups such as Asian Americans and Latinos, histories in the United States that make it difficult to take advantage of the economic opportunities made possible by the globalization of banking.

CONCLUSION

Korean immigrant entrepreneurship is informed by the globalization of Korean banking and contributes to Korean small business ownership. These institutions were established to promote Korean economic development in the United States, employ Korean cultural norms when delivering services, and emphasize educating clients in terms of the banking and business norms of the United States. Whereas the research conducted by Li et al. (2001) and Li et al. (2002) do not mention Chinese American banks' collaboration with U.S. government organizations, I examined how Korean banks work with federal institutions, such as the FDIC, to promote financial literacy among clientele and potential small business owners. Korean banks also negotiated the issue of financial risk by engaging in relationship lending and taking advantages of opportunities facilitated by the globalization of Korean banking.

Disparities among minority banks and different racial histories may make it difficult to simply "replicate" the example of Korean banking among non-immigrant people of color. In other words, policy recommendations may not easily remedy the fact that minority groups have different inroads to transnational economic exchanges. What findings do suggest is that immigrant-based groups will most likely benefit from the minority banking industry given their networks and the globalization of banking. Such preliminary conclusions, however, deserve more empirical and theoretical attention, especially as the current U.S. financial crisis continues to affect the financial industry and put at risk loans, mortgages, and business ownership.

REFERENCES

Federal Deposit Insurance Corporation. (2008a). FDIC mission, vision, and values. Retrieved January 1, 2008, from http://www.fdic.gov/about/mission/index.html.

————. (2008b). Money Smart—An adult education program: Money Smart adult financial education curriculum. Retrieved January 1, 2008, from http://www.fdic.gov/consumers/consumer/moneysmart/overview.html.

————. (2008c). FDIC List of Minority Institutions—Excel (6/30/08 Performance Information). Retrieved November 25, 2008, from http://www.fdic.gov/regulations/resources/minority/MDI.html.

Federal Financial Institutions Examination Council. 2007. Retrieved January 10, 2008, from http://www.ffiec.gov/.

Li, W., Zhou, Y., Dymski, G., & Chee, M. (2001). Banking on social capital in the era of globalization—Chinese ethnobanks in Los Angeles. Environment and Planning A, 33: 1923–1948.

Li, W., Dymski, G., Zhou, Y., Chee, M., & Aldana, C. (2002). Chinese-American banking and community development in Los Angeles county. Annals of Association of American Geographers, 92(4): 777–796.

Packaged Facts. (2005). Market trends: Opportunities in the "unbanked" consumer market. Retrieved from Market Research.com Academic.

Peek, J. (2007). The Value to Banks of Small Business Lending. Retrieved June 1, 2007, from http://www.sba.gov/advo/research/rs301tot.pdf.

U.S. Bureau of the Census. (2006). Survey of business owners—Korean-owned firms: 2002. Retrieved January 30, 2007, from http://www.census.gov/csd/sbo/koreansummary offindings.htm.

24

Black and White Fraternity Men's Approaches Toward Women

RASHAWN RAY AND JASON A. ROSOW

We revisit a privileged upper stratum of society in this chapter about fraternity men. Ray and Rosow offer an interesting twist on this familiar topic by contrasting the culture of sex and romance among Black and White fraternity members. This chapter complements Boswell and Spade's portrayal of low- and high-risk fraternity rape cultures and delves into several aspects of the fraternity influence. While most people are familiar with cultural depictions of fraternity men that describe their pursuit of sexual conquests as single-minded and often manipulative, this chapter takes you further inside this elite world. As you read this chapter, you may want to think about the influence of structural factors, such as housing, on fraternity men's behavior. How do you think men living in close quarters acclimate to such sexual intimacy? Why were the behaviors of the Black fraternity men so different? How are the Black men portrayed here affected by their class position, and how does this build on the child-rearing practices described by Lacy for Black middle-class families? What does this tell us about the relative power and status of White versus Black college women? Why is one group treated so much better than the other by their fraternity dates? Do you attribute these differences in the behavior of the Black and White fraternity men to individual differences, cultural differences, racial/ethnic differences, class differences, structural differences, or other factors? How and why might these men change their behavior when they get out of college and away from their fraternities?

SOURCE: From *Men and Masculinities*, 12(4), pp. 523–546 © 2010 by Sage Publications. Reprinted by permission of Sage Publications Inc.

D espite the proliferation of research on collegiate gender and sexual relations, we know little about one of the key groups within this institutional arrangement—fraternity men. Meanwhile, we know even less about differences and similarities in Black and White high-status men's relations with women. Because fraternity men typically are situated on top of the peer culture hierarchy, a comprehensive understanding of the organization of collegiate social life must take into account how these specific enclaves of men understand and perceive gender relations and sexuality, and whether these understandings and perceptions vary by race.

Scholars have offered three competing explanations regarding racial differences in men's approaches toward women: (a) Black and White men objectify women similarly; (b) Black men objectify women more than White men; (c) White men objectify women more than Black men. The first possibility contends that most men, irregardless of status or race, sexually objectify women in the same manner. Thus, Black and White men's performances in masculinities are expected to be similar. In patriarchal societies, men control sexual and romantic environments by promoting sexually aggressive behavior among men. Through the emphasis of the importance of sexual prowess, cultural mandates concerning gender encourage men to "sexually objectify" women and appear "sexual." However, such mandates encourage women to stress relationship viability and appear "romantic." Hence, men are often authorized to express themselves sexually, while women who act this way are shunned. This possibility suggests that gender trumps race and status concerning men's interactions with women.

A second possibility is that Black men exhibit more sexually objectifying approaches toward women than do White men. This explanation is most in line with scholarship on Black men's relations with women. More specifically, cultural motifs like the "cool pose" portray Black men as culprits of sexual violence. However, this perspective, which is echoed with public discourses and much scholarly research, gives the impression that all Black men are part of the same cultural spaces, thereby neglecting the fact that Black men may be part of different sociocultural spaces that yield distinctly different structural consequences for their treatment of women. It should also be noted that the stereotypical nature of Black men as the Mandingo—overly aggressive, sexually promiscuous, physically superior yet intellectually inferior—has long been purported in mainstream discourses. Race scholars assert this is a problematized, dramatized, and monolithic perception of Black men that is often exacerbated in the media.

Finally, the third possibility asserts that White men are more sexually objectifying than Black men. By virtue of their presumed greater status and esteem, White men are more likely to control social environments and accept, and even normalize, sexual objectifications of women. This perspective echoes the sentiments of women who claim sexual harassment in high-status institutions such as law, academia, and corporate America where White men are typically the controllers of social environments. In contrast to the aforementioned "cool pose," some extant literature finds that Black men's gender attitudes, compared to their White counterparts, are more supportive of gender equality because of a shared oppression and subordination with women.

In this article, we assess these three predictions by analyzing 30 in-depth, individual interviews and surveys and two focus group interviews with Black and White high-status fraternity men. We find evidence that White men are more sexually objectifying than their Black counterparts, in support of the third prediction. However, we also find that the reasons behind this pattern go beyond the explanations typically asserted by this prediction and the first two predictions. Collectively, the three explanations noted above neglect the extent to which cultural and social norms are embedded within and shaped by the structure of institutions, and in turn, how structure shapes men's approaches toward women and the performances of masculinities. Accordingly, we contend that "normative institutional arrangements" are one of the key factors that underlie racial differences regarding how men interact with women romantically and sexually on college campuses.

NORMATIVE INSTITUTIONAL ARRANGEMENTS
IN HIGHER EDUCATION

Normative institutional arrangements are boundaries that shape social interactions and establish control over social environments. Normative institutional arrangements identify social contexts (e.g., social environments in fraternity houses), whereby certain behaviors are more or less acceptable and certain structures hold individuals more or less accountable for their treatment of others. Here, we showcase the implications of the intersecting forces of race and status by examining two normative institutional arrangements that are common themes in Black and White men's understandings and perceptions of gender and sexual relations: (a) small Black student and Greek communities; (b) living arrangements including a lack of on-campus fraternity houses.

The Black student community at most Predominately White Institutions (PWIs) is small and insular. There is also a limited amount of social interaction between Black and White fraternities and between Black and White students overall. Similar to patterns at the societal level, interracial dating is infrequent. As a result, even high-status Black fraternity men are mostly invisible in White social arenas.

In contrast, the relatively small number of Black students and limited interactions with Whites indicate that Black fraternity men are much more visible in the Black community. In fact, this group of Black men aligns with the ideals of what DuBois (1903, 1939) conceptualized as the "Talented Tenth." Such members of the Black elite are expected to sacrifice personal interests and endeavors to provide leadership and guidance to the Black community. However, being part of the Talented Tenth signifies the monitoring of this group's behavior, particularly actions that are inconsistent with a greater good for the Black community. This monitoring by others on Black fraternity men is intensified in a structural setting with a small community size, and in turn, increases the likelihood that their treatment of women will be publicized and scrutinized by members of their own social community and the broader college and off-campus communities. Although White fraternity men may also be visible, the sheer number of White students leads to them being held less accountable, and consequently, able to perform masculinity in a manner that Black fraternity men cannot.

Not only is the Black community relatively small but Black Greeks have very different on-campus living arrangements than White Greeks. There is a historical legacy of racial discrimination, both within and external to the university, that has traditionally precluded Black fraternities and sororities from gaining equal access to economic resources such as Greek houses and large alumni endowments. To date, most Black Greek Letter Organizations (BGLOs) do not have fraternity or sorority houses on university property. If they do, these houses normally are not the same size or stature of those of their White counterparts. To the extent that the structure of living arrangements facilitates a certain treatment of women, racial differences in access to housing on-campus may have implications in potential racial differences in approaches toward women…. We pose two essential questions (a) Regarding high-status fraternity men's relations with women, are there racial differences in romantic versus sexually objectifying approaches? (b) How do "normative institutional arrangements" structure men's approaches toward women? Because there has been limited empirical research on elite men, we privilege their accounts and voices to gain an insider's perspective into the intersections of masculinities, status, sexuality, and race.

SETTING AND METHOD

We conducted 30 in-depth individual interviews and surveys, along with two focus group interviews, from 15 Black and 15 White fraternity men at a PWI that we call Greek University (GU). Enrolling approximately 30,000 undergraduates, GU is ideal for this study because of its strong academic reputation, vibrant social life, and party scene. GU's emphasis on Greek life facilitates the examination of gender relations among high-status men. About 20% of GU undergraduates are members of Greek letter organizations, which is larger than similar universities. For members, the Greek system normally offers a home away from home, friendships, and social and philanthropic activities. There are approximately 25 White fraternities with memberships around 100, some with on-campus and some with off-campus status, and five Black fraternities with memberships around 10 and all hold off-campus status. Although approximately 25% of White students are members of Greek organizations, less than 10% of Black students are members of Greek organizations. Black and White fraternities are operated by two different governing bodies, the National Pan-Hellenic Council (NPHC) and the Interfraternity Council (IFC), respectively. Although none of these fraternities appear to explicitly discriminate on the basis of race, there is virtually no overlap in race among members of these organizations….

As a Black and White team of male researchers, we note that gender may elicit certain responses with participants de-emphasizing romanticism. We also conducted interviews with the authors matched with participants by race to elicit candid responses about the other racial group. Based on the data presented throughout this article, we are confident that we limited methodological biases. For example, one White respondent states, "Blacks will fuck anything." Another says, "Yeah, my friends at home are Black. They like to put it in girls' asses." Based on our experiences interacting with these respondents, we believe they would not have made these comments if they were being interviewed by a Black interviewer ... for a Black quotation about Whites. These quotations show that respondents did not hesitate to make derogatory statements about the other group.

All the men in our study report being family-oriented and having lofty career goals. Most participants are active on campus and have higher GPAs than non-Greeks. However, a substantial class difference exists between Blacks and Whites. The Black men's self-reported family household income is lower middle class, whereas the White men's self-reported family household income is upper-middle class. Many of the Black fraternity men have scholarships, student loans, and/or jobs to pay for tuition and housing costs, whereas most of the White fraternity men have scholarships and/or their parents pay a substantial portion of their tuition and living expenses. All respondents self-identify as heterosexual....

RACIALIZING GENDER RELATIONS ON CAMPUS

The interviews suggest that Black fraternity men exhibit more romantic approaches than White fraternity men. Although both groups sexually objectify women, Black men emphasize romanticism more than their White counterparts. They indicate that women are physical objects of enjoyment but should also be respected. White fraternity men make few romantic references and primarily view women as sexual objects.

The following quotations exemplify sexually objectifying approaches. This participant suggests that romance is unnecessary in the quest for gratification.

> Pretty much you do not need to do all that wine and dine them and all that. You can skip all that and just bring them back to the house and do what's important to you. (White)

In two different parts of the interview, a participant explains which factors affect how far he will go with a woman.

R: If they [women] were decent or just okay, I'll just mess around with them ... Get head.

I: When she gives you head, do you go down on her?

R: Honestly, I don't like that ... I do it every once in a while. Honestly depends how hot the girl is. If I'm drunk and into the girl, I probably would. But other girls, I just make out with them for a little bit.

R: We were talking for about a week and we started messing around. She starts giving me head, and when I took her shirt off, I put my hand on her stomach and this girl had abs. I think that's the most disgusting thing. Like, girls with abs, its like … too masculine. So that like turned me off and I couldn't get off and I never called her again. (White)

Nine of the 15 White participants report engaging in sexual behavior that they do not prefer including performing oral sex because of a woman's desirable physical characteristics. They also rarely describe "hot girls" in terms of social competence and popularity. Reports from Black men also contain sexually objectifying approaches. While describing what he desires in a woman, a participant compares women to cars as he explains why his standards for sexual encounters are lower than for relationships.

I: Are your standards lower for a hookup than a committed relationship?

R: I use this analogy. Some people say it's corny, but whatever. When you have the title of a car, you want it to be nice, but you'll jump in your friend's car. You'll ride, you'll ride anything because it's not your title. But if I'm going to have the title to you, you've got to be nice because you represent me! But now I'll ride in a pinto, but I just won't buy one. (Black)

Although both groups exhibit sexually objectifying approaches, romantic approaches in quality and content are far more prevalent among Black men. They respond when asked to "describe ways you or your friends respect women on campus."

I definitely think my fraternity brothers do a lot of stuff that make them [women] feel appreciated like getting them flowers; whether write them a poem, whether it's just tell them they look beautiful. (Black)
I think you have to treat women with respect. I think because of how society is I think a lot of males have been misshapen to be like the world leader; the dominant figure in the relationship. They wear the pants in the relationship. I feel like I would treat a woman the way that I would want to be treated. (Black)

Conversely, many White men describe a very different notion of respecting women.

We respect women. We won't take advantage of them if they're wasted. If she's puking in our bathroom, one of the pledges will get her a ride home. (White)
One way that I respect women? A lot of ways. I'll never ask if she needs a ride home after we hookup. I'll let her bring it up or let her spend the night. You respect a girl more if you let her stay. (White)

Black and White differences are also evident in responses to "what do you consider a serious relationship?" White men understand a serious relationship primarily in terms of physical monogamy, whereas Black men define serious relationships in terms of socioemotional exchanges.

R: If you're in a serious relationship, you shouldn't be making out ... that's wrong.

I. So serious relationships are when you don't cheat on a girl.

R: No. You shouldn't be making out in front of people. If you have a girlfriend you can't be like all over girls at parties. (White)

> Serious relationship is pretty much a basic understanding that two people are together. You have somebody to talk to; somebody who is going to be there on the other end of the phone call. When you leave that message they're calling back. Maybe at night you got somebody to cuddle with. Somebody that could possibly cook for you. Somebody that might be taking you out, picking you up. Somebody that is worrying about what you're doing. (Black)

Twelve of the 15 Black fraternity men explain that having someone to "share" and do "special" things with is the best thing about being in a serious relationship.

> I'd say you get the companionship, the love. You've got somebody there in daytime hours, not just in nighttime hours. The nine to five hours they're going to be there to go out with you. They might send you out with some stuff, take you out to eat, go see a movie, and like it's that constant companionship. (Black)

Comparatively, only 7 of the 15 White fraternity men mention that this is a benefit of a committed relationship. Instead, 12 of them explain that having a "regular hookup" is the best thing about being in a relationship.

> Lots of sex. You can have it everyday without having to go out and get it. It's a lot easier, but you do have to put up with shit occasionally. (White)
>   The best thing is you don't have to use a condom. It feels better and you can go right to it. And you got someone to call that you know what they want and knows what you like. (White)

In contrast, only 3 of the 15 Black men mention sexual convenience as a benefit of being in a committed relationship.

Perhaps most revealing are the responses to "describe a romantic evening." Black men volunteer specific details without hesitation and reveal intimate knowledge of their partners, thoughtful planning, and intricate execution.

> I try to do romantic things on occasion, not just on occasions. On her birthday I surprised her. I told her we were going out to dinner. There is a whole day of events. I left a dozen roses in front of her door. I have a key to her apartment just because she likes to have that kind of security just in case I need to go over there and do something for her. When

> she came home I had prepared a dinner for her. I cooked her favorite dinner which was spaghetti and she was really surprised. It was a candlelight dinner, lights were off, food all served, salad, and spaghetti. She really liked that and I gave her some more gifts, but the last thing I got her was a ring that she loves. (Black)

Of course, "romantic" does not necessarily imply equitable gender relations. "Romantic" can also have negative connotations for gender relations (e.g., women need to be taken care of, pampered, put on a pedestal). Comparatively, most White men's narratives imply less thought and planning. Only three of them could describe a romantic evening, two of which were descriptions encompassing "dinner and a movie," preferably an "expensive establishment."

> I clean up, shave, put on a nice shirt with a nice pair of pants, you come out of the car, you wait for her, open the door for her. Nice expensive restaurant; something with a good reputation. Maybe somewhere someone's parents would take them, because that lets them know you're dropping some cheddar, you know, your [sic] dropping some money. Have some easy conversation, then come back have a few more drinks, and then, you know. {laughs}. (White)

Another White participant says, "Well, on her birthday I got her an ipod [sic]. She loved it. I took her out to dinner, an expensive dinner."

Finally, the language used by White fraternity men to describe women in gender interactions suggests sexually objectifying approaches, whereas the language exhibited most frequently by Black fraternity men implies more romantic approaches. White men commonly refer to women as "chicks," "girls," and other belittling terms. Conversely, Black men generally use more "respectful" terms like "women," "ladies," and "females" or refer to individuals by name. As seen in many quotations throughout the article, the examples below illustrate the role of language regarding gender relations.

> I know this one time I was real drunk, a little too flirtatious with a female who was actually a friend of mine. I did—I was not trying to hook up with her. Actually, she was trying to hook up with me. And when the alcohol mixed with the flirtatious lady, mixing with me not driving having to be at her house that night. I did regret it when I woke up the next morning {laughs}. (Black)
>
> When the booze settles in you can make mistakes and you'll screw up with a "frat rat" or something. (White)

Collectively, Black excerpts normally acknowledge women's agency, whereas the White accounts typically display the use of the passive voice, whereby a woman is always acted on and never acting. Black men emphasize more romanticism in their accounts regarding experiences with and attitudes about women. However, these differences are not solely related to race.

NORMATIVE INSTITUTIONAL ARRANGEMENTS STRUCTURE APPROACHES TOWARD WOMEN

We find that differences in men's approaches toward women are structured by normative institutional arrangements centering on community size and living arrangements. Participant accounts suggest that the size of their respective racial communities on campus and the presence or absence of a fraternity house underlie racial differences in romantic versus sexually objectifying approaches toward women.

Greek and Racially Based Communities

Fraternity men are concerned about their individual and group reputations when making gender relation choices. Thus, they aim to steer clear of certain social scenes to preserve their status as elite men. In the following accounts, participants indicate that to maintain their reputations they normally will not "hookup" with low-status women. We asked, "Are there any women you wouldn't hook up with?"

> Fat girls. I stay away from them. Sluts too. They're disgusting. I don't need to hookup with that, that's not *our* [his fraternity] style. (White)
>
> Everyone has the one, two, or three girls that they're like what the hell was I doing? But you don't want to have too many. I mean its good to hookup, but you don't want to do it with a girl that's easy. If it's a girl that every guy wants and you bring her back its like, "Wow! You hooked up with that girl? That's impressive!" It feels good. If you hookup with an ugly girl, you're friends will give you shit for it. (White)
>
> Yes. They're not attractive. [laughs] That might sound mean, but that's what they are. Not attractive girls. I don't think there's no woman here that's higher than what we think we can reach. And then lower, yes there is a group of people that you should just not touch. I hear a lot of guys in other fraternities say, "Man I wish I could get a girl like that." Instead, we just get the girl like that. (Black)

… The Greek community seems to hold White fraternity men accountable to sorority women. As the quotation above alludes, it is only for reasons of saving face that will allow them access to other women in the same sorority. Another White participant describes how cheating could result in a bad reputation if he got caught.

I: What would happen if you got caught cheating?

R: The way I could see it [cheating] affecting something is if it's a sorority girl you fuck over.

I: You can't screw with a sorority sister?

R: You could, but you could get the name, you're an asshole, you're a player, or something like that. I mean it might. It could spread around the [her sorority] house. Then you're Blacklisted. (White)

Although repercussions exist with sorority women, there are an abundance of non-Greek women with no reputational constraints. The large number and high percentage of White students give White fraternity men an ample pool of women not connected to the Greek community. When relating with non-Greek women or GDIs,[1] White fraternity men do not have to worry about "the word" getting around. Moreover, White fraternity men can disassociate from the fraternity, blend into the crowd, and interact as they please.

R: GDIs come here and it's like sensory overload. They are like in awe. If you're in awe, it's like so easy. {laughs} You can say anything to a GDI. Adam makes girls cry.

I: He wouldn't do that to a sorority girl?

R: You make a sorority girl cry she's going to tell all her friends. "I was at XYZ [fraternity] and this guy made me cry and he's such an asshole." If you say it to a GDI, she's going home and you're probably never going to see her again and she's not going to tell all of XYZ [sorority] that you said this and you're not going to have a whole sorority that hates you. (White)

Black fraternity men face a different organizational structure. These men feel that they cannot "do things like other guys."

Because there's only seven [Black Greek] organizations on campus, we have a huge impact on the Black race here. Where there's like 750 different [White] organizations, their impact is not as severe. It's not as deep, especially cause they have more people than our race. (Black)

It's kinda like being on the basketball team or being on a football team. You know what I'm saying? Its kinda like "Eta" [his fraternity name] puts you on the next level. Like you're Black Greek but you are like the … you are supposed to be representing the Black Greek. It's kinda hard to get that out, but when we do something we are suppose to be setting the bar for everybody else. It's like a known thing that we suppose to be setting a bar. You know what I mean? (Black)

Black fraternity men, and many Black students, cannot overcome the reputational constraints of the small Black population. Black men report being very conscious of their behavior when interacting with women. Although White fraternity men can generally be anonymous and "get off" safely, Black fraternity men perceive themselves to be constantly visible and therefore continuously held accountable for their treatment of women.

The Fraternity House

The organizational structure of "the house" facilitates sex, discourages intimacy, and is used as a resource, which affords White fraternity men control of sexual environments. For instance, these men report that women normally engage in

---

[1] "GDI" is an acronym for "God Damn Independent," which is a derogatory term used to describe non-Greeks.

relationships to be associated with a particular fraternity and to have access to fraternity functions and/or alcohol.

College-aged women younger than 21 years old seem to rely on fraternities for basic ingredients of the mainstream version of the college experience—big parties and alcohol. In fact, an interview with a White participant was interrupted twice because of orders for alcohol placed by an ex-girlfriend and another woman from the dorm.

"The house" also facilitates a convenient means of engaging sexual behavior. A participant discusses the difference between living in "the house" and living off-campus.

> You meet a lot more girls in the house. The frat [house] is easier, a lot easier too in that sense cause coming back from the bars, it's not necessarily like "let's go back to my place." Instead it's like, "Let's go back to the frat [house] and have a couple more drinks." It's like you don't sound like you're trying to hook up with them. "Let's go back to my house and just … get it on" … [laughs].… It's easier. (White)

"The house" also constrains men's gender relations. Although the fraternity houses at GU are impressive mansion-like structures, they are chaotic, nonprivate spaces that promote nonromantic activities. Most White fraternities require a "live-in" period. In the first year, members sleep in cold dorms, which are rooms composed of dozens of bunk beds.

> It's like fifty of us sleep together. But you put your beds together and have all these sheets and stuff. It's like a bungalow. But sometimes you can hear other people having sex. (White)

If members earn enough "house points" for representing the fraternity well through activities like philanthropy or sports, they then typically live with three roommates in a tiny bunk-style room.

White fraternity men indicate that they could never "get away" with having romantic time. A White participant says, "There's so many people running around that house, someone's bound to see or hear something." The public nature of fraternity living arrangements is also confirmed in our field notes. While entering a fraternity room to do an interview, the interviewee and interviewer interrupted a roommate who was masturbating. The interviewer reported surprise that the masturbator seemed only slightly uncomfortable with the interruption. This suggests that interruptions like these are commonplace.

Although privacy would intuitively be linked to more sexually objectifying approaches, in the context of Greek social life, a lack of privacy facilitates these approaches by preventing intimacy. While having other people as witnesses should reduce the degree of exploitation, Greek social life is a normative institutional arrangement structured by hegemonic masculinity with sexual prowess as one of its essential ideals. Thus, men who engage in public displays of sexual objectification are applauded. A participant describes one evening in the cold dorm.

> Lunch on Fridays are the best. It's like all the stories from Thursday night. It's pretty funny. It's a good time. For instance, Tom came into the cold dorm and he was with his girlfriend and they were really drunk. And he's like, "We're having sex." I was like, "You should have heard him. He punished her." (White)

Romantic displays, because they are not in concert with hegemonic ideals, are sanctioned. For example, participants indicate that men who make romantic displays like saying "I love you" or opting for alone time with a woman over "hanging with the guys" will quickly be referred to as "pussy-whipped."

> You don't want to be known as pussy-whipped. Guys that are pussy-whipped are wimps. They just let their girl tell them what to do. You can't count on them. They'll tell you one thing, but if the girl says something different, they're doing what she says. (White)

Another White participant characterizes how public displays of romanticism are considered to be uncool by the general Greek community.

> I don't know how romantic it gets. Am I like going to set up a table in my frat room and light a candle? It'd be cool, if I had the balls to do it. (White)

When asked directly "why don't you and your friends have romantic evenings," a White participant explains.

> Frat houses aren't the place for that [romantic behavior]. Have you looked around? The place is filthy and you have no privacy. None. I shower with five guys; people always coming in and out. You're never alone. I used to feel weird about it [sex], but now I don't. Like I used to try to be quiet, but I'm having sex less than four feet from where my roommate, whose [sic] having sex with his girlfriend. You're going to hear something. So you don't worry about it. (White)

Conversely, Black fraternity men's off-campus status offers private space for romantic relations. Most members are scattered across two- to four-person apartments and rental houses. Interviews conducted in bedrooms at Black residencies were devoid of interruptions, whereas interviews with Whites had three to four interruptions, on average....

As an added point of emphasis, focus group and ethnographic field note exerts highlight participants own awareness of the importance of normative institutional arrangements. When Black men were asked in a focus group if things would be different if GU was not a PWI, and instead a Historically Black College or University (HBCU), they unanimously responded, "Yes." If Black men had a house, they think their behavior would be similar to their White counterparts. They perceive "the house" as a place to socialize in large groups that is free from police contact and potentially hostile strangers. If they were the "majority," they perceive being free from the incessant scrutiny of the general campus community. Black men explain how "nice it would be" to not have to represent "every Black man on the planet."

When White men were asked in a focus group to imagine life without "the house," they replied, "It would be like being a GDI." They further explained that "the house" is "like a face" which enables them to "meet girls." White men also mirror issues of safety indicated by Black men. For instance, they are concerned that if they lose "the house," they would have to go to bars, small house-parties, or third-party vendors and would then have to worry about drinking and driving, public intoxication, and police breaking up parties. In other words, White fraternity men perceive that losing "the house" would make them "just like everybody else."

In sum, normative institutional arrangements—the presence or absence of a fraternity "house" and the size of the Greek and racially based community in the larger student population—afford Blacks and Whites different opportunity structures for romantic and sexual relations. The small, highly visible and insular Black communities normally force Black fraternity men to be conscious about their positions as leaders and role models, thus affecting their experiences with and treatment of women. This consciousness often leads Black fraternity men to conveying more romantic approaches toward women. Because of the size of the White student population, White fraternity men often find relief from reputational constraints. "The house" facilitates White fraternity men's relations with women by putting them in control of sexual environments. At the same time, however, the "public" nature of fraternity houses constrains gender relations by providing only nonprivate and unromantic spaces, thereby promoting more sexually objectifying approaches.

CONCLUSION

… Our findings suggest that both Black and White fraternity men sexually objectify women; however, Black fraternity men exhibit more romantic approaches in their perceptions of their relations with women. Black college social scenes, particularly Black Greek scenes, are often more gender egalitarian. Although the small size of the Black community and the organizational structure of the Black Greek system generally force Black men to be more conscious about their treatment of women, the organizational structure of the White Greek community facilitates sexually objectifying approaches toward women. White fraternity men also have a larger pool of non-Greek women to engage; therefore, they are held less accountable for their relations with women because of a hyper level of anonymity. Although the presence of a fraternity house enables White fraternity men to be in control of sexual environments, it also constrains gender relations by offering nonprivate and nonintimate spaces.…

Our research further argues that structural settings shape how actors perceive others (e.g., as sexual, romantic, and/or holistic others) and reflect the racial and gender dynamics of college campuses including racial segregation and skewed gender ratios. Along these lines, hegemonic masculinity is about much more than gender beliefs and masculine performances. Hegemonic masculinity is also about normative accountability structures and the preservation of normative personal, social, and institutional resources. Privileges across gender, race, and status divides afford White fraternity men less accountability when performing a hegemonic masculine self during interactions with women. As shown here, under certain institutional arrangements, racial disadvantage, as with Black fraternity men, can decrease gender inequality and reduce a traditional hegemonic style of engagement toward women. However, race and/or class advantage, as with White fraternity men, and disadvantage, as with Black men, can increase gender inequality and propel a hegemonic presentation of self such as the "cool pose."

REFERENCES

DuBois, W. E. B. 1903. *The souls of black folk*. New York, NY: Dover.

DuBois, W. E. B. 1939. *Black folk, then and now: An essay in the history and sociology of the Negro race*. New York: Henry Holt.

25

Racialization, 'Flexible Ethnicity,' Gender, and Third-Generation Mexican American Identity

JESSICA M. VASQUEZ

Latinos have recently surpassed Blacks as the largest minority group in America and are poised, by mid-century, to end the majority status of Whites, giving no single racial/ethnic group a numerical plurality in the population. Latinos have dispersed all over the country, although they are most highly concentrated in the Southwest. With increasing time, Latinos have higher rates of naturalization, English fluency, greater presence in high-status occupations, higher labor market wages, and lower rates of poverty. As Latinos' presence in America extends over time, we are able to see the longitudinal effects of their assimilation. Vasquez captures some of the advantages and dilemmas third-generation Mexican-Americans face as they blend and intermarry with the existing population.
In fact, multi-racialism is one of the fastest-growing racial/ethnic categorizations, forcing the U.S. Census, as Kilty and Vidal de Haymes show in Chapter 4 on Census categorization, to modify their categories and terminology. Vasquez's study illustrates the variations Latinos may encounter in this country according to their appearance and associates. We also see differences in the ways multi-racial people choose to self-identify or self-present in various groupings or contexts.

SOURCE: From "Blurred Borders but Not 'Others': Racialization, Flexible 'Ethnicity,' Gender, and Third-Generation Mexican American Identity by Jessica M. Vasquez. *Sociological Perspective,* Vol. 53, No. 1, 2010. All rights reserved. Reprinted by permission of the University of California Press Journals.

As you read this chapter, you may want to think about how media or politically based presentations of Latino immigrants compare to the image presented here. How have Latinos, as a group, been upwardly mobile compared to other immigrant populations? What are some of the factors that help or hurt this? What do you think about public policies geared toward Latino immigration? How might you devise a solution to the issue of illegal Latino immigration?

Globalization and migration trespass the U.S.-Mexico border and create a distinct border culture. Mexican-origin people in borderland zones are conceptualized as "juggling cultures" and having "plural personalities." How, then, do third-generation Mexican Americans in these territorial and symbolic borderlands navigate their "plural" identities? How is this process both racialized and gendered? This article explores the dynamics of third-generation Mexican Americans' social position. Mexican Americans live in a "borderland" that is both racialized and gendered. Race and gender remain "foundations for systems of power and inequality" and have "overlapping and cumulative" effects on people's experience (Anderson and Collins 2007:5). Third-generation Mexican American men and women confront specific racialized and gendered stereotypes. Race is critically important in understanding the social construction of gender. Perceived by others as exotic or docile women and/or villainous men, third-generation Mexican Americans are subject to "controlling images" against which they are challenged to define themselves. "Controlling images" demarcate the boundary of belonging and are "major instrument[s] of power" in that they create and maintain race, class, and gender oppression (Collins 1991:68).

Mexican Americans in the Southwest exist within a vast geo-political and socio-cultural borderland. The physical space, part of Mexico prior to the Treaty of Guadalupe Hidalgo of 1848, extends along a 2,000-mile land border where distinct cultures, economies, and widely disparate standards of living converge. The Mexican-origin population in the United States—including immigrants and their descendents—has nearly doubled in size in every decade except for 1930–1960. In 1910, the Mexican-origin population was 385,000 as compared to the 2000 population of 20.6 million. Considerable differences exist between the generations of the Mexican-origin population. Generational status affects self-labeling practices, acculturation processes, and life outcomes. Currently, nearly seven million people are third-plus-generation Mexican Americans, yet the complexities of their borderland identities and experiences are rarely addressed. To date, the relationship between structural assimilation (education, occupations) and the absence of prejudice and discrimination has been inadequately studied, a hole that my research on middle-class third-generation Mexican Americans fills.

Using intensive interviews with third-generation Mexican Americans living in California, I argue that two poles of a *racialization process continuum* shape the identity experiences of third-generation Mexican Americans: "flexible ethnicity" and racialization. The *racialization process continuum* encapsulates the terms "flexible ethnicity" and racialization (which I use here to refer to enforcing a non-white racial status) in a larger context of racialization dynamics. Racialization, in the broad sense, includes the possibility of whitening, being identified and treated as non-Hispanic white. "Flexible ethnicity" is predicated upon being "whitened," that is, racialized as non-Hispanic white. As supporters of racialized assimilation assert, Latinos can be racialized as non-Hispanic white, Hispanic/ Latino, or non-Hispanic black.

"Flexible ethnicity" refers to the ability to deftly and effectively navigate different racial terrains and be considered an "insider" in more than one racial or ethnic group. Although "flexible ethnicity" may enable individuals to access resources and privileges, it cannot always be wielded to attain specific ends. "Racialization" refers to the process of imposing racial assignments on others and linking those racial ascriptions with differential expectations and value assessments. To borrow from Omi and Winant (1994:64), racialization, a fundamental categorization process, is "the extension of racial meaning to a previously racially unclassified relationship or group." Racialization of Mexican Americans as non-white or "off-white" anticipates and enforces difference, creating and reproducing social distance and unequal power dynamics.

In this article I argue that although many third-generation Mexican Americans exhibit "flexible ethnicity," racialization hampers their efforts to define the terms of their own race/ethnicity. U.S. society racializes Mexican American men more stringently than Mexican American women; also, women are allowed greater degrees of "flexible ethnicity." The gendered stereotypes to which Mexican Americans are subjected have major implications for their structural assimilation: men, figured as violent gangsters, encounter a more rigid barrier to mainstream acceptance than women, who are typecast as exotic.…

METHODOLOGY

This article is based on interviews with twenty-nine third-generation Mexican Americans. For this study I conducted interviews in one large metropolitan area in Northern California and another large urban region in Southern California. I located my research in California because it is the state with the largest Hispanic population and the highest numbers of in-migration from Mexico. According to U.S. Census Bureau estimates about the California population in 2007, 36.2 percent were of Hispanic origin, compared to the national average of 15.1 percent.

Thirteen women and sixteen men comprised my interview sample. Seven individuals were multiracial and twenty-two were monoracial, multiraciality in my sample meaning that a respondent has at least one non-Hispanic European-descent parent. While the tendency in my sample was for multiracials to be phenotypically ambiguous and/or bear similarity to whiteness, I do not intend to conflate multiraciality with light skin and European-like physical features or monoraciality with dark skin and indigenous somatic characteristics. Where appropriate, I refer to skin tone, and thereby acknowledge the range of skin colors that both multiracials and monoracials can possess. I always use pseudonyms with reference to my interviewees to protect their anonymity. Pseudonyms are assigned to correspond with interviewees' Hispanic or non-Hispanic first and last names.

The age range of respondents is from seventeen to forty-five years old, with twenty-eight as the median age. The vast majority of respondents were at least college-educated or, if high school students, were on an educational track to college. Among the non-full-time students, virtually all held white-collar jobs, such as an architect, teacher, college counselor, police officer, and sales associate. By measures of individual or household income, educational level, and occupation, all respondents were middle-class and structurally integrated into U.S. society....

RACIALIZATION AND "FLEXIBLE ETHNICITY"

Third-generation Mexican Americans experience a *racialization process continuum* that runs the gamut from racialization, an outsider-imposed racial identity, to "flexible ethnicity," an identity that effectively crosses and engages both Mexican and American social circles and communities. In both racialization and "flexible ethnicity," there can be a mismatch between asserted and ascribed identity, limited by enforcement of racial boundaries. In what follows, I present a section on racialization, succeeded by a section on "flexible ethnicity." Both data sections are organized identically: I present the experiences of men and women who are monoracial, and then I turn to an analysis of multiracial men and women.

Racialization: Forced and Enforced Racial Identity

"All ethnicities are not equal, all are not symbolic, costless, and voluntary" (Waters 1990:160). Indeed, the "symbolic ethnicity" or "ethnic option" that white ethnics enjoy—they may practice elements of their ethnicity with no detrimental effects—is not available to members of all groups. Racialization prohibits racial minorities, marked as nonwhite and thereby devalued, from having a symbolic or optional race/ethnicity. Racialization refers to the expectation that "differences in skin color, or other racially coded characteristics, explain social differences.... Temperament, sexuality, intelligence, athletic ability, aesthetic preferences, and so on are presumed to be fixed and discernable from the palpable mark of race" (Omi and Winant 1994:60).

Monoracial Men Ongoing immigration from Mexico creates a particular social context for later-generation Mexican Americans in California to negotiate. At times, Mexican Americans sympathize and bond with Mexican immigrants, and at other times, they distinguish and separate themselves from the newcomers. A motivation for maintaining social distance from the new arrivals is the threat that stereotypes against immigrants will adversely affect Mexican Americans born in the United States. Mexican immigration, coupled with the inability of mainstream society to notice differences between immigrants and natives, complicates the situation of Mexican Americans. In many respects, as Mexican immigrants are racialized, so too are Mexican Americans.

In a context of sustained Mexican immigration where "dark skin becomes a frequently invoked indicator of foreignness" (Jiménez 2008:1548–9), natives assume that people of Mexican descent are foreign *and* unauthorized. This often-erroneous assumption provokes nativist fears of the "browning" or "Latin Americanization" of the United States and engraves inter-ethnic boundaries, often drawn and guarded via discrimination. A gendered stereotype—or "controlling image"—that ensnared dark-skinned monoracial men most often was that of the potentially violent Latino male.

RICK TORRES: I think when people look at me they see my scars and the moustache ... a lot of times people get the impression that I'm a gangster or a *cholo*. That really bothers me because that is just totally what I'm not about.

MOISES RAMOS: This lady saw me [in a bar with friends] and she said out loud, "oh, he's mean looking." I was thrown by it. For some reason I just smiled and she was like, "oh my god, he has a dimple." We started talking and she asked what I did and I told her [a high school career counselor]. "Oh my God, I never would have guessed." I was like, "obviously you shouldn't judge a book by its cover." She had this image of us—Mexicans—[as] bald and mean and involved in gangs.

Racialization can occur even if one's race is incorrectly perceived. "Multiple passing," where an individual "can easily pass and blend in and out of their cultures and ethnicities, as well as others not associated with their background" (Guevarra 2003:84), can lead to "cross discrimination," when a minority may suffer from discrimination aimed at a *different* minority group. This happened to Tom Acevedo, a slim monoracial teenager who has dark skin, black hair, and dark brown eyes, and is sometimes mistaken for an Arab. The worst instance occurred several months after the September 11, 2001 terrorist attacks when Tom was thirteen years old and was stopped by a police officer as he was walking home from school:

TOM: I got stopped because I was carrying a suitcase with my trumpet in it. I kinda looked like I was Arab—the cop said so. He pulled me over ... and questioned me.... And this is after September 11th, too, so they got a little more suspicious about that.

JMV: What did you think was going on there?

TOM: I believe he thought I had a bomb in my hands. [Laughs] But I told him "nah." I opened up the suitcase ... and I showed him the trumpet.... He's like, "OK, I see." I told him, "I'm not a terrorist or anything, sir, so don't take me for that."

JMV: How did you respond to that internally?

TOM: That kind of freaked me out, because you know, most of my friends they do kind of look Arab but they are actually Mexican. It just freaked me out. He might think I actually have a gun and he might not trust [me]. He might take me to jail.... *I didn't want to get arrested for being Mexican.* [Laughs] [my emphasis]

Monoracial Women   In contrast to Mexican American monoracial men, Mexican American monoracial women were included by mainstream institutions until their Mexican heritage was discovered. At this point, they were excluded, experiencing withdrawal of access to resources, support, or jobs. Phenotype, surname, and "cultural toolkit" (such as language, dress, and cultural knowledge) are markers of difference that people use to target others for discrimination. It is easy to see the process of racialization at work in situations when a Mexican American initially "passes" as non-Hispanic white and is later "outed" as Mexican American. The change in how people are defined racially often leads to a corresponding change in treatment—loss of jobs, withholding of support or friendships, or withdrawing of resources.

The blonde, light-skinned sister of a woman I gave the pseudonym "Veronica Guzman" was mentored by the high school Vice Principal until she was "discovered" to be Mexican American. Veronica and her similarly brown-haired, dark-skinned brother had markedly different relationships with their high school Vice Principal than did their more Anglo-looking sister:

> My sister, she's light skinned, she looks American, but she's a Mexican American. The Vice Principal thought she was white. They [the administration] didn't know that my brother and sister and I were related [despite sharing the same surname]. They would send information for her in English and for my brother they would send it in Spanish. She's blonde and according to them she's a *guera* [white woman] and he's *mexicano,* Mexican. The vice principal treated her differently [better] because she was a blonde, she was a *guera.* When they found out she was Hispanic, she was a Mexican, it wasn't the same anymore.

Caitlyn Benavidas similarly experienced racialization when she was "found out" to be Mexican American. While Caitlyn often "passes" as non-Hispanic white or Persian, she becomes vulnerable to racialization at the moment when others realize they have misread her racial identity:

> I've had some ... really uncomfortable situations with people thinking that I was not Mexican.... I was hired as a waitress by a Middle Eastern family and they hired me thinking I was Greek or Persian. I started speaking Spanish to the busboys once and they were like, "Why the hell are you speaking Spanish? You can't be Mexican." And basically went off on me about how they probably wouldn't have hired me if they had known I was [Mexican].

Multiracial Men   In contrast to their monoracial peers who routinely experience overt racialization, multiracial Mexican American men are racialized to the extent they are perceived as non-European-descent white. Despite the ability of some to "pass" as non-Hispanic white, the "controlling images" for multiracial men are the same as for their monoracial counterparts: villains and gangsters. Two of the four multiracial Mexican American men interviewed claim that they have never suffered discrimination, both men crediting not looking stereo-typically "Mexican" as being a shield (one man also has a non-Hispanic-sounding surname).

Pierre-Mecatl Ramirez, a master's-degree-holding twenty-nine-year-old Mexican American and French multiracial man with a two-adult household income of approximately $100,000, serves as a prime example of how "'authentic' black and ... brown identity is imagined as lower-class, urban, often violent ... and male" (Bettie 2003:48). Pierre-Mecatl, whose pseudonym mirrors his French and indigenous heritage, explains how he feels he was profiled according to a racialized and gendered "controlling image" of a threatening, potentially violent criminal by a police officer when he was a teenager:

> I went to a little park downtown.... This cop comes up to me, just out of the blue, harassing me, asking me these questions about this piece of graffiti next to me that I hadn't even seen.... When I was a teenager, I affiliated with the Gothic subculture, so I was dressed in a velvet blazer and a bowler and this guy's talking to me about this graffiti. "Well this is Mexican graffiti." ... I caught that he was basically saying, "This is Mexican gang graffiti and your last name is Ramirez." Because [he] had my ID. He gave me back my license and told me to get the fuck out of there.

Multiracial Women Like multiracial men, multiracial women undergo racialization when they are perceived as non-European descent white. Racialization can involve the assumption that Mexican Americans possess superior and specific cultural and Spanish language knowledge. This pervasive assumption of ethnic cultural and language knowledge obscures the diversity within the Mexican-origin community and reifies a particular version of Mexican ethnicity as authentic. Jillian Rosenberg, whose fictional name reflects her Jewish heritage, experiences a "magnified moment" of racialization when she reveals her Mexican heritage to her college counselor. Jillian's mother is Mexican American and her father is Jewish; she has light skin and brown hair and is presumed to be non-Hispanic white. In a situation with significant long-term ramifications, a college career counselor told Jillian that the program she was interested in was for minorities only. When she told him that indeed she is Mexican American, he retorted that she would have to "prove" that she is a "real Mexican" on the application. Jillian recounts the conversation:

> In college ... I went to speak with a pre-med advisor about my plans and mapping out my courses. He's known for being a really big jerk.... I asked the counselor about a minority program I heard about. "Oh, that's just for minorities."
>
> I said, "Oh, actually I am Mexican."
>
> "Oh, well, if you want to take advantage of that, on applications—I don't want to sound un-P.C. [politically correct] here or anything—you are really going to have to prove that you are a real Mexican."
>
> I was like, "Excuse me?"

"Well, I mean, are you part of any cultural organizations here?"

"No, don't really agree with a lot of their purposes so I chose not to join them."

"Well, have you gone to cultural events or organized Latino heritage month?"

"No, like I said, I don't really agree with their purposes."

"Well, do you speak Spanish fluently?"

"Actually, it was my first language, but I lost it."

"Well, are you taking any Spanish classes at least?"

"No, it doesn't fit into my schedule."

"Well, you are really going to have do something here to prove that you are a real Mexican."

The college counselor's effort to racialize Jillian is notable: he challenges her authenticity (expressing skepticism about whether she is "really" Mexican), commands her to fit a preconceived and static notion of Mexican American identity, and tries to limit her access to programs for minority students.

Racialization in schools can involve withholding administrator support and information, as seen above, or can entail embarrassing students in the classroom by wrongfully assuming they possess superior knowledge of Spanish. In her Spanish class, Amalia Ruiz, of Mexican and Italian descent, felt she was presumed to be conversant in the language:

> I'm not [fluent in Spanish]. In high school … I got really pissed off because a teacher asked me a question in … second-year Spanish. I just barely passed first-year Spanish. And I didn't know the answer. And in front of the whole class he said, "You know, just because you speak Spanish at home doesn't mean that you don't have to study for this class." What?! I felt so angry and humiliated. First of all, I didn't know Spanish so I don't speak it at home! He assumed that I did. And he scolded me in front of the whole class…. He thought I was just … a lazy Mexican, basically.

A "controlling image" with which Mexican American multiracial women in my sample were confronted is the gendered stereotype of being exotic, erotic, and sexual. Even if not explicit, "there is a sexual message imbedded in ethnic stereotypes and categories" (Nagel 2003:255). Renata Contreras, a mixed race half non-Hispanic white and half Mexican American woman, feels stereotyped by both Caucasian and Mexican-origin men. She described white men's gendered stereotype of Latinas as sexually attractive or promiscuous:

> [From] the Caucasian perspective, being half-Latina is pretty sexy. [She] is more fiery and willing to do things. The perception of "Oh, spicy Latina!" … I wonder if that has to do with the stereotype of the servile woman who is willing to do anything. I think that's what the Caucasian guys think about Latin women…. Maybe they've seen lots of Latin

women with lots of babies so they think they are promiscuous....
Or what they see in movies about the way Latinas are played as the
seductress....

She contrasts this with the reception she often receives from Mexican men:
"From Mexican guys' perspective, I think they just see me as, 'Oh, you look
like a *gringa* [white girl].' That's attractive too because you are different from
our women, like *gringa* is somehow better ... like it's classier or something."
Renata refers to how the lure of marrying into an equal or more prestigious
social group has influenced Mexican men's reactions to her. Renata's gendered
borderland is impacted by stereotypes from the two racial/ethnic groups to
which she belongs. On the one hand, white men see Renata as appealing
because she is "spicy" and "willing," and on the other hand, Mexican men see
her as a desirable marital partner due to her "classier" non-Hispanic white
heritage.

   Like Renata, Jillian Rosenberg, introduced earlier, is multiracial. Despite
often "passing" as non-Hispanic white, since racialization and "flexible ethnicity"
are not mutually exclusive categories, she is subjected to racialized and gendered
stereotypes. Growing up in a Southern California town in her youth, Jillian was
ashamed of her Mexican heritage because she felt it carried negative, lower class
connotations. It was "dirty" and something from which to actively estrange herself: "I had really awful stereotypes of what Mexican people were like too, like
all Mexican people are gardeners and maids.... It was always ... the half of me
that I had to hide.... I was soiled in some way because I had Mexican blood in
me." In contrast, her peers at Yale University, who were enthusiastically looking
for some "uniqueness" to mark them as "not just white," exoticized her. In her
new social context, Jillian experienced an "identity crisis" about the different
attributions of meaning to her Mexican origins:

> Now, going to Yale, it's a total reversal. I'm having a completely new
> identity crisis. Because suddenly it's cool to be Mexican, it's exotic, and
> "I'm of a different race, I'm not just a white kid" [her voice gets
> breathless and sexualized]. It's a really diverse campus and people defi-
> nitely put a lot of emphasis on diversity. It's just so funny. Now all the
> white kids are like, "Oh my god, I wish I were half of something like
> you are, at least, to make me exotic." I'm like, I never thought of being
> Mexican as exotic. I always thought it was gross and dirty and lame and
> not exotic. "Oh, your eyes are so Mexican, they are so exotic and
> pretty." ... All of growing up I was trying hard to be white and I was
> too Mexican....

"Flexible Ethnicity": The Privilege and Predicament of Being "White-xican" and Living in Two Worlds

Many third-generation Mexican Americans live "between" Mexican and
American social worlds. Racial liminality is due in part to the physical appearance

of Mexican Americans, particularly those of "mixed race" who are offspring of non-Hispanic white and Mexican American partnerships. Multiracial (and to varying degrees, monoracial) Mexican Americans tend to have pale skin and less stereo-typically indigenous features that allows them to be perceived and treated as non-Hispanic white at times. For some, the ambiguity of racial liminality leads to crises, while for others it leads to opportunities.

Many third-generation Mexican Americans display ... the ability to navigate two different social worlds, that is, mainstream U.S. culture and a Mexican-oriented community. Their U.S. citizenship and their possession of "cultural toolkits" and skill sets marks the third generation as American, allowing many to play out flexibly their racial background, magnifying or minimizing aspects of their heritage. The volitional aspect of "flexible ethnicity," however, is limited by how others perceive, treat, and racially mark third-generation individuals. Mexican Americans have a variety of experiences along the *racialization process continuum*, with racialization as nonwhite and "flexible ethnicity," which involves whitening, at the poles....

Monoracial Men   A man I assigned the pseudonym "Auscencio Dos Santos," a monoracial child of two Chicano activist parents, finds "flexible ethnicity" an advantage as well as a source of identity struggles. He tells a story about living in a borderland where he is both Mexican *and* American. Auscencio is a heavy-set twenty-eight-year-old man with light skin and short dark hair. He has deep brown eyes, a short goatee, and wore a modern shirt reminiscent of a *guayabera* with intricate threadwork embroidery. In describing himself, Auscencio said he is "as much hamburger as taco," thus portraying himself as living in the middle of a cultural borderland. Auscencio, who goes by Ceño among friends and family, reviews his bicultural, borderland status:

> I think it's clear that I'm probably more Americanized than I think I would like to admit. I've got my Internet.... My whole lifestyle is pretty American. But I do feel, at times ... when I read history books or I see the horrible things that this country has done, I'm like, "Oh, man, I'm so glad I'm not 100 percent American." I don't feel I fit in anywhere. I'm right in between—and it's okay. Especially here in California. I think my attitude would be very different in somewhere like Nebraska. Or Montana. Your environment dictates who you are, and almost how you act.... It depends where I am too, because when I was living in El Paso, Texas ... I felt much more Mexicano there than I did American. Because I was "Ceño" and my friends were Julio and Oscar. I would go to the store and the woman would swipe my card and say, "Thank you, Mr. Dos Santos" [correct Spanish pronunciation]. I'd be like, "Yeah!" ... There is something to be said for being completely accepted and understood.

Auscencio is "structurally assimilated" (education and occupation), "linguistically assimilated" (language), but not "identificationally assimilated" (identification). In essence, Auscencio is "in" mainstream culture yet he is not "of" mainstream culture. Note that Auscencio distinguishes between being "an American" and being a "white American." Auscencio resists the "fusion of race and nationality" that suggests that the American "imagined community" exists exclusively of non-Hispanic white Americans.

Reflecting the process of "racialized assimilation," monoracial men were more inclined than any other category to affiliate themselves with Mexico. "Flexible ethnicity" can be an uncomfortable social space. Dillon Castillo, a tall, slim, lanky monoracial seventeen-year-old with dark hair and light skin, wards off the discomfort of "flexible ethnicity" by making efforts to *racialize himself:* "I'm usually perceived white. So I hear all the jokes. That's not an advantage." As a coping mechanism, Dillon wears his heritage around him, displaying a Mexican *zarape* (blanket) on the seat of his low-rider car and *la Virgen de Guadalupe* (patron saint of Mexico) on his window. Here, attempts to racialize one's self are attempts to disarm racist comments that are regularly aired when individuals are assumed to be non-Mexican American.

Monoracial Women Women did not explicitly mention attempting to racialize themselves. A common female response is expressed by monoracial Samantha Diaz, who exists in a liminal racial space, neither "Mexican Mexican" nor "American." Samantha has medium-brown skin and long, straight, black hair: "I see myself as Mexican, but a little more whitewashed. I'm part of the generation that is a little lost. Some people consider me not Mexican enough but I'm not American enough. So I'm really stuck in the middle." Samantha considers how she can morph into different social contexts because of her "flexible ethnicity" and her biculturalism: "When I'm around my Mexican friends, I'm Mexican. When I'm around people at work [Anglo dominant law office], I'm less Mexican. Honestly, I slip into it when it's convenient. When I have to fit whoever I am around my personality changes." Samantha is more outgoing among Mexican-origin peers and more restrained among non-Hispanic whites. While Samantha can call on her "plurality of selves" to "fit" a number of different social situations, the voluntary nature of personality or cultural shifts should not be overstated because, depending on the situation, her flexibility can be limited by racialization. For example, one day when Samantha left her high school, she found a racist epithet spray-painted on her car because she was perceived as racially "other."

Alternatively, demonstrating the significance of skin color and physical features, Carmina Dos Santos, who is monoracial with light hair, light skin, and green eyes, tells me about the privilege of "passing" for non-Hispanic white:

It's an advantage. I have the privilege of blend[ing] in…. I think there are certain privileges to looking not stereotypically Mexican. Things I take for granted like not being followed in a store, not being labeled as somebody who doesn't have money…. The other thing is that … because I am so light I see things and I hear things that other people say just assuming that I'm on their side.

Multiracial Men   Multiracial men find "flexible ethnicity" to be both uncomfortable and advantageous. All multiracial men interviewed found they were "an insider to jokes about Mexicans," and they would confront the speaker if s/he "mattered" to him. Andrew Rosenberg noted that being both Mexican and Jewish in racially tracked high school classes was distinctly disturbing, asking rhetorically, "There was always a divide between the white kids and Mexicans—what was *I* supposed to do?" Feeling conflicted about his nebulous position in a racially stratified high school, he commented, "It's scary because you don't know where you fit in." Yet despite these emotional challenges, Andrew delights in belonging to two cultures: "It's kind of fun, though, because you get a little bit of both. So even though it's been hard it ... it's cool." Being in "two worlds at the same time" allows for some benefits such as "helping him to understand other people." Andrew lives at the crossroads of borderland, American, and Mexican cultures—his flexibility providing insight into race relations by allowing him to "see two completely different cultures, but seeing them get along, at least in the context of his family."

Multiracial Women   Multiracial men were reflective about their experience of ethnic flexibility, but they did not express the same heightened self-consciousness about emphasizing or de-emphasizing their racial/ethnic identity as did the females. Multiracial women made calculated decisions about the identity they wished to project. Renata Contreras, easily perceived as non-Hispanic white because of her blonde hair and pale skin inherited from her Caucasian mother, makes thoughtful decisions about her racial claims depending on her social context:

> If it's the Junior League or something like that I ... probably would put white and ignore the Hispanic part. Because I just feel like the people there would judge me, "Oh, a Hispanic, how nice, what diversity" [sticky sweet and singsong voice]. In high school I played tennis a lot and we'd go to the tennis club in Montecito [high-class neighborhood], I wouldn't highlight the Mexican part.... I don't need that kind of judgment. In those situations, I'd probably just put white. Then white-slash-Mexican American probably for job applications or [if] I feel like people really would have an open mind or encourage diversity.

Most respondents who experienced "flexible ethnicity" were women and/or multiracials (or light-skinned monoracials), though they too were subject to racialization. Recall that while "flexible ethnicity" was a common experience, two women and four men claimed not to experience the phenomenon. Mexican American women experienced more "flexible ethnicity" and less racialization than men. Mexican American women, viewed as "soft," desirable, docile, and sexual, are seen as less threatening, and therefore more acceptable than Mexican American men who are viewed as violent, aggressive, and threatening. While men attempted to racialize themselves and affiliate with Mexico, reflecting racialized assimilation, women did not vociferously claim connection to Mexico. This suggests a gender division regarding a feeling of comfort, inclusion, and acceptance in mainstream America. In answer to a question on respondents' relationship to what they consider "mainstream American society," all respondents except one man said they felt a part of mainstream culture. Exemplifying their racialized, borderland status, nine men and three women qualified this affirmation by asserting their bicultural status. Of those who asserted their bicultural status, ten were monoracial and two were multiracial Mexican Americans. Those who were granted flexibility circulated in the U.S. mainstream, suggesting that the dominant racial group is more tolerant of women than men and multiracials than monoracials.

CONCLUSION

… These third-generation Mexican Americans who are middle-class and structurally integrated into U.S. occupations, institutions, and mainstream culture live at an identity and cultural "crossroads." Being both Mexican and American, yet not entirely either one, the image of a nimble and pluralistic "mestiza consciousness" holds true for some third-generation Mexican Americans. Respondents who are either part European-descent or have lighter skin and hair colors have "flexible ethnicity," traversing multiple racial terrains with dexterity. Some found this "two world" perspective to be an advantage of being bicultural, in part because their experiences of cultures meeting on equal ground in their homes and families facilitated an openness and comprehension of other cultures. Others, however, were beset with identity crises or a "struggle of borders" (Gueverra 2003:82) that were neither positive nor voluntary. While "flexible ethnicity" affords individuals room to negotiate a variety of racial identities, this is not entirely an agent-centered process or state of being. "Flexible ethnicity" posits that a variety of ethnic scripts or presentations of self are available to actors, but that the array of options is constrained by racialization.

REFERENCES

Anderson, Margaret L. and Patricia Hill Collins. 2007. "Why Race, Class, and Gender Still Matter." Pp. 1–16 in *Race, Class, & Gender,* edited by M. L. Anderson and P. H. Collins. Belmont, CA: Wadsworth.

Bettie, Julie. 2003. *Women without Class: Girls, Race, and Identity.* Berkeley: University of California Press.

_____. 1991. *Black Feminist Thought,* Vol. 2. New York: Routledge.

Guevarra, Rudy P., Jr. 2003. "Burritos and Bagoong: Mexipinos and Multiethnic Identity in San Diego, California." Pp. 73–96 in *Crossing Lines: Race and Mixed Race across the Geohistorical Divide,* edited by M. Coronado, J. Rudy, P. Guevarra, J. Moniz, and L. F. Szanto. Santa Barbara, CA: Multiethnic Student Outreach.

Jiménez, Tomás R. 2008. "Mexican-Immigrant Replenishment and the Continuing Significance of Ethnicity and Race." *American Journal of Sociology* 113(6):1527–67.

Omi, Michael, and Howard Winant. 1994. *Racial Formation in the United States: From the 1960s to the 1990s*, 2nd edition. New York: Routledge.

Nagel, Joane. 2003. *Race, Ethnicity, and Sexuality: Intimate Intersections, Forbidden Frontiers.* New York: Oxford University Press.

Waters, Mary C. 1990. *Ethnic Options: Choosing Identities in America.* Berkeley: University of California Press.

26

How Models Manage
Bodily Capital

ASHLEY MEARS AND WILLIAM FINLAY

Models represent the ultimate embodiment of femininity. Through television, magazines, movies, and other elements of our media culture, women learn that their looks symbolize their value, and that they must be young, thin, and pretty to have worth. Aging is a traumatic experience for women as they peak early and progressively lose their cultural capital; men, in contrast, earn more money and increase their cultural capital as they age. The use of cosmetic surgery by broadening numbers attests to the widespread pressures felt by women to conform to demanding yet ultimately unattainable norms of appearance. Mears and Finlay's study shows how models, who appear successful in their body management, suffer intimidation and loss of self-esteem on a far greater scale than typical women, as their worth is even more highly tied to their appearance. Far from experiencing the glamorous and luxurious life that most people imagine, models work long and often physically grueling hours in an occupation characterized by unsteady employment. More than most, models become objectified and commodified, a process in which they willingly engage to earn a living. To what extent do you see some of the attitudes and behaviors of these models in the women around you? Why do you think women are so forced to live in their bodies while men can earn their identity through their achievement? What kinds of things might free women from the tyranny of being judged so heavily on their appearance? Is the trade-off they make with men, exchanging their looks for financial security, still relevant in today's society where so many women work?

SOURCE: Ashley Mears and William Finlay, "Not Just a Paper Doll: How Models Manage Bodily Capital and Why They Perform Emotional Labor," from *Journal of Contemporary Ethnography*, 34(3), June 2005, pp. 314–43. Copyright © 2005 by SAGE Publications. Reprinted by permission of SAGE Publications, Inc.

Modeling is an uncertain and stressful occupation. Although a handful of models achieve "superstar" status, most of them work intermittently during the course of a career that peaks in their late teens and terminates in their mid-twenties. During their brief careers, they are rejected for jobs, they are humiliated on the job, they struggle to maintain their physical appearance, and they face continual competition from younger and thinner models. It is an occupation that requires them to be passive and silent at work while paying careful attention to the management of their "bodily capital" (Wacquant 2004)....

In this article, we address how models manage their physical appearance and argue that models manage their bodily capital—the expectation that they be young and thin—by lying about their age and by turning their body into an object that they rigorously evaluate and monitor. We argue that they embrace emotional labor—by attempting to charm agents, clients, and photographers and by defining their work as acting—because they believe it makes it more likely that they will be hired and because it enables them to find dignity in work that mostly consists of the passive display of physical beauty....

WHY STUDY MODELS?

Modeling is a valuable occupation for sociological study for at least three reasons. First, modeling is both similar to and yet different from other occupations, such as stripping and exotic dancing, in which people make their living by displaying their bodies. Although models perform their work by wearing rather than removing their clothes, nevertheless, the bodies of models, like those of strippers, are observed, evaluated, criticized, and touched. An analysis of models' work, therefore, is a way of exploring issues that have been raised in studies of sex workers: contemporary definitions of beauty and desirability, the objectification of the body, and workers' efforts to assert their autonomy.

Second, modeling provides a good example of how workers learn how to monitor and manage their bodily capital. Wacquant (2004) uses the concept of *bodily capital* to describe the way in which boxers meticulously monitor all parts of their bodies in preparation for a fight. This includes training, conditioning, and most crucial of all in the days immediately before the bout, getting their weight down to the required weight limit. Managing bodily capital is, of course, a major concern for all athletes, although few are required to be as obsessed with weight—with some exceptions such as wrestlers, gymnasts, and lightweight rowers—as are boxers.

The attention that models bring to their weight and general physical state makes them closer to boxers than strippers. Like boxers, they are routinely measured and weighed. Models who deviate more than slightly from the ideal height (between five feet eight inches and 6 feet), physical dimensions (34-24-34 inches), and age (25 or younger) will have considerable difficulty in obtaining employment, whereas strippers do not have to meet equivalent norms of bodily conformity. Although strippers monitor their weight and are attentive to their physical appearance, they vary widely in age and appearance, which reflects in part the fact that customers' tastes are diverse. Unlike boxers and other athletes, however, models have to learn how to manage their bodily capital without coaching and away from their work sites. It is a solitary activity—every model has to learn how to do it by herself. If she fails, her agent or client will simply pick someone else for the job....

STUDYING MODELS

The first author used participation observation and interviews to collect the data for this study. Having modeled in Atlanta and abroad for four years, she had access to and familiarity with the model's occupational world. She kept daily field notes during her final semester in college working as a model, totaling an average of fifteen hours of participant observation per week during the course of four months, from May through August 2002. She also conducted fifteen semi-structured interviews with female Atlanta-based models in June and July 2002. Both authors conducted independent analyses of the field notes and interviews.

Only women were selected for this study for two reasons. First, modeling provides greater opportunities for women than men—the well-known supermodels are overwhelmingly female, for example. Naomi Wolf (1991) has pointed out that modeling is one of the very few occupations in which women routinely earn more than men. Second, a career as a model remains an enduring fantasy for girls, much as many boys may dream of becoming a professional athlete. Moreover, modeling is glorified in American culture as an exciting and prestigious career for young women, especially by popular culture outlets such as teenage fashion magazines.

Respondents ranged in age from sixteen to thirty-three years old. Eleven respondents were white, two black, and two of mixed racial backgrounds. They were asked questions about their careers, their earnings, their relationships with agents, clients, and other models, and what they liked and did not like about modeling. The interviews lasted one to two hours in length and were tape-recorded....The researcher invited models to participate during "bookings" or jobs that they had together. Having modeled in Atlanta for several years, the researcher was well acquainted with many of the few hundred women who work fairly regularly in this city, especially if they did fashion shows. At two of these shows, while waiting backstage for hair and makeup to be done, models were asked if they were willing to be interviewed for this project. The researcher deliberately selected models with varying years of work experience to make the sample as diverse as possible....

MANAGING BODILY CAPITAL

If a model works regularly, it means that she has learned to manage her bodily capital effectively. This consists of two processes. First, she... has to be able to take the criticism and rejection that is a constant feature of modeling, even for models who have enjoyed considerable success. Second, she must maintain her bodily capital as she gets older and heavier....

TAKING CRITICISM

Rejection is guaranteed in the modeling industry because there are too many models chasing too few jobs. A model may travel for an hour to a casting, and spend another thirty minutes waiting in line, only for the client to glance her over and make an immediate decision that she is not right for the job. A model is rejected for any number of reasons. Her look may not be exactly what the client wants or the client may consider her nose too big or her bust too small. She may be too tall or her hair may be too dark or too short. Most damning of all, she may be dismissed as too fat. She may never even know why she was rejected, but will be left to wonder, as many respondents did, "What did I do wrong?" Whatever the reason, models must harden themselves to a barrage of rejections, as Amelia (twenty-six, seven years of modeling) explained:

> I mean, you hear a hundred "no's" before you hear a "yes" with everything. So you know, everyone thinks that we have these huge egos and, in reality, it's probably quite the opposite.

Rejection is common to a number of occupations, particularly those that require occupants to persuade others to buy their products or service....

Compounding models' sense of failure, however, is the criticism, often detailed and explicit, that frequently accompanies these rejections. One occupational hazard of modeling is that everyone—clients, agents, hairstylists, photographers, makeup artists, and designers—feels entitled to make brutally pointed comments about models' physical deficiencies. Another is that this rejection is taken very personally because it challenges a model's conception of her own beauty. Disparaging remarks that could be construed as sexual harassment in most other job settings are taken as a given in the model's daily routine.

The most common criticism that models hear is that they are overweight. Of course, the average model, at five feet eleven inches and 117 pounds, is underweight relative to the average American woman who is five feet four inches and weighs 140 pounds. Yet even the thinnest of models in our sample constantly fretted about their body weight, which is consistent with Brenner and Cunningham's (1992) study of female fashion models.

Kim (twenty-three, five years of modeling) mentioned she had put on ten pounds while living in Germany. She had gone there to work but had received few modeling jobs, so she had spent the time hanging out with friends— "I drank, I ate bread, I ate cheese." She described what happened when she returned to Atlanta:

> I had a shoot the day after I came back. I went into the agency two days after that shoot. First thing I hear is, "Kim, uh, Sammy, the photographer,

called. He said your ass is too big. You need to lose weight." I just burst into tears.... I couldn't deal with that. And then I called them back and I was like, "I'm sorry. I understand that I, I know that I gained weight. I'm very aware of that. That just, I just took that really to heart." And they were like, "you can't take this to heart, this is a business. You just need to understand. Ten pounds gone, you're working again."

Sophia (thirty-three, ten years of modeling) has a number of stories about the scrutiny to which her eating and weight have been subjected. In Milan, a booker told her not to eat fruit:

"No fruit! People in concentration camps didn't have fruit." I swear to you. That's I think one of my all-time favorite quotes that someone in the business said to me, that people in concentration camps don't have fruit. I was like, "Well, I didn't know that I was in a concentration camp."

In New York, she had an agent who would conduct unannounced "weigh-ins" at the agency. She would take each model into the bathroom where she would weigh her and measure her waist.

Another way that models are humiliated about their weight is by being asked to try on clothes that are too small. Models have to fit into clothing, ranging from size zero to size eight, to book their jobs. Fitting into the clothes is generally not a problem in catalogue shoots because these usually come in the larger sizes (i.e., six to eight) and can be pinned on the models if the fit is not perfect. Runway shows, on the other hand, often present a challenge because these clothes have to be worn properly and because they come in far smaller sizes. As Simone (twenty-four, seven years of modeling) observed

For runway jobs … they come in with their sample sizes that have been made on these really ultra skinny girls in New York, and in order to book the show you have to be able to fit in the clothes and look good in them.

If a model does not fit into them, she risks summary public dismissal from the show to the accompaniment of disparaging comments about her being "too big." Even if she is not fired, the designer's attempts to alter the clothes to fit her are likely to cause considerable embarrassment.

Fitting into clothes is an inherent problem in modeling because of the age-body relationship. Most models begin their careers when they are young, barely out of their teens in some cases, an age at which they still retain much of the shape of their skinny, prepubescent bodies. As they age, their bodies develop their mature shape and no longer fit into small sample sizes. Kelly (eighteen, six years of modeling) explained that her agents and clients had always known me as little size zero, fourteen-year-old, cute little girl, and all of a sudden, I was getting some curves on me and they didn't know what to do. They thought it was because I was getting fat and really it was just because I was growing up.

These models only have to look around them to see their competition—younger and thinner models who do fit into the tiny clothes. Heather (twenty-two, seven years of modeling) said,

I know that I used to be fifteen, sixteen, and I used to be skinny-minny and I used to be the fresh face. And I'm not anymore.

Models are rejected so frequently and so quickly that it is easy for them to become disheartened, especially if they are newcomers to the business. It is hard for new models to get used to a system in which clients make instant and seemingly arbitrary decisions about which models they do want and which ones they do not want.

A model has to be able to handle rejection to succeed. First, she must accept it without becoming too discouraged. This means not taking it "personally." Although rejection is in fact deeply personal, models try to remind themselves that it is just their external appearance that is being turned down. Cameron (nineteen, two years of modeling) explained that she had to tell herself to accept rejection without considering herself a failure:

It probably isn't just because of you…. You're gorgeous, but they might want a different look, and I'm not the right look right now. So you have got to say to yourself, it's not you….

The second way in which models handle rejection is by working on their external appearance. Even though they believe that luck and timing shape careers, they are not fatalists. Instead, they spend considerable time and effort on managing their bodies to increase their chances of employment.

TAMING THE BODY

Agencies and clients prefer models who are young and thin. If a model can conceal her true age and keep her weight down, she is more likely to book jobs. Both, with some effort, are possible. Models, of course, cannot literally prevent themselves from getting older, but they can and do lie about their age. When the first author met with agents in New York, her Atlanta agency instructed her to bump her age down from nineteen to eighteen. She lied to bookers, clients, photographers, and even to other models if they asked her age. Josie, who is twenty-six, claims to be twenty-two, and she says she has altered her date of birth in her passport to conceal her true age from agents when she works abroad.

Although age is a disqualifying factor for models, it can be circumvented, particularly if a model looks younger than she really is (this is what enables models older than twenty-five to continue working). More difficult to hide and more damaging to a model's career is weight gain. Models fear the effects of time on their figures. Simone (twenty-four, seven years of modeling) does sit-ups because she feels that her stomach is "starting to get a little bigger than it used to be." The weight gain that accompanies aging is a threat to the livelihoods of even the youngest and thinnest models—changes to their bodies may be slower than is normal for someone of their age, but they are inevitable. Amanda (seventeen, two years of modeling) has gone from a size zero to a size four during the period she has been modeling. Despite having thirty-four-inch hips and a perfectly flat stomach, younger models make her feel insecure about her body:

> You'll go to a job when there will be a younger girl who is like thirteen or fourteen or something and is really tiny and I'm like, "Oh, man, maybe I need to be skinnier."

Faced with the twin threats of demography and younger rivals, models go to considerable lengths to maintain their size and weight. Some monitor themselves and what they eat very closely. Kelly (eighteen, six years of modeling) keeps a tape measure on her kitchen table, which she uses daily to check her waist and thighs. She is working with a personal trainer to reduce her hips to thirty-six inches, as instructed by her agents. Since the age of fourteen, Elizabeth (sixteen, two years of modeling) has been trying to lose weight. She had even tried the Slim Fast diet ("That about killed me!" she said) but now counts her caloric intake to make sure she does not exceed 1,200 calories in a day. Amelia (twenty-six, seven years of modeling), after first declaring her aversion to dieting, admitted that she would diet during the fall bathing-suit season:

> I'm just like every other girl and I want to drop three pounds before the show. I'm only eating salad for a week because, you know, I want to fit in that swimsuit.

In addition to dieting, models exercise rigorously and lift weights.

The effect of these pressures has been to make models the permanent overseers of their own bodies who continually compare themselves to an industry standard that becomes increasingly difficult for them to maintain. Furthermore, models have to do this on their own, unlike boxers who are monitored by trainers to ensure that they are managing their bodily capital correctly (Wacquant 2004). At most, a model might be told by an agent to keep her hips down to a certain size or to lose a few pounds, but little, if any, guidance is provided as to how to do this. Instead, models must internalize the expectations of agents, clients, and designers to become the harshest critics of their own bodies. It is striking how readily they disparage their own physical appearance by comparing it to what they consider to be normal—the ideal of the perfect thin body. This self-objectification leads them to one conclusion: they are abnormal or deviant because they are not thin enough....

Strippers and exotic dancers are also conscious of how they compare to others in their occupations, but their efforts are directed toward transforming their bodies through technology in addition to taming them. For example, Wesely argues that the various body technologies that strippers use to make themselves attractive to customers—including breast implants, dying of their hair, nose jobs, chin surgery, and liposuction—represent an attempt to "further the doll-like image of the female body while hiding the realities of biology" (2003, 654). The available technology is too crude, however, to transform models' bodies into a physical state that would meet the stringent facial and bodily standards of the fashion industry (aside from the near impossibility of making someone taller). Models depend on their genetic predisposition to tallness and thinness, which is then cultivated and maintained through exercise and dieting....

REFERENCES

Brenner, Jennifer B., and Joseph G. Cunningham. 1992. Gender differences in eating attitudes, body concept, and self-esteem among models. *Sex Roles* 27: 413–37.

Wacquant, Loic. 2004. *Body and soul: Notebooks of an apprentice boxer*. New York: Oxford University Press.

Wesely, Jennifer K. 2003. Exotic dancing and the negotiation of identity. *Journal of Contemporary Ethnography* 32: 643–69.

Wolf, Naomi. 1991. *The beauty myth: How images of beauty are used against women*. New York: William Morrow.

27

Varieties of Masculinity
in Male Cheerleading

ERIC ANDERSON

A stratification hierarchy exists among men, with some forms of masculinity holding greater status than others. At the top is hypermasculinity, characterized by highly traditional gendered behavior, sexual conquest, competition and dominance, strength and athleticism, drinking and partying, and devaluing of all things feminine. Below this are various types of mid-range masculinities, where men acknowledge and incorporate some positive aspects of femininity, including fashion sense (metrosexuals), feelings (sensitive men), and equalitarianism (feminist men). At the bottom of the masculinity hierarchy are homosexuals; below them are women of all kinds. Masculinity is fragile and ephemeral for men, as they find themselves constantly called upon to prove themselves in the face of masculinity challenges. This is particularly the case when they cross over into a feminized terrain such as cheerleading. Anderson uses this setting to contrast two different masculinity styles: hypermasculinity, which he calls orthodox masculinity, and mid-range masculinity, which he calls inclusive. His descriptions of the culture of these two arenas showcase some of the differences among these strata of manliness. To what extent and in what locations do you tend to see these different types of masculinity displayed? What are the benefits of one versus the other? What are the forces shaping men into these different varieties? Do you think these distinctions are changing or are they merely replacing themselves?

This research uses sixty-eight in-depth interviews of collegiate male cheerleaders and participant observation from four selected cheerleading teams to examine the construction of masculinity among heterosexual men in a feminized

SOURCE: Eric Anderson, "Orthodox and Inclusive Masculinity: Competing Masculinities among Heterosexual Men in a Feminized Terrain," *Sociological Perspectives*, Vol. 48, No. 3: 337–335. © 2005, Pacific Sociological Association. Used by permission. All rights reserved.

terrain. Previous studies maintain that a hegemonic process of masculine dominance and submission influences most heterosexual men in feminized arenas to bolster their masculinity through the approximation of orthodox masculine requisites, including the expression of homophobic and antifeminine-acting attitudes. This research shows, however, that men in collegiate cheerleading today exhibit two contrasting and competing forms of normative masculinity, each supported by organizational and institutional culture and each with near-equal membership.

The first category of masculine performance is labeled as *orthodox*. The men categorized into this group are shown to perform masculinity in a manner consistent with previous studies of men in feminized terrain: they attempt to approximate the hegemonic form of masculinity, largely by devaluing women and gay men. The performance of masculinity among men in this group is influenced by a number of factors, including the institutional culture of one of the two major cheerleading governing bodies, the "Orthodox Cheerleading Association."

The second category of masculine performance is labeled as *inclusive*. The men in this group view orthodox masculinity as undesirable and do not aspire to many of its tenets. Particularly important to the study of men in feminized terrain, this research shows that inclusive masculinity is based less on homophobia and antifeminity than orthodox masculinity. Men in this group willingly embrace the feminized underpinnings of their sport and largely value their gay teammates. Notably, the construction of inclusive masculinity is influenced by the institutional culture of the other major governing cheerleading body, the "Inclusive Cheerleading Association."

Whereas previous studies of heterosexual men in feminized terrain found that men almost unanimously attempt to align themselves with orthodox masculinity, this research finds that heterosexual men in collegiate cheerleading are nearly evenly split between these two normative forms of masculine expression. The emergence of a more inclusive form of masculinity is attributed to many factors, including the structure of the sport; the reduction of cultural, institutional, and organizational homophobia; and the resocialization of men into a gender-integrated sport.

BACKGROUND

David and Brannon (1976:12) have categorized four basic tenets that "seem to comprise the core requirements" of American masculinity. These include: (1) no sissy stuff, (2) be a big wheel, (3) be a sturdy oak, and (4) give 'em hell. Although all four rules are important in understanding the construction and stratification of masculine power and privilege, this research is mostly concerned with the "no sissy stuff" principle because of a durable sociological understanding that contemporary masculinity largely exists in opposition to femininity. Kimmel (2004:97) says, "While different groups of men may disagree about other traits and their significance in gender definitions, the antifemininity component of masculinity is perhaps the single dominant and universal characteristic."

The marginalization of men in feminine fields has been shown effective in deterring heterosexual men from engaging in these settings, perhaps because men who enter these fields find their sexuality publicly scrutinized. Accordingly, it has also been shown that North American masculinity is based in a disassociation from homosexuality, something particularly true to athletic settings. When one adds homophobia to David and Brannon's (1976) tenets, it produces an orthodox form of masculinity. Orthodox masculinity is reflected in and reproduced through an institutionalized, gender-segregated, and highly competitive sporting culture in North America.

Previous investigations into the social construction of masculinities have shown a relationship between the dominant form of masculine expression and subordinate forms. Connell (1995:77) has described hegemonic masculinity as a social process in which one form of institutionalized masculinity is "culturally exalted" above all others. Key to understanding the operation of hegemony in relationship to masculinity, Connell (1987, 1995) maintains that most men exhibiting a subordinate form of masculinity actually desire to obtain the hegemonic form. Essentially, the process of hegemony influences the oppressed to maintain the rightfulness or naturalization of their oppression. Indeed, it is this aspiration that makes the process hegemonic. Thus, if hegemony applies to masculinity, one would expect most who transgress masculine-defined boundaries to preserve, if not inflate, their position by adopting as many tenets of orthodox masculinity as possible. These identity management techniques might include: (1) being homophobic, (2) devaluing femininity, (3) increasing masculine bravado, and (4) claiming masculine space within the larger feminized arena.

Examinations of feminized terrain have consistently shown men to approximate orthodox masculinity and to justify their transgression as consistent with hegemonic expectations of masculinity. For example, men who occupy feminized space are quick to defend their transgressions as consistent within normative boundaries of orthodox masculinity, maintaining that they have not transgressed masculine acceptability, rather that the space has been inappropriately gendered. Or, if men do acknowledge the feminine underpinnings of their field, they attempt to select a particular role within that space and define it as masculine. Accordingly, heterosexual male cheerleaders have been shown to emphasize that certain tasks within cheerleading (such as lifting women above their heads) are masculine, believing women lack the strength to perform these tasks as well as men. Conversely, they designate other tasks (such as erotic dancing) to be exclusively feminine....

METHODS

This research uses in-depth interviews and participant observations to examine how heterosexual men in collegiate cheerleading construct masculinity through micro and macro social processes....

The informants are sixty-eight self-identified heterosexual men who used to play high school football but became collegiate cheerleaders because they were unable to make their university football teams. Although a self-selection process cannot be ruled out (i.e., it is possible that men most affected by the masculinization process of football do not become cheerleaders), most of the informants reported that upon entering cheerleading, they held orthodox notions of masculinity, including sexist views and overt homophobia. The men, between eighteen and twenty-three years of age, come from diverse regions from throughout the United States, but 80 percent of the informants were white, middle-class men, so generalizations can only be made for this group....

While having lunch with a group of male cheerleaders, I asked, "How many of you would rather be on the football team?" All six resoundingly answered, "I would." After indicating a fondness for the cultural power that came to him as a football player, Richie said, "Yeah, I wish I could have made the football team; I really miss football." He added, "But I wasn't going to make any other team, so cheerleading was a way of getting back into the game. Well, as close as I could anyhow." This was the leitmotif among men who were recruited into cheerleading after playing football. To these men, cheerleading became an acceptable last effort to return to sport. For them, being in a feminized athletic arena was judged to be better than being outside it altogether....

PERFORMING ORTHODOX MASCULINITY

During the first day of cheerleading, Randy (a college senior) instructed the younger men:

> It is really important for you guys to give these women a lot of respect.... You are going to be putting your hands in certain places, and catching them when they fall, so be sure to be respectful.... Remember that we do things better; we pick up on things faster than women do, so don't rub that in by telling them, "We are better than you." Be respectful of the fact that guys are better. Just as important, when you are out in the field you have to portray a masculine image. When you are on the field, you must be the king of masculinity. We don't care what your sexual orientation is; that is none of our business. What you are at home is none of our business. But when you are here you have to be masculine. And if anyone gives you shit, and says you are gay or whatever, remind them that while they are out there playing with guys, you are out here with all these beautiful women.

Randy's talk illustrates the institutional and cultural attitudes of masculinity, homophobia, and sexism among men categorized as belonging to the orthodox group. This speech, in some variation, is traditional for veteran male cheerleaders to give to new men in the Orthodox Cheerleading Association. It is something they call "guy talk," and it best exemplifies orthodox masculinity because it maintains that men should strictly avoid activities culturally determined to be feminine and portrays women as less intelligent and less athletic than men.

The form of masculinity promoted in Randy's speech also subjugates homosexuality and sends a message to gay men that to be accepted they must downplay their sexuality and act in accordance with dominant notions of orthodox masculinity....

Those who valued orthodox masculinity often relied on the same identity management techniques discussed in previous investigations of collegiate cheerleading. Namely, they maintained that their role within the sport was consistent with orthodox expectations of masculinity and that the feminized nature of their terrain was falsely attributed. Daren said, "I know that people don't think that this is a masculine sport, but I challenge them to throw a girl up in the air and then catch her as she falls. Besides, the original cheerleaders *were* men." Daren, like many men in cheerleading, postulates that not only is cheerleading a sport in which men and women are polarized into masculine and feminine roles, but that it is also a sport that requires men to be "really" masculine. "Yeah, most of the guys are *really* masculine.... We aren't a bunch of fairies out here dancing in skirts."

There was variance about attitudes toward women among those belonging to the orthodox group. Although some held misogynistic attitudes, viewing women as purely sex objects, most maintained that they respected women as athletes but usually considered them to be inferior to men in their athleticism. Patrick agreed:

> I was asked to be on the team because the women needed me to help them do their routines better. They just can't throw girls as high as guys can. I'm not saying that they can't throw girls, but I am saying that the best cheerleading comes from the coed squads, and that's because we give them a better show. We can do what women can't.

In addition to the maintenance of sexist attitudes, many of the men categorized as belonging to the orthodox group also expressed varying degrees of homophobia. Although this was sometimes found in overt expressions, it was most often expressed covertly. Perhaps much of this group's homophobic and/or heterosexist sentiment was used to challenge cultural assumptions that male cheerleaders are gay. Indeed, it is common for heterosexual men to confront (or displace) accusations of homosexuality with homophobia ... and these men certainly found themselves inundated with homosexual suspicion. In fact, the suspicion of homosexuality may be even more prominent among male cheerleaders today than among male cheerleaders of yesteryear....

However, the reduction of cultural homophobia and the increased presence of openly gay men seems to have made the expression of homophobia somewhat outmoded, even among many members of the orthodox group. The expression of homophobia is therefore largely accomplished through covert mechanisms of heterosexism. One such manner comes in the form of *defensive heterosexuality*.

Defensive heterosexuality is characterized by the expressive signaling of heterosexuality through a variety of repeated mechanisms. For example, the most common narrative heterosexual male cheerleaders used to explain their transgression into feminized space was a well crafted and collectively constructed story about men lusting for their female teammates. One male cheerleader said, "Yeah, there are all these hot chicks in cheerleading. That is why I came out for the team." Another said, "Who wouldn't want to be out here with all these beautiful women?" The story was common: so compelled to be around hot women, heterosexual male cheerleaders were sexually drawn to the feminized arena of cheerleading.

These heterosexist and objectifying remarks even dominated the Orthodox Cheerleading Association's Web site. In one discussion forum, the question was asked, "How do I get more guys to cheer in high school?" Fifteen of nineteen responses included the "tell them about the girls" pitch. It was also common for men classified in the orthodox group to make heterosexualized and/or objectifying comments about women, often talking about whom they slept with or would like to sleep with. One night, five heterosexual male cheerleaders were sitting in a hotel room when one said, "Let's order a prostitute." The men then talked about this for the better part of an hour. It was doubtful this would happen, however, for they had had the same conversation the previous night....

In contrast to previous studies of masculine construction in cheerleading, however, only about half of these men in collegiate cheerleading were categorized as subscribing to the orthodox form of masculinity. The other half were shown to perform masculinity in a surprising, fascinating, and theoretically important manner.

PERFORMING INCLUSIVE MASCULINITY

With their competition finished, dinner eaten, and the movie over, a group of seven men (five straight and two gay) walked back to their hotel. Howie said, "Time for some drinking games. I've invited over the guys from Lincoln." When asked if he was worried that the cheerleaders from the other team might think him gay because he was not only sharing a room with a gay man, but also sharing a bed, he responded, "No. Why would I?"

After an hour of drinking games, one of the heterosexual men said, "Hey guys, do you want to see if coach will drive us to a club?" Howie responded, "You guys know of any around here?" To which his best friend, Steve, answered, "There is Gold Diggers, the Slush House, and then of course there is the Phoenix; it's a gay club." Howie interrupted, "Let's go there," and the others agreed. When asked why they would rather go to a gay club than a straight one, Howie answered, "The vibe is better, the music is better, and there are still good-looking women, so why wouldn't we want to go there?" When asked, "Aren't you worried about being thought gay?," the five heterosexual men shook their heads no and Howie asked, "Why would we?" They rounded up more teammates, piled into two vans, and headed for the club. Once there, heterosexual men danced with both women *and* gay men, two heterosexual men even "freaked" each other (a term used by these men to describe two people dancing with their groins together).

Although there was not a universal position on homosexuality among men in the inclusive group (just as there was not in the orthodox group), these men had few inhibitions about homosexuality. Their attitudes ran from tolerant to celebratory. Typical comments included, "I don't care what people think of me" to "Why is it necessary to have a label?" One male cheerleader even said, "I used to go to gay clubs all the time, and then I actually got a job at a gay club. I got hit on all the time. It was flattering." Still another said, "Why should I care? Why should people care if I'm straight?"

These attitudes are particularly unusual given that these men previously played high school football. "I used to hate gays," one inclusive cheerleader told me. "But now I don't care. I've gotten over it." His teammate added, "Yeah, most of my teammates used to just hate gays. I mean, what football player doesn't?" Perhaps most telling, another cheerleader said, "To be honest with you, I used to be homophobic. I used to be one of the guys calling the cheerleaders on my high school team fags." He continued, "Now, I'm on the other side. I mean, I'm not gay, but others sometimes think I am because I cheer, and if that's what they want to think, I don't bother to try to tell them different."

Data suggests that this shift in attitude from homophobic to gay-friendly might be made possible for several reasons. First, gay male cheerleaders seem to have strong support from female and older male teammates. For example, Dan said, "Oh yeah, you learn not to be homophobic real quick. I mean, you can't be. The women and coaches in cheer would never stand for that." Another said, "I made some homophobic comment when I first joined, and one of the guys pulled me to the side and schooled me on it." Second, heterosexual men generally befriended at least one gay male teammate. Jeffrey, a fourth-year cheerleader, said,

> I grew up in a town of 2,000. I never met a gay person. In my town, you were just taught to hate them, even though we didn't know who it was we were supposed to hate. So I did … until I met Jaime [who was the only openly gay member on his team]. I mean, I used to call guys fags all the time, but I'd never call him that. He was a real cool guy, and now I think that gay people are just really cool people.

A third possible reason for the shift in attitude is that institutional support has helped in shaping a new understanding of homosexuality. Whereas overt homophobia was generally replaced with heterosexism in the Orthodox Cheerleading Association, homosexuality and femininity among men was institutionally supported in the Inclusive Cheerleading Association. This has led many gay men who cheered in high school to self-select into colleges that compete for the Inclusive Cheerleading Association. Thus, there are more openly gay athletes in this association, and this makes it easier for closeted men to come out.

The heterosexual men classified as belonging to the inclusive group were not only less concerned with mitigating homosexual suspicion through homophobia and heterosexism, but they were also less concerned about associating with femininity. For example, Jeff, a heterosexual cheerleader, practiced with another male teammate for nearly half an hour, trying to put a female into the air with perfect form. But after growing bored he said, "My turn." The athletes switched positions, and Jeff stood atop the hands of a male and a female.

According to those with an orthodox understanding of masculinity, this position is one of the most feminine things a man can do in cheerleading. Jeff, however, was unconcerned. He willingly embraced the coded femininity of such tasks. In this respect, men in this group were shown to be less concerned about performing consistently with some of the tenets of traditional masculinity, challenging the bifurcation of gender.

When men in this group were asked about their masculine identities, many indicated that they considered themselves to be "metrosexual," a recent popculture term they understand to describe a gay-friendly heterosexual male who presents himself with the style-conscious behaviors otherwise attributed to gay men. Thus, just as "guy talk" was a useful defining construct for men in the orthodox group, the term "metrosexual" was useful for men in the inclusive group....

DISCUSSION

The Orthodox Cheerleading Association institutionalizes masculinity along the lines of other masculinized team sports. It values a bifurcation of gender and views homosexuality as a threat to this polarization. In the maintenance of orthodox masculinity, this cheerleading association uses covert and overt techniques to police masculine behaviors. For example, cultural norms stigmatize men (or entire teams) that perform in ways contrary to orthodox masculine perspectives. Men who dance complicated or erotic choreography (similar to women) find that their showmanship lowers their team's score. Because of this, men move in a rigid fashion, leaving the hip-swinging and erotic choreography exclusively to women.

In the Inclusive Cheerleading Association, however, men are expected to dance as competently and erotically as women. Men often take center stage, thrusting their pelvises and caressing their bodies to the thunderous approval of the audience, while their female teammates wait on the side. In the Inclusive Cheerleading Association, men even throw other men into the air, the strictest taboo in the Orthodox Cheerleading Association. Although this does not happen with regular frequency (largely because it is more difficult to throw and catch a 180-pound man than a 100-pound woman), one squad concluded their national championship routine in the Inclusive Cheerleading Association by having a man fly over a two-person high pyramid and land safely into the arms of four other men. This finale brought cheers of deafening approval that carried on longer than is customary. Because of this kind of activity, the Inclusive Cheerleading Association's competitions are widely recognized as being more dynamic and daring than those of the Orthodox Cheerleading Association, whose members, in turn, feel that they uphold the traditional form of cheerleading.

The institutional variance of gender roles between these two associations is fiercely contested and politically charged. The Orthodox Cheerleading Association even bans (for three years) any collegiate team that participates in an Inclusive Cheerleading Association event. In this manner, the Orthodox Cheerleading Association relies on traditional tools of marginalization, stigmatization, and institutional punishment for associating with femininity. The Inclusive Cheerleading Association makes no such demands of its participants.

The two leading cheerleading associations also maintain near-opposite perspectives on homosexuality. Highlighting the institutional perspective on homosexuality in the Orthodox Cheerleading Association, men who are out or who act in less than masculinized ways are equated with being unprofessional. Accordingly, it was harder to find openly gay cheerleaders in the Orthodox Cheerleading Association. Similar to the "don't ask, don't tell" policy found among openly gay athletes in highly masculinized team sports, their status was often privately recognized but not publicly discussed. Conversely, in many cases, gay members of the Inclusive Cheerleading Association talked more freely about their sexuality....

Previous investigations of masculine construction among men in feminized terrain have shown that hegemonic processes serve to stigmatize the expression of masculinity in ways that do not meet orthodox perspectives. This is attributable to the hegemonic understandings of masculine construction that requires cultural and institutional punishment for those who fail to meet the mandates of the dominant form, in whichever form it currently exists. The hegemonic privileging of one form of masculine expression and the subordination of all others are particularly salient among men in feminized terrain because these arenas have been shown to be more permissive of marginalized men than masculinized arenas. Thus, there is a greater variance of masculinities exhibited among men in feminized terrain than among men in highly masculinized fields....

REFERENCES

Connell, Robert. 1987. *Gender and Power*. Stanford, CA: Stanford University Press.

_____. 1995. *Masculinities*. Berkeley: University of California Press.

David, Deborah and Robert Brannon. 1976. *The Forty Nine Percent Majority: The Male Sex Role*. Reading, MA: Addison-Wesley.

Kimmel, Michael. 2004. *The Gendered Society: Sociological Perspective on Sex and Gender*. London: Oxford University Press.

28

Grinding on the Dance Floor: Gendered Scripts and Sexualized Dancing at College Parties

SHELLY RONEN

The previous two chapters examined gender roles, capital, hierarchies, and dilemmas for men and women individually. Here, we use Ronen's study of co-ed sexualized dancing to take a glimpse into the scripts and interactional dynamics of cross-gender relations. Bumping and grinding on the dance floor is, no doubt, a behavior familiar to many readers (see also Boswell and Spades' Chapter 18 on fraternities). But this study of a college scene offers a breakdown of the way these kinds of dancing interactions proceed from start to finish. Along the way, Ronen shows the influence of outside factors on these dynamics between couples, from the feedback of their specific friends to the larger reputational grapevine they have to negotiate. Have you ever grinded on the dance floor or been part of a "grind line?" How do people you know think about this behavior? Do you believe that it leads to sex? Is it considered a form of "hooking up" itself? How does this form of non-verbal sexualized "courtship" compare to verbal interactions?

The lights are dim and the music loud, and university students at a campus-wide party engage in behaviors not to be mistaken for superficial social interaction and fun. The night goes on and the alcohol flows, students meet, flirt, and perhaps

SOURCE: "Grinding of the Dance Floor: Gendered Scripts and Sexualized Dancing at College Parties," *Gender & Society* 24(3), pp. 355–377, © 2010 by Sage Publications. Reprinted by permission of Sage Publications Inc.

"hook up." The hookup is a common practice among contemporary university students that has received increasing attention in the past decade. As students use the term, a hookup can mean any sort of sexual encounter from kissing to intercourse. Some hookups occur on dance floors at parties, while others are initiated there but concluded in private. Hookups that occur on the dance floor, facilitated by alcohol consumption, are often preceded by the sexualized dancing known as "grinding."

Grinding is also known as "freaking," "freak-dancing," "dirty dancing," "bumping," and "booty-dancing." This style of dancing is associated with club culture and hip-hop, the type of music commonly played at mainstream college parties. When using the word "grinding," I mean to capture a common set of dancing behaviors that may nonetheless differ slightly across regions. Though different terms exist to refer to grinding, the sexualized nature of the dancing, and accompanying explicit gestures, are the defining characteristics.... While some might argue grinding is *just* dancing, I will argue grinding on the college dance floor is a kind of sexual act that carries significant, socially constructed meaning and may influence behavior in other settings.

Grinding is more than just erotic dancing, just as Schwartz and Lever (1976) consider a social gathering to be more than merely fun sociation. In their study of the college mixer, they argue that dating scripts "carry great meaning for the individual and the group" (1976, 414). In an ostensibly recreational setting, participants strive for a balance of vulnerability and self-protection. Grinders on the dance floor encounter the same challenges of conveying both their openness to grinding and protecting themselves from potential damage to their self-image. They negotiate this status-grading social process and strive to conserve "interaction order" using stable interactional scripts (Gagnon 1990; Goffman 1983; Simon and Gagnon 1973, 1986). Grinding is sociologically interesting because it is a public manifestation of contemporary heterosexual scripts, which are a crucial site for the reproduction of gender inequality....

Though we might separate different sexual and romantic practices for some purposes, scripts governing interaction in one context, such as grinding, may continue into others, such as hooking up in private. Since Simon and Gagnon's introduction of script theory (1973, 1986), it has been applied to sexual behaviors, such as dates, consensual heterosexual intercourse, and coercion. According to the theory, cultural scenarios, "the ordering of representations of self and other," organize interpersonal scripts (Simon and Gagnon 1986, 97). In turn, these interpersonal scripts construct individual identities because sources of sexual interest depend upon intrapsychic scripting "embedded within the stereotyped interpersonal script" (Simon and Gagnon 1986, 118)....

Research on college sex presents evidence for a double standard in which sexually agentic women are judged more harshly than men. Little research has examined double standards enforced on the dance floor.... I present an interactionist analysis of the heterosexual scripts in this sexual arena. I systematically describe grinding, as well as the gendered behaviors used to initiate it on the dance floor at college parties. I argue that gender beliefs, primed by heterosexual and heterosocial interaction, lead to a scripted interactional dynamic that as a gendered social process reinforces gender inequality in sexuality. Though the culture of hookups implies consensuality and possibly even egalitarianism in the pursuit of short-term encounters and unattached intercourse, like other studies of hookups, my analysis reveals that interactional scripts are subject to enduring gendered rules that restrict behavior and disadvantage women.

METHOD

During the fall of 2006, I and a group of other undergraduate students at an elite, private university attended campus parties that were open to all undergraduate students and engaged in participant observation. Participant observers received several hours of training prior to their fieldwork through class lectures and online training regarding the rights of human subjects. Each investigator spent at least 90 minutes in the field and wrote field notes immediately upon returning home. The data on which I draw in this article are made up of multiple investigators' ($N = 72$) accounts of one time in the field, similar to Grazian's ethnosurvey method (Grazian 2008). Students were instructed not to change their appearance or behavior from their norm and to respect the privacy of those being observed by avoiding recording identifying details. The assignment allowed both participant observation and observation without participation. Fifty-three percent of participant observers were men and 47 percent were women.... Typically, there were four observers at a single party. They blended in as participants at parties with estimated attendance ranging from 100 to 350. We do not have any information about the racial makeup of students attending these events, but they were open to all undergraduates on campus. The racial makeup of the university in which this research took place at that time was 40 percent white, 25 percent Asian, 10 percent Black, 10 percent Hispanic, 11 percent other groups, and 4 percent unknown when this fieldwork was conducted (Common Data Set 2006)....

I acknowledge the limitations of these data as collected within a specific portion of college social life at one school, at one time. This university is a private institution, on the West Coast, with a relatively privileged and mostly white and Asian student population. These patterns I observed may or may not apply in other university settings, though "grinding" has been described as a relatively ubiquitous part of heterosexual culture among college students....

THE LIFE CYCLE OF GRINDING

In this section, I use accounts to portray a life cycle of the most common grinding interaction. Behaviors on the dance floor appear fluid and transition seamlessly. However, in this analysis they have been separated into stages: pre-grinding behaviors, initiation, acceptance or rejection, and heterosexual dyadic grinding....

Two Is Company: Heterosexual Dyadic Grinding

Single-gender dancing in groups and grinding heterosexual dyads filled these dance floors. Successful instances of grinding dyad formation resulted from initiations in which a man approached a woman with the outcome of the pair dancing closely....

Grinding, as it appeared at these parties, generally involved a woman rubbing her buttocks into a man's groin and her back against his torso in a repetitive motion to the beat of the music. One account describes the most common formation, "they were generally grinding so that the girl's back was to the guy's front, and both parties were rolling their bodies in unison." An alternative configuration involved the two partners facing one another and repetitively pushing their groins and chests into each others' bodies. These two formations were sometimes interchangeable, but grinding dyads seemed to prefer the former configuration. Depending on the form of initiation, men were sometimes behind their grinding partner and sometimes facing them; "they started dancing facing each other then the guy put his hands on the girl's waist and spun her around to start grinding." Grinding movements were accompanied by touching. In one account, "the boy began moving his hand up the side of her leg," while in another, "his hands started wandering over her breasts and waistline." Sometimes the dyad leant against a wall while grinding, and "for the most part men are allowed to touch women in such ways that would otherwise be deemed inappropriate." In both positions—whether face to face or not—grinding involved miming, alluding to, and representing intercourse. "There [was] a lot of hip thrusting and in the pelvic regions, and the man and woman [were] always touching." Observers commonly noted pelvic thrusts, repetitive vertical movements in friction with a partner's body, and pressure on hips, buttocks, or breasts. One movement involved women leaning forward, while men placed their hands on their partner's lower back, pushing her buttocks into his groin. One account made the sexual nature explicit: "they were dancing with girls like they were having sex ... guys lift[ed] girls off the floor but most of the time they just thrust on the girls' butts."

Men were predominantly the active partners. Women's parts in grinding descriptions were limited to response. However, women generally appeared receptive to the movements once grinding had been initiated. According to one observer, "A lot of girls appeared to like it." ...

As a demonstration of public sexuality, these behaviors were particularly fraught for women. One woman, having been grinding with and kissing a man, talked to her nearby friend about the interaction.

> "Oh my god ... I can't believe I just did that ... I feel like a slut." Her friend responded, "No you're not, don't worry you're having fun!" The girl continued, "Everyone's looking at me ... they think I'm a slut," and the friend replied, "Who cares? ... No they're not ... Just have fun." After speaking with her friend, the girl began to make out with the guy again.... While the two students were making out, individuals in the vicinity gave them disapproving looks.

The distress this woman expressed implies that kissing this man on the dance floor was sexually meaningful; it carried significance enough for her to be judged as a "slut." Furthermore, while she was worried by the prospect of being labeled a slut for her behavior, she then returned to the activity. Both this woman and the disapproving onlookers are enforcing a double standard. The woman worried, probably with considerable justification, that she would be judged more harshly than her hookup partner. We can presume this woman was not in a prior relationship with the man she was kissing, by her expressed surprise to be kissing him at the party. No doubt, this aggravated the situation because she was in violation of the relational imperative, which holds that women should only want to engage in sexual behaviors with someone in an attached relationship. This instance was a pressure point that revealed some of the conflicting gendered rules that structure behavior on the dance floor....

Showing Off and Looking On: Gendered Display and Coded Space in Pregrinding Behaviors

Gendered scripts designated appropriate preinitiation behavior at these parties. The life cycle of a potential encounter on the dance floor generally progressed in several steps. Women entered the dance floor with their friends, mostly other women. The dance floor initially held "mostly girls [dancing] with other girls." These women were not invited onto the dance floor by anyone. Verbal requests to dance were not necessary, unlike the conventions of Berk's (1977) singles dance. Nor were women expected to depend on male-initiated "protocols" such as those Bailey (1988, 35) discusses in a history of American courtship. On these dance floors, unattached women were not just granted access; they assumed ownership of the space. These women would be best described as dancing *near* one other, rather than *with* one other....

Women's group dancing was nevertheless suggestive. Women moved their "hips seductively," drawing attention to their bodies while dancing without a partner. Their movements were "independent of each other" but in close proximity. This form of dancing differed from heterosexual grinding in that most women stayed further away from each other than in grinding dyads, and avoided prolonged eye contact with their fellow dancers. Women often moved among their groups, rotating through closer proximity with different women throughout.

Pairs or groups of women sometimes mimicked heterosexual grinding and "danc[ed] on each other" or in "mini circles." They synchronized movements dancing closely back to back, or in the formation normally taken by a man and a woman. These displays between women both already on the dance floor, however, were temporary and "often accompanied by jokes and laughing with each other so they were not as sexually charged," suggesting they were acts of pretence—display—not genuine interest in grinding with other women....

Group dancing seemed designed to display these women's *sexiness* for onlookers, rather than sexual interest between them. All group dancing, including mock-grinding dyads, involved suggestive gestures and open body language that welcomed initiation from men. These displays were visible enough such that men who were "awkwardly standing around the perimeter" could see them.... Men talked while at the outskirts, standing and facing in towards the dance floor. "There were about a dozen males leaning against the wall of the dance room either watching dancers or talking among themselves in small groups." Others filed through the dance floors looking for women they knew or wanted to dance with; "guys seemed to prowl the exterior of the dance floor peering in at the different groups of girls searching for a girl to dance with." Women used their bodies provocatively to encourage initiation. But they depended ultimately on men to approach; and far from being a transgressive indication of same-sex interest, women dancing with other women reinforced the heterosexuality of grinding conduct (Hamilton 2007).

While women appeared to enjoy the dancing, entering the dance floor without grinding partners, men never danced together in groups. Men depended on the women for motivation to leave the perimeter of the floor. Goal-oriented grinding initiations were the predominant reason for a man's entrance onto the dance floor. The threat of being labeled as gay was too great and presumably damaging to heterosexual grinding initiation prospects. At one event, groups of men sought confirmation that there were women on the dance floor whom they knew before they joined in with the dancing.... Men sometimes looked for acquaintances on the dance floor, presumably to increase their chances of successful grinding initiation. In contrast, women danced without particular targets. The difference between single gender dancing for men and women highlights the gendered nature of homophobia. Men presumably avoided dancing with other men because this might signal homosexuality, while women dancing with each other did not signal lesbianism....

Making a Move

Encouraged others to join the dancing simply by being there, as well as gesturing to friends and pulling them onto the floor....

Direct requests to dance, verbal or otherwise, were extremely unusual. When women initiated, they risked punishment for their gender incongruent agency. One example of such initiation reveals the boundary for women's behavior, in an exceptional account that underlines the rule.

> A very slutty sailor girl tried to pull a guy from the courtyard to the dance floor ... it was clear that she was quite intoxicated. Initially it seemed like the guy was enjoying himself but as her dancing became more and more sexually intense, he bolted.

In this case the man was put off by this woman's direct initiation. The student observer, also a man, was judgmental of her behavior, describing her as "slutty" despite the instruction to be objective in his field notes. This ethnographer and the man involved in the grinding are enforcing the implied gendered double standard. Women should be dancing, mimicking sexualized movements with friends, and displaying their sexiness as expressions of desirability. They should desire grinding initiations, but they should be objects rather than agents. A woman's direct initiation is a behavior that must be excused with reasons such as intoxication or understood as the result of her loose morals.

Get Close and Anything Goes: Acceptance

Since erotic movements were the standard form of grinding, mere engagement in the dancing did not necessarily imply grinders' sexual interest in dance partners. Rather, students signaled interest by allowing proximity to others and sustaining grinding for longer. Some women "danced provocatively, but refused to dance with a partner." The incongruence between their dancing style and their actual interest in grinding highlights the ambiguity of sexualized dance displays. They are a meaningful mode of signaling to encourage initiations and yet provide women with a safety of a ubiquitous dancing style. In this sense, women are protected from the responsibility of directly encouraging an initiation. Also at play is the restrictive relational imperative, making grinding with a stranger a riskier choice than dancing provocatively with friends.

Observed initiation scripts were in some ways similar to a traditional request to dance. In the traditional formal dance the initiator is a man, and once he begins dancing with his partner or cuts in, he is "responsible for the woman until she [is] taken over by another partner" (Bailey 1988, 31). At our observed events, "requests" took the form of an approach rather than the traditional verbal invitation. There was "no clear form of asking for a dance … people simply pressed up on each other."… In virtually all cases observed, initiation was instigated by men, and verbal requests were unusual. Men entered the dance floor and danced their way towards women. Men did not speak, but attempted close physical contact with their chosen targets. Initiators never approached a woman if she was grinding with another man, presumably because this was a sufficient tie-sign. Entrance into a woman's immediate surrounding space was interpreted as interest in forming a grinding dyad. In most cases, one man approached a group of women and selected one woman with whom to make contact. Alternatively, a group of men approached a group of women.

To form a grinding pair, men and women communicated nonverbally by use of body language and facial expressions, a characteristic feature of the contemporary hookup culture (Bogle 2008). During this tense period of negotiation, women deliberately controlled their space and attention. Loaded gestures included drawing closer to or further from an initiator and avoiding or reciprocating eye contact and smiles. Women "would signal acceptance by laughing or touching lightly."… The line for women's agency was drawn at initiation, as they could only be powerful by submitting or withholding their bodies from initiating men. Once approached, therefore, women would choose compliance or cessation.

Conversely, men's scripts offered little to no signaling prior to forming grinding dyads. In rare instances when women approached and initiated, men "convey[ed] their intentions with the women by scaling up or down the sexuality of the dancing." Those few men who were disinterested in their grinding partners demon-strated nonverbal behaviors that revealed the rule.

> If the guy was only dancing for fun, he would often raise his arms from the girl's hips in order to show that he was only having fun and didn't want it to be sexual [or] didn't want to get aroused.

Such men avoided grinding in a remote part of the dance floor, preferring instead to stay close to other friends. Any increased privacy might have signaled interest; "Guys talked to nearby friends more when not interested." Conversely, they were "quiet and focused on their body movements" when interested. Observed scripts seemed inflexible: When men violated expectations, such as not wanting to grind with will-ing women, they did not have a clear course to reject. Finding themselves dancing with women, men danced passively so as to perform the expected scripted behavior, while at the same time rejecting or preventing further pursuit.

Men approaching a woman controlled how much warning she received before the enforced proximity. When approaching from the front, men could seek some degree of permission, or at least provide women with a warning. Despite loud music and dark settings, some men sought a woman's attention. These approaches sometimes included "minimal conversation," but most often men sought eye contact and open body language that "demonstrated a woman's availability." If approaching from the back, men gave women little warning. In a surprise initiation, there was no opportunity to give permission, directly or oth-erwise. This surprise approach was, in fact, the most common one: most women were not aware of the initiation until physical contact. Men "just crept up on girls and started thrusting." Some men were sometimes labeled "creepers" and were frowned upon. Men labeled as creepers often elicited looks of "disgust as women moved away." Men who went "behind a female and grabb[ed] her and [started] grinding" were not as successful as those who initiated some eye contact and "small talk." Women more likely rebuffed them with "defensive tactics" (Snow, Robinson, and McCall 1991), by "pull[ing] their unknowing friend towards them and shift[ing] away from the guy."…

Getting It On: Acceptance as Compliance

Regardless of approach, if the interaction progressed beyond initiation, then a grinding dyad formed. Acceptance was seldom verbal. Most often, it took the form of a woman allowing the initiator to continue dancing in proximity and allowing him to touch her during dancing—that is, not rejecting him. One char-acterization described acceptance as the lack of opposition, saying "[the woman] didn't oppose physically or verbally and began to dance." Another observation cast the woman's acceptance as more active: "If a woman accepted a man's approach, she would shift her motions slightly so that it became clear that she was dancing *with* the man rather than merely next to him."…

Women's friends also played a part in grinding acceptance. "The girl exchanged a look with the friend directly in front of her, and when her friend nodded the girl moved back and began grinding with the man."… Women who were dancing with only one other friend were less likely to be receptive to initiations, because this would leave the friend on her own. If initiations separated pairs of friends, women were more likely to reject. Indeed, men increased their chances by approaching in groups: "two or three males who were friends walked up to two or three females." This approach ensured that groups of women all had a partner, and no one woman was left alone.

Standing alone was uncommon for women, presumably because it was undesirable. "The only people standing alone were the undesirable guys hanging out along the edges of the dance floor."... As Schwartz and Lever discuss, women might stand slightly apart so as to be available for approach, or stand together so as to be seen as desirable and indifferent to approach. Presumably standing alone made women uncomfortable because they seemed desperate, "alone and needing someone" (1976, 419). Alternatively, women might have preferred standing with friends to make use of their assistance during initiation....

Calling It Off and Cooling Out: Rejection as Deference to Outside Forces

Dancing initiations did not always lead to grinding pairs, particularly those following surprise approaches. Rejection behaviors followed consistent scripts to a greater degree than observed initiations; women used certain behaviors for rejection more consistently than men used particular behaviors for initiation. Just as the most common form of initiation was nonverbal, rejections were also mostly nonverbal. In the case of rejections too, women often relied on their friends for decision-making assistance. Dancing in a group better positioned women to cool out men when rejecting unwanted approaches.

> One woman mouthed to her single friend dancing across from her "Is he hot?" The female friend mouthed back "No" and the woman grimaced. She then gently disentangled herself from the man's hands around her hips and began provocatively dancing with her friend.

The presence of friends worked as a preemptive avoidance tactic.

Women most often used avoidance as a cooling out strategy. Flight was the most noticeable and immediate following an approach.... Women shifted attention away from initiators and toward their friends. Most often, a woman drew her friends nearer while still dancing; sometimes she stopped altogether and walked away with her friends.... These rejections protected the woman from culpability and the man from embarrassment by having others disrupt the privacy of the proffered dyad and giving women an excuse to move away.

Women who could not preemptively avoid an approach had to respond to an initiation while already grinding. In these circumstances, women employed a number of avoidance strategies such as nonverbal cues of disinterest or excuses. Women danced disinterestedly—even limply—for some period of time with unwanted grinding partners before leaving or being left. Some women waited for the end of a song to break off and excused themselves; others acted immediately and left before a natural pause....

Regardless of specific cooling out strategy, women showed consistent preference for rejections that suggested contravening excuses. Rather than polite refusals, which required conversation, women nonverbally cooled out their initiators. They avoided direct expression of their (lack of) desire. They avoided gesturing to any real preference to grind or not. These women had a vested interest in preventing humiliation and conflict with their initiators. These college parties were open only to university students, meaning some chance existed that these students would encounter one another in other contexts. While this is no doubt a strong motivation to preserve initiator's face, the gendered nature by which face-saving is distributed reveals a systematic inequality in these cooling out strategies. Men initiated and women cooled out. This is the dynamic that reproduced gender inequality in this sexual script. Rather than risking commandeering the masculine role of agency and power, women sought feminine—deferent, submissive, communal—ways to express their own agency, and their refusal, while avoiding embarrassing men.

DISCUSSION AND CONCLUSION

The heterosexual grinding script at these parties reveals a gendered interactional dynamic that reproduces systematic gender inequality. The interactional dynamic is one in which men directly initiate, while women indirectly cool out unwanted approaches. In exceptional gender violations—when initiations do not follow the prevailing script—participants compensate with gender congruent behaviors....

This gendered dynamic biases grinding towards male pleasure; women do not have equal access to direct agency and unpunished sexuality. Women reckon with the heterosexual relational imperative in which they are judged if they do not confine sexual behavior to committed relationships, but men are expected to pursue sexual behavior of all kinds. We can see this as interactional evidence of the relational imperative: Women on the dance floor should prefer to grind (and kiss or do anything sexual, for that matter) with a boyfriend or a man with whom a relationship is likely. Though women are ostensibly entitled to desire sexuality, signaling an increasingly liberalized sexual script, double standards identified in college courtship endure in this setting. Women face the Madonna-whore dichotomy in which they must present themselves as sexy and sexually competent to be desirable for grinding but, in taking part in grinding, risk being seen as immoral....

REFERENCES

Bailey, B. L. 1988. *From front porch to back seat: Courtship in twentieth century America.* Baltimore: Johns Hopkins University Press.

Berk, B. 1977. Face-saving at the singles dance. *Social Problems* 24:530–44.

Bogle, K. A. 2008. *Hooking up: Sex, dating, and relationships on campus.* New York: New York University Press.

Common Data Set. 2006. Office of University Communications. http://ucommstanford. edu/cds/cds_2006.html.

Gagnon, J. H. 1990. The explicit and implicit use of the scripting perspective in sex research. *Annual Review of Sex Research* 1:1–43.

Goffman, E. 1983. The interaction order. *American Sociological Review* 48:1–16.

Grazian, D. 2008. *On the make: The hustle of urban nightlife.* Chicago: University of Chicago Press.

Hamilton, L. 2007. Trading on heterosexuality: College women's gender strategies and homophobia. *Gender & Society* 21:145–72.

Schwartz, P., and J. Lever. 1976. Fear and loathing at a college mixer. *Journal of Contemporary Ethnography* 4:413.

Simon, W., and J. H. Gagnon. 1973. *Sexual conduct: The social sources of human sexuality.* Rutgers, NJ: Transaction Publishers.

Simon, W., and J. H. Gagnon. 1986. Sexual scripts: Permanence and change. *Archives of Sexual Behavior* 15:97–120.

Snow, D. A., C. B. Robinson, and P. L. McCall. 1991. "Cooling out" men in singles bars and nightclubs: Observations on the interpersonal survival strategies of women in public spaces. *Journal of Contemporary Ethnography* 19:423.

PART IV

Social Institutions

In this penultimate Part, we look at the core social institutions forming society. Social institutions are not primarily social organizations, such as schools, businesses, or churches, but more abstract concepts, once again, describing social features that cannot be seen or touched. Some of the major social institutions in our society are politics, economics, religion, education, and the family. These social institutions have organizations, groups, and concrete aspects operating within them that carry out the business of each sphere. For example, politics consists partly of governing bodies, political parties, elections, and campaigning. Within the institution of economics we find the stock market, banks, businesses, and the way people manage their budgets. Religion encompasses houses of worship, spirituality, dogma, sacred objects, and beliefs. In education we find the classrooms, books, tests, ideas, and the passing on of knowledge from one generation to the next. The family has extended kinship networks, new family forms, inheritance, dominance, geographic location, and the strength of blood relations. Therefore, though the idea of a social institution is abstract in the ideal, it translates into concrete and highly relevant realities that profoundly affect our lives. Social institutions contribute to the structure within which we live, play, and work.

Sociologists sometimes differentiate between primary and secondary social institutions. We regard the economy, the political system, and the military as three primary ones. Their leaders make the decisions that most profoundly affect our lives through their influence on society. In thinking about the attacks of 9/11, we can see that the airplanes of Al Qaeda were aimed at Wall Street (the economy), the White house (the political system), and the Pentagon (the military). If these institutions changed, we would undoubtedly be different people. Due to America's place in the world, major changes in these institutions would also affect people in other countries.

Secondary social institutions include the family, religion, education, work, medicine, law, sports, and the mass media. These serve and are influenced by the primary ones, providing support for them in good times and distraction from them during bad times. Like the primary social institutions, these operate at the most abstract level but translate down to the nuts and bolts of concrete everyday life.

At the same time, each social institution and its organizations and component parts have *statuses* and *roles* that people play within them. Politics has politicians, speechwriters, press secretaries, fundraisers, elected officials, and managers. Economics has market traders, brokers, bankers, tellers, investors, salespeople, and cashiers. Religion has clergy, the faithful, disbelievers, heretics, congregation presidents, and youth group leaders. Education has teachers, students, principals, administrators, writers, publishers, and secretaries. The family has mothers, fathers, children, grandparents, aunts, uncles, and pseudo-kin. All of these roles and statuses are filled by people whose position in relation to each other is defined by the structure of society and its organizations. We learn the social scripts and we enact the behaviors associated with them, occupying our position in society.

The *family* represents the foundational *primary group* of society, within which children are born and raised. It also serves as the fundamental economic unit, with adults working to support the family and arranging for the food, clothing, and shelter of its members. The family is a *micro-social system*, sanctioning the legitimate conduct of sexuality and pairing off people for reproduction. Within modern society, we generally observe cultural values of monogamy (fidelity among two partners), freedom of partnership choice, bilateral kinship descent (lines of inheritance following through both men and women), neolocal residence (neither with the groom's nor bride's family, but in a location of their own), moderate gender equality in interpersonal power relationships, and a nuclear rather than extended family structure.

Several important norms guide the character of family formation: *endogamy* (that marital partners should be similar to each other), where we expect some degree of similarity between partners in terms of age, race/ethnicity, class, and religion; *exogamy* (that marital partners should be different from each other in other ways), where we expect people to marry outside their gender and family (the incest taboo); and *homogamy* (that marital partners should be similar to each other in respect to personal characteristics), where we expect some overlap in personality. These are in a constant state of evolution as the definitions of acceptability shift with new social developments. Since the middle of the twentieth century, our society has seen several changes in existing family forms that have shaken the family as an institution and raised serious questions about its functioning and future. These include divorce, family violence, remarriage, blended families, single-parent families, gay families, dual-career families, commuter families, childless families, and non-married families.

In our first selection, Kathleen Gerson considers the rise of several different nuclear family forms we have seen proliferate over the last decades and the way children raised in these arrangements view them. She not only focuses on divorce and the varieties of remarriage but also discusses the role of working mothers on these households and children. You may find it interesting to compare your own and your friends' experiences with the adult children presented here. But new family forms go beyond what Gerson considers, as we see in Maura Ryan and Dana Berkowitz's chapter on gay and lesbian families. Although gay marriage is still nationally debated, gay families appear to have taken off without this blessing. Ryan and Berkowitz look at some of the difficulties same-sex parents face in adopting or birthing their own children and how they navigate the biological imperative people commonly associate with parent-child relationships. Confounding this are issues of racial/ethnicity diversity, which add to their already non-traditional family forms.

Religion serves as another building block of society. All religions offer a *dogma,* a set of tenets, or a *belief system,* which explains the meaning of life. People accept these convictions on faith and use them as guiding principles. Religious beliefs separate objects and behavior into the realms of the *sacred* and the *profane,* with the former encompassing the divine or supernatural and the latter the earthly or secular. Sometimes these overlap in the area of *civil religion,* where church and state come into union. Countries characterized by civil religion have religious beliefs guiding their policies and often think that their nation is divinely favored, is guided by moral standards of behavior, and it has a mission to perform in the world. Religion is one of the major sources of international and domestic conflict with groups battling over territory, legitimacy, tolerance, and extremism.

In our lead chapter here, Lori Peek offers a study of Muslim American students after the crisis of 9/11. Although many of these young people had assimilated ethnically and religiously to American society, she traces how the attack by Muslim terrorists leading to the subsequent ostracism of identifiable Muslims drove these young people to more closely identify with their religion. What do you think the role of religion should be in our government? How does it unite people or drive them apart, even in a society that purports to be as diverse as ours? How do you think 9/11 affected Americans' views about the religious fundamentalism of culture and politics? Jesse M. Smith provides us with a portrait of the opposing process in his chapter on atheism. Non-believers are raised in American society, surrounded by religiosity. He shows how many break away from this trend, disillusioned by the dogma of organized religion. Just as with Peek's chapter, we see an identity change process whereby people go through patterned stages in their religious transformation.

We turn, next, to the role of *education*, conceived by many as the great route of upward mobility. America is one of the most highly educated countries in the world, and people from all over come here to study in our colleges and universities. Over the years, we have turned increasingly to the school system to teach our children about morals and values, health and sex education, independent living in society, and a host of other matters that were once handled in the home and place of worship. We also want our schools to teach diversity, conflict resolution, tolerance, and to impart the American culture to new immigrants. Yet public funding for education remains a controversial issue, as we vie between investing in our next generation or in existing ones, and between supporting public schools or letting people transfer their property tax dollars into the private domain. Curricular matters are also a battlefield where sacred and secular views clash, as witnessed by the ongoing debates over teaching evolutionism versus creationism. The same vehemence can be witnessed in the way our citizens approach forced busing, integration, prayer in schools, and school violence, topics that divide liberals and conservatives across the nation. Thus, we are increasingly asking our schools to do more, but supporting them less.

With so many important issues abounding in this field, we can address only a limited few. Jonathan Kozol poignantly depicts the different structural opportunities children from diverse geographic and economic groups find when they get to their public schools. As a nation, we are torn between our principles of equality of opportunity and the very real costs of making this a reality. Elizabeth Aries and Maynard Seider examine critically important questions that parents and children from working-class families must face as they strive for upward mobility. Comparing state universities with elite colleges, they trace the difficult experiences that young people from less affluent backgrounds face in college and their different social and educational challenges compared to youth from more privileged backgrounds. They then address the often unspoken irony that occurs when young people manage to assimilate into higher social status and ultimately find themselves alienated from those who helped them make this upward leap.

Work in America is a critical topic as people are laboring more than ever before, outpacing the hours toiled of nearly every other industrialized country. With free trade agreements, the outsourcing of jobs to foreign countries, and the decline of the manufacturing base, America's economy has shifted dramatically to an intellectual, entertainment, and service economy. The relationship between workers and their employers, formerly heavily slanted toward full-time, career-long commitment, has been eroded by the competitive pressures of the global economy. People are more likely to find themselves employed as contingent workers, without health benefits or pensions, and to move between different jobs or to experience periods of unemployment. Union membership has declined over the years. Small businesses, restaurants, and retail outlets have given way to large chains. Government regulations protecting workers' overtime wages have eroded, and the general climate has shifted toward favoring businesses as a way to stimulate the economy. As a result, the earning power of many Americans has declined. The internationalization of work has led to greater competition and interdependence between countries, so that swings in the supply or demand for goods or services, whether crude oil or call centers reverberate from China to the Middle East, to the European Union, to the United States. American workers have thus entered the global labor market, and it has affected everyone's jobs, their financial security, the industries they are likely to enter, and the length and character of their working lives.

In this section, we offer two contrasting views of work in America: entrepreneurial and industrial. Hernan Ramirez and Pierette Hondagneu-Sotelo take us inside the former realm, describing the working communities and lives of Mexican immigrant gardeners. New to the country and ensconced within their extended networks of kin and race/ethnicity, members of this community rely on each other to extend a helping hand. Laborers benefit from having a ready-made job and the opportunity to acquire skills and move up the ladder, while longer-term residents gain employees with a hard work ethic and guaranteed reliability. The patterns by which people move into and through this entrepreneurial venue are very applicable to other forms of entrepreneurial work by new immigrants. On the other side of the coin, many blue-collar Americans coming from generations of entrenched manufacturing laborers have found themselves cast adrift by the globalization of work, through which their jobs were taken by people in countries with lower labor costs. Elizabeth Mikyla Legerski and Marie Cornwall describe the multi-faceted crises members of unemployed families face when confronted with serious and likely lasting job loss. Husbands and wives are dislocated occupationally, emotionally, and interpersonally by this profound transformation, which forcefully tears them out of their traditional and preferred gender roles.

Like our educational system, meaningful aspects of *health care* delivery function at all levels, from the doctor-patient relationship to the hospitals, to medical delivery organizations, to the economics of medicine. We live in an era of modern medicine, where many diseases have been eradicated, science and technology are battling others, and sterilization and sanitation procedures are highly advanced. As a result, people are living longer and enjoying a better quality of life into their later years. The science of *epidemiology* has flourished, with advanced research being conducted into the patterns by which diseases thrive and spread throughout populations. Other issues in the sphere of public health include prevention (wellness) versus treatment, harm reduction versus criminalization (drugs), and medical insurance coverage. Compared to some countries, the United States does not make medical treatment accessible to all its citizens, but the level of health care available is better here than in many nations. Like other dimensions we have examined, however, we see grave inequalities here, with some people getting the most advanced treatment while others suffer deprivation. Another important trend in medicine has been the expansion of *medicalization* into previously non-medical spheres. We have seen many behaviors redefined as medical issues, such as addiction (physical and psychological), childbirth, anxiety, classroom misbehavior, and learning differences. All of these swell the influence of medical professionals and expand the domain of the *sick role* in society. Though this may make us a more humane nation, it may also lead to the over-medicalization of our population: doctors may be prescribing more drugs now than ever before.

William Marsiglio considers a largely under-addressed health issue for the country's population in his examination of the effect of male culture and behavior patterns on the health of men and their children. He offers a wealth of information, in his review of the literature, on how men view and treat their health, and how the male gender role interferes with their listening to the messages they get from their bodies or taking action to ameliorate them. Marsiglio suggests that they are transmitting these dangerous health patterns to subsequent generations through the vehicle of male culture. Next, Jonathan Leo directs our attention to a powerful medical trend by examining the widespread practice of prescribing Ritalin, a drug designed to combat ADD and ADHD, for children. With advertisements promoting this phenomenon pervasive on television and in the print media and school officials joining with doctors to press drug therapy on parents, many mothers and fathers are regulating their children's moods and behavior through this new, potentially problematic, control paradigm.

29

A New Generation Negotiates Gender, Work, and Family Change

KATHLEEN GERSON

Family forms and parental roles have undergone seismic changes from the post-war image of the stable, single-breadwinner, suburban family with two cars, two children, and the picket fence. Changes in the labor market have necessitated the entry of women in massive numbers into the workforce, leading to the gradual decline of the traditional family role of stay-at-home mother. Intact, lifelong nuclear families have been supplemented by divorced, remarried, and blended families, with definitions of "parent" and "siblings" undergoing dramatic transformation. When feminists urged middle-class women to self-actualize through careerism in the 1970s, our nation went into a collective anxiety over the traumas that would be caused by daycare. Anti-feminist backlash led many young women in subsequent generations to return to traditionalism, or, as Gerson puts it, to search for a "neotraditionalism." As you read Gerson's chapter, you may want to reflect on how our assumptions, fears, and attitudes toward family forms and parental roles have evolved over these last few decades. How do you feel about divorce and remarriage? Do you believe that parents should stay together until the children grow up? How many of you come from reforged families with step-parents and step-siblings or have close friends who do? How would you compare these families with traditional nuclear families of origination? What kinds of arrangements are best for the adults involved, and what provides the most favorable situations for children? How would you rate the advantages versus the

SOURCE: Kathleen Gerson, "Changing Lives, Resistant Institutions: A New Generation Negotiates Gender, Work, and Family Change." *Sociological Forum*, Vol. 24, No. 4, 2009. Copyright © 2009 by Wiley-Blackwell Publishers and the Eastern Sociological Society.

disadvantages of young people growing up with divorced and remarried parents? How should divorced parents assess their relative needs compared to those of their children? How would you assess the pulse beat of your generation with regard to your expectations for male and female roles within marriage?

• • • At the outset of a new century, sociology's enduring concern with explaining the links between individual and social change has never been more relevant. We are poised at a moment when changing lives are colliding with resistant institutions. On the one hand, growing demographic and cultural diversity has given people new ways to live, work, and build families. Dual-earner, single-parent, and same-sex couple homes now greatly outnumber the once-ascendant homemaker-breadwinner family. Legions of work- and career-committed women, including married and single women with and without children, have taken their place alongside and now outnumber home-centered mothers. The "traditional" career, where male workers of all classes (though not all races) could gain economic security through loyalty to their employers and earn enough to support wives and children, has been supplanted by a myriad of time-demanding but insecure jobs. The life course has become more fluid and unpredictable as people travel new paths through work and family in adulthood.

Yet these intertwined social shifts—revolutions in family life, gender arrangements, work trajectories, and life-course patterns—face great resistance from institutions rooted in earlier eras. At the workplace, employers reward "ideal workers" who provide uninterrupted full-time—often overtime—commitment, an ideal that workers now perceive as not just a requirement to move up but even to keep their place. In the home, privatized caretaking leaves parents, especially mothers, facing the seemingly endless demands of "intensive parenting." The ideal of permanent marriage persists for relationships despite the fluid and uncertain nature of intimate commitment.

The tensions between changing lives and resistant institutions have created personal dilemmas for women and men alike. Even though children increasingly depend on their mother's earnings, women remain primarily responsible for caretaking. And however much men would like to be involved fathers, their success in the job market remains the prime measure of their "marriageability" and social status. To explain and develop effective strategies for transcending the impasse between public demands and private needs, we need a deeper understanding of how these structural and cultural conflicts play out in the lives of young women and men.

"CHILDREN OF THE GENDER REVOLUTION" AS A LENS FOR MAPPING CHANGE

The lives of young adults growing up in a period of large-scale institutional restructuring provide a fulcrum point of social change. It is they who face the most intense conflicts and they who will be forging new directions and strategies.

Norman Ryder aptly termed young adulthood a strategic phase in the "demographic metabolism" of birth, aging, and death. Poised between the dependency of childhood and the irrevocable investments of later adulthood, this life stage represents both a time of individual transition and a potential engine for social change.

Each generation's choices are both a judgment about the past and a statement about the future, but the life strategies of today's young adults are especially consequential. They came of age in an era of unprecedented change, and they are building lives in a world that bears little resemblance to that of their parents and grandparents. Their experiences illuminate the ways that diverse work, family, and gender arrangements shape life chances and how, in turn, people use their experiences to craft strategies that influence the trajectory of change.

Members of this generation are "children of the gender revolution" in two senses. They grew up watching their parents cope with new family forms, unexpected economic insecurities, and expanding options for women. Facing dilemmas about whether and how to craft their own ties to partners, offspring, and jobs, they are also negotiating their own transition to adulthood. To discover the experiences of this strategically situated generation, I interviewed a carefully selected group of 18 to 32 year-olds about their experiences growing up, their current work and family strategies, and their outlooks on the future. Their lives provide a window through which to view the consequences of social change and its future prospects.

The Ambiguity of Family Structure: Divided Views on Parents' Work and Marital Choices

The young women and men who were interviewed experienced the full range of changes that have been taking place in U.S. homes. Most lived in some form of "nontraditional" household before reaching 18. About 40% experienced a parental separation or divorce at some point, and among the 60% who grew up with both biological parents, more than half of these parental couples relied on two paychecks to keep the family afloat. Although the remainder lived in a more traditional home, where mothers worked intermittently, secondarily, or not at all, many of these marriages also changed in significant ways as children grew to adulthood.

How do these children view their parents' diverse arrangements and choices? While the conventional wisdom argues that children do best in families with two biological parents and a home-centered mother, the young women and men in my study hold more complicated and divided assessments. Among those who lived in homes where mothers did not work for pay in a committed way, almost half (48%) wished their mothers had pursued a different alternative. When domesticity appeared to undermine their mother's satisfaction, disturb the household's harmony, or threaten its economic security, a child concluded it would have been better if his or her mother had pursued a more sustained commitment to paid work.

In contrast, almost 8 out of 10 of those who grew up in a home with a work-committed mother believe this was the best option. Although a minority concluded that long working hours, blocked opportunities, and family-unfriendly workplaces made their mothers feel overburdened and time-stressed, most focused on the increased economic resources, financial stability, and personal self-confidence that employment provided to mothers as well as fathers.

Young people are even more divided about whether their parents should have stayed together. Among those whose parents broke up (or never married), a slight majority wished they had stayed together. Yet close to half concluded that, while not ideal, a parental separation was better than living in a conflict-ridden or silently unhappy home. More surprising, although most children whose parents did stay together thought this was the best arrangement, 4 out of 10 felt their parents might have been better off apart. Whether their parents remained together or broke up, young people drew lessons from the long-term consequences. When married parents appeared to grow more distant and unhappy, children developed doubts about the wisdom of sticking it out. When divorced parents were able to get back on their feet and create a better life, children developed a positive outlook on the decision to separate.

In all these circumstances, we cannot deduce children's outlooks and reactions from the *form* of family arrangements. Instead, children focused on how well their parents (and other caretakers) were able to meet the twin challenges of providing economic and emotional support.

FROM FAMILY STRUCTURES
TO FAMILY PATHWAYS

Family life matters, but family "type" provides a limited and ambiguous framework for explaining children's perceptions. Such bulky categories as traditional, dual-earner, and single-parent mask more complex and subtle variations *within* family types. These static categories also draw attention away from how families *change* in both form and functioning as children grow up. In contrast to the image of a static family "structure," my respondents recounted dynamic processes that unfolded in unexpected ways, creating "family paths" that often involved dramatic shifts in a child's sense of support.

Even when children did not experience a clear change in the composition of their household, such as a parental breakup, most recalled living in different "families" as their parents' relationships, economic circumstances, and job statuses shifted over time, thus expanding or eroding their sense of support. About a third described families that remained generally stable and supportive, while less than 1 in 10 faced chronic domestic conflict and insecurity; but almost a quarter believed their families improved over time, while more than a third experienced eroding support as family life seemed to unravel. Despite their diverse destinations, these trajectories underscore how families are fluid and dynamic. Family life is a film, not a snapshot.

Family paths are crucial for charting children's views, but they cannot be reduced to changes in family form. Among those whose families remained or became traditional, slightly more than half report that their homes became more supportive while slightly less than half report eroding support. Similarly, while the majority of children exposed to a parental breakup described an eroding family path, 44% experienced improving domestic circumstances after a parental breakup. Even though most children in dual-earning families recounted stable or expanding support, a quarter of these children disagreed. Expanding and eroding support could occur in the context of either lasting marriages or parental separations as well as in both dual-earner and single-earner households.

Family Paths and Gender Strategies

If trajectories in family form do not explain a child's perceptions of support, then what does? The answer lies in how parents developed strategies for breadwinning and caretaking in the face of unexpected economic contingencies and interpersonal crises. *Gender flexibility* helped households meet children's financial and emotional needs, while *gender inflexibility* left them ill-prepared to cope with unpredictable economic squeezes and declining parental morale.

Gender Flexibility and Expanding Support Gender flexibility can take different forms and unfold in a variety of ways. For children who grew up in a two-parent home, a mother's decision to take a job and build a career created more sharing and parental satisfaction. When Josh's depressed mother went to work, her morale improved and his father became a more involved caretaker. When Chris's mother agreed to become the family's main breadwinner, his father was able to leave a frustrating job and retrain for a more satisfying career. In these cases, two-parent homes became more egalitarian, cohesive, and financially secure.

In one-parent homes, parental separations sometimes reduced domestic conflict, enhanced a custodial parent's morale, and helped households establish greater financial stability. For Danisha, divorce reduced the turmoil swirling in her household and prompted her parents to better collaborate in caring for her and her siblings. After Miranda's mother's left her father, a kind but "stubborn" man who could not keep a job but did not want his wife to work, it marked a turning point when Miranda's family life became more economically stable. When Mariela's philandering mother moved out, her father remarried someone who became, in Mariela's words, a "real mother" who gave her more attention and also contributed much-needed financial resources.

Family support also expanded when parents were able to rely on a broader network of caretakers and breadwinners. When their single working mothers moved near their grandparents, Nate and Isabella both gained "another parent." In the face of his parents' job setbacks and losses, Ray's family relied on his grandparents' financial contributions to weather their economic storms.

Though the changes took different forms, these families encountered un-expected crises that prompted their custodial parents and other guardians to transgress traditional gender boundaries and create new ways of earning and caring. For everyone with an improving family trajectory, more domestic equality in lasting marriages, beneficial parental breakups, new—and better—remarriages, and expanded care networks all encouraged rising parental morale and increased economic stability.

Gender Inflexibility and Eroding Support Gender inflexibility left other families ill-equipped to cope with unavoidable, but unanticipated, challenges. Parents in some two-parent homes were unable to develop new ways to share working and caretaking, despite dissatisfaction with rigidly gender-divided arrangements. Joel's parents, for example, "got stuck" in a traditional division of responsibilities even though his mother grew increasingly unhappy at home and his father felt trapped in a dead-end job. Sarah's mother became depressed and "overinvolved" when she faced an unexpected pregnancy and relinquished a teaching career to stay home.

Dual-earner homes do not guarantee happier outcomes, however. When parents became mired in power struggles or overwork, domestic contexts were also demoralizing. Michelle's father, for example, opposed her mother's career and refused to help at home. Patricia's mother insisted on doing all the cooking and cleaning, even though she provided the lion's share of the family's income. Michelle's parents ultimately parted after she left home, and Patricia wished her mother would do the same.

It is not surprising that parental morale and financial resources decline in the wake of a breakup, but these consequences also reflect parental difficulties in transcending gender boundaries. Nina's home fell into poverty not just because her father abandoned them, but because her mother never held a job and feared joining the workforce after his departure. When Hank's father "walked out," his stay-at-home mother not only resisted going to work but, even worse, turned to alcohol. On the other side of the gender divide, William's mother had no trou-ble supporting the family on her banker's income, but his once-involved father left to marry a younger woman and no longer spent time caring for him and his brothers.

Finally, the loss of support from other caretakers contributed to an eroding family path, especially when it coincided with the loss of parental support. When Jasmine's grandmother died shortly after her parents' breakup, it felt like a larger loss than her father's departure. Whether the problem was a marital impasse, a problematic parental breakup, or a smaller care network, when rigid gender boundaries prevented mothers from taking jobs or fathers and others from becoming or remaining involved caretakers, declining family support followed.

Families and Gender in a Changing Social Context The nature of parents' and other guardians' gender strategies shaped children's perceptions of their fam-ilies' pathways. Gender flexibility in breadwinning and caretaking helped parents and others meet children's economic and emotional needs. Gender inflexibility left them poorly prepared for a host of unavoidable challenges to a traditional division of tasks and responsibilities. Although these challenges were unexpected, they are not random. They reflect widespread and inexorable social shifts that have undermined the "family wage" and the organized "male" career, raised expectations for marital happiness and provided new opportunities to remain single or leave unhappy marriages, and fueled women's growing need and desire to pursue a life beyond domesticity and dead-end jobs.

In the context of rising economic uncertainty, expanding options and fluidity in intimate relationships, and the rising work aspirations of women, most families will confront economic, social, and interpersonal contingencies that encourage, and often force, family change. Homes that are flexible in their strategies for breadwinning and caretaking are better equipped to cope with—and prevail over—these unpredictable, but inescapable, challenges. Homes that are unable or unwilling to transgress gender boundaries are, in contrast, ill-prepared to cope with economic squeezes, a mother's declining morale, or a father's inability (or refusal) to provide support.

Negotiating the Future: High Hopes, Guarded Strategies

So what do young adults want for themselves? Despite their diverse childhood experiences, they have notably similar aspirations for adulthood. The overwhelming majority (95%) hope to create a life-long intimate bond with one partner. It would be misleading, however, to equate the ideal of a lasting relationship with the desire for a traditional one. To the contrary, 80% of women and 68% of men wish to build an egalitarian partnership with room for considerable personal autonomy. Not surprisingly, three-fourths of those who grew up in a dual-earner home want their spouses to share breadwinning and caretaking; but so do more than two-thirds of those from a more traditional home and close to nine-tenths of those with a single parent. Whether reared by traditional, dual-earning, or single parents, the overwhelming majority of men as well women want to forge a committed bond where both partners share paid work and family caretaking flexibly and equally.

When it comes to their aspirations, young women and men are thus more alike than different. Both hope to integrate family and work in their own lives and to balance care and autonomy in their relationships. Yet they are also worried that seemingly insurmountable obstacles block the path to integrating work and family life in an egalitarian way. They hold deep and realistic fears that time-demanding jobs, a dearth of child-care and family-leave options, and their own high standards for an intimate relationship will place their ideal scenarios out of reach.

Confronted with so many obstacles, young women and men are preparing for less ideal circumstances by pursuing fall-back strategies that offer some insurance against their worst fears. Despite the large overlap in women's and men's aspirations, their second-best strategies point to a new gender divide that differs starkly with the one touted by media analysts and social critics. In contrast to the popular argument that young women are "opting out" of the workplace, almost three-quarters of the women are preparing to fall back on "self-reliance." They see work as essential to their survival and marriage as an appropriate option only if and when they can find the right partner. Men, however, worry that equal parenting will cost them at work, which they believe must remain their first priority. Seventy percent of men are planning to fall back on a neotraditional arrangement that leaves room for their partner to work but reserves the status of primary breadwinner for themselves. These fall-back strategies are not only different but also at odds. Despite the shared desire to strike a balance between work and caretaking in the context of an egalitarian relationship, "self-reliant" women and "neotraditional" men are on a collision course.

Women's Search for Self-Reliance   In contrast to the media-driven message that young women are "opting out" of the workplace for marriage and caretaking, almost three-quarters of the interviewed women are reluctant to surrender their autonomy in a traditional marriage and are determined to seek financial and emotional self-reliance. To these young women, the fragility of marital bonds makes relying on a husband for economic security seem foolhardy. Self-reliance offers protection against the dangers of fragile relationships, economic dependence in marriage, and the social devaluation of domesticity. Accordingly, they seek both economic self-sufficiency and a separate identity by establishing strong ties to paid work. Danisha, an African American who grew up in an inner-city, working-class neighborhood, declared:

> Let's say that my marriage doesn't work. Just in case, I want to establish myself, because I don't ever want to end up, like, "What am I going to do?" I want to be able to do what I have to do and still be okay.

Jennifer, who was raised in a middle-class, predominantly white suburb, agreed:

> I have to have a job and some kind of stability before considering marriage. Too many of my mother's friends went for that—"Let him provide everything"—and they're stuck in a very unhappy relationship, but can't leave because they can't provide for themselves or the children they now have.

These young women do not believe their search for a nonnegotiable base in the world of paid work precludes having a life partner, but they are determined to set a high standard for a worthy relationship. Economic self-reliance and personal independence make it possible to resist "settling" for anything less than a satisfying, mutually supportive bond. This outlook, in turn, encourages them to postpone commitment and to view marriage as both optional and reversible. Rachel, whose own parents separated, explained:

> I'm not afraid of being alone, but I am afraid of being with somebody's who's a jerk. I want to get married and have children, but it has to be under the right circumstances, with the right person.

Self-reliant women also concur that if a worthy relationship ultimately proves out of reach, remaining single need not mean rejecting motherhood or becoming socially disconnected. Just as they are redesigning relationships, they are also redesigning motherhood. These women would prefer to raise children with a committed partner, but they are willing to do so without one. They see breadwinning as an aspect of good mothering; and they hope to create a support network of kin and friends with whom to share care and who, if needed, can substitute for an intimate partner....

Men's Search for Neotraditionalism   Men, in contrast, are more inclined to fall back on a more traditional relationship, although in a modified form. Faced with escalating time pressures, rising insecurity at work, and a cultural paradigm that sees men's earnings as the core measure of their "marriageability," 70% of men concluded that equal sharing, however appealing, is too costly. Yet these men also felt torn between their desire to succeed—or at least survive—in the marketplace and growing pressures for egalitarian sharing in their relationships. To reconcile these conflicts, they hope to soften the boundaries between earning and caring without relinquishing their claim to breadwinning prerogatives.

In a variety of ways, these men seek to create a neotraditional alternative in an age of women's work. Breadwinning remains an integral, nonnegotiable aspect of their own identity and thus forms the bedrock of their family commitments. Involvement in caretaking, though crucially important, must nevertheless take second place. As a corollary, they distinguish between a woman's "choice" to work and a man's "responsibility" and "right" to do so. Jim believed he needed to work "full-time, all the time," which meant that his wife would be the one to "fit work in" when a child arrived.

> How are you gonna get ahead if you're not at work? [So] if somebody's gonna be the breadwinner, it's going to be me. I always feel the need to work.... This may sound sexist, but she'll just have to take time off.

By shifting the meaning of equality from equal sharing to "women's choice," this outlook makes room for an employed partner without undermining men's position as specialists in breadwinning. Because this strategy frames women's—but not men's—work as "optional," it converts belief in a child's need for intensive parenting into an injunction for "intensive mothering."...

Reaching Across the Gender Divide?

Young women and men face different dilemmas, but shared uncertainties. The social and economic shifts that have pushed and pulled women into the paid workforce have also eroded stable career paths for men in both white-and blue-collar occupations. These changes have prompted young people to develop new strategies for career and family building.

Young workers are losing faith in the "career mystique" that once promised steady, predictable movement up a structured occupational trajectory. Fearful that a career tied to an organizational hierarchy may leave them on a ladder that collapses before they reach its higher rungs, young men as well as women hope instead to personally tailor their own careers by shifting jobs and even occupations as new opportunities arise or older ones are foreclosed. After watching his father suffer a career-ending layoff after 25 years of service, Joel rejected the notion that remaining a loyal employee would ensure job security or a rising income:

> He was figuring to retire with that place. It came as a shock. I used it as a learning experience—that things aren't as stable as you might think, and not to make a choice just because of security. Consciously or subconsciously, I don't want to fall in that situation.

...Young workers also hope to integrate their public and personal lives by balancing work and family. Men as well as women hoped to cross the spatial and temporal boundaries that separate work and care, both by bringing work home and taking children to work. William sought a small biotechnology firm with a relaxed, child-friendly environment after finishing his chemistry degree.

> I'm hoping to work in a small company which is really informal, so I can bring the kids in the office and play around, work odd hours that make me able to do it all.

...Women and men also refashioned definitions of ideal parents and partners. Rather than focusing on differences between mothers and fathers, they emphasized that a good parent provides *both* financial support and devoted care.... Rather than focusing on differences between husbands and wives, they argued that a good partner is *both* an earner and a caretaker. Although Ken was reared in a traditional middle-class family, he knew searching for a work-committed partner implies providing not just moral, but practical, support at home.

> [I'm looking for] the opposite of what my parents have—someone who's professional, with mutual admiration and support. Showing respect for what the other person does. Not just saying that you love somebody, but showing it through actions. So I hope we split things right down the middle.

...In sum, these children of the gender revolution seek to personally craft their own careers, to transcend the spatial and temporal boundaries between home and the workplace, to redefine the meaning of an ideal parent and an ideal partner, and to reject rigid judgments about "better" and "worse" family forms. These behavioral and ideological strategies are responses to changing social and economic contingencies, but they also offer new templates for enacting more flexible, egalitarian gender ideals.

Institutions and Changing Lives: Reframing the Theoretical and Political Debate

Growing up during the gender revolution has prompted young adults to see family life as an unfolding process that responds to changing social and interpersonal contingencies. It has encouraged them to develop a set of shared aspirations, including wanting to forge a life-long intimate partnership, to balance committed work with devoted parenting, and to craft flexible, egalitarian ways of sharing earning and caretaking. Yet encounters with resistant institutions have also convinced young women and men to develop fall-back strategies to survive in the (all-too-likely) event that their options fall short of their ideals....

Gender Flexibility as the Key to Family Resilience   Postindustrial life poses risks and challenges to *all* types of households. Single-parent and dual-earning homes may face difficulties balancing paid and domestic work, but sole-breadwinner homes may face equally perplexing dilemmas about how to survive on one paycheck or avoid the perils of feeling stuck in rigid gender "roles." Since few contemporary families are immune from some type of crisis, a household's ability to resolve the specific conflicts it faces are more consequential than the form it takes at one point along the way. Why, then, did some children conclude their homes became more supportive and stable while others recounted a cascade of destabilizing events?

Only a gender lens can make sense of these divergent family pathways. Across diverse family types, the flexibility of parents' and other caretakers' gender strategies shaped a child's perception of expanding or eroding support. When families encountered unanticipated contingencies, flexible approaches to breadwinning and caregiving helped them overcome economic uncertainties and interpersonal tensions. When mothers, fathers, and others could not transcend gender divisions that prevented them from providing financial support or gaining personal satisfaction, children watched their caretakers endure the difficulties of unhappy marriages, dissatisfying jobs, and a dwindling safety net. Amid a social and economic landscape that is undermining clearly drawn divisions between earning and caring, flexible gender strategies help families meet children's economic and emotional needs, while rigid gender boundaries leave them ill-prepared to cope with twenty-first-century contingencies.

Social Change as a Clash Between Changing Lives and Resistant Institutions   Analysts of family and gender change have posed starkly divergent and contradictory scenarios for the future. Some argue that a rising tide of "opt-out" mothers foreshadows a return to tradition, especially among educated women. Others see a deepening "decline of commitment" in the growing number of single adults. Although profound changes are undoubtedly occurring, this debate poses a false dichotomy between turning back to a more stable but unequal family order versus moving toward a society of frayed social bonds. Instead, the experiences and strategies of today's 20- and 30-somethings reveal a growing clash between new needs and intransigent institutions. While most seek to blend the traditional value of commitment with the modern values of gender equality and work-family balance, they face workplaces and communities that expect them to choose between sustained work commitment and intimate caretaking. The direction of social change thus depends on whether the structures of work and caretaking can change to support the revolutionary and irreversible shifts in individual aspirations and family needs....

Gender flexibility needs to be a centerpiece of collective efforts to restructure work and caretaking. Not only do most young adults want to create flexible, egalitarian partnerships, but a mother's earnings and a father's involvement are both increasingly integral to the economic and emotional welfare of children. We can thus achieve the best family values by creating flexible workplaces, providing equal economic opportunity for women, outlawing discrimination against all parents, and building child-friendly communities with plentiful, affordable, and high-quality care. Amid new economic and marital uncertainties, institutional support for flexible, egalitarian options to blend earning and caring provides the key to fostering both individual and collective well-being. Gender flexibility and equality are not in conflict with family well-being, but are necessary ingredients to achieve it. The answer to twenty-first-century work and family conundrums is to finish the gender revolution, not turn back the clock.

30

Constructing Gay and Lesbian Parent Families 'Beyond the Closet'

MAURA RYAN AND DANA BERKOWITZ

Beyond step- and reforged families, we have seen the rise of gay families. The first real wave of gay families and the children they adopted or birthed occurred in the 1980s. Since that time families, the courts, and members of society have had to think about how they feel about this nontraditional family form and the children reared within it. People have struggled over whether children raised by gays are more likely to become gay, if they experience sexual or gender role confusion, or if they lack certain advantages held by traditional heterosexual families. But most studies have found that outcomes for children of gay and lesbian parents are no better—and no worse—than for other children, whether the measures involve peer group relationships, self-esteem, behavioral difficulties, academic achievement, or warmth and quality of family relationships. The children of gay and lesbian families may encounter anti-gay sentiments, and most report feeling angry, upset, or sad about these experiences. Gay parent support groups and supportive environments generally help to alleviate some of these problems. Like families headed by heterosexual parents, lesbian and gay parents and their children are diverse groups. Ryan and Berkowitz's study ably illustrates this point, offering data on gays and lesbians, on those who are racially homogeneous or mixed, and on those who birth as opposed to adopt their offspring. They also offer data on societal preferences for blood family ties. After reading this article, how do you feel about gay families? Do you think that attitudes about gay marriage are changing or will remain steady? It was

SOURCE: Reprinted by permission of Springer Publishers.

only in 1967 that interracial marriage became legalized; how do you think people will look back at gay marriage bans fifty years from now?

Gay men and lesbians today have more opportunities to create families than ever before. Such opportunities are an amalgamation of the modern gay and lesbian movement, transformations in cultural ideologies, broad changes in families, and revolutions in medical technology. Still, the creation of a gay or lesbian parent family inevitably requires the reproductive assistance of parties outside of the same-sex couples who wish to create families. This paper provides a multi-layered analysis of how lesbian mothers and gay fathers construct their families where negotiating non-biological relatedness is a necessary aspect of their endeavors. We highlight that while such negotiations also hold true for many infertile and adoptive heterosexual couples, gay and lesbian parents differ in that they face an added layer of complexity as they confront heterosexism in family-building bureaucracies and traverse a social terrain where they are visible as non-biologically related families.

To explain this, we employ our 40 interviews with lesbian mothers and gay fathers as another piece of empirical evidence that sustains Seidman's (2004) notion that for many gay and lesbian individuals, subjectivities and experiences are forged "beyond the closet," in that many gays and lesbians do not have to live closeted lives anymore, but where their lives are defined by minority sexual orientation status in a system of heterosexual dominance. We stress how our participants' family-building experiences are somewhat similar to other non-biologically related families, but qualitatively distinct due to heterosexual dominance.

Gay and lesbian parents offer a unique opportunity to think about how some parents must juggle the limitations of their physiologies *and* homophobia in child-granting bureaucracies *and* a constant response to their visibly non-biological family. Relying on a qualitative methodology of in-depth interviews with 18 lesbian mothers and 22 gay fathers who constructed their families through non-hetero-normative means, we generate new insights about how the privileging of biological relatedness coupled with heterosexual dominance shapes the family experiences of lesbian mothers and gay fathers in institutional settings and interpersonal interactions.

STANDARD NORTH AMERICAN FAMILIES (SNAF) AND NON-BIOLOGICALLY RELATED FAMILIES

The idealized notion of the standard nuclear family holds a sanctified place in the hearts and minds of the American public and it is embedded in our social, religious, and legal institutions. The hegemonic standard North American family (SNAF) of two heterosexual married persons parenting their biologically produced children is more than a privileged model for families; it has become an "ideological code." Despite the heterogeneity of families in contemporary Western society, those that deviate from the norm are judged to be deficient and inadequate.

The past decades' culture wars over homosexuality have meant that both popular media pundits and policy makers concern themselves with the rights granted (or not granted) to gay and lesbian citizens. In recent years the political attention directed at gays and lesbians has been particularly concerned with issues of gay headed families, focusing on gay marriage and same-sex adoption. Despite these legal obstacles lesbians and gay men have continued to build families and consequently the definition of "the family" has changed to incorporate them. Over the past two decades, some gay men and lesbians have "turned the adoption world on its head" while others are utilizing their own physiological capabilities and employing the assistance of surrogate mothers and sperm donors in unprecedented numbers (Lev 2006, p. 73). By using these emerging opportunities and creating planned families, lesbians and gay men challenge normative definitions of family and parenthood.

A key dimension of SNAF's ideological code is the presence of biologically related children. Although many factions of society might disagree that biological relatedness is enough to constitute a family, the ideological code of SNAF has historically defined a real family as such. Schneider's (1980) classic research confirms that the majority of Americans consider the crucial defining elements of kinship to be genetic. More recently, Bartholet (1999) has referred to the privileging of biological relatedness as a "blood bias" or "the assumption that blood relationship is central to what family is all about" (p. 7). Similarly, Nelkin and Lindee (1995) have argued that America's recently burgeoning fascination with genetics has allowed the biological family to gain unheralded strength. Thus, the ideology of genetic family superiority has had significant consequences for the institution of adoption in that all non-biological family forms are rendered pathological and deviant. Adoptive families are viewed as second-best and infertile couples are urged to seek treatment so that they can have the opportunity to conceive a child who is a part of their genetic union.

It is this genetically-biased portrait of family that has led to the "social construction of infertility as a problem requiring high technology medical treatments to produce a biologically related child" (Miall 1996, p. 310). The question of whether or not gay and lesbian couples (and individuals) should be regarded as infertile is a controversial philosophical and political matter. Nevertheless, we argue that a "relational infertility" does accompany gay and lesbian relationships to a certain extent.

Similar to many heterosexual infertile women, heterosexual adoptive parents, and single mothers-by-choice, gay and lesbian parents often carefully labor to construct families that mirror dominant families. Oftentimes this includes a focus on a genetic link. After all, "genetics is both an idea and a roadmap of identity" (Hertz 2002, p. 3) and for many gay and lesbian parents, genetics is one of the few blueprints they have to work with as they mentally sketch their future families. Drawing upon the stories of 40 lesbian mothers and gay fathers, we attend to the nuanced parallels and distinctions between gay and lesbian parents, adoptive heterosexual parents and infertile heterosexual couples. Hereafter, we emphasize the simultaneous importance of institutional heterosexism and the interactional dynamics of constructing gay and lesbian headed families beyond the closet....

DATA COLLECTION

Semi-structured open-ended face-to-face interviews were conducted with one of the 18 lesbian mothers and 18 of the 22 gay fathers; the rest were conducted over the telephone... All of the lesbian birth mothers and gay fathers in our studies identified themselves as white. They were also overhelmingly middle-class and upper-middle-class. Although all of our participants were white, some of them had constructed multi-racial families. Of the 18 birth mothers, three of them were in interracial relationships during their pregnancies (two with African American women and one with a Latina woman). The white woman with a Latina partner and one of the women with an African American partner chose a sperm donor who was Latino/African American, respectively, in order to reproduce bi-racial children. Of the 22 fathers, eight (three couples and two single fathers) were raising children of color, all of whom were adopted....

Although the lesbian participants' intentions toward and experiences with becoming pregnant varied slightly, the majority of them—14 of 18—acquired sperm (or in one case, an embryo) from a sperm bank. Gay participants became fathers in the following diverse ways: 12 through adopting or fostering; five through employment of a surrogate; three through co-parenting with lesbian couples; and two (one couple) through both the employment of a surrogate and an egg donor.

INSTITUTIONAL AND INTERACTIONAL HETEROSEXUAL DOMINANCE

According to Herek (1990), heterosexism is an ideological system that denies, denigrates, and stigmatizes any non-heterosexual form of behavior, identity, relationship or community; it occurs at the cultural and individual level and can be observed in institutions and customs. Similarly, Seidman (2004) refers to heterosexual dominance as the legal, cultural, and social privileging of heterosexuality and persons who are defined as heterosexual.

THE INSTITUTIONAL DIMENSIONS OF HETEROSEXUAL DOMINANCE: NAVIGATING INSTITUTIONS THAT PRIVILEGE BIOLOGY AND HETEROSEXUALITY

All participants were concerned with the extent that their reproductive pathways would be impeded by homophobic individuals and heterosexist norms and as such, navigated fertility clinics and adoption agencies with caution. Although not every participant experienced homophobia and heterosexism in their family-building paths, most did report encountering an experience with at least one of these. Importantly, heterosexism is more than a benign oversight of gay and lesbian experiences resulting from the great number of heterosexual people in society; it is a form of social control which can negatively affect the life chances of gays and lesbians. In family building bureaucracies, heterosexism has the potential to block gay and lesbian opportunities to have children. For example, consider the ways in which the following stories illuminate institutionalized heterosexism in participants' paths to creating families with children:

When Leonard, his partner Ariel, and two women embarked on their co-parenting agreement, the first step was a known-donor insemination process at the neighborhood sperm bank in a suburb of Boston. Leonard was escorted into a small room and given a Playboy magazine as a visual aid to assist him in the masturbation process. He laughed at the heteronormativity of the process and managed to find his own way of filling the plastic cup. However, after completing this process, he was told to fill out a form about his personal, familial, and sexual history. On the last page, was a question that read, *have you ever had sex with a man?* And, *might you ever have been exposed to a person with AIDS?* He was honest on the application and, although Leonard has tested negative for HIV multiple times, the next day his application was denied. Denied access to the clinic's procedures, the four future co-parents resigned themselves to insemination at home.

Donna, a lesbian residing in a small city in Florida, attempted to receive insemination at the only local clinic in the city that provided such a service. Forewarned by friends that the physicians would not inseminate lesbian women she concealed her sexual identity (and her partnered status) by describing herself as a single woman. However, the physicians still denied her request claiming that, "they did not inseminate single women." Such restricted access to fertility treatments is neither uncommon nor undocumented (Robinson 1997). With few options left, Donna chose to have heterosexual intercourse with a male friend and ultimately became pregnant.

Where these stories represent extreme cases, in that Leonard was only one of four fathers who resigned themselves to at home insemination and Donna was the only lesbian mother who chose to have heterosexual intercourse to get pregnant because she had no other options, we share these anecdotes to elucidate how gay men and lesbians seeking parenthood via fertility clinics navigate an arena laden with homophobia and institutionalized heterosexism. For example, although not all of our lesbian participants experienced discrimination, as Donna did in being denied insemination procedures, all of them anticipated institutional discrimination and took steps to avoid it. Cassandra said she was "afraid of finding a doctor who would work with her and her partner." Many other participants cautiously interviewed doctors or toured hospitals before they embarked on pregnancy so that they could guarantee fair treatment during their pregnancies and births.

Although many individuals today can choose to live beyond the closet, they must still reside in a world where most institutions maintain heterosexual domination. Heterosexual dominance is deeply rooted in the institutions and culture of American society and must be understood as not simply a product of laws or individual prejudice, but institutionalized pervasive dominance.

Those men and women who choose not to navigate fertility clinics can employ various forms of adoption to construct their families. However, the field of foster care and adoption remains one in which homophobic practices frequently surface. Yet, 39% of all adoption agencies in the United States did report placing a child with gay or lesbian adopters in 1999-2000. 12 men in author's Berkowitz's sample took advantage of such opportunities in adoption and employed private or public adoption agencies to construct their families. Akin to many adoptive parents, gay adoptive fathers have the dilemma of negotiating bonds with their future children when a birth parent or institutional agent could decide to back out of the pre-birth agreement of their adoption. Often, because of these considerations, gay fathers-to-be discover innovative ways of securing their emotional investment.

Craig and Darrel, an interracial couple who became fathers of their two young girls through fostering by use of the public adoption system discussed the risks involved in this route that surfaced as a direct consequence of biogenetic dominance. Although biogenetic dominance affects all non-biologically related families, the privileging of biological ties is a consequence of the heterosexist arrangements that naturalize heterosexuality. Because of legal privileges granted to the biological mother, Craig and Darrel did not have any law-binding tie to their first daughter until after they fathered her for 2 years; at the time of the interview they still did not have legal rights to their second daughter who they had been fathering for two and a half months. Like many foster parents, regardless of sexuality, they choose to deal with the impending risk of the birth mother returning to claim her children, yet traversed their way through the foster care system in a highly cognizant and conscientious manner. Their first daughter was the eighth child in succession placed in adoptive homes by her biological mother and their second child is the third. By consciously choosing to foster daughters who have mothers that have relinquished their other biological children, these men believe they are minimizing the chances of their parental rights being challenged by their children's biological parents. Similarly, Drew and Nico discussed their apprehensions of creating a family through adoption because of the risks of the birth mother returning:

> The thing about adoption is…that even though that child or those children are legally yours, they are never your children. And that is very frightening to me. That [we] would have this wonderful child or children through adoption and then at some point, something could happen, either through the courts or a change of the birth mother's mind…it is very unsettling to me and scared me. It scared me that the family we would create would be shaken by the birth mother or the genetic father coming back into our lives or the baby's life.

Because of this fear, and because their privileged class status permitted them to do so, Drew and Nico chose to construct their family through the employment of a surrogate mother. With the legal authenticity of a blood-tie, the couple felt more able to invest affection into their children without concern that they could one day be taken from them. All families interested in adoption or foster care must navigate the legal privileges granted to biological families. However, gays and lesbians enter into such investments differently than heterosexual families because of the conscious efforts to juggle the limitations of physiology and discriminatory legal practices.

For instance, Josie, who would later become the birth mother of two boys, experienced a failed adoption before deciding to get pregnant. An acquaintance of hers became pregnant with a child she neither wanted to abort nor raise. Josie's pregnant friend agreed to let her adopt the future child. Much later in the pregnancy, Josie's pregnant friend had a chance encounter with the biological father and told him she was pregnant with his child. He was willing to relinquish his genetic rights to the child, but his parents were not. On the grounds that the child should not go to *someone like* Josie, based on her sexual orientation, she had a failed adoption which propelled her to become pregnant herself. Although Josie is the only birth mother Ryan interviewed who had a failed adoption, the possibility of legal difficulties and fear of the eventual loss of child custody were cited by other lesbian participants as the foremost reason in the decision to produce a biological child.

For Susan, pregnancy may have been exciting, but the prospect of attaining the goal of having children overshadowed her view of pregnancy. While many of Ryan's participants also discussed excitement regarding the bodily experiences of pregnancy, this was secondary to the perceived legal protection of one parent having biological ties to their child. Note, for example, Blanche and Maura (Ryan)'s conversation below:

MAURA: Why was it that you and your partner thought about having children through pregnancy rather than other options?

BLANCHE: Adoption. We've talked about adoption, but in Florida, it's impossible. If it ever became legal for us to adopt in Florida, we'd do it in a heartbeat.

MAURA: So it wasn't really that you wanted to experience pregnancy, but that it was the most—

BLANCHE: Yeah, I guess we did. We both wanted to know what it was like to be pregnant and to give birth and to be able to hold that against people when they say something—to say, I've given birth! [laughs] I'm kidding. We did definitely wanted that whole experience.

Notice that although Blanche and her partner wanted to experience pregnancy, their investment in experiencing pregnancy was secondary to the legal restraints posed by Florida's ban on gay adoption. For Ryan's lesbian participants, the "whole experience" includes the experience of pregnancy, a biological connection between one parent and the couple's child, and a legal claim to their children....

INTERACTION DIMENSIONS OF HETEROSEXUAL DOMINANCE: NAVIGATING SOCIAL SITUATIONS THAT PRIVILEGE BIOLOGY AND HETEROSEXUALITY

The everyday production of family life in the United States privileges biological relatedness and blood-ties in institutional and symbolic ways. Our participants' use of biological ties in creating their families is a strategic reconciliation of homophobic discriminatory legal practices, we also maintain that it points to ideological preferences for blood-ties.

Dominant family ideology establishes biological relatedness as critical for defining family. As such, being a parent is often understood as being a *biological* parent. Billy and Elliot, recent fathers of a set of twins decided to mix their sperm before inseminating their chosen egg donor. They maintain that because there are two children and two "fathers," each man is the biological father of a twin. Although these men are uncertain about their biological paternity status for each twin, their story illuminates how meanings associated with aspects of the reproductive sphere emerge out of a social process with the "blood bias" at the forefront of these men's consciousness. Further, their explanation allows their family to appear like other "normal" families: They are both biological parents.

It is perhaps true that biology can contribute to an individual's physical and mental health, the way they look, or the way they behave. Still, its strongest power lies in what people believe it can do. Folk knowledge understandings of biology posit that biology is nearly fully responsible for making people who they are and that on its own it can create bonds between people. For this reason—because of the things society believes it to be—biology is also social. The social aspect also becomes salient in our participants' description of manipulating biology for an audience who is judging the authenticity of their families. In short, biology can be finessed in order to procure better social treatment.

USING BIOLOGY TO SYMBOLIZE FAMILY

Individuals or couples who construct non-biological families have a unique opportunity to consciously create what their families will look like. In her study on British lesbian donor choices, Jones (2005) found that her participants preferred donors who resembled the non-biological mother in racial and ethnic characteristics. Similarly, when the choice was available, all of our participants focused on reproducing children who would blend into their families. Many participants evaluated paper or web-based documents about potential donors, and others engaged in face-to-face interactions with the future donor or co-parent(s) before embarking on constructing their families. Billy spoke about how he and his partner, Elliot, evaluated a series of egg donors from a catalog. He explains, "It's funny how you can read these profiles. After you read a couple of them, you sort of really hear the voice of the person…it was like a yearbook. You know a photograph with a site description."

This same phenomenon of a preference for homogeneity also manifests itself in the context of planned lesbian headed families. Sameness in lesbian planned families refers to the idea that the donor is preferably very similar to either the birth mother or the non-biological mother in the characteristics of race, ethnicity, hair color, eye color, complexion, and other physical characteristics. Amongst participants who looked for similar characteristics of the non-biological mother in their chosen donors, it was evident that *intra*racial white couples chose ethnic markers (e.g. a donor who was Irish if the non-biological mother was Irish) and hair color preferences which matched the non-biological mother; however, this preference was more salient for participants in interracial partnerships. Interracial lesbian couples face the double burden of sanctions for their sexual orientation and their interracial romantic commitment. When having children, some research suggests that interracial lesbian couples may choose to incorporate the race of the non-biological mother in their donor decision so that their family oppression may be lessened slightly by their resemblance to each other (Jones 2005). This was true for two of our three lesbian participants in interracial relationships. For example, Susan, a white participant who had an African American partner said, "We're together, we're a biracial couple, doesn't it make sense for the child to be more a representation of us than just of me?" As with Jones' (2005) participants in an interracial couple, Susan and her partner chose to reproduce a biracial child to reflect the non-biological mother's race. Importantly, while the non-biological mother is *not* genetically related to the child, the choice of an African American donor allows their family to *look* genetically linked.

Some mothers chose donors who looked like them. For instance, Carol, another white birth mother in an interracial relationship, chose a white sperm donor even though she was going to be raising the child with an African American woman. While she had finalized the donor decision before meeting her partner, she was adamant in her decision to use a donor who resembled her features. She elaborated:

> I want when he's going through school, you know, I want...I don't know. I might be selfish I guess. But I just wanted him to look like me to say, *Oh, yeah, he looks like you*, you know? Like I wanted someone to know that he is mine.

Whether the desired characteristics of the sperm donor lie in the sameness of the non-biological mother or the birth mother, the guiding reason is analogous: to create the appearance of a genetic family. For instance, Lydia, birth mother of one young son, said that although she knew there would be no interweaving biological tie between her, her partner, and their child, they could create the appearance of a genetic link. She said, "We're a family and if we have the chance to increase our chances of our kid looking like both partners, of course we would do that." Rhonda, birth mother of one young daughter, similarly said that the decision to make their child characteristically similar to her and her partner might make them "a little bit more of a close family." In these ways, our lesbian participants are responding to culturally pervasive heterosexist family understandings. If the leading mainstream logic is that families must look similar to be taken as families (because children are products of the two heterosexual people involved in creating the child), lesbian mothers may negotiate this assumption by striving to construct families that appear genetically linked. Further, the manipulation of biology—making a family appear genetically linked when it is evident that they are not—is a particular strategy in softening meso-level heterosexual domination. For example, Drew and Nico spoke to Berkowitz about the conversations that occurred as they navigated their surrogacy agency of choice's website that depicted hundreds of potential surrogate mothers. Drew explained:

> Well, on the website a lot of the women were 4 foot 2, Guatemalan women; it just wasn't going to work for us....We wanted to find a surrogate who was white and like get rid of one other problem that these children, or child would have to deal with, you know, to be mixed race.... We wanted someone who was fairly young, who had done it before and who was remotely attractive.

Drew and Nico did not explicitly state that they needed their child to resemble them (e.g. we wanted her to have a nose like Drew and eyes like Nico); however, in explicitly stating the importance of a white surrogate mother (who would produce a white child) they highlight how their child's whiteness would help their family appear more like the ideological code of SNAF.

Carol, the white birth mother who chose to raise a white child with an African American woman, explained to Ryan that when her African American partner is out alone with their white child she is regularly asked if she is the child's nanny. When family compositions deviate from the ideological code of SNAF whether it be because the family is multi-racial or headed by same-sex parents, the family is more susceptible to questions about its formation. For this reason, the doing of family becomes more of a performative accomplishment....

DISCUSSION

Today the choice of having a child is available to those lesbian women and gay men who are able to financially and interpersonally navigate the bureaucratic apparatus of insemination clinics, surrogacy, adoption and fostering agencies, or who choose to traverse co-parenting arrangements. They can choose to be single parents, coupled co-parents, or co-parents with platonic friends. Lesbian mothers and gay fathers have more choices than ever before in constructing their families. Although lesbian and gay parent families clearly push the boundaries of SNAF, it is pertinent to keep in mind that gay and lesbian parent families do not occur in isolation. To a large extent, these new options for gays and lesbians to build families are still shaped by social institutions and by dominant ideologies about families. Gay headed families are resisting dominant constructions of family by their very existence. However, in this article we have suggested that the larger social processes that surround gay and lesbian parent families, independent of their individual will, construct their knowledge about how to appropriately do family. The binary categorization of gender, heteronormativity, and the ideological code of SNAF compose some of these larger processes. The complex ways that these socio-cultural institutions shape society and the individuals in them are critical in shaping gay and lesbian procreative consciousnesses, reproductive decision-making, and parenting experiences.

REFERENCES

Bartholet, E. (1999). *Family bond: Adoption, infertility, and the new world of child production.* Boston, MA: Beacon.

Herek, G. M. (1990). The context of anti-gay violence: Notes on cultural and psychological heterosexism. *Journal of Interpersonal Violence, 5,* 316–333.

Hertz, R. (2002). The father as an idea: A challenge to kinship boundaries by single mothers. *Symbolic Interaction, 25,* 1–32.

Jones, C. (2005). Looking like a family: Negotiating bio-genetic continuity in British lesbian families using licensed donor insemination. *Sexualities, 8,* 221–237.

Lev, A. I. (2006). Gay dads: Choosing surrogacy. *Lesbian & gay psychology review, 7,* 73–77.

Miall, C. E. (1996). The social construction of adoption: Clinical and community perspectives. *Family Relations, 45,* 309–317.

Nelkin, D., & Lindee, M. S. (1995). *The DNA mystique: The gene as a cultural icon.* New York: Freeman.

Robinson, B. (1997). Birds do it. Bees do it. So why not single women and lesbians? *Bioethics, 3,* 217–227.

Schneider, D. M. (1980). *American kinship: A cultural account.* Illinois: University of Chicago Press.

Seidman, S. (2004). *Beyond the closet: The transformation of gay and lesbian life.* New York: Routledge. Reprinted by permission of Springer Publishers.

31

Muslim Self-Identities
after 9/11

LORI PEEK

*Americans got a jolt about Islam, a religion that had been flying below the radar
screen, after the tragedy of 9/11. Suddenly we were focused on Muslims around
the world and in our own country. Muslims in America were jolted into a new
and harsh reality. Were they the enemy? Could they be trusted? Were they
working undercover for terrorists? Overnight we went from a country that had
embraced multiculturalism to one that practiced racial profiling. In this poignant
paper, Peek traces the experiences of Muslim college students in the aftermath of
9/11. While the students she studied in Colorado and New York City came
from diverse ethnic and national origins, they were united by their religious piety
and observance. The young men prayed seven times a day and wore beards; the
young women covered their faces with the veil. This made them very apparent to
their suddenly hostile neighbors. Peek traces the development and evolution of
their religious identity as it came under fire. You will see the effect of the threat
from an out-group on in-group solidarity and identity. How would you feel if an
identity of yours came under attack from a majority of the people? To what
extent do you think these young people's attitudes and behavior might have been
different without 9/11? Do identities forged in crisis last? How important is your
religious identity to you and others around you?*

The religious landscape of the United States has changed markedly over the past
four decades with an unprecedented diversification of the American population,
as millions of immigrants and refugees arrived from Africa, Asia, the Caribbean,
Eastern Europe, Latin America, and the Middle East. The "new" immigrants are
racially, ethnically, linguistically, and religiously more heterogeneous than the

SOURCE: Lori Peek, "Muslim Self-Identities Since 9/11." Reprinted by permission of the author.

immigrants of a century ago, making the United States the most religiously diverse nation on earth (Eck 2001).

Muslims constitute an important part of this increasingly diverse religious landscape. Estimates vary regarding the current population, though it is generally accepted that approximately five to seven million Muslims live permanently in America and the community is growing steadily. In fact, Islam is the fastest growing religion in the United States, and is poised to surpass Judaism and become second only to Christianity in the number of adherents. The Muslim community includes large percentages of African Americans as well as many first-, second-, and third-generation immigrants of South Asian and Arab descent. Additionally, an increasing number of Whites, Latinos, and Native Americans has converted to Islam over the past several decades. Thus, the adherents to the faith represent a broad range of ethnicities, cultures, nationalities, and Islamic ideologies.

As the body of knowledge regarding the traditions and experiences of Muslim Americans continues to expand, few empirical investigations have specifically explored the process of religious identity formation among Muslims in the United States. Based on a qualitative study of a sample of 127 young, mostly second-generation, Muslim Americans in New York and Colorado, this article examines the role that religion played in the lives of the participants. In particular, the analysis focuses on why and how religion became the most salient source of personal and social identity for this group of Muslim Americans.

Various theories have been advanced regarding why certain individuals and communities highlight and develop religious identities, as opposed to other forms of personal and social identity such as race, ethnicity, or nationality. According to Smith (1978:1175), immigration itself is often a theologizing experience, with new arrivals to a country turning to religion. Second, religion may become an important basis for identity due to the functions religion plays in society. In addition to meeting spiritual needs, membership in a religious organization offers many non-religious material, psychological, and social benefits, including community networks, economic opportunities, educational resources, and peer trust and support. A third explanation maintains that religious identity and expression serve to ease the tensions caused by immigrant and ethnic diversity, so that as individuals define themselves, first and foremost, in religious terms, their ethnic variation and national differences become less problematic. A fourth explanation contends that religion may be used to maintain personal and social distinctiveness in the multicultural American context, serving as an important identity marker that helps promote individual self-awareness and preserve group cohesion.

A MODEL OF MUSLIM RELIGIOUS IDENTITY DEVELOPMENT

My research with Muslim university students revealed three stages of religious identity development: religion as ascribed identity; religion as chosen identity; and religion as declared identity. As the participants moved through each of the stages, their faith became more intense and their religious practice increased, as did their identification with the religion of Islam and the individual characteristic of being Muslim. The identity formation model presented below is based on three broad assumptions: that identity is acquired through a social and developmental process; that the length of time taken to proceed through the stages differs from person to person; and that this model applies to a particular group of individuals in a specific social and historical context and is not meant to serve as a universal model for all Muslim Americans during all time periods.

Religion as Ascribed Identity

Of the 127 individuals interviewed for this study, 117 were born into Muslim families (the other 10 converted to Islam as young adults). Because almost all of the interviewees were raised in Muslim homes, most participants viewed religion as an ascribed characteristic of their individual selves and social worlds. During this first stage of identity development, the students reported that they engaged in very little critical reflection when they were children regarding the meaning of "being Muslim" because their religious identity was taken for granted as part of their everyday lives. However, practices required by their parents such as dressing modestly or attending religion classes at a mosque nonetheless reflected a Muslim religious identity. Individuals absorb and internalize many norms, values, and behaviors when they see them exemplified by their parents, peers, and others long before they understand them intellectually.

Although questions regarding religious identity were not important when the participants were younger, they sometimes had difficulties when their religion prevented them from participating in peer activities. Many of the students pointed out that just because they were born into a certain religion did not mean that they comprehended or appreciated the belief system and practices, particularly when they were children. Ali, whose family moved to the United States from Indonesia when he was seven years old, said:

> Junior high school I think there were some Muslims, but I wasn't really practicing, so I didn't know anything. I had many friends. You didn't think about religion that much. It was sports or cartoon characters or whatever. In high school, I think there were at least two more Muslims in my graduating class. I didn't know them that well. I wasn't really practicing as a Muslim. It wasn't my main concern. It was my own personal thing.

Almost all of the participants stated that religion was just one of the many aspects that defined who they were during their formative years. Because most students were the children of immigrants, they often identified according to their parents' national or ethnic backgrounds, at least when they were younger. After some reflection, some of the interviewees concluded that they were more likely to identify themselves this way because of social norms and external pressure in the United States to define oneself by race, nationality, or ethnicity rather than religion. Salma, a native of New York City, related her thoughts on this subject:

> People would say to me, "Where are you from?" Then I'd say, "I was born and raised here. I was born here. I'm an American." And they would say, "No, where are you *really* from?" Once I told my mother that and I said, "I just say I'm Indian." She said, "Why do you say that? You're not Indian." I'm like, "You guys are Indian and so I'm Indian." She's like, "We're Indian because we were born and raised in India. We're Indian citizens. You were born and raised here. You're American. You have Indian background. Your culture is Indian. Your religion is Islam. You're an American citizen." Nowadays, I just say, "I was born here. My parents are from India." It clears up every question.

The interviewees may not have consciously contemplated their religious identities during this stage, but they were well aware of the pressure to assimilate to "American" values and norms. It is not surprising that some of the interviewees said that during their younger years they cast off their religious identity in an attempt to "pass" as part of mainstream society. This is not something they were proud of, but they explained their behavior as a result of their need to fit in or their lack of understanding of the true meaning of Islam. Maryam, a second-generation immigrant from Trinidad who was wearing a headscarf when I interviewed her, discussed her prior fears of being mocked for wearing religious attire:

> In the beginning, when I was younger, I was like, I'm not going to be seen with the *hijab*. The kids are going to make fun of me. I was completely against all of this. I had to dress in the newest jeans that came out, have the nicest sneakers. In the beginning, when I was younger, it was really a big deal.

Religion as Chosen Identity

As children age, they begin to develop a more concrete conception of their religious identity. As the participants in this study matured, they began to view religion not as an unquestionable, ascribed characteristic, but as a chosen identity. When they told stories about their identity development, they frequently mentioned various factors that led to their choosing first and foremost to be identified as Muslim.

Some of the interviewees believed it was "only natural" that as they grew older, they would begin to contemplate important life questions and their religious backgrounds, and hence re-examine that aspect of their identities. Asma, who was originally from Guyana, explained:

> Even though on the exterior I was practicing since I was young, it doesn't mean necessarily that I was spiritually, I don't know, thinking about God, into the faith. It doesn't really happen until you've become an adult, until you learn about the world. You never really get to live on your own, to get a chance to think about what you want for yourself

and what kind of person you are until you get to college. In high school, you always have people telling you what to think.

Asma touches on an important factor in this stage of religious identity development: entering college. Leaving home for college represents a point of transition when rites, rituals, institutional expectations, or regulations cause individuals to reflect on their behavior and identity. For the subjects in this study, the campus setting provided space and time to explore their identities and make choices about who they wanted to be and how they wanted to live their lives. Ultimately, it enabled them to further construct a Muslim religious identity. Zoya, who was raised in New York City, became a lot more religious during college. When I last interviewed her in the spring of 2002, she had decided to quit her business school honors program so that she could focus on serving the on-campus Islamic Center where she had recently been elected president. Her parents were not pleased with this decision, as they had consistently stressed the importance of receiving a good education and getting a high-paying job. However, Zoya felt that it was much more important that she devote her time to the religious community, even though it was difficult for her to defy the wishes of her parents.

The peers that students found in college represented another factor that contributed to their increased religiosity. While growing up, the participants were often one of only a handful of Muslim students, with few Muslim friends. However, all spoke about connecting with other Muslims in college, most frequently through the Muslim Student Associations (MSAs) on their campuses. It became easier to identify as a Muslim simply because they had discovered a larger Muslim peer group to associate with and new friends with whom they could relate. Much of the pressure to conform to non-Muslim society seemed to subside as they found others with similar interests who adhered to the same religious tenets and social norms. Ali discussed the impact of finding Muslim friends at his university:

> In high school I was not very religious. It was in college that I developed a renewed interest in Islam. Being with other Muslims was a factor. People are open as Muslims. In high school they weren't. In college you have more freedom. You're exposed to different ideas and cultures. You're encouraged to experiment. I experimented with Islam. The importance of Islam, being a Muslim, is my main identity and prioritizing that in my life came after I came to college, when I was more integrated in the community here. I knew more Muslims and became more active, learned more about Islam and myself.

Peers and close friends played a significant role in constructing, reinforcing, and affirming the strong emerging religious identity of almost all participants. The interviewees often told me that they began to learn about Islam with their friends since they were going through a similar process of religious exploration. Some individuals even turned to their on-campus religious organizations and peers because they felt they were not receiving enough religious education at home. Their parents were the ones who taught them the fundamentals of Islam, but the interviewees sometimes complained that their mothers and fathers were "too cultural" because they mixed cultural norms and values with religious practices. As they matured, it became increasingly important to develop their own religious identity. Henna, a second-year law student who was raised in the United States, stated the following:

> Definitely in my generation, I've noticed that there's that return to Islam among Muslim Americans. Maybe it's because where our parents grew up, they didn't have to think twice about their identity. They were Muslims. They took it for granted. They never had to question it or explore it. It was a given. Whereas here, for us, it's something we have to work at because there's no one in our high school or elementary school who teaches us about Islam. We have to go and find out about it on our own. Maybe that's the motivation that's making our generation more religious.

Most also felt that it was not only important to maintain their Muslim identity for themselves, but to teach the religion to their children in the future.

Religion as Declared Identity

For the participants in this study, the third stage of religious identity development occurred in response to the events and aftermath of 9/11. Within hours of the terrorist attacks, an unprecedented rash of xenophobic incidents began. In the months following 9/11, thousands of Muslims and Arabs (as well as South Asians, Latinos, Sikhs, Hindus, and other individuals who were mistakenly perceived to be Arab or Muslim) endured discrimination, harassment, racial and religious profiling, and verbal and physical assault. Despite this reaction, most of the Muslim students that I interviewed continued to publicly affirm their religious identities. Indeed, many participants reported that their religious identity became even stronger during this time.

The overwhelming magnitude of the events of 9/11 led many students to pray more often and increased their need for a spiritual anchor, just as the events did for numerous other Americans of various faiths. Many of those interviewed reported becoming more reliant on God as they became more cognizant of their own mortality. The increasingly negative portrayal of Islam following the attacks and participants' perceptions of how non-Muslims now viewed them also caused some students to identify more closely with Islam. In public, they were frequently questioned about their faith and religious beliefs. To help them respond to these inquiries from both friends and strangers, they studied the *Qur'an* and other religious documents. As they searched for answers, they felt they were becoming "better Muslims" and drawing even closer to their faith. Thus, the 9/11 tragedy had the inadvertent effect of causing many Muslims to learn more about Islam, which over time further strengthened their religious identities. Sanae, who had recently moved to the United States from India, described how she had come closer to Islam:

> After 9/11, Muslims were so criticized for it, we all turned in. We were like, "What's happening? Let's read up." After 9/11 when we were

subjected to the questions, like, "Tell us, what the hell is *jihad?*" Then we started reading more. We turned in and came together and started attending more meetings. We read up more and more and became stronger Muslims.

Along with learning more about the religion, the interviewees noted the increasing importance of positively representing Muslims and Islam to others. Because the students believed that their religion was now viewed negatively by many of their fellow citizens, they felt a need to both explain and demonstrate their faith more strongly than before. Many of the participants remarked that if they could just show people what a "good Muslim" and the "true Islam" were, some of the stereotyping and antipathy would end.

Many students also reported encountering confrontations with hostile people who stared them down with dirty looks and made hurtful remarks. Sara, who came with her family to the United States from Pakistan, discussed how she managed an incident with a police officer so as to diffuse negative stereotypes about Muslims:

> This is a time when the patience of Muslims is being tested. We're not supposed to react how we want to. I got stared down by a cop, I've never been stared down. I don't mean to say racism or anything, maybe he's from a place where he didn't grow up with any minority people or with an ethnic background. I took that into account. He looked at me a couple of times. People had to look back to see who he's looking at. I was very scared, embarrassed, ashamed. I thought, "Why are these people looking at me?" I looked at him and said, "Hi, everything's okay?" He got caught by the moment that I asked him, he thought I wouldn't ask him. He said, "Hi, how are you doing?" I said, "Fine."

A number of the interviewees reported that they had become more likely to approach others and offer information about their religion. Some of the students said that if they saw someone staring at them or giving them "looks" they would walk directly up to that person, introduce themselves, and ask if he or she had any questions. This gave them the opportunity to explain what it means to practice Islam, including the importance of proper dress and other aspects of the religion. Conversely, others were not willing to confront people, even if they felt harassed, because they feared for their safety.

Much of the discourse surrounding 9/11 involved dualities such as "good and evil" and "us and them." Just as the students were aware that they had been cast as "the Other" immediately following 9/11, they also believed that there was some expectation for them to choose between their American and Muslim identities. Natasha, a second-generation immigrant of Egyptian descent, talked about this duality:

> I think September 11 made me feel forced to choose between identities. It was big. You've got to choose one or the other, and they're not going to accept you at all. At first people weren't going to accept us.

I was like, if they're not going to accept me as an American, meaning that I'm going to have to choose one or the other, then I'm going to have to choose Islam. If they're not going to accept me as an American, if they're going to tell me I don't deserve to be here, when I am an American, if they're going to try to make me feel that way, then, hey, I'm going to be a Muslim. No one's going to ever tell me, "You can't be a Muslim. You're not a Muslim. Go back to some other planet where there isn't Islam." I felt like I had to choose then. I don't think I felt like I had to choose before then. But after that it was like, well, fine. If I have to choose, I choose to be Muslim.

With their greater self-identification as Muslims, many students decided to make a stronger visual representation of their faith. Young men grew beards and young women began to wear headscarves, despite significant disapproval from their parents who were worried about their safety. They ignored their families, hurt that their mothers and fathers would make requests that they felt violated their religious beliefs. Anna, a native of India, was quite distressed as she discussed what happened with her family:

My mom wants me to take my headscarf off because she's terrified that somebody is going to hurt me. This is my religion. I can override whatever my parents want. I have to do what the religion says. I told them, "Leave me alone. I want to wear it. You can't tell me to take it off." But it's very hurtful, growing up in an Islamic home and then having your parents say this to you because they're scared. It makes me feel very sad. I lock my door and think about how they're so religious and they pray all the time. What is going through their minds? How do they feel when I leave the house in the morning? They know that they can't stop me. If I want to do something, I'm just going to do it. It's hard that this happened and the situation that it puts us all in as Muslim women.

A small number of the interviewees reported changing their appearance following 9/11, mostly against their will, as they tried to appease their parents. A few of the men shaved their beards, and five of the women stopped wearing headscarves for a short time. Still, most participants were unwilling to alter their appearance, despite possible danger. The students contended that it was more important during this time than ever before to declare their Muslim identities and faith in Islam.

CONCLUSION

Secularization theory predicted the diminishing significance of religion in modern society (Cox 1965; Parsons 1963), yet this study illustrates the continued importance of religion as a basis of personal and social identity. Moreover, this research reveals that the formation of a religious identity is a dynamic and ongoing process and that religious identity itself is not a static phenomenon. As the data show, religious identities are actively constructed by individuals and groups in our social world, in addition to being defined, challenged, accepted, or rejected by other people, communities, and institutions. The development of a strong religious identity involves heightened reflection and self-awareness, individual choices, and the acknowledgment of others. Religious boundaries and meaning are constructed both from within and without, in response to internal conflicts and choices and external pressures and rewards that drive identity formation. Religious identities are ultimately achieved identities, which can be affirmed or denied.

REFERENCES

Cox, H. 1965. *The secular city*. New York: Macmillan.

Eck, D. L. 2001. *A new religious America*. New York: HarperCollins.

Parsons, T. 1963. Christianity and modern industrial society. In *Sociological theory, values, and socio-cultural change,* edited by E. A. Tiryakin, 33–70. New York: Free Press.

Smith, T. L. 1978. Religion and ethnicity in America. *American Historical Review* 83: 1155–1185.

32

Rejecting God,
Adopting Atheism

JESSE M. SMITH

Non-believing is on the rise worldwide, particularly in developed countries, with poorer countries tending toward high levels of religiosity and richer ones having lower rates. Some people who reject religion have moved themselves into the category of "agnostics," or doubters; a small but growing number are willing to express their atheism, or full non-belief in the existence of God. As a fairly religious country, the United States ranks low among other wealthy, industrialized countries in its rate of professed atheism. Particularly high growth can be found among college students. There are many reasons today's youth have chosen to either shy away from or shun religion altogether. For one, the number of students attending college is much greater than it was in the baby boomer generation. Campuses generally tend to be liberal strongholds that promote free thinking and debates on controversial topics. Religion is among those topics discussed, whereas it was taboo for older generations to even question religion or belief in God. Giving people forums where their opinions can be freely expressed will increase ideological diversity and promote open-mindedness. As a result, both agnosticism and atheism are increasingly notable among the "Millennial" generation, who have a greater skepticism toward the dogmatism of organized religion. Campus atheist groups are growing as well, as non-believers who arrive on campus often find themselves dismayed by the array of religious groups with which to affiliate, while they have found no such home.

In this chapter, Smith examines the identity progression individuals often follow in their movement away from theism toward non-belief. What kind of religious upbringing did you have growing up? Do you think people generally tend to follow their families' religious beliefs and identity, or is this changing?

SOURCE: Used by permission of the author.

Does atheism represent a form of rebellion, or simply individual liberation? Do you see any trends at the societal level to suggest that the extreme pressure toward religiosity in America might be diminishing, or is the reverse possible? What might be the consequences for our society of becoming less religious? How does the trend toward greater agnosticism and atheism in the developed nations and more extreme religious fervor in the poorer countries affect international relations?

Sociologists of religion have increasingly taken interest in the topic of irreligion, and several scholars have turned their attention to examining those who claim no religion and/or lack theistic beliefs. Through studies of religious "nones" we now know more about the demographic characteristics of those Americans who are unaffiliated with traditional religion (Hout and Fischer 2002). For instance, studies show that younger males with high levels of education are more likely to be religiously unaffiliated. Further, Hout and Fischer (2002) have helped explain the recent increase in the numbers of Americans claiming no religion. They argue that the political differences between the religious right and religious liberals have impelled some of the latter to disaffiliate themselves from traditional religion.

In this study I examine one subset of the irreligious: self-identified atheists. Although we know more about the social characteristics of atheists than previously, we continue to know little about how or why people actually *become* atheists. Yet, understanding this seems particularly relevant as the latest surveys show that the number of self-identified atheists is on the rise. Of course, even if their numbers continue to increase in the years to come, those who express an explicitly atheist viewpoint remain a very small segment of the population—around 2 percent. Undoubtedly, the United States continues to be a heavily theistic society. For example, one major survey reports that about 84 percent of Americans claim a religious identity; nearly 92 percent report they believe in the existence of a God or "universal spirit"; and 60 percent of these believe in a personal God, many of whom also believe God intervenes in their everyday lives (Pew Forum 2007). This point is more than just observation. As the current study will show, the formidable influence of theism in America is closely connected to the identities of those who reject it.

There have been relatively few qualitative studies that examine atheists themselves directly. One such study by Heiner (1992) explores the deviant status of atheists in the Deep South. Based on interviews with ten atheist activists affiliated with the Freedom From Religion Foundation, he discusses some of the strategies atheists use to counter the prejudice and discrimination they face in a highly religious part of the country. He finds that atheists, like theists, are engaged in the "othering" of those who do not share their views. A more recent qualitative study by Hunsberger and Altemeyer (2006) investigates why it is some Americans come to claim atheism. Their findings in part suggest that despite the normative status of belief in God, as well as the various benefits associated with religious affiliation, atheists are simply those who "could not make themselves believe" (2006:42).

This study seeks to contribute by examining the process by which people construct an atheist identity in a theist society. I consider the social psychological aspects of rejecting the culturally dominant and normative belief in God. I begin with a discussion of the methods I used to study atheists. I then present a four part model that discusses and illustrates the process of how these participants came to reject theism and adopt the label "atheist."

SETTING AND METHODS

I discovered that the mountain west university town in which I currently reside is an ideal setting for the study of atheists. I discovered several well-established atheist organizations in the surrounding area. To gain entrée, I attended a picnic at a local park where members of three atheist groups came together for informal socializing. After a friendly conversation with the president of one of the groups, I disclosed that I wanted to conduct a study of atheists. The organizer was very receptive, and she introduced me to group members.

DATA COLLECTION

My primary on-site data gathering methods included conducting individual in-depth interviews and engaging in participant observation. The interviews lasted from one to two hours and took a life-course approach, which focused on the personal biographies, relationships, social interactions, and life transitions that were important to forming an atheist identity. Because of the nature of the setting, and the fact I disclosed that I was a researcher, my role with relation to those I studied was that of an overt, participant-observer with "peripheral membership" (Adler and Adler 1987:36).

In total, I conducted 40 semi-structured in-depth interviews with atheists ranging in age from 18 to 92 years old. The sample consisted of 17 males and 23 females. Most respondents were white, and highly educated. Although some of the characteristics of the sample align with the previously mentioned demographics of the unaffiliated, neither the sample, nor the study itself was designed to be generalizable to all atheists. This study's focus is on depth and process, rather than breadth and representativeness. However, given the focused concern with the issue of identity formation, the discussion below may nevertheless be *suggestive* of a generic identity process undergone by many atheists.

CONSTRUCTING AN ATHEIST IDENTITY

My research revealed four major elements in the construction of an atheist identity in America: (1) the starting point: the ubiquity of theism, (2) questioning theism, (3) rejecting theism, and (4) "coming out" atheist.

The Starting Point: The Ubiquity of Theism

In addition to the 92 percent of people expressing *belief* in God, about 60 percent are *certain* that God exists (Chaves 2002). This reality plays a fundamental role in the construction of an atheist identity. For most participants, belief in God and religious practice had characterized their lives at some point. Thirty-five of them described their upbringing as being somewhere between somewhat religious and extremely religious. They each expressed at least some belief in God; they each identified with a particular religion or spiritual philosophy, and they each participated in some kind of religious activity. Though not *all* of those in this study came from overtly religious backgrounds, I discovered that they too underwent a similar process in rejecting God, and that the ubiquity of theism and religion itself played an important role in their becoming atheists.

Much like Peek's (2005) study of the identity processes of Muslims, most atheists viewed their early religious identities as ascribed statuses. Participants used statements such as, "I had no choice about church as a kid. You know, kids just do what their parents tell them to do." "Religion was imposed on me when I was young … you just don't have a choice when you're a child." Many participants made similar surface interpretations of their early experiences with religion. But when asked to elaborate, more thoughtful reflections were offered. For instance, Chris, an ex-Army paratrooper and recent father—who at one time even aspired to be a priest—expressed how deeply he felt about his faith growing up:

> I was raised Roman Catholic. I was *very* strongly religious as a kid. Did everything, the plays, did catechism … I'd sit there [in church] and—wow!—actually as a child I was pretty emotional about it. You know, I'd think about Jesus on the cross and I'd cry.

Similarly, Kelly, a 36 year old graduate student and teacher commented: "I was *very* religious, and church was a *very* big part of my life. I went every Sunday, Sunday school and all. I liked church, I liked the ritual … and I enjoyed going." Clearly, despite their own acknowledgement of their lack of choice in the matter, for both Chris and Kelly, religion was of central importance to them growing up.

Participants' level of belief and religiosity varied by degree, and for most, their involvement was more tempered than this. Still, all but a few discussed some level of religiosity before becoming atheists. This is evinced by the fact that participants began sharing their stories by first naming and identifying with the particular religious organization or belief system with which they grew up. And even when there was low religious *participation*, the normative *belief* in God was still present. Jenny, for example, was a young and bright student at a local university. Her comments are indicative of the ubiquity of belief (especially in childhood), and were typical of many, "I grew up religious, I was Lutheran; but we didn't go to church all that much. By late high school, I don't think I was really going anymore at all … but it wasn't that—I mean, I *still believed* in God."

From her comments, Jenny's belief in God seemed almost a "default" position. And as has been observed, many individuals—especially in their formative years—simply accept the normative theistic viewpoint with which they are accustomed; sometimes passively internalizing belief in God based on their experience within their social environment, where belief is not just accepted, but often expected. Indeed, because belief is so pervasive, most people born and raised in the United States experience socialization toward theism, and the atheists I studied were no exception to this socialization process.

As social psychologists have argued, the transmission of cultural values and beliefs through primary socialization within the family context greatly influences the values and beliefs that children will adopt and carry with them into adulthood. However, I found that even those few who were raised in families without much religion (or no religion at all), nevertheless encountered, and were influenced by, the high levels of religiosity and belief present in the general milieu of American culture. Even the least religious were not invulnerable to the influences of their communities, their neighbors and their peers, their schooling, and other aspects of socialization when it came to belief in God and religious behavior.

This certainly was Helen's experience. Helen was in her early thirties and had recently moved to start a new job. She discussed how her immediate family was entirely non-religious, yet she ended up participating in religion anyway by attending Christian church services with her friends or other neighborhood children. She puts it this way, "I was definitely raised in a secular household ... but I'd still go to church with some of my cousins and stuff; all the rest of the people I knew were religious, so I had lots of exposure to it. Plus one of my best friends for a long time was Mormon. So I went with her to the Mormon Church several times."

All participants, whether raised with religion or not, were surrounded by theist friends, neighbors, and others who influenced them significantly.

QUESTIONING THEISM

If most participants were raised with religion, and had once believed in God, what caused them to reject this, and claim an atheist identity? The ways participants began to understand God and religion as they carefully contemplated their beliefs and interacted with others became a determining factor. Respondents, over time, and in new social contexts, began questioning what they had learned growing up, which included the religious meanings they had often taken for granted. They developed new outlooks and commitments as their interactions took on new meanings.

Most started experiencing significant doubts about the existence of God when they left their homes for the first time to begin college. This was a critical time in life, as they encountered other points of view, met people from other backgrounds, and started new relationships. Jason, a 61 year old former Catholic, who had worked various odd jobs throughout his life before recently retiring, described it this way:

> I didn't really question it at all until I got to college and started studying.
> I remember reading a book by Bertrand Russell, *Why I'm Not a Christian*

and that just opened my eyes. It's like, wow!... And when I got to college I met other people ... it was my first exposure to people outside my social milieu. They [fellow students] didn't go to church on Sunday, and nothing bad was happening to them (laughs), and God wasn't striking them down for not going to church ... I mean I wasn't an atheist at this point, I still thought you know, maybe there is a God out there.

New relationships with friends, professors, and others prompted and drove this questioning. Some developed friendships with others who were themselves questioning God, or were critical of religion. One respondent remarked how a new friend she considered "very smart," and who made a considerable impression on her, would send her secularist podcasts that critically engaged religion and debated the "God question." Another felt "inspired" by a brilliant professor who "seemed to be an atheist." Ironically however, for many participants, it was their interactions with believers that compelled such questioning. For instance, 34 year old Amber related how one of her good friends—a strong Catholic—"basically believed I was going to hell for questioning whether there is a God ... [yet] she was still willing to hang out with me." Although Amber thought this aspect of the relationship was strange, it led her to recognize the "illogical" aspects of religious belief, and drove her to question more.

Research has found that going off to college tends to be an important stage in individuals' lives in terms of experimenting with, and developing new identities and establishing a more autonomous sense of self. This process can also sometimes happen relatively abruptly. But for these participants, the process of questioning their beliefs and the existence of God tended to take time, because as has been observed, the effects of religious socialization—and especially *early* religious instruction—can be quite strong. And because religious beliefs and practices are socially learned, there must take place an *un*learning process for those who eventually come to reject it. Thirty year old Joel was raised Episcopalian. He related his experience with this gradual process:

In college I still believed in God, but I no longer had a desire to go to church or anything. And then I guess it was just a slow progression, as every year went by my belief was getting less and less, and as I started getting into astronomy, and reading, and studying evolution ... it [God and religion] just started sounding more and more implausible.

This "slow progression," which was described by many participants, is consistent with Ebaugh's description of the "doubting stage" in her study of the process of role-exit: "The doubting process is usually gradual in that the individual first experiences an overall dissatisfaction in a generalized way ..." (1988:41). Indeed, these participants experienced a generalized and non-descript discontent with their religious beliefs. But through ongoing and particular interactions they began to specify and articulate their doubts in ways that opened a space for an atheist viewpoint.

As participants sought more education, they became increasingly skeptical of the religious teachings they grew up with. Once they viewed their religious beliefs as being challenged by scientific and secular explanations of the world, the religious ideas and objects (e.g. the afterlife, the Bible) began to take on different meanings. As Ebaugh argued: "The doubting stage is essentially one of reinterpreting and redefining a situation that was previously taken for granted. Events and expectations that had been defined as acceptable begin to take on new meanings. [These] doubts involve a reinterpretation of reality, a realization that things are not what they had seemed."(1988:41). However, as significant as this was, doubts about God alone were not sufficient for participants to adopt an atheist identity. More active engagement with substantive moral issues and specific interactions with others were necessary.

The Centrality of Morality   Morality was of central importance to these atheists, and specific moral issues progressively drove their questioning of God and religion more intensely. Forty-nine year old Matt, a former Lutheran, described it this way:

> I was reading the old testament, and what really got me started on a different path was that God was telling Joshua to go in and destroy the city, destroy every man, woman and child, every goat, every chicken, I mean leave nothing! And I remember thinking "what's going on here; why does God want to kill cows?" And as I continued reading and finding more troubling things I just started questioning more and more. And then it dawned on me that the God I was worshipping and reading about all this time was really just a sinner himself. Here he is being jealous and vengeful and human.

Matt had read the Bible before in its entirety as a young adult. But it wasn't until he had left the country to serve in the Air Force, and later attend college, where he made new friends and experienced other views, that he underwent this "re-reading" of the Bible and engaged his beliefs in a critical way. Many participants raised similar issues addressing scriptural teachings. This is consistent with Hunsberger and Altemeyer's (2006) study that reported that serious doubts about religion were often caused by reading the Bible itself. But more than just having problems with scripture, these respondents also pointed out what they saw as the hypocrisy of religious people they knew. Kelly, for instance, reflected that, "In college, I started to think about some issues from when I was growing up, that people [church members] were not nice. They preached about it [the Bible and morality], but then they were not good to each other. That really frustrated me." As participants confronted these issues, they increasingly desired to distance themselves from religion, and to extricate morality from its traditional association with religion. What emerged for them was a new outlook on morality: its origins, meaning, and implications for behavior.

An important outcome of this is that a tension arose in which participants saw themselves as good and moral individuals, but were beginning to doubt that this came from observing religious practices or adhering to religious "truths." They began to construct a cognitive and symbolic boundary between morality and religion, and asserted themselves as moral individuals against what they increasingly viewed as a false connection between being religious and being moral. They each in some way observed—and criticized the idea—that people *need* religion to be moral and good. As one participant remarked, "I don't think having morals has anything to do with religion. I know right from wrong. I think it benefits me in the long-term to do good things and be a good person; you know, not to lie, cheat, steal or kill; and it just makes my life a lot easier. It frustrates me that people think that one has to do with the other."

Atheism as a Rejection Identity

The earlier ideas of God and religion began to be replaced with new meaning as participants moved from the questioning stage, to a more deliberate and active stage of rejection. Sometimes intimate relationships with other non-believers facilitated this process. For instance, 30 year old John commented, "I was on the fence, borderline, you know, for a long time, like an agnostic type thing until I met Angie (now his spouse) and she brought me 'over' ... and I said, okay, I really don't believe anymore." But for participants, it was not simply a matter of reaching a place where they happened not to "believe anymore." Rather, a veritable *atheist identity* was only possible in the context of having *explicitly* rejected religion and the notion of God itself, and then elaborating in this space that which they saw as the more accurate way to view and interpret the world: through the lens of science and secular thinking. Respondents experienced an increasing commitment to this new viewpoint, which is apparent in their discussions of their developing world-views. Brittany put it this way:

> If people could become more educated and more capable of making reasonable, peaceful decisions that would support the smooth functioning of a society without anybody over them holding a whip ... that would be better.... That's how it is with religion. As we acquire the education and the sophistication to begin to choose an ethical system for [its] advantages, in a scientific, rational, social context, we can move away from the crutch of a carrot-and-stick approach, which is what I consider most religions [as] basically being.

The generalization that religion is an uncritical and unsophisticated attempt at interpreting the world, that a rational scientific approach is more accurate and desirable was one echoed by many participants. Helen, for instance, elaborates this at greater length:

> My interest is in the *natural world* ... it's a scientific worldview ... if you think about the world in a scientific method kind of way, I think it becomes increasingly difficult to reconcile that with any kind of super-natural belief. A scientific worldview is the idea that you approach the world and ideas in your life in a way that mirrors the scientific method. You go out and you gather data and see whether that refutes your ideas and your hypothesis or whether it supports it. And that might sound kind of funny for everyday life, but it's not. The idea that a supernatural God created the Earth either happened or it didn't. It's real or it's not

real … I mean, there are good theories and bad ones, and they [religion and science] are not equally valid.

This new commitment to scientific and secular thinking was accompanied by an explicit rejection of the supernatural. Although most made a point to observe that they cannot *prove* God does not exist, when probed further, most comments were of this variety: "I find no compelling reason to believe in a god," "There is *nothing* magical or supernatural about the universe or anything in it." Some remarks were even a bit sarcastic: "No spooks here." Or, "I just don't believe in the 'great sky fairy.'" This language is telling because it indicates a strong rejection and distancing from what participants view as the ideas of theists.

This process of dissociating from theism and drawing symbolic boundaries is not just an abstract or intellectual affair. Developing a self that includes *not* being a theist had real consequences for participants' social relationships. For instance, though they generally viewed themselves as highly tolerant of others, many discussed how they could not date or have a serious relationship with a strong believer. For example, James, a college student in his late twenties, said "I couldn't date a true believer; somebody that was just like 'this is how it is,' because I couldn't respect how they approached the universe … I mean, I respect them as people, but that line of thinking is not something that I would want in my intimate life. Someone who makes religious claims like that. Anybody who, for instance, can buy the Bible as truth, I'm going to have a problem with that, and it's just not going to work out."

"Coming Out" Atheist

Claiming an atheist identity vocally and using the label had important implications for respondents. Different factors impelled participants to "come out of the closet." Having "known" and thought of themselves as atheists for a period of time, the desire to *claim* the identity grew stronger as they interacted with theists and sought to dissociate themselves from religion. This is in line with Hout and Fischer's (2002) findings, in that claiming atheism can serve as a practical, and symbolic statement against traditional religion (and its often politically conservative ideas). Finally, despite any initial reticence, as interviewees began to claim atheism overtly in social interactions, a concomitant sense of empowerment, confidence, and new sense of self emerged.

Applying the Label   These atheists thought of themselves as such before they applied the label in the presence of others. Unlike some stigmatized and marginalized identities, atheism is not something that is readily apparent to others, or that has physical or even social indicators. Further, on some level declaring an atheist identity is similar to the coming out process gays experience. That is, though difficult in the beginning, for some, publicly adopting the label and coming out as an atheist was an important step toward a new self-concept and a feeling of independence and empowerment. And like the social stigma faced by those who adopt a marginalized sexual identity for these respondents, claiming their atheism not only influenced their self-concepts, but shaped their future interactions. Importantly, it was at this stage that an atheist identity began to crystallize and acquire greater salience.

The setting in which participants came out—that is, first applied the label atheist to themselves in the presence of others, ranged from the virtual and abstract (e.g. MySpace, Facebook, e-mail) to the concrete and personal (e.g. face-to-face interactions with family members). James for instance, remarked, "It was on MySpace. You know, under the 'religion' thing, I put 'atheist.' It just made it very public. It's kind of cool too, because I actually started to realize that some of my friends were atheists too, and I didn't even know it." For James, declaring the identity online led to important validation from friends. He went on to talk about how some of his friends too, ended up posting "atheist" on their MySpace accounts after they saw his. Further, he discussed how he and one of his religious friends got into a "huge debate" after he made his atheism public, which changed the dynamic of their friendship. When asked why he decided to come out on his account, James replied that he had been thinking about it for some time, and that he eventually just worked up the "nerve" to do it. Clearly, he had an internal conversation with himself, knowing that this label was appropriate for his identity, he was preoccupied with these thoughts until he was able to "put it out there" online. Several others shared experiences with claiming their atheism on their profiles of similar social networking websites. For most however, their coming out occurred during face-to-face interactions with family and/or friends. Moreover, like James, many had experienced a period of time where the "internal pressure" built up to the point where they felt a strong desire to adopt the label publicly, and reveal their "true" selves, even if in an environment that may disapprove. And despite any disapproval, being open about their atheism and discussing it with believers and non-believers alike, ultimately led respondents to become more convinced that their position was correct. That is, much like the gay Christians O'Brien (2004) studied, the challenges that arose from *claiming* the identity gave further substance and meaning *to* the identity. And although the social setting varied, the application of the label shaped, and in some cases, fundamentally changed the nature of their relationships with important others. This in turn, shaped their future interactions, and in most cases reinforced their identities as atheists.

Liberation   These atheists felt their coming out was an overall positive experience. For example, Jason remarked, "It felt good you know? Like this is what I *really* think … It was liberating; liberation from the whole religion thing." With even greater enthusiasm, Matt put it this way, "I'd reached this breakthrough, it's like I'd broken the chains, and I felt free! It was such a relief; it's a huge weight off your shoulders." And Dennis, a college student majoring in philosophy said, "it was very empowering, because—I mean—when you lose the idea of God, it *could* be like you're lost, like there's no higher purpose … but for me it was the idea of setting your own purpose, that it's just you, [that] it comes from within; that's pretty exciting!" Further, this initial excitement tended to produce greater confidence in claiming the identity in the future, and with an expanding group of people.

Terms such as "liberation" and "freedom" were used repeatedly, and it was clear from their accounts that although they acknowledged their views were non-normative, and that atheists are a heavily stigmatized group in America, each described a sense of satisfaction and confidence with having labeled themselves atheists. That is, they felt good about labeling themselves with a term that best represented not only their point of view, but their "real" selves. And yet, during this process, atheism became more than simply an identity label. It became an increasingly salient and meaningful aspect of the self. Framing their coming out as a necessary and positive milestone in their biographies contributed greatly to their self-conceptions. Through the development of this personal identity, they carved out their own sense of self and suffused it with meaning based on constructing "a sense of difference and separateness" in society. But this also provided a "sense of location and meaning within society" based on the perception that it was their individual autonomy and personal biographical experience that got them there (Hewitt 2000:98). Rather than simple non-belief, participants constructed coherent identities with their own meanings and boundaries, which challenged the norms of American culture. This speaks to the power of religiously based identity in America, from which, participants could not simply stand apart. Rather, they had to specify and articulate just what their departure from this dominant way of thinking means.

REFERENCES

Adler, Patricia and Peter Adler. 1987. *Membership Roles in Field Research.* Newbury Park, CA: Sage Publications.

Chaves, Mark. 2002. "Abiding Faith." *Contexts* 1:19–26.

Ebaugh, Helen R. F. 1988. *Becoming an Ex: the Process of Role Exit.* Chicago, IL: University of Chicago Press.

Heiner, Robert. 1992. "Nones on the Run: Evangelical Heathens in the Deep South." *Deviant Behavior* 13:1–20.

Hewitt, John. 2000. *Self and Society: A Symbolic Interactionist Social Psychology.* Needham Heights, MA: Allyn & Bacon.

Hout, Michael and Claude S. Fischer. 2002. "Why More Americans Have No Religious Preference: Politics and Generations." *American Sociological Review* 67:165–190.

Hunsberger, Bruce E. and Bob Altemeyer. 2006. *Atheists: A Groundbreaking Study of America's Nonbelievers.* Amherst, NY: Prometheus Books.

O'Brien, Jodi. 2004. "Wrestling the Angel of Contradiction: Queer Christian Identities." *Culture and Religion* 5:179–202.

Peek, Lori. 2005. "Becoming Muslim: The Development of a Religious Identity." *Sociology of Religion* 66:215–242.

Pew Forum on Religion and Public Life. 2007. *U.S. Religious Landscape Survey.*

33

Savage Inequalities

JONATHAN KOZOL

Education has long been considered, rightly or wrongly, one of the dominant venues for upward mobility in the United States. We pride ourselves, as a nation, on having a system of public education that is available to all, one of the cornerstones of our value of equal opportunity. Over the years, we have seen many struggles in the educational arena that address this situation. For a long time, students were bussed out of their neighborhoods into other school systems to reduce segregation, based on the philosophy that separate education for different racial groups does not promote equality. We have been in the midst of debates for a while over such issues as school vouchers, discussing whether students who voluntarily withdraw from public schools to attend private schools should receive educational rebates to take to their new schools. In this selection, Kozol shows us how unsuccessful we have been in trying to achieve equality of education. He compares an inner-city African-American school with a suburban White school and shows us the dramatic differences. How are children's life chances affected by attending a poor, inner-city school as compared to a suburban, well-funded one? Are there any solutions to the differentials that Kozol identifies? Where does the school you attended fall as compared to the ones described by Kozol?

"East of anywhere," writes a reporter for the *St. Louis Post-Dispatch*, "often evokes the other side of the tracks. But, for a first-time visitor suddenly deposited on its eerily empty streets, East St. Louis might suggest another world." The city, which is 98 percent black, has no obstetric services, no regular trash collection, and few jobs. Nearly a third of its families live on less than $7,500 a year; 75 percent of its population lives on welfare of some form. The U.S. Department of Housing and Urban Development describes it as "the most distressed small city in America."

SOURCE: *Savage Inequalities* by Jonathan Kozol, Copyright © 1991 by Jonathan Kozol. Used by permission of Crown Publishers, a division of Random House, Inc.

Only three of the 13 buildings on Missouri Avenue, one of the city's major thoroughfares, are occupied. A 13-story office building, tallest in the city, has been boarded up. Outside, on the sidewalk, a pile of garbage fills a ten-foot crater.

The city, which by night and day is clouded by the fumes that pour from vents and smokestacks at the Pfizer and Monsanto chemical plants, has one of the highest rates of child asthma in America.

It is, according to a teacher at Southern Illinois University, "a repository for a nonwhite population that is now regarded as expendable." The *Post-Dispatch* describes it as "America's Soweto."

Fiscal shortages have forced the layoff of 1,170 of the city's 1,400 employees in the past 12 years. The city, which is often unable to buy heating fuel or toilet paper for the city hall, recently announced that it might have to cashier all but 10 percent of the remaining work force of 230. In 1989 the mayor announced that he might need to sell the city hall and all six fire stations to raise needed cash. Last year the plan had to be scrapped after the city lost its city hall in a court judgment to a creditor. East St. Louis is mortgaged into the next century but has the highest property-tax rate in the state....

The dangers of exposure to raw sewage, which backs up repeatedly into the homes of residents in East St. Louis, were first noticed, in the spring of 1989, at a public housing project, Villa Griffin. Raw sewage, says the *Post-Dispatch,* overflowed into a playground just behind the housing project, which is home to 187 children, "forming an oozing lake of ... tainted water."... A St. Louis health official voices her dismay that children live with waste in their backyards. "The development of working sewage systems made cities livable a hundred years ago," she notes. "Sewage systems separate us from the Third World."...

The sewage, which is flowing from collapsed pipes and dysfunctional pumping stations, has also flooded basements all over the city. The city's vacuum truck, which uses water and suction to unclog the city's sewers, cannot be used because it needs $5,000 in repairs. Even when it works, it sometimes can't be used because there isn't money to hire drivers. A single engineer now does the work that 14 others did before they were laid off. By April the pool of overflow behind the Villa Griffin project has expanded into a lagoon of sewage. Two million gallons of raw sewage lie outside the children's homes....

The problems of the streets in urban areas, as teachers often note, frequently spill over into public schools. In the public schools of East St. Louis this is literally the case.

"Martin Luther King Junior High School," notes the *Post-Dispatch* in a story published in the early spring of 1989, "was evacuated Friday afternoon after sewage flowed into the kitchen.... The kitchen was closed and students were sent home." On Monday, the paper continues, "East St. Louis Senior High School was awash in sewage for the second time this year." The school had to be shut because of "fumes and backed-up toilets." Sewage flowed into the basement, through the floor, then up into the kitchen and the students' bathrooms. The backup, we read, "occurred in the food preparation areas."

School is resumed the following morning at the high school, but a few days later the overflow recurs. This time the entire system is affected, since the meals distributed to every student in the city are prepared in the two schools that have been flooded. School is called off for all 16,500 students in the district. The sewage backup, caused by the failure of two pumping stations, forces officials at the high school to shut down the furnaces.

At Martin Luther King, the parking lot and gym are also flooded. "It's a disaster," says a legislator. "The streets are under water; gaseous fumes are being emitted from the pipes under the schools," she says, "making people ill."

In the same week, the schools announce the layoff of 280 teachers, 166 cooks and cafeteria workers, 25 teacher aides, 16 custodians and 18 painters, electricians, engineers, and plumbers. The president of the teachers' union says the cuts, which will bring the size of kindergarten and primary classes up to 30 students, and the size of fourth- to twelfth-grade classes up to 35, will have "an unimaginable impact" on the students. "If you have a high school teacher with five classes each day and between 150 and 175 students..., it's going to have a devastating effect." The school system, it is also noted, has been using more than 70 "permanent substitute teachers," who are paid only $10,000 yearly, as a way of saving money....

East St. Louis, says the chairman of the state board, "is simply the worst possible place I can imagine to have a child brought up.... The community is in desperate circumstances." Sports and music, he observes, are, for many children here, "the only avenues of success." Sadly enough, no matter how it ratifies the stereotype, this is the truth; and there is a poignant aspect to the fact that, even with class size soaring and one quarter of the system's teachers being given their dismissal, the state board of education demonstrates its genuine but skewed compassion by attempting to leave sports and music untouched by the overall austerity.

Even sports facilities, however, are degrading by comparison with those found and expected at most high schools in America. The football field at East St. Louis High is missing almost everything—including goalposts. There are a couple of metal pipes—no crossbar, just the pipes. Bob Shannon, the football coach, who has to use his personal funds to purchase footballs and has had to cut and rake the football field himself, has dreams of having goalposts someday. He'd also like to let his students have new uniforms. The ones they wear are nine years old and held together somehow by a patchwork of repairs. Keeping them clean is a problem, too. The school cannot afford a washing machine. The uniforms are carted to a corner laundromat with 15 dollars' worth of quarters....

In the wing of the school that holds vocational classes, a damp, unpleasant odor fills the halls. The school has a machine shop, which cannot be used for lack of staff, and a woodworking shop. The only shop that's occupied this morning is the auto-body class. A man with long blond hair and wearing a white sweat suit swings a paddle to get children in their chairs. "What we need the most is new equipment," he reports. "I have equipment for alignment, for example, but we don't have money to install it. We also need a better form of egress. We bring the cars in through two other classes." Computerized equipment used in most repair shops, he reports, is far beyond the high school's budget. It looks like a very old gas station in an isolated rural town....

The science labs at East St. Louis High are 30 to 50 years outdated. John McMillan, a soft-spoken man, teaches physics at the school. He shows me his lab. The six lab stations in the room have empty holes where pipes were once attached. "It would be great if we had water," says McMillan....

Leaving the chemistry labs, I pass a double-sized classroom in which roughly 60 kids are sitting fairly still but doing nothing. "This is supervised study hall," a teacher tells me in the corridor. But when we step inside, he finds there is no teacher. "The teacher must be out today," he says.

Irl Solomon's history classes, which I visit next, have been described by journalists who cover East St. Louis as the highlight of the school. Solomon, a man of 54 whose reddish hair is turning white, has taught in urban schools for almost 30 years. A graduate of Brandeis University, he entered law school but was drawn away by a concern with civil rights. "After one semester, I decided that the law was not for me. I said, 'Go and find the toughest place there is to teach. See if you like it.' I'm still here...."

"I have four girls right now in my senior home room who are pregnant or have just had babies. When I ask them why this happens, I am told, 'Well, there's no reason not to have a baby. There's not much for me in public school.' The truth is, that's a pretty honest answer. A diploma from a ghetto high school doesn't count for much in the United States today. So, if this is really the last education that a person's going to get, she's probably perceptive in that statement. Ah, there's so much bitterness—unfairness—there, you know. Most of these pregnant girls are not the ones who have much self-esteem...."

"Very little education in the school would be considered academic in the suburbs. Maybe 10 to 15 percent of students are in truly academic programs. Of the 55 percent who graduate, 20 percent may go to four-year colleges: something like 10 percent of any entering class. Another 10 to 20 percent may get some other kind of higher education. An equal number join the military...."

"I don't go to physics class, because my lab has no equipment," says one student. "The typewriters in my typing class don't work. The women's toilets...." She makes a sour face. "I'll be honest," she says. "I just don't use the toilets. If I do, I come back into class and I feel dirty."

"I wanted to study Latin," says another student. "But we don't have Latin in this school."

"We lost our only Latin teacher," Solomon says.

A girl in a white jersey with the message DO THE RIGHT THING on the front raises her hand. "You visit other schools," she says. "Do you think the children in this school are getting what we'd get in a nice section of St. Louis?"

I note that we are in a different state and city.

"Are we citizens of East St. Louis or America?" she asks....

In a seventh-grade social studies class, the only book that bears some relevance to black concerns—its title is *The American Negro*—bears a publication date of 1967. The teacher invites me to ask the class some questions. Uncertain where to start, I ask the students what they've learned about the civil rights campaigns of recent decades.

A 14-year-old girl with short black curly hair says this: "Every year in February we are told to read the same old speech of Martin Luther King. We read it every year. 'I have a dream....' It does begin to seem—what is the word?"

She hesitates and then she finds the word: "perfunctory."

I ask her what she means.

"We have a school in East St. Louis named for Dr. King," she says. "The school is full of sewer water and the doors are locked with chains. Every student in that school is black. It's like a terrible joke on history."

It startles me to hear her words, but I am startled even more to think how seldom any press reporter has observed the irony of naming segregated schools for Martin Luther King. Children reach the heart of these hypocrisies much quicker than the grown-ups and the experts do....

The train ride from Grand Central Station to suburban Rye, New York, takes 35 to 40 minutes. The high school is a short ride from the station. Built of handsome gray stone and set in a landscaped campus, it resembles a New England prep school. On a day in early June of 1990, I enter the school and am directed by a student to the office.

The principal, a relaxed, unhurried man who, unlike many urban principals, seems gratified to have me visit in his school, takes me in to see the auditorium, which, he says, was recently restored with private charitable funds ($400,000) raised by parents. The crenellated ceiling, which is white and spotless, and the polished dark-wood paneling contrast with the collapsing structure of the auditorium at [another school I visited]. The principal strikes his fist against the balcony: "They made this place extremely solid." Through a window, one can see the spreading branches of a beech tree in the central courtyard of the school.

In a student lounge, a dozen seniors are relaxing on a carpeted floor that is constructed with a number of tiers so that, as the principal explains, "they can stretch out and be comfortable while reading."

The library is wood-paneled, like the auditorium. Students, all of whom are white, are seated at private carrels, of which there are approximately 40. Some are doing homework; others are looking through the *New York Times*. Every student that I see during my visit to the school is white or Asian, though I later learn there are a number of Hispanic students and that 1 or 2 percent of students in the school are black.

According to the principal, the school has 96 computers for 546 children. The typical student, he says, studies a foreign language for four or five years, beginning in the junior high school, and a second foreign language (Latin is available) for two years. Of 140 seniors, 92 are now enrolled in AP [advanced placement] classes. Maximum teacher salary will soon reach $70,000. Per-pupil funding is above $12,000 at the time I visit.

The students I meet include eleventh and twelfth graders. The teacher tells me that the class is reading Robert Coles, Studs Terkel, Alice Walker. He tells me I will find them more than willing to engage me in debate, and this turns out to be correct. Primed for my visit, it appears, they arrow in directly on the dual questions of equality and race.

Three general positions soon emerge and seem to be accepted widely. The first is that the fiscal inequalities "do matter very much" in shaping what a school can offer ("That is obvious," one student says) and that any loss of funds in Rye, as a potential consequence of future equalizing, would be damaging to many things the town regards as quite essential.

The second position is that racial integration—for example, by the busing of black children from the city or a nonwhite suburb to this school—would meet with strong resistance, and the reason would not simply be the fear that certain standards might decline. The reason, several students say straightforwardly, is "racial" or, as others say it, "out-and-out racism" on the part of adults.

The third position voiced by many students, but not all, is that equity is basically a goal to be desired and should be pursued for moral reasons, but "will probably make no major difference" since poor children "still would lack the motivation" and "would probably fail in any case because of other problems."

At this point, I ask if they can truly say "it wouldn't make a difference" since it's never been attempted. Several students then seem to rethink their views and say that "it might work, but it would have to start with preschool and the elementary grades" and "it might be 20 years before we'd see a difference."

At this stage in the discussion, several students speak with some real feeling of the present inequalities, which, they say, are "obviously unfair," and one student goes a little further and proposes that "we need to change a lot more than the schools." Another says she'd favor racial integration "by whatever means—including busing—even if the parents disapprove." But a contradictory opinion also is expressed with a good deal of fervor and is stated by one student in a rather biting voice: "I don't see why we should do it. How could it be of benefit to us?"

Throughout the discussion, whatever the views the children voice, there is a degree of unreality about the whole exchange. The children are lucid and their language is well chosen and their arguments well made, but there is a sense that they are dealing with an issue that does not feel very vivid, and that nothing that we say about it to each other really matters since it's "just a theoretical discussion." To a certain degree, the skillfulness and cleverness that they display seem to derive precisely from this sense of unreality. Questions of unfairness feel more like a geometric problem than a matter of humanity or conscience. A few of the students do break through the note of unreality, but, when they do, they cease to be so agile in their use of words and speak more awkwardly. Ethical challenges seem to threaten their effectiveness. There is the sense that they were skating over ice and that the issues we addressed were safely frozen underneath. When they stop to look beneath the ice they start to stumble. The verbal competence they have acquired here may have been gained by building walls around some regions of the heart.

"I don't think that busing students from their ghetto to a different school would do much good," one student says. "You can take them out of the environment, but you can't take the environment out of *them*. If someone grows up in the South Bronx, he's not going to be prone to learn." His name is Max and he has short black hair and speaks with confidence. "Busing didn't work when it was tried," he says. I ask him how he knows this and he says he saw a television movie about Boston.

"I agree that it's unfair the way it is," another student says. "We have AP courses and they don't. Our classes are much smaller." But, she says, "putting them in schools like ours is not the answer. Why not put some AP classes into their school? Fix the roof and paint the halls so it will not be so depressing."

The students know the term "separate but equal," but seem unaware of its historical associations. "Keep them where they are but make it equal," says a girl in the front row.

A student named Jennifer, whose manner of speech is somewhat less refined and polished than that of the others, tells me that her parents came here from New York. "My family is originally from the Bronx. Schools are hell there. That's one reason that we moved. I don't think it's our responsibility to pay our taxes to provide for *them*. I mean, my parents used to live there and they wanted to get out. There's no point in coming to a place like this, where schools are good, and then your taxes go back to the place where you began."

I bait her a bit: "Do you mean that, now that you are not in hell, you have no feeling for the people that you left behind?"

"It has to be the people in the area who want an education. If your parents just don't care, it won't do any good to spend a lot of money. Someone else can't want a good life for you. You have got to want it for yourself." Then she adds, however, "I agree that everyone should have a chance at taking the same courses...."

I ask her if she'd think it fair to pay more taxes so that this was possible. "I don't see how that benefits me," she says.

34

Lower Income College Students at Elite vs. State Schools

ELIZABETH ARIES AND MAYNARD SEIDER

Attending college has been preached as one of the strongest routes to upward mobility for the poor in our society. The American dream of opportunity suggests that children from lower income families who earn college degrees can ascend into the middle class or better. Yet much less has been written about the relative value of different types of higher educational experiences. Parents and their children often heavily ponder the competing merits of private versus state schools in deciding on where they can get the best value for their financial and life investments. Where parents are affluent and come from elite college backgrounds, this decision is less problematic. Families of less wealthy means, whose children aspire to attend college and have the academic records to be accepted at elite schools, find this choice more difficult. Aries and Seider offer some unexpected insights in their comparison of these two genres of colleges and the way students from lower income backgrounds fare in them. They trace students' progression through school from their initial discomfort, through their acclimation, and the resulting disconnect they often feel from their families. Looking around you, how do you see the friendship patterns of people you know? Do college students tend to form friendship groups of homogeneous class backgrounds, or do students from different levels of affluence mix easily? What kinds of mixing or segregation among students do you see with regard to urban versus rural backgrounds? In-state versus out-of-state? Private

SOURCE: Springer, *Qualitative Sociology*, 28(4), 2005, pp. 419–43, "The Interactive Relationship and the College Experience: The Case of Lower Income Students," by Elizabeth Aries and Maynard Seider. Used with the kind permission of Springer Science and Business Media.

versus public high school? How many students do you know who hold full- or part-time jobs while they attend college? How many have to work during school vacations? How do these constraints affect the formation of friendship groups? Do you think students who have to work their way through college get more or less out of their education? How would you evaluate the trade-off that these students and their families make in sending their children to elite colleges?

Prestigious colleges and universities in the United States have become increasingly concerned that the students they are educating come disproportionately from upper-income families. The top colleges draw three-quarters of their students from the wealthiest quarter of families in America, and only three percent of students from the poorest quarter. Lower-income white students in particular are underrepresented at highly selective colleges and universities, and with only 1% of white matriculants coming from families with incomes below $22,000 with two parents who did not attend college. Apprehension is growing at elite colleges and universities that they are "reproducing social advantage instead of serving as an engine of mobility" (Leonhardt 2004). Presidents of elite schools argue that these schools have a "profound responsibility to help meet our national challenge of achieving equal opportunity" (Summers 2004), and that if opportunities are not increased for low income students to attend such schools, "we will neither prepare any of our students for the world, nor will we serve our role in that world" (Marx 2004). These issues, while somewhat unique to the more mass-based tradition of higher education in the United States, may well strike a similar chord in societies like Great Britain and that have been widening class access to colleges and universities over the past three decades.

As prestigious colleges commit themselves to seeking greater economic diversity among their student body, attention must be paid to the challenges as well as the benefits for low income students who enter elite colleges. A college education indeed broadens an individual's employment opportunities and opportunities for self-development, but comes not without personal costs. For lower income students, many of whom represent the first generation in their family to attend college, entrance to college means an encounter with faculty and students from very different class backgrounds than their own. Baxter and Britton (2001, p. 93) suggest the changes lower income students undergo have "significant effects on their sense of self, as well as on relations with friends and colleagues who still inhabit the 'old' world."...

In the present research we sought to understand how class-based aspects of identity shape the college experience, and how the college experience influences the class-based aspects of identity of lower income students. This interactive relationship between class-based identity and the college experience occurs within a particular social context. Thus we highlight the ways that institutional variability (i.e., the type of college one attends) shapes that relationship. It is not uncommon for students entering college to feel intimidated, uncertain of the competition and concerned about how they will adjust to their new circumstances. But lower income students at prestigious liberal arts colleges face an additional set of challenges, as they lack the financial resources (or "economic capital") and cultural capital (Bourdieu 1977) of their more affluent counterparts....

METHOD

We interviewed a total of 30 lower income students, including 15 students who attended a highly selective liberal arts school (labeled Little Ivy), and 15 students who attended a state college (State College), as part of a larger study of social class and identity. We recognized that age, race and gender would influence the college experience of lower income students. Given that the experiences of lower income students who are African American, Hispanic, Asian and white may differ from one another, we decided to hold race constant and to restrict our sample to students who were white. Students at Little Ivy were all of traditional college age, so we limited our sample at State College to students of traditional age as well. The groups were equally balanced by gender....

Difficulties Faced by Lower-Income Students Little Ivy

Differences in the economic capital of the affluent and lower income students became salient to the lower income students upon arrival at college.... The economic capital of the affluent students stood out in their possessions, in their electronic equipment (e.g., large screen televisions, stereos, computers), dorm furnishings (e.g., leather couches, designer bedding), designer clothes, expensive cars, and in the money they spent on possessions (e.g., CDs, DVDs, videos, clothes), meals off campus at expensive restaurants, and vacations (e.g., to Europe, Caribbean resorts, family estates). For the lower income students, their lack of such possessions and the money required for such lifestyles seemed clear markers of their difference. The lower income students made many references to the possessions and lifestyles of the affluent students. Despite the fact that many of these students had a parent who had completed college, their lack of economic capital proved highly salient to them.

Carl grew up in a single parent home with his mother who had a college degree but worked as a secretary. Carl had attended a public magnet school that included students from some very wealthy backgrounds, and as Carl described himself:

> I was always the different kid who didn't have the right equipment....
> I felt I didn't have the right clothes, and I was made fun of for having
> dirty old holey speakers, or just not the right clothes, especially in
> elementary school and middle school.... [At Little Ivy] I came with
> just clothes and a toothbrush where others came with computers,
> televisions, video games, etc.... I didn't have a winter coat. You don't
> need a winter coat in [my home town] and those are expensive. I didn't
> have the money for that. I didn't have very warm clothes at all.
> I didn't have a computer. I didn't think I needed a computer. I can
> use the ones here in the computer lab....

Inadequacy, Inferiority, and Intimidation

A number of the lower income students worried about their self presentation and focused in particular on inadequacies of their linguistic competence, e.g., their inability to articulate their ideas clearly, deficiencies in their grammar, their regional accents. These problems seemed particularly acute for the first generation students lacking in cultural capital. When speaking up in class some students worried that their speech would mark them as less intelligent than the other students. Ann, the daughter of blue collar immigrants and a first generation student, commented on her own anxiety: "In class I have all these ideas sometimes and saying them is like, 'What did I just say? I must have come across like an idiot!' So I don't think I'm very eloquent."

The language of home was often not the language spoken in the academy. Sarah, a first generation college student whose parents worked blue-collar jobs remarked. "The people I'm used to at home are very different in just grammar." Several lower income students became aware of their regional accents, which they found to be devalued in the college context.... Speech, then, proved to be a marker of class background that could cause others to look down upon lower income students, and aspect of self that made them feel inadequate.

Lower income students, and in particular first generation students, came to view their own parents through a new lens as they compared their parents to the parents of the affluent students, and the differences seemed associated with deficiency. Their parents had less education, and their parents' occupations held less status and prestige. Sarah spoke of having a hard time telling people about her dad's blue-collar job as a mechanic "when they say [their] father is a doctor, or [their] father is a writer." She described her parents as "not like" Little Ivy in "their level of education, the way they speak," and confessed that "it was always kind of weird for them to come here. It seemed like they didn't belong." ...

While many of the affluent students had taken numerous trips abroad and within the United States while growing up, most of the lower income students had not had such experiences. As Mark, who had one parent who had completed college, stated, "My personal experience is very narrow really. I grew up in a rural county. I went to high school in a small city but I've never had the experience of a major city like New York or Chicago or anything like that." Marie noted that:

> Most of the kids in high school that I would have known definitely had never left the country. I think that in one class [at Little Ivy], I think it was French, they went around the classroom and most everybody had been to France. If you had asked that question at my high school it wouldn't have been like that at all.

Such occurrences reminded lower income students that they were outsiders from a culture and set of experiences shared by other students. In addition, lower income students felt inferior because the cultural practice of travel abroad helps affluent students with the acquisition of foreign language skills, while they lacked this capital.

Several lower income students spoke of feeling intimidated by the wealthy students, especially by those who had attended prep schools, bastions of social privilege to which most lower income students had no access, and which provided a breeding they had not acquired. Some were made anxious more generally by the possession of wealth, which to them signaled superiority. As Paul admitted, "I feel intimidated by money. This is like psychology. Rich people are better. Rich people are successful. I don't have money so I'm not successful. That's generally how I think of it to an extent. It's required a lot of conscious effort to overcome the initial hesitancy." Despite the fact that Paul's parents had graduate degrees, they had worked low wage jobs, and Paul recalled: "My parents fought every single day about money for every single day of my life until they were divorced."...

Exclusion

Many lower income students felt their class backgrounds made it difficult for them to connect to the wealthy students. While the first generation students were more likely to express feelings of inadequacy, inferiority and intimidation, feelings of exclusion were not exclusive to first generation lower income students. As Rob phrased it:

> I would love to be friends with [wealthy students], but I do feel I am
> excluded based on class.... One of the first nights I got here I tried
> to meet people. There was a party in my dorm on the second floor.
> I didn't realize it was a rich white kids' party. Everyone was giving
> me one word answers, turning their face and that made me feel kind
> of crummy that night.

It was not just that more affluent students failed to include them; the lower income students often found it difficult to bridge the gulf. For Paul, simply knowing he was around wealthier people caused him to be "deaf and dumb." He went on to say:

> There is just such immaterial knowledge. [Wealthy students] know that
> they make up a certain subset of the population, and that they have a
> certain niche, and it fosters in them a really unique bond between each
> other and excludes you. They can be totally friendly with me, not
> hostile or anything like that, but it excludes me in ways. It's attitude and
> personality. The way you react to everything. The way that we look at
> a sporting event on television, like a basketball game. I know a lot about
> basketball and I enjoy it. I am like this passive fan where my roommate
> and his friends talk about prior games that they have been to, players
> they have met, training in a gym with one player on the opposing team.
> Just being part of that and feeling much more involved in that team or
> event in a way that I wasn't....

Like the social geography of the elite boys preparatory school studied by Kuriloff and Reichert (2003, p. 760) where an exclusive "center" existed made up of upper class students with lower income and non-white students on the margins, so too at Little Ivy there appeared to be an unspoken "center" from which many lower income students felt excluded. The exclusion felt by lower income students also resembles the experience of African American girls in a predominantly white, elite, all-girls preparatory school. The African American girls at that school felt they were outsiders, that they did not fit or belong and were not recognized as full members of the school community....

The ongoing lifestyles of the affluent students required resources that the lower income students did not possess, and that also led to their exclusion. Lower income students could not afford to accompany their friends on trips abroad for spring break, or to go to dinners at high-priced restaurants. So, too, the African American girls at the predominantly white, elite, preparatory school studied by Horvat and Antonio (1999) could not afford to accompany the affluent girls to their frequent lunches off campus. Some Little Ivy students reported that they never went out to eat, or if they accompanied others to restaurants, they hid their lack of resources by not ordering anything. While the cultural signals of the affluent students may have been sent unconsciously, the unintended consequences remained.

Powerlessness

One lower income student spoke of being rendered powerless socially because of his lack of economic resources, a requirement for shaping social life. Carl described:

> the system of drinking [at Little Ivy], that you host and you provide and nobody ever gives you any money towards it. You host and people get some sort of power by buying the keg. People who have a lot of money to throw lots of parties and have provided all the resources of alcohol and marijuana, and people get a sense of power from that.

While lower income students might be included in these parties, the organization of social life required resources they did not possess.

Lower income students also felt greater powerlessness relative to the affluent students in determining their own futures. The affluent students could plan on future careers that required graduate degrees without having to worry about the costs of further education, while the lower income students felt constrained by their financial need. Karen, a second generation college student, had planned to go to graduate school, but her father had recently gone on disability and faced a future of expensive medical care. She said: "It makes me feel that these financial concerns are shaping my future goals and not necessarily shaping the rest of [the other students' goals]."

However, class-based difference did not always bring feelings of inadequacy, intimidation and exclusion for the lower income students. Friendships did, in fact, develop across class lines. Marie felt that "people at [Little Ivy] do a pretty good job of hiding [class]. It has surprised me to discover that some of my friends do have a lot of money." Several times students referred to the ways in which others might display or "perform" a class identity different than their own. As Carl, a Little Ivy student, put it, "It's hard to tell who's from that background, because class is something that you can hide. You can perform your identity and pretend to be part of another class. So it's really hard to tell." When asked whether there had ever been an incident on campus where they felt left out or put down due to their class position, for half the students no incident came immediately to mind.

State College

Because our State College sample was not surrounded by highly affluent students, class-based differences did not generally seem salient to them. The State College students did not make comments about markedly different material possessions from other students, and made no mention of students who went out to expensive dinners, made extravagant purchases, or took trips abroad for spring break. They did not seem particularly aware of their speech (e.g., ability to express themselves, grammar, or regional accents); most had not gone out of state to college and thus were not in contact with students speaking in different regional accents. They had less reason to feel excluded and intimidated, and never mentioned feeling power-less. In the few instances where working class students at State College spoke of interactions with more affluent students, they were cognizant, however, of being judged as less adequate. Chris, a first generation college student whose father was disabled and often unemployed, and whose mother was unemployed due to illness, spoke of having a roommate from an "upper-level class" who, because of his money, "thinks he's just way better than everyone."

While many State College students volunteered that their clothes reflected their social class position, they did not see their clothes as a marker differentiating them from other students on campus. Many State College students spoke of not having "a lot of new clothes," or designer clothes. Elena, a second-generation student said, "I don't really dress in name brand clothes. I go to Goodwill a lot." Craig, a first generation student, reported, "I don't wear Gucci or Armani or what not." While their clothes marked their exclusion from the affluent classes, they did not symbolize their exclusion from other students on campus.

New Forms of Capital that Lower Income Students Develop at College Little Ivy

Since coming to college, many Little Ivy students described changes that had occurred in their language, dress, and behavior, in their self-confidence and self-respect, and in their understanding of their position in the world.

Language, Dress, Behavior

Regardless of the educational level of their parents, the experience of being at Little Ivy was marked for lower income students by the assimilation of aspects of speech, attire and behavior commonly associated with the middle and upper classes, a process reported on earlier among lower income boys at an elite board-ing school (Kuriloff and Reichert 2003). As Carl remarked:

> Part of what [Little Ivy] does is socialize people into the wealthy elite. And so I dress differently. I speak differently. I'm more articulate than [my mother].... My education has given me the ability to adjust to situations rationally.... Freshman year it was people telling me on my floor, friends of mine, what to wear, how to dress, how to act in certain social circumstances. And there's this whole culture that goes along with [Little Ivy] that' s very different from my friends at the state university and that culture. And so there is a tension between who I have been and who I want to be. I think I'm part of a different social class than I came in here with, I was socialized to be part of.

The changes were motivated, in part, by students' desires to win acceptance in this new environment. Rob spoke of being highly aware of "people's criti-cisms, the desire to fit in and please people." Sarah commented, "I would prob-ably say that people of my class would say that I look kind of preppy or probably not in their social class. I think I kind of try to fit in to what a [Little Ivy] student is like to some extent."

The speech style of lower income students showed evidence of convergence to that of their higher income counterparts, consistent with the predictions of speech accommodation theory that individuals may move to another's speech style to evoke social approval, and that subordinates tend to show greater con-vergence to the speech style of superiors Naomi, the daughter of college edu-cated immigrant parents, for example, described herself as using "intellectual SAT word speak."

Self-Confidence, Self-Respect, and Position in the World

Perhaps the most significant changes that accompanied the acquisition of middle class cultural capital were the attainment of new forms of self-confidence and self-respect that came along with diminished feelings of difference, inadequacy and exclusion. Based on their academic performance, some students in fact recognized that their academic talents measured up to that of the affluent students. Rob, who spoke of feelings of inferiority when he initially arrived on campus, discovered that his academic work was equal to that of other students. He noted with pride that he no longer thought of himself "as less capable than anyone on this campus." Peter came to realize also that "poor people are just rich people without a lot of money."

Little Ivy opened up opportunities that put lower income students on a new trajectory into society. Several lower income students spoke of feeling they inhabited a new position in relation to the world around them. Peter recognized that "I am one of the people that might be making big decisions in the future." Paul reported having "more self respect and optimism. I don't use drugs. I look more up instead of down." Reflecting on how things might have been different had he gone to a state college, Allen said, "I think that my ambitions and prospects would probably be a lot less than they are now. There wouldn't be as much opportunity to get into medical school; there wouldn't be connections with the alumni.... I would be thinking a lot more practical in a state college." It is interesting to note that the increased feelings of self-confidence are more frequently voiced by students who had a parent who graduated from college, who arrived at Little Ivy with more cultural capital than their first generation lower income counterparts.

State College

State College students had less to say about the acquisition of new forms of cultural capital (e.g., changes in language, dress or behavior) although some students articulated these changes. Tom, whose parents both had college degrees, knew that his "vocabulary has grown as a side effect of all that I've learned at college. I think I talk with bigger words." Tom went on to say, "I feel that I can present myself decently well now in a situation where I would need to." Craig, a first generation student, noted the changes that had occurred in the way he constructed an argument. His opinions were now grounded in knowledge he had acquired.

> Me and my dad sometimes if we get into a discussion, we almost always have different opinions on the subject. And I feel my opinion comes more from my education in college than from anything else. I feel everything I opinionate is from what I've learned from courses I've taken, and I think it's different for him. I think he makes his opinions based on his experiences.

Explicit statements about increased self-confidence and self-respect were lacking, although such feelings are implicit in the above statements....

DISCUSSION AND CONCLUSION

Our interviews with lower income students revealed that the type of college one attends differentially influences the relationship between class identity and the college experience. Lower income students at a prestigious private college faced more class-related challenges and difficulties than students at the state college, despite the fact that they came to college with greater cultural capital than the state college students. The disparities of wealth between students at the prestigious college heightened awareness of social class, whereas greater homogeneity in class backgrounds at the state college made class less salient. The differences between the college experiences of the elite and state college students cannot be explained by differences in residential versus commuting status, as 80% of the state college students were living on campus as were 100% of the elite college students. While previous work has documented the difficulties working-class students face in a range of university settings, our direct comparison of elite and state college students demonstrates the importance of social context (i.e., of the type of institution one attends) in shaping the class-based experience of identity for lower income students and in understanding the meaning that cultural capital may have for individuals.

The first generation students who entered the elite college with the least cultural capital evidenced the greatest initial feelings of intimidation, discomfort, inadequacy, and deficiency. As to the issue of exclusion and powerlessness, lower income students at the elite college displayed roughly similar feelings regardless of the educational background of their parents. Our study points to the importance of looking at both economic and cultural capital in an analysis of social class in a college setting. Simply differentiating students by family income or parental "collar" ignores the attitudes, skills, and levels of confidence that distinguish adolescents within similar income brackets. Upwardly mobile second-generation students may still be subject to the "hidden injuries of class" but their added cultural capital helps them to buffer the class issues they face and lessens the conflict within their "straddler" identities. Perhaps that is why the students at the elite college with a college educated parent showed the greatest gains in self-confidence and self-respect.

Our data also highlight the importance of returning to Bourdieu's original definition of cultural capital as encompassing not just "highbrow" aesthetic culture, but also skills and abilities. When speaking of their lack of cultural capital, students at the elite college referred to competencies and "immaterial knowledge" such as not possessing the "right" linguistic skills, the proper forms of dress, knowledge of how to act in certain social situations, and the strategies and contacts used to procure summer jobs. These were important aspects of cultural capital that differentiated them from affluent students upon entry to college....

REFERENCES

Baxter, A., and Britton, C. (2001). Risk, identity and change: Becoming a mature student. *International Studies in Sociology of Education, 11,* 87–102.

Bourdieu, P. (1977). Cultural reproduction and social reproduction. In J. Karabel, and A. Halsey (Eds.), *Power and ideology in education* (pp. 487–511). New York: Oxford University Press.

Horvat, E. M., and Antonio, A. L. (1999). "Hey, those shoes are out of uniform": African American girls in an elite high school and the importance of habitus. *Anthropology Education Quarterly, 30,* 317–342.

Kuriloff, P., and Reichert, M. C. (2003). Boys of class, boys of color: Negotiating the academic and social geography of an elite independent school. *Journal of Social Issues, 4,* 751–769.

Leonhardt, D. (2004). As wealthy fill top colleges, concerns grow over fairness. *New York Times,* April 22, 2004, Section A, Column 5, p. 1.

Marx, T. (2004). Amherst College 183rd Commencement Address, Amherst College, Amherst, MA, May 23, 2004.

Summers, L. (June 10, 2004). Harvard University President's Commencement Address. Harvard University, Cambridge, MA. 2004.

35

Mexican Immigrant Gardeners: Entrepreneurs or Exploited Workers?

HERNAN RAMIREZ AND PIERRETTE HONDAGNEU-SOTELO

Work in America takes many different forms. One of the mainstays of the American dream is to open a small business and be your own boss. This entrepreneurial vision has captivated immigrants and existing residents alike. Ten percent of Americans own their own business, making the United States one of the most entrepreneurial nations in the world. Immigrant entrepreneurs play a large part in helping to drive a new wave of globalization. American immigration policy and the outcome of the current immigration debates will affect how this segment performs over the next decade. These entrepreneurial ventures include small businesses and personal businesses, composed of one person companies with no employees. Immigrant entrepreneurs are the fastest-growing segment of small business owners today. Immigrants are increasingly turning to entrepreneurship to steer around traditional barriers of entry to the workplace. Although immigrants often bring education, professional experience, and a developed network with them to their adopted country, their professional assets do not always translate into value across cultures. Immigrants often lack corporate contacts, English-language skills, and the larger company experience required to find a traditional corporate job. Starting a business is often easier than finding a job.

> *In this chapter, Ramirez and Hondagneu-Sotelo introduce us to the world of Mexican immigrant gardeners. Landscaping services represents one of the top ten Latino immigrant industrial niches on the basis of the number of workers employed (restaurant and other food services, as discussed in Chapter 11 by Gaytán on Mexican restaurants, also falls into the top ten). They discuss the way workers are recruited to this business and how they move up through the ranks of apprenticeship and acquire their own businesses. In reading this chapter, how would you assess Ramirez and Hondagneu-Sotelo's question of whether this represents opportunity or exploitation? How do you think the ethnicity of these workers affects their occupational patterns? Are these jobs likely to be held by domestic residents? Do you see any parallels between the occupational mobility of these Mexican gardeners and immigrants in any other occupational arenas, either legitimate or illegitimate?*

Concentrated numbers of Latino immigrant workers are now working in unregulated, informal economy jobs in U.S. suburbs and cities. These include income generating activities that are not illegal, but which occur outside of state regulations. Los Angeles is perhaps the mecca of Latino immigrant informal economy, with vibrant sectors of day laborers, street vendors, garment assembly workers, nannies, domestic workers, and gardeners. Many observers note that Latino immigrants wind up being exploited in the informal economy because they have low levels of literacy, English fluency, and job skills. However, all informal sector jobs are not alike, and many include degrees of informality and formality. More importantly, some of these jobs afford disadvantaged workers opportunities not otherwise available to them in the formal economy. To disentangle some of these job characteristics, in this article we step inside the social and economic world of Mexican immigrant maintenance gardeners in Los Angeles.

Throughout the twentieth century and into the present era, Latino, and particularly Mexican immigrant gardeners, have transformed the landscape of Los Angeles, enabling the lush, leafy, suburban visual character of the city and surrounding areas. From Pasadena to Beverly Hills, and from the Hollywood Hills to Santa Monica and the modest-sized yards of Culver City, small crews of brown-skinned, Spanish-speaking men drive around in pick-up trucks packed with lawn mowers and other landscaping equipment. They circulate and descend on residential neighborhoods—and not just the toniest ones—six days a week (and some even work on Sundays). Usually working in teams of two or three, they restore clean, orderly greenscapes by quickly mowing lawns, blowing leaves, and trimming bushes. They typically do several gardens in one neighborhood before moving along, allowing them to maintain routes that may include working in 10 to 15 different yards in one day. They also perform "extras," such as pruning trees and planting annuals, for extra pay.

What is the occupational status of these *jardineros*? On the one hand, Mexican immigrant gardeners appear to be highly exploited cheap labor, toiling and sweating under the scorching sun in residential corners of Los Angeles. Newspaper articles refer to them as "leaf blowers" and "brown dirt cowboys," and lump them in phrases like "maids and gardeners," suggesting servant-like, racially oppressed, unskilled, subordinated labor. If gendered dirty work involves men working with dirt and grime and women with bodies, this is classic masculinized and racialized dirty work. Gardeners cultivate plants growing out of soil, and dust and dirt literally swirl around them when they use the blowers. Visually, Mexican immigrant gardeners certainly look like workers: they are dressed in boots, caps, and work clothes, and lug heavy mowers and blowers in and out of their trucks to do hard, manual labor.

On the other hand, many of them operate as independent contractors, and in this regard, today's Mexican immigrant gardeners may be following in the footsteps of Japanese American gardeners, acting as small entrepreneurs (Huerta 2007; Pisani and Yoskowitz 2006). Route-owner gardeners own their own trucks and tools, they bill their clients for services rendered, they strategize how to minimize risk, they enjoy a degree of autonomy in when and how they do their work, and most of them employ one or sometimes a few employees. During the twentieth century, Japanese American men established residential maintenance gardening as a skilled occupational and entrepreneurial niche, one that allowed them to capitalize on their background working in agriculture and to negotiate racial discrimination and exclusion. Gardening jobs afforded them and their families routes to upward socioeconomic mobility, and were pathways to their children's professions. Are Mexican immigrant men following in their footsteps?

Based on interviews conducted with 47 Mexican immigrant gardeners, this article examines the occupational structure of this informal sector job in order to assess contemporary possibilities for socioeconomic mobility. We ask, are these men exploited workers, toiling in a low-wage, dead-end, dirty, dangerous informal sector job, or are they acting more autonomously as small business owners who employ co-ethnics, strategizing risk and opportunity, and thereby enabling socioeconomic mobility? What organizes the pathways into this job, and how do gardeners move up the ladder? As we address these questions, we gauge the range of informality in the job and the gendered aspect of job demand and organization, and we show how the local labor market conditions in Los Angeles have changed during the last 20 years through labor immigration, economic restructuring, and social and cultural shifts in the consumption of services....

FINDINGS

Occupational stratification and an informal organizational system mediate residential gardening jobs. In part, this reflects specialization and division of labor, with some gardeners entrusted with particular tasks that other gardeners do not do (driving the truck, tree pruning). Most gardeners, however, do a variety of tasks, including driving, mowing, blowing leaves, pruning, clipping, planting annuals, fertilizing, and so on.

The primary differentiation among gardeners is determined not by the tasks they do or the hours that they spend working at one particular residence, but according to occupational social relations. Here, the job breaks down into three categories. These include *ayudantes* (helpers), or waged employees who work for an independent gardener. This is the point of entry into the job. Next are the independent, self-employed route owners, who maintain a route of residential customers for whom they provide regular gardening services. Some of these independent gardeners work alone, but most employ, or have in the past employed, at least one or two ayudantes. Finally, some Mexican gardeners become licensed landscape contractors. Degrees of informality and formality characterize each of these positions, with the ayudantes usually paid in cash and tilting almost exclusively into the informal economy, landscape contractors veering toward more formal, contractual economic transactions, and the route-owner gardeners somewhere in between.

The *Ayudante* Apprenticeship

There is a linear progression of mobility in the job, and while not everyone becomes a financially solvent route owner and fewer still become landscape contractors, all newcomers begin by working as ayudantes or employees. These are not advertised job positions. Rather, gendered social networks provide an "in." The men typically start off working for male family members or for acquaintances from their ranchos or towns of origin. A few are hired from day laborer hiring sites, but generally, social networks and social capital assure the independent route owners of a trusted, loyal work force.

Social network hiring and social capital are institutionalized mechanisms of immigrant occupational niches and ethnic enclave employment and social networks may in fact help regulate informal sector occupations. It is not just solidarity among co-ethnics, however, but a familial and localized sense of trust and obligation that allows newly arrived Mexican men to work as gardening helpers for their family members and close acquaintances. At the point of occupational entry, informality and social capital rule.

Wages are generally paid in cash. During the summer of 2007, when these data were collected, the daily wages averaged between $75 and $80, with drivers earning slightly more ($100 to $110). Most of them earned between $450 to $480 for a six-day workweek. Whenever possible, the ayudantes earned additional pay (typically $100 a day) by working weekend "extras." Tree trimmers, a job that involves more danger and skill, might earn $200 a day. Informality prevails, and part of their wages may even be paid in kind, as room and board to men who are newly arrived from Mexico.

When discussing their earlier experiences as ayudantes, some of the men recounted narratives of gratitude while some expressed resentment, but they all agreed that working as a gardener's helper served as an apprenticeship and an important entry into residential maintenance gardening. As one man recalled: "I started working (in 1974) as a gardener's helper with some relatives, but really, it was practically out of appreciation for food and everything, and I worked that way for four months. And that's what allowed me to learn (about gardening)." Newcomers to the job may work as helpers for several months or years before they break out on their own.

Migrant social networks among family and friends constitute powerful channels pulling the men into gardening. Thirty-seven out of the 47 gardeners reported finding their first job with family members or friends. One interviewee said that he had initially shown an interest in construction work, but this proved impossible to break into because his brothers and uncles all worked in gardening jobs. "So, well, okay, gardening it is," he conceded. Another gardener explained the magnet of family networks this way: "All of my *paisanos* (countrymen) that are here, those from the rancho, my friends, the brothers of my friends, my ancestors, all of them came here to work in this (gardening)." Recognizing the powerful tug of these ties, but the arbitrary job sector to which they connect, he explained: "If they had all worked in restaurants, I would assure you that today we'd all be working in restaurants."

Mexican immigrant gardeners have low levels of formal education, few job skills, and backgrounds rooted in rural, peasant agriculture in central western Mexico. The majority of our interviewees grew up in ranchos, or rural villages, and as adolescents many of them cultivated and harvested corn, beans, and other crops in small plots of land, working alongside their fathers. All of the gardeners interviewed, except one, had previous experience working in small-scale agriculture in Mexico (the one exception was an engineer who was pulled into gardening, he said, because "my relatives were gardeners"). Some had also worked in the agribusiness fields in California. Residential maintenance gardening is hard physical work, but the gardeners were unanimous that it was easier and *menos matado* (less backbreaking, but literally, less killing), than either peasant cultivation on Mexican ranchos or field work in California agriculture. When asked about what they liked best about their job, they said it was the ability to work outdoors, in *aire libre* (open air), as they had been accustomed to doing in Mexico. They saw working outdoors, and among plants, as preferable to the limited opportunities and stifling work sites available to them in factories, sweatshops, or restaurant kitchens.

Economists and sociologists maintain that a background without formal schooling in rural, preindustrial agricultural societies is a hindrance to economic incorporation in the city. For gardening jobs, we propose that a background rooted in ranchos and peasant agriculture serves as a form of positive human capital, as it provides the gardeners with experience that allows them to withstand and perhaps even enjoy a job that others might reject. This constructs a particularly gendered, masculine human capital. Mexican immigrant gardeners' background of working in the fields, in small crews composed of brothers, sons, and fathers, endows them with gendered human capital well-suited for the way that residential commercial gardening is organized in the United States.

Mexican gardeners express an affinity for working outdoors and among plants, but residential maintenance gardening involves a different set of skills than plowing fields or harvesting crops. As gardeners in suburban California, they must learn to tend to ornamental shrubs and flowers, mow, edge, and apply fertilizers to lawns, and to prune according to homeowners' specifications. Working as an ayudanate or employee for an established gardener provides newcomers with an active apprenticeship. They learn the job by doing it alongside their brothers and uncles. On the job, they are exposed to the daily rigors and seasonal rhythms of tree pruning, planting annuals, or applying rye grass seed to lawns. A few of the men said that a customer had provided instruction on how to prune a rose bush or deadhead a plant, but the majority of them learn by doing, and over time, job skills increase.

Becoming an Independent Gardener

The potential for higher earnings increases for independent gardeners who maintain a "route" of regular customers. The independent gardeners act as worker/ entrepreneurs. They continue to do the manual work of gardening maintenance, but they also own their own trucks, machinery, and tools, and importantly, they negotiate the price and collect fees for services from the customers. In essence, they own a route of paying jobs. Most ayudantes eventually try to venture out on their own as route owners, and they need four things to do so: human capital, social capital, legal capital (legal status and attendant papers), and a modest amount of financial capital.

Mexican immigrant men become independent gardeners by building on the gendered human and social capital they have already accumulated, and by cobbling together help from family members and friends who give or lend them equipment, and who sometimes give them *casas,* or customers. The route of regular customers is the primary business commodity that these gardeners own, and the gifting and sharing of routes poses an interesting sociological dilemma. The gardeners talk about the size of their business not in relation to how many ayudantes or employees they have working, nor how much they own in tools and machinery, but rather by reference to the amount they gross from their route of paying customers each month. *"Traigo una ruta de $7,000"* or "I've got a route worth $7,000 (a month)" is the common lingo they use to discuss their businesses. Why would they give away part of their business? Some well-established gardeners want to downsize their routes, or they want to get rid of their lowest paying clients, but they also do it to help newcomer relatives. An uncle or father may give his nephew or son a portion of a route as a wedding gift. One gardener offered this analysis of the social network chains that initially absorb men as ayudantes or employees, and then spawn new, independent gardeners by giving them portions of routes:

> For example, my nephew arrived and I got him in working with me.
> One day soon he'll want to become independent, and I'll help him.
> I'll say, "Here are a few houses," and then it will be one more who is
> separating (to become an independent route owner) and then his
> brother will come, and he'll put him to work. And that's how it
> grows (Adrian, 47).

Even with assistance from friends and family, the men need financial capital to become independent route owners. According to our interviewees, breaking into independent gardening in 2007 required a financial investment of about $5,000. Gardeners need a truck, tools, and equipment, and a list of paying customers. Gardeners agree that becoming an independent route owner is harder than it used to be because there are too many gardeners, and because equipment is now more expensive. Careful calculations of investment are in order. As one gardener said:

> If you're going to start from the bottom, you need a truck. You need your equipment, and all of that. Just your *maquina* (mower), how much is that going to cost you? Brand new, it'll cost more than a $1,000. That's just to cut, and then maybe another $500 for the blower. And then the edger, let's say another $400 and some. And you have to buy your truck, your rakes, all of it—hoes, rakes, hoses, oil, and gasoline everyday. You have to spend a lot (Salvador, 38).

Before inflation, it was cheaper. In the 1980s, one gardener recalled spending $1,100 on a used Datsun pickup, and going to the swap meet to buy used equipment. "Back then, 20 years ago, it wasn't that expensive," he said. "You could buy a lawnmower for $160, a weeder for $60, a blower for $40" (Antonio, 49). Although no one spoke about buying stolen machinery or equipment, just about every gardener told a story of having had a blower or power mower stolen from his truck. This suggests a lively underground market at swap meets for low-priced gardening equipment, a market that lowers the entrance fee for becoming a route owner. With a strong route, a successful route owner can gross $5,000 monthly, so the initial costs can be quickly recouped.

Routes are also bought and sold, usually for two to three times what they generate in monthly earnings, and this too is another informal practice. Typically the transaction occurs between friends and acquaintances, but at least one L.A. lawnmower repair shop features a corkboard with notices of routes for sale. These are delicate negotiations. While social capital and trust prevail, the buyer still risks purchasing a route where the customers may be reluctant to accept a new gardener, or he may risk getting jobs with few opportunities for earning "extras," where the clients are difficult because they are habitually late payers or too picky, or where the homes are distantly located. Buyers entering into transactions also risk dishonest route sellers who may reappear after several months to get their clients back, a practice referred to as "route snatching." This practice still continues. One interviewee even reported that a route buyer had murdered an unscrupulous route seller "who was knocking on doors, trying to get his clients again." We have no corroboration of this event, but this kind of information flows among the social networks, informing the negotiations and decisions of jardineros, encouraging them to keep these transactions limited to well-known, trusted friends and family.

Finally, becoming a route owner requires legal capital. We use this term to refer to legal work authorization, and the range of particular job permits and credentials dependent on it. Lack of legal work authorization exacerbates immigrant worker exploitation and deters job mobility and depresses wages. Here, we emphasize that mobility in gardening is enabled by legal status. Legal status is technically required for route owners because they own and drive trucks, and since 1993, applicants for driver's licenses in California must present Social

Security numbers. All of the ayudantes we interviewed were undocumented, and most of the route owners (32 of 36) were legal permanent residents or naturalized U.S. citizens. Legal status is practically a necessity for route owners.

Still, a few interviewees had ventured into independent gardening in spite of undocumented status, and like many of the estimated 12 million undocumented immigrants in the United States, they lived and worked in fear. They fear not only deportation, but also having their trucks impounded and losing their investments. "I don't have a driver's license so I must drive very carefully," said one man who had his truck registered under his wife's name. Besides driver's licenses, gardeners are also required to have work permits to do gardening in particular municipalities. These are generally not enforced, and gardeners seek permits if they have many houses in one city, and forego the permit fees if they only have one or two houses in the city limits. Undocumented gardeners also fear inspectors who may issue tickets for using loud, gas-powered blowers. Gas-powered blowers are banned in 20 California cities, many of which are in Southern California, but this is haphazardly enforced and blowers are widely used.

Legal capital also enables business growth. One gardener claimed that undocumented status held him back from expanding his business. He regretted being unable to compete for large landscaping jobs that required insurance. "I feel pressured right now," he said. "Maybe if I had my *seguro* (Social Security number) I would venture to open more doors ... get bigger jobs, get bigger trucks. With my seguro, I could get credit, invest in machinery. I know that with my seguro I could place ads in the yellow pages." Without legal status, those opportunities were closed to him.

The fees for residential maintenance gardening vary considerably, and reflect local labor markets, the size of the property, the extent of work involved, and the number of weekly visits. The lowest rates in Los Angeles, found in neighborhoods with modest-sized yards, are $40 to $80 a month, but the typical monthly fee is $150 to $200. At the high end are larger estates that yield monthly fees between $600 and upward of $1000 for garden maintenance.

Managing the Route: Shades of Grey

Once they own a route, Mexican immigrant gardeners rely on different strategies for managing it so they can thrive in an increasingly competitive environment. They innovate these business strategies along a continuum of economic informality and formality. It is instructive to consider the particular contours of this grey zone.

First, it is important to note the many practices of formality in which the gardeners engage. Most of the independent route owners are legal permanent residents or naturalized U.S. citizens. This means that they can, and usually do, abide by the rules that regulate the roads. They generally hold valid California driver's licenses, pay state registration fees for their pickup trucks, buy auto insurance, and carefully abide by traffic laws. They also pay income tax as well as the annual municipal permits in cities where they have a concentrated number of customers. Moreover, they act as formal business agents by submitting monthly bills in writing to their customers. The customers pay by check, not cash. The gardeners cash these checks at banks, and they pay income tax, although they may underreport their earnings.

Still, informality characterizes other dimensions of their jobs, particularly those involving their paid helpers and paying customers. Most hire informally, through family and migrant networks, and they pay these ayudantes in cash. Some even try to maintain good relations with their workers by providing lunch for them, a paternalistic practice of informality. While the gardeners bill their customers in writing, mailing the bill or leaving it in the mailbox, they do the jobs based on verbal agreements. There is typically no signed contract between the gardener and client. When the gardeners encounter clients who are months late in paying their bills, they do not penalize these late bill payers or go to small claims court, but they informally handle it by patiently waiting for payment. It is not uncommon for a gardener to keep working at a home where he has not been paid for three months. Similarly, when they are terminated due to a house sale or disagreement, or when they decide to stop carrying a particular customer because they are giving that yard away to a newly independent gardener, a verbal conversation handles the transaction. Independent route owners thus navigate their businesses by abiding by some rules and practices of the formal economy, but they also rely on informal occupational practices.

Managing Competition

The biggest complaint, repeated by every independent route owner interviewed, is that customer fees have remained stagnant while competition and underbidding from new independent route-owner gardeners has become fierce. *"Estamos entre la espaday la pared* ... We're caught in between a rock and a hard place," is how one gardener summarized the situation of being caught between rising costs and stagnant fees. Also, more gardeners willing to work for less are crowding the field. Independent route owners who had been in business for many years said that the Immigration Reform and Control Act (IRCA), which offered amnesty/legalization to many formerly undocumented Mexican immigrants in 1986, had acted as a catalyst for helpers to venture out on their own as route owners. Amnesty/legalization freed the ayudantes from the yoke of working for someone else, and it emboldened those who remained as helpers to be more demanding about their working conditions and pay. Labor costs, and the cost for equipment and maintenance had steadily increased, and during the summer of 2007, gasoline soared to over $3 a gallon. All of this created competitive pressures.

> The clients don't go with the first estimate they get, but they get two, three, maybe four estimates from different jardineros. They tend to go with the lowest bidder. If you charge $300 (a month) for a place, there will be people who will charge $250 (Salvador, 38).

> Things have gotten tougher for jardineros. Clients today look for the lowest price, and they almost always go for the lowest bid. This has been the change I've noticed over the last 15 years. Lots of new

jardineros have entered the field ... They charge lower prices, and as result, the work is going downhill for all jardineros (Raul, 46).

I think jardineria is still a good line of work. It's good to be your own boss. But I don't like it when younger jardineros undercut each other ... clients take advantage of the situation and pay the lowest possible price (Fernando, 62).

Veteran gardeners who have been working for the last 30 years in Los Angeles said the market was now flooded with competitors. "For every jardinero who retires, there must be two new guys who enter jardineria," said Alberto (63). "In the 1980s, there were half as many as today." While their periodization varied, many of these veteran gardeners portrayed the 1970s and 80s as the "golden age" of gardening in Los Angeles.

These perceptions of competitive pressures mesh with growth in Mexican immigrant labor markets in Los Angeles. The number of Mexican immigrants in Los Angeles quadrupled between 1980 and 2000, and by the 1990s there was rapid growth in all kinds of immigrant self-employment. Consequently, Latino immigrant wages in Los Angeles declined between 1980 and 1990, and Mexican immigrant wages were found to be lowest in workplaces saturated with co-ethnics. While the demand for gardening services increased during this period as well, the new jobs were not necessarily better paying ones, but ones that responded to the new terrain of underbidding. The fee floor seemed to be moving down.

Managing the route of customers is the top business skill required of successful gardeners, and Mexican immigrant gardeners innovate various approaches to doing this in the increasingly competitive environment. Route size varies, but most keep a route of about 40 to 60 clients. One route-managing business strategy is to simply increase the route size, to obtain and keep as many clients and jobs as possible, regardless of what they pay. One interviewee maintained a route between 200 and 250 customers, but he owned four trucks, each manned by different crews. This gardener's strategy was to build the business and route size to the maximum, and his customers included ones that he'd had since he first began in the business. "I'm not going to leave them just because they pay little," he reasoned. Rather, he said his business philosophy was: *"No te fijes en lo que te da uno, sino fijate en lo que juntes de todos* ... Don't pay attention to what you make from one, but rather to what you make from all of them." This jumbo-sized route yielded him gross monthly earnings of $18,000 to $20,000, but after labor costs, machine maintenance and gas, he said he reported $120,000 to the IRS.

Maximizing route size, however, was not a favored business strategy with most gardeners because it involves managing more customers and a larger work force, and both may entail problems. The route owners generally reported that they had good employees, and they said they were thankful for their many good clients, but they also complained that clients have gotten cheaper, and the ayudantes have become less reliable. Here, their thinking follows that of many business people—they want to keep their labor costs down and they want to increase the price of their service. Instead, the opposite seems to be happening and they expressed complaints such as the following:

> If you used to have an ayudante, you would put him to work, and he'd work. There wasn't a problem. But after that (1986 amnesty/legalization), they got finicky. They'd say, "Oh, I'm not going to start work at such and such time," or "I don't want to work late" (Ramon, 54).

> Nowadays, ayudantes don't want to be paid too little, but clients don't want to pay well either … I've always tried to carry worker's compensation insurance, in case something happens, but it has become an added burden (Horacio, 50).

> The clients are sometimes too demanding. They ask for too much. They pay too little. And sometimes the workers—there are days they know how to do the work, and there are days that they don't (Juan, 48).

In this context of stagnant fees and higher labor costs, most gardeners do not want to build jumbo-sized routes. Once they reach what they consider an optimal route size, they pass jobs onto friends or relatives who are starting out in the business. They also try to minimize risk and costs by guardedly, and cautiously, taking on new customers. These route owners are selective about whom they will accept and keep as their paying customers. Some gardeners deliberately "downsize," doing away with ayudantes altogether in order to save on labor costs, or labor management headaches.

It would seem economically rational that Mexican immigrant gardeners would prefer a route made up of the very highest paying customers. One gardener reported that he had a customer who paid $3,500 monthly, and 24 out of the 36 route owners interviewed reported having at least one customer who paid $600 or more a month for gardening services. But some gardeners remained wary of taking on a big job, because it makes a large percentage of their total earnings dependent on only one customer. If that customer disappears, then a big portion of their total monthly earnings disappears as well. As entrepreneurs, gardeners make constant calculations of risk.

Gardeners complained of stagnant fees, but by asking the clients for a raise, they risk being fired and losing the job. In this aspect, they are in the same position as domestic workers who may work at the same house without a raise for many years. As workers, they feel the inflationary pressures. As one gardener explained: "You get home and each month you've got bills to pay … but they (customers) don't ask you if your (gardening service) fees are going up. They don't ask that." He had concluded, as had other route owners, that "then you have to resort to other tactics … not to depend solely on maintenance gardening." Long standing customers, he said, *"ya no dejan* … no longer leave a profit."

The proven strategy cited by all the route owners involved taking on extras. They keep their route of steady residential garden maintenance jobs, even if fee levels remain flat, in hopes that the clients will approach them for the more lucrative extra jobs. These extras involve special tasks, such as the seasonal pruning of big trees, laying sod, cleaning brush on hillsides, putting in sprinklers or walkways, or planting annual flower beds or bulbs. Many gardeners said this is where they earned real money.

> What generates a profit is (extra) planting … I've had the same fees for a long time. But like I tell you, what helps me out are the extra jobs I do for them. They never, never haggle (over prices for extras) (Juan, 48).

> With the route you make enough to get by, you make enough to pay the rent and to cover your basic expenses. And the extras, you know, are the ones that generate profit… (the route) is where you make enough to sustain yourself, but the extras are the ones that leave you enough money for savings (Miguel, 53).

Extras can yield the route owners anywhere from $2,000 to $4,000 month. Good ones might pay $500 to $1,000 a day. "It's better business," and "you make more money and you work less," were common refrains about the extras. One gardener said he even targeted the earnings from extras to his savings account. "I notice that when I get checks from (the route), I don't deposit anything into my savings account. When I *do* make deposits to my savings is whenever I do extra jobs."

Keeping the maintenance route allows the gardeners access to the more lucrative extras, and it allows them to balance the reliability and predictability of steady earnings with the opportunities of the extras. In lieu of asking for raises or charging higher fees, they count on the extras. When they charge for extras, gardeners position themselves as independent contractors, such as professional painters, roofers, or appliance vendors. They name a price, and most clients go for it. Also, the fees may involve selection, purchase, delivery, and installation of plants or materials. The gardener charges the clients more than he paid for it, so he profits from both labor and the mark up on material. This is how they navigate the turbulence of saturated labor markets.

Still, many route owners expressed verbal defeat from competition. They hadn't quit maintenance gardening, but some believed that "it's no longer a good business. We're just surviving because of the extras, tree trimming, and all that. But from gardening? No." The defeatist attitude was not expressed by all the gardeners, but was succinctly stated by one man whose frustration had replaced *raza* immigrant optimism *and ganas* (the will to triumph). *"Ya no se puede* … It can no longer be done," he concluded.

Formalizing the Informal: Landscape Contractors

A few route-owner gardeners make the transition away from maintenance and tilt their occupation towards exclusively doing extras. They become landscape contractors, but this requires overcoming many hurdles and largely stepping out of informality into practices of formality. It also requires enhanced, occupation-specific legal capital. Landscape contractors are required to work with a state-issued specialty contractor's license, and this necessitates successfully passing a comprehensive written exam administered by the California Contractors State License Board (CSLB). The Board's landscaping examination is rigorous, covering every facet of the occupation, from landscape design and job estimation to contracts and business practices, irrigation system installation, landscaping maintenance, and job site safety. It also requires more financial capital. In 2007, it cost $400 in fees to obtain a landscaping contractor's license. To qualify for a license, applicants must also undergo a criminal background check, submit their social security number, and possess more than $2,500 in operating capital. Contractors must file a $10,000 bond and present proof of worker's compensation insurance coverage as a condition of licensure.

When they bid on jobs, and accept work, landscapers issue formal written contracts. Unlike informal route owners, who must be wary of advertising their unlicensed services with cards or flyers, the landscape contractors are able to take out ads through conventional means (yellow pages, flyers, etc.). The landscape contractors do big one-time jobs, installing new lawns, irrigation systems, hardscape walkways, or covered patios, and planting mature trees and entire yards. Often they continue overseeing a gardening route in addition to their work as landscapers. Keeping a maintenance route on the side allows their landscaping firms to flourish with referrals. Customers are not willing to pay much for their regular garden maintenance, but when it comes to improving their yards they often invest thousands of dollars. One man described obtaining contracts worth between $70,000 and $80,000:

> I've been working on my own for about 30 years. And about 12 or
> 14 years ago, I got a ticket for putting in sprinklers without a license.
> So I got my (landscape) contractor's license. Yes, I had to get it.
>
> I figured it was cheaper to get it than to pay fines … So, it's been
> 30 years, and I've got six kids, and all of them, I think, have had
> a good education. And you can see that our neighborhood is nice
> (Teodoro, 53).

Over the course of the years, the income generated by his landscaping business allowed him to purchase a home in a mostly white, upper middle class section of Los Angeles and put his children through college. Two of his children were in law school, and another one worked at a Spanish-language television network. His story, like those of countless other route owners and landscapers, is a testament to the ways in which jardineria involves entrepreneurship and can facilitate intergenerational socioeconomic mobility.

CONCLUSION

This internal occupational mobility is nothing short of stunning. Within several years, a newcomer rookie can gather his apprenticeship knowledge, a driver's license, and truck and modest savings, and use these to become an independent route-owner gardener. Acting as both worker and entrepreneur, he will need to strategically manage the route and to negotiate rising costs and stagnant fees, but if he does well, he might increase his earnings anywhere from between three to ten times greater than what the paid employees earn. Affluent homeowners create backyard sanctuaries and status yards, and they seek to protect their residential investments by paying for home improvements and maintenance. Gardeners are poised to benefit from this situation. By strategically managing their routes, and by working long hours and six and seven day workweeks, some Mexican immigrant gardeners earn six figure incomes. Few occupations in the contemporary, post-industrial service economy offer Mexican men with less than primary school education and limited English fluency this opportunity.

In the context of Los Angeles, where low-skilled immigrant workers from Mexico and Central America face dead-end jobs in downgraded manufacturing, construction, and services, the economic integration of Mexican immigrant gardeners looks far more promising than other job sectors. Future research, however, will need to evaluate the extent to which the sons and daughters of Mexican immigrant gardeners are benefiting from the gardening business. A segment of them appear to be following the children of Japanese gardeners into the more lucrative professions....

36

Working-Class Job Loss, Gender, and the Negotiation of Household Labor

ELIZABETH MIKLYA LEGERSKI AND
MARIE CORNWALL

For over a half century, American manufacturing dominated the globe. It turned the tide in World War II and hastened the defeat of Nazi Germany; it subsequently helped rebuild Europe and Japan; it enabled the United States to outlast the Soviet empire in the Cold War. At the same time, it met all the material needs of the American people. During this period, many American icons were born. Companies like General Motors, Ford, Boeing, Maytag, and Levi Strauss became household names. American manufacturing became synonymous with quality and ingenuity. On the back of this industrial output, America's middle class arose. High-paying manufacturing jobs, in turn, helped spur a robust and growing economy that depended little on foreign nations for manufactured goods and armaments. However, manufacturing as a share of the economy has been plummeting. In 1965, manufacturing accounted for 53 percent of the economy. By 1988 it only accounted for 39 percent, and in 2004 it accounted for just 9 percent. The decline of the manufacturing industry manifests itself most clearly in job losses.

Manufacturing loss is occurring because of globalization and outsourcing. Globalization is the increased mobility of goods, services, labor, technology, and capital throughout the world; outsourcing is the performance of a production activity in another country that was previously done by a domestic firm or plant. At the dawn of globalization, the elimination of trade barriers opened up access to

SOURCE: From *Gender & Society*, 24(4), pp. 447–474, copyright © 2010 by Sage Publications. Reprinted by permission of Sage Publications Inc.

foreign markets for American manufacturers in return for building factories abroad. In due course, more and more manufacturers set up shop overseas, producing goods to be sold to Americans. Today, the trend is so severe that analysts predict that in some industries, a quarter to a half of all jobs are likely to migrate. In this chapter, Legerski and Cornwall examine the impact of this pattern of job loss on families, looking specifically at how husbands' and wives' unemployment affects their gender role enactment and household division of labor. How does the plight of these families compare with the image you have formed about laid-off industrial workers? How do their experiences compare with those from your family's economic sector? How do their gender role adaptations compare with those of the families described by Gerson in chapter 29? To what do you attribute some of the similarities or differences? What does this make you think about America's role in the world economy over the next half-century? How do you envision positioning yourself occupationally to fit into reduced American job niches? How do you imagine yourself negotiating a household and gendered division of labor with your partner?

The 2009 economic recession has generated interest in the consequences of rising unemployment among men. Although the media have highlighted rising numbers of professional dads attending PTA meetings and cooking dinner, do these observations represent real, lasting change in the household division of labor? Furthermore, do difficult economic times have the same effect on working-class dads? Could the social structural changes brought about by the recession undermine gender inequalities across class boundaries and reignite the stalled feminist revolution? Or will change at the macro level be thwarted by resistance at other levels, preventing long-term change in the gender order?

Observing "when husbands and wives are doing gender traditionally and when they are undoing it" (Risman 2009, 82) requires examining "social relational contexts...." The household is one such arena where structural and cultural changes may affect gendered selves and identities, influencing the negotiation of household labor and the doing of gender....

Despite the increase in dual-earner families, the economic necessity of women's paid employment, and trends toward greater attitudinal gender egalitarianism, a significant gender gap in housework persists. Two puzzling findings remain: men who earn less than their wives or are unemployed may be less inclined to do housework, and women whose earnings are higher than their husbands' tend to do more. Nevertheless, this research often focuses on the experiences of the middle-class. The result is a limited understanding of the division of household labor among couples at the bottom of the earnings pyramid.

Our purpose is to explore the possibility of change in gender arrangements among white, heterosexual, religiously conservative, working-class couples experiencing the long-term unemployment of the primary breadwinner by examining the gender processes that underlie and reinforce the division of household labor. Specifically, we ask, do couples faced with men's unemployment negotiate new household and caregiving responsibilities? And what are the gender processes that structure the gendered division of household labor in this context?...

We approached our study of unemployed steelworkers with the assumption that gendered cultural processes are best studied during "unsettled times," when the adequacies of set "strategies of action" are tested. During "settled times," gender ideologies and gender behaviors are well established, institutionalized, and naturalized. During unsettled times, when well-worn strategies of action are no longer available or prove inadequate, the institutional-, interactional-, and individual-level processes that reinforce gender become more visible and more identifiable. We argue that data collected from middle-aged, working-class, religiously conservative couples who are facing long-term unanticipated unemployment offer insights into the doing and undoing of gender....

In the following analysis, we define household labor as the activities, tasks, and interactions that couples engage in to maintain both family members and the household and focus on tasks traditionally defined as women's work. This includes both routine physical tasks (shopping, cooking, and cleaning) and nurturant and emotional aspects of providing care for household members. Because the sample consists of middle-aged families whose children are typically older, we do not focus on the child care component of household labor....

METHOD

Data Collection

In the fall of 2001, a large steel manufacturing plant closed in a metropolitan area of the western United States, displacing more than 1,200 steel-workers. The closure occurred during an economic downturn in a region where steady employment in well-paying jobs had become hard to find. Working with the local union in June 2002, we mailed each of the displaced union workers $(N = 1,177)$ a 16-page survey to assess the impact of the plant shutdown on families. A total of 622 steelworkers replied to the survey (valid response rate of 53.6 percent). In the survey, respondents were asked if they would agree to be interviewed. The 385 who consented were categorized by their reported level of financial difficulty and age. A randomly selected subset of steelworkers from each of nine categories was then identified for in depth interviews. Interviews typically took place in the respondents' homes and lasted between 45 and 120 minutes. Participants were compensated with 35-dollar gift certificates....

FINDINGS

Changes in Household Labor Following Unemployment

Although couples reported some changes in housework, close examination of these accounts revealed that change in the gendering of tasks was relatively minimal, and respondents often reinforced gender as they described the daily practices of their family life. Only 2 women insisted that they split the housework evenly, while 24 described a gendered division of labor in their homes. Despite acknowledging the inequality of the situation, more than one-third of these women felt the division of labor was fair.

Women's paid employment status at the time of our interviews was directly related to their housework contributions; women who worked full-time and had children in their homes reported reducing the time spent doing household chores such as cleaning and cooking. The loss of wages reduced available money, and 19 of the women reported cutting costs on food. Joan, a full-time seamstress at age 50 and the sole breadwinner for her family, described her experience: "You try to find more sales. That's what it's all about—you have to cut out a lot of unnecessary foods and different things so that we can afford them. But then that's still scrimping. We go without that, and make with what we do have, so that we can buy something else." All categories of women spent more time clipping coupons, looking for sales, and planning meals using less expensive items, and employed women were more likely to report eating out less—strategies that created more work for them.

Overall, 23 of the women described how their spouses increased their efforts to "help out" around the house. In addition, 11 of the women who had children or other adult family members living in their home reported help with housework from their kids (usually daughters) and able-bodied mothers. For example, both Barbara, age 43, and Linda, age 45, found full-time office work after the layoff. Both found it hard to keep up with things given their new schedules, so Barbara's teenage daughter and Linda's adult daughters helped out by cleaning the bathrooms occasionally. Nancy, age 34, was also able to increase her hours at a bakery to full-time. She was fortunate in that she was able to pick shifts opposite her spouse's so that they could avoid child care expenses for their young children. She said her spouse George, age 36 and self-employed in construction, was also good at helping with housework after the change: "Once in a couple of weeks it will be totally clean—not like scrubbing toilets and mopping floors—but it will look all clean. He'll have it all picked up and stuff like that."...

Even when men increased the amount of time they spent doing housework, the task itself remained gendered. This was apparent in the language used to describe the adjustments that were made. Husbands helped with meal preparation, but they did not "cook-cook." Men did not do the grocery shopping; they "picked something up" on their way home. The wives planned the meals, made the grocery lists, and shopped for the best prices. Men "straightened" the house, but they did not "clean-clean," mop, or scrub floors. When women got help from family members, it was usually daughters and mothers who stepped in. Thus, the help women received maintained gender distinctions so that household work did not become men's work and the most time-consuming, most physical, and least desirable tasks remained women's work.

Institutional Processes

In most cases, new opportunities for market work were highly limited for both partners. Women's educational attainment and employment histories constrained opportunities for finding well-paying jobs that would replace their husbands' lost wages. Most women had substantial work histories, but their labor force participation was often intermittent as the majority had taken time out of the market for childbearing and mothering. During the interviews, they made clear distinctions between "working" and "having a career." Most did not see themselves as the "career type" and reported having few expectations for advanced education as young adults. Those who had started college did not finish—usually dropping out at marriage or with the birth of their first child. Those who hoped to go back to finish their degrees found it difficult to do so because of limited financial resources.

Of the 23 employed women, 22 felt that their wages were necessary for their families' survival. Nevertheless, 19 described their market work as secondary to that of their husbands. Thus, 14 reported taking on most of the housework early in their marriages with the view that this division of labor was reasonable and fair since their husbands had better-paying jobs. Although 23 of the women expected to be stay-at-home mothers when they married, only 6 were able to avoid the labor market. Change in their husbands' employment status was often the impetus for women's reentry into the labor force. This instance of unemployment was no different. While 12 of the women reported looking for jobs or trying to increase their hours when the layoff seemed imminent, only 4 women were able to secure new employment or more hours....

Where women did earn good wages, respondents adapted their traditional ideals to economic realities, resulting in symbolic support for traditional ideals but practical support for a "pragmatic egalitarianism"—an ideology that recognizes the necessity of women's employment without abandoning traditional gender beliefs. Women emphasized that "kids need someone home" but acknowledged that sometimes, "you have to do what you have to do to make ends meet." Caroline described the importance of mothers' being available for children: "You should be there to help your kids with school, or keep on them about homework and stuff like that. I think that's part of being in a family." But she then added, "I mean, I'd like to stay home, but you know, [it takes] two incomes to make ends meet and stuff."...

Even substantial strain did not create real change in housework allocation. Julie, age 57, was working full-time in document control while her husband Dave, age 58, was unemployed. When asked why they maintained a gendered division of housework, she explained that it had been that way since the beginning of their marriage and she could not explain the lack of change: "My mom was the one to do housework and my dad always did the job. That's how I was raised, so I had no problem with the fact that I did the housework and he'd do the job." Even women who had working mothers often described them as traditional. For example, Kay, age 50 and employed full-time as a waitress, described her decision to do as her mother had done:

> My mother didn't really, you know, she worked a little bit, different times whenever my dad would get laid off, then she would go to work. But other than that, she was a stay-at-home mom. And then as we got older she worked just to get ahead and stuff. But I thought that I would do the same thing I'd been taught.

Kay's spouse Richard, also age 50, had found full-time work as a baker in a grocery store. Although she was equally employed, Kay described her work in the home as fundamental to her understanding of self. Yet Kay understood that younger women were more likely to do things differently:

> We still do the same. I'm from a different generation than the young girls, and I still feel that my job is in the home and the meals. I think that the new generation is more like, "I work, you work." I know that my girls do that. They'll come home and my son-in-law will come home and she hasn't fixed supper and she goes, "If you want something, fix it." My generation just didn't do that....

Individual Processes

Interviewees drew on strong gender identities as they described how they coped. Women's gender identities were embedded in responsibilities for care that extended beyond mothering young children and included the care of husbands, adult children, and grandchildren. In at least three cases, women intentionally avoided market work to have time to maintain family relationships. Donna, age 47, who was one of only two women in our sample with college degrees, opted out of the labor force to be a stay-at-home mom. Her spouse Ron, age 42, had enrolled in retraining and was still out of work. Finances were tight, but Donna insisted it was important for her to be home:

> Even though I don't have any little preschooler people, I still manage kids in and out and around, do grocery shopping, and I manage the home, and we don't want me to go to work. I enjoy doing things for me, and for my kids, and for my husband and family. I don't feel like I have done anything less valuable to the world than someone who goes out and has a wonderful career.

We found similar sentiments among women who had entered the labor force. Even when women were working full-time, both husbands and wives reinforced gender identities in language that described the women as full-time homemakers and the men as only temporarily without jobs. In fact, 20 of the women (both in and out of the labor force) extolled the virtues of caregiving and described such work as central to who they were....

For most couples, women's employment was secondary to their identity as caregivers. Because their employment was often temporary, poorly paid, and unskilled, it was viewed as a means for providing for family but not a core feature of their self-definitions. In fact, despite acknowledging the importance of their wages, 16 of the employed women in our sample did not describe their employment as a core feature of their identity. When women described their role as mothers they often explained it was something they always wanted to do; when they talked about their role as wage earners, they often described their employment as temporary or just to secure the "extras" (i.e., family vacations and activities for children)....

Thus, housework was rarely only housework; it was imbued with care and formed a central component of women's identities.

Although some men may enjoy active fathering, household labor did not provide a similar sense of identity for these unemployed steelworkers. While unemployment benefits in some cases were sufficient to make women's wages feel secondary, they also could not replace the source of identity and control paid work provided men. Debbie, age 52, for example, was quite aware of the connection between her husband Al's work and his identity:

> Men have a tendency to identify with what they do as who they are. To Al, he was a steelworker. That's who he was. I don't know whether that makes sense or not. It's kind of like saying he's Al, but to him, he's Al the steelworker. Not only did it provide for his family, but it gave him a sense if identity, and when he was severed, it took him several months to figure out that he was somebody apart from that.

While women worried about their inability to keep up with both housework and caregiving, men were distressed about their inability to maintain their bread-winner status.

We found some support for a felt need to neutralize "deviant identities," but more significantly for these men, job loss left them adrift and psychologically vulnerable. For working-class men, being stably employed provides a sense of control. So while women's employment was additive to their other responsibilities, men's job loss created a vacuum, a loss of something definitive of manhood and necessary to sustain the self. Taking on tasks understood to be women's work did not replace the loss of occupationally reinforced gender identities....

Interactional Processes

Relations between husbands and wives were strained and tense. When asked to explain how they made decisions regarding their families' division of labor, 17 claimed no formal decisions were ever made and that family members just did what needed to be done. Further probing resulted in frequent admissions that current arrangements were inequitable and often overwhelming for the women. Working women—especially those with children—were most likely to report feeling exhausted by their household labor demands....

Women's concerns about the mental health of their husbands (evidence of the strength of their caregiving identities) shaped spousal interactions.

Most women acknowledged that their husbands' senses of self had been radically shaken, and 12 confessed that they were worried about their spouses' mental health. When we asked if they ever experienced marital conflict over the unequal distribution of labor, Louise, age 46 and employed part-time in a warehouse, said,

> Um, sometimes, [but] very rarely. I figure it's just easier to do it. I mean, it'd be nice to have a perfect marriage that you can say, well you know, honey, if you'll do this and this and this, I can do this and this. But like I

say, with his depression and that, it's like I walk on water, walk on eggshells, not sure if I say something if he's going to take it wrong.

... Despite their worries, eight of the women confessed that they also fought more since the layoff. Linda described how she "confronted" her spouse: "I ask him, I jokingly tease [him], 'Don't you think you should make the bed?' [Laughs] You know, things. And he likes to grill. He'll grill for us, but we don't have meat as much as we used to." Even after venting her frustration, Linda changed the subject and praised her husband for the help he provides—a behavior common among the women. Although they teased and sometimes yelled, with few resources to remedy the identity challenges faced by their husbands, women mostly worried and rarely renegotiated tasks. In most cases, women seemed more concerned about preserving their husbands' health and sense of masculinity than adjusting demands on their time. In this way, the women were "held hostage" by their caregiving responsibilities. Cognizant of their spouses' fragile psychological states, they resisted bargaining for more equitable changes in the division of labor. They were in a very real sense "prisoners of love."

Although they rarely encouraged their husbands to talk about their emotions and feelings, the women were keenly aware of their spouses' emotional states. They noted shifts in diet, sleep, and mood. Six women revealed that their spouses were taking antidepressants. Debbie, age 52, for example, said, "He finds it very easy, when he gets really blue, to sleep 14, 15 hours." Marci, age 57, described her response to such episodes: "One day a couple of weeks ago I could tell that he was starting [to get] down and I said, 'You've had a couple of bad days, haven't you?' And he said, 'Yeah.' And I said, 'Well, you know that you'll get a job when the time's right.'" The women provided an important source of emotional support for the men, and few were willing to add to the emotional strain by demanding greater participation in household tasks....

Our data suggest some women did not renegotiate the division of labor primarily because they were concerned about the mental health of their husbands. Nevertheless, it is possible that in some cases resistance to pushing husbands to do more also may have been because of concerns they might lose their temper or do something violent. At least eight women reported that their spouses were more irritable and susceptible to losing their tempers following the layoff. For example, Mary Ann, age 52, explained that her husband Bill, age 52, had become "short-fused, short-tempered, not particularly at anybody, just with things. If something he's doing isn't going right—that's cause for an explosion. And it's usually loud but not too dangerous." While we found no direct evidence of domestic violence, research shows higher rates of domestic violence among couples in which husbands are unemployed, and the local newspaper did report cases of violence following the plant closure.

CONCLUSION

... In general, we found no significant "undoing" of gender. Instead, efforts to develop new egalitarian strategies were constrained by four underlying gender processes: (1) the lack of well-paying jobs for women that would allow them to fully replace the lost wages of husbands, (2) the lack of new ideologies to support men's equal participation in household labor and the persistent gendering of household tasks, (3) the reinforcement of traditional gender ideologies that framed the wives' employment and income as secondary to the husbands' role as breadwinners despite their unemployment status, and (4) the maintenance of gender identities through interactions between spouses. The results illustrate how each level of gender processes does not exist independently of the others; for example, features of men's and women's identities (i.e., the differential significance of care work and paid work for men and women) played a central role in forestalling changes in the distribution of household labor at the interactional level....

REFERENCE

Risman, B. J. 2009. From doing to undoing: Gender as we know it. *Gender & Society* 23: 81–84.

37

Healthy Dads, Healthy Kids

WILLIAM MARSIGLIO

*Men are historically notorious for ignoring their health, a finding that has serious
implications for their physical and mental welfare. Help-seeking rates are even lower
for men who conform to traditional masculine norms. They tend to follow an "If it
ain't broke, don't fix it" attitude, and even when symptoms appear, they are likely
to delay or avoid going to the doctor. Interestingly, men are more likely to refer
others for medical attention, while ignoring their own similar symptoms or problems.
Yet American men generally die approximately five years earlier than women and
also have higher mortality rates for 14 of the 15 leading causes of death. In this
chapter, Marsiglio examines broad patterns of men's health, from their main health
problems to the effects of the male gender role on their likelihood to silently tolerate
illness, pain, and injury. He considers the effect of these tendencies on a broad
segment of the population: their children. In so doing, he draws some conclusions
about the consequences of the way men address their health that are both surprising
yet commonsensical. What health messages have you received as you were growing
up? How have these compared to your siblings or others of the other gender? How
did your parents handle their health issues? Outside of the family, what groups or
institutions may have affected the way you think about how you should define and
treat your body? Do you buy Marsiglio's argument that the way fathers treat their
health affects their children? If so, what factors are likely to help change men's
health patterns or to hold them constant?*

Far too many babies and children in the United States today struggle with
myriad conditions that negatively affect their emotional, mental, and physical
health. We initially see this tragedy in the high rates of infant mortality and low
birth weight babies and then in the discouragingly high numbers of youth who
are obese, get pregnant, contract a sexually transmitted infection, smoke, binge

SOURCE: William Marsiglio, "Healthy Dads, Healthy Kids," *Contexts*, Vol. 8, No. 4, 22–27. © 2009,
The American Sociological Association. Used by permission. All rights reserved.

drink, abuse drugs, develop an eating disorder, or attempt suicide. Sadly, when young people look at adult men they often find poor role models who are ill-equipped to help them avoid or correct unhealthy behaviors.

One third of American men are obese and one in 10 will become an alcoholic in his lifetime. Men are more likely than women to smoke, eat fatty foods, drink and drive, use guns, play violent sports, and not get enough sleep, and they're less likely to use seatbelts. Men are also less likely than women to seek medical attention for either routine physicals or when more serious problems occur.

Many factors contribute to children's poor health, but one we hear little about is how fathers act and what they do and don't say about health. Indeed, what men say and do can help prevent or minimize some of their children's health problems and effectively manage the adverse effects when problems do arise.

However, social pressures and financial struggles limit the choices men can make to prioritize health for themselves and their children. Understanding how fathers make decisions, as well as their social networks and diverse experiences over their lifetimes, is essential for cultivating a more engaged, health-conscious style of fathering that will, in turn, positively affect their children's health.

The theory of "constrained choice," developed by health policy experts Chloe Bird and Patricia Rieker, can guide efforts to help fathers do a better job in this regard. Touted as a "platform for prevention," the theory suggests that individuals' opportunities to pursue healthy options are shaped by decision-making processes at multiple levels: nation/state, community, workplace, family, and individual. By paying attention to gender-based health disparities, this framework also highlights the diverse social forces that organize men's and women's lives differently. These conditions, along with biological processes and other social realities like socioeconomic status, expose men and women to specific stresses, burdens, and health risks. Consequently, men in general, and fathers in particular, face unique challenges to assert themselves as more positive role models for healthy behavior.

We're entering a propitious moment in history to foster real changes in how fathers perceive, practice, and promote health. President Barack Obama's commitment to reforming health care on multiple levels while promoting preventative care and wellness resonates with Bird's and Rieker's idea that individuals' health experiences are shaped by a multilayered social context and their call for prevention strategies. Indeed, as an outspoken public advocate of getting men to step up and become more involved fathers, Obama and his administration are poised to spearhead cultural, policy, and programmatic changes that could link health promotion with good fathering.

CONNECTING FATHERS' AND CHILDREN'S HEALTH

Since the 1980s scholars have explored extensively how fathers from all kinds of families are involved with their children and how that involvement affects their children. Currently, a nascent and promising multidisciplinary research agenda (involving sociologists, nurses, pediatric psychologists, and public health experts) considers how fathers' health consciousness, practices, and outcomes relate to their children's quality of health and care. Attention has even been given to how men's actions prior to a child's conception or birth can influence that child's health.

Political scientist Cynthia Daniels argues that men's behaviors can indeed make a difference. Prevailing cultural conceptions of how masculinity is related to reproduction, she observes, have compromised scientific assessments of how sperm damaged from toxicity at work and at war, as well as from men's smoking, drinking, and drug habits, produces negative outcomes for fetuses and infants. She also notes that scientists and funding agencies apply a higher level of scrutiny when reviewing reproductive studies of male sperm compared to those examining female reproductive issues. Such scientific bias has perpetuated the public's disproportionate interest in how women's allegedly bad behavior increases fetal health risks while largely ignoring how men's preconception experiences can negatively contribute to fetal and children's health. Indeed, Daniels argues that many have resisted human and animal research that suggests connections between males and both birth defects and childhood diseases, in part because it "places men closer and closer to culpability for the health problems of their children."

In addition to direct biological influences, various studies show that fathers may indirectly affect children's health outcomes by influencing the mother's prenatal and early postnatal behavior and stress levels. Researchers can't say definitively whether and how men make a difference, but as sociologist Rachel Kimbro's work from the Fragile Families and Child Well-being study indicates, women in more stable and supportive relationships do exhibit more positive prenatal health behaviors. They're more apt to receive prenatal care and less likely to smoke, drink, or abuse drugs during pregnancy. Women with partners who have completed at least some college have lower odds of smoking and using drugs during pregnancy. (On the other hand, these women have higher odds of drinking occasionally during pregnancy and, surprisingly, fathers' education is not related to the women's inadequate prenatal care.)

Health policy researcher Laurie Martin and colleagues also show that fathers with at least a high school education, first-time fathers, and those who want a pregnancy are more likely to be involved prenatally. And, health researchers Manoj Sharma and Rick Petosa argue that partners' views consistently play a major role in whether mothers start and sustain breastfeeding. Finally, Craig Garfield, a pediatrician at the Evanston North-western Healthcare Research Institute, confirms that fathers' poor postnatal mental health brings on negative consequences for both children's developmental outcomes and maternal mental health.

Resident and nonresident fathers have many opportunities to affect their children's health beliefs and practices after the infant and toddler years as well. Though relatively few studies have explored these possible links, sociologist Chadwick Menning's longitudinal research with a national sample suggests that nonresident fathers' greater involvement with their adolescent children reduces the likelihood the children will begin smoking regularly. This outcome changes with fathers' level of involvement. Furthermore, children are more likely to smoke when their fathers smoke.

Menning's earlier research with family demographer Susan Stewart paints a similar portrait of resident and nonresident fathers' contributions to their children's obesity. Children are more likely to be obese if their fathers (or mothers) are obese, they found. Among nonresident fathers, those more highly educated and more involved with their children tend to have children at lower risk for obesity. Similar trends are evident among white, rural families in Iowa, where fathers' lack of exercise, poor eating, excessive drinking, and smoking predicts the same behaviors among adolescents. Lower family social status, as measured by education, increases fathers' chances of exhibiting health-risk lifestyles, partly explaining the path of intergenerational transmission of poor health behaviors from fathers to adolescents.

Fathering may also have significant health consequences for fathers themselves. For example, Garfield, the Northwestern pediatrician, describes how stressful experiences associated with fathering can accumulate and debilitate fathers' health over time. Presumably, fathers at risk of experiencing this pattern include those frustrated by their inability to fulfill their breadwinning role because of low wages or job loss, as well as those working stressful, high-risk jobs for the money.

MASCULINE BODY IMAGES, CAREGIVING IDEALS

How men see their own bodies and their role as caregivers shapes the role fathers play in influencing their kids' health. Boys and men are regularly exposed to media, family, peers, and other sources that transmit messages about health. How they think about gender and social class, for example, affects how they construct images of manliness, perceive their own and others' bodies, manage friendships, and approach fathering.

Discussions about health are influenced by how males think about gender and their individual exposure to stress and risk, factors themselves that are affected by various social processes and limited constrained choices, be they cultural, structural, or interpersonal. For example, the messages permeating organized sports, friendships, and workplaces often encourage males to assert a stoic, risk-taking, and "hard" image that rejects expressions of vulnerability and femininity.

Sociologist Michael Messner asserts that contrary to popular wisdom, boys' and men's sports activity often breeds "unhealthy practices, drug and alcohol abuse, pain, injury, and (in some sports) low life expectancy." Referencing the "pain principle," Messner observes that if boys don't learn to "'shake it off,' ignore their own pain, and treat their bodies as instruments to be used—and used up—to get a job done ... [then] they may lose their position on the team, or they may be labeled as 'women,' 'fags,' or 'pussies' for not being manly enough to play hurt."

The masculine ideology that perpetuates bodily harm extends well beyond the sports world, whether it's excessive drinking, drug use, fast driving, fighting, or some other display of a potentially self-destructive behavior. Much of this is tied to and supported by males having friends and acquaintances who take unnecessary risks.

For example, in his recent book *Guyland,* gender scholar Michael Kimmel discusses the disturbing way peer pressure fuels hazing rituals. In fraternities, young men seduced by the masculine status that flows from belonging to a tightly knit, all-male group, subject themselves to humiliation and sometimes untold health risks. A 2008 study of hazing in more than 50 schools found drinking was involved in the hazing of 31 percent of the men, and 17 percent "drank until they passed out."

Consistent with sociologist Erving Goffman's view that "men must be prepared to put up their lives to save their faces," men who work dangerous jobs as loggers, miners, construction workers, police officers, firefighters, EMTs, and the like are engulfed in an atmosphere that defines the drama of manhood as physical, fearless, and full of risk. Recent studies, such as the compelling ethnography of wildland firefighters by sociologist Matthew Desmond, suggest a much more complex picture, where the skills and dispositions children and adolescents acquire from their blue-collar upbringing prepare them to view as unthreatening the high-risk work many will perform as young men. What's fascinating, Desmond and others have observed, is that men don't avoid, but instead actively pursue, jobs that threaten their bodies and health.

In recent decades, diverse tactics have been used to persuade adult men to adopt a more attentive self-care philosophy. Men have been encouraged to become more body conscious, embrace healthier life practices, and develop closer ties with the health-care community. Magazines launched since the late 1980s like *Men's Health* found a niche among an expanding segment of professional men eager to learn the latest developments in nutrition, fitness training, and body care. Increasingly, too, a range of books, newsletters, magazines, websites, and other media outlets have driven the boom industries to educate men about a host of issues including prostate, colon, and heart care; testosterone therapy; hair replacement and surgical implants; and, of course, erectile dysfunction therapies.

Just as men's health advocates try to transform negative perceptions of self-care as feminine, they must wrestle with the notion that providing care for the ill or disabled is women's "work." Women are more likely than men to practice caregiving, but as social worker Betty Kramer and sociologist Edward Thompson illustrate in their edited volume *Men as Caregivers,* many men are effective caregivers. Estimates indicate that between 14 percent and 18 percent of men informally provide various forms of caregiving for needy friends and family.

Yet, mainstream cultural messages downplay fathers' caregiving capacity and ability to address children's health-care needs. Parenting magazines and books are commonly tailored to informing moms more so than dads about the newest and best ways to care for and protect their children. This gender bias has been so engrained in the public's mind that a book published in 2004, *From Boys to*

Men: A Woman's Guide to the Health of Husbands, Partners, Sons, Fathers, and Brothers, seems intuitively marketable whereas serious doubts would accompany its hypothetical counterpart, *From Girls to Women: A Man's Guide to the Health of Wives, Partners, Daughters, Mothers, and Sisters.* Whether it's wiping noses in a childcare facility or wiping bottoms in a nursing home, societal images depict women as best suited for these positions, and the social networking patterns that shape fathers' everyday lives reinforce these stereotypes.

In her book *Do Men Mother?* sociologist Andrea Doucet interviewed Canadian primary caregiving fathers, most of whom were single fathers or stay-at-home dads. She reports that even highly motivated fathers find it difficult to feel comfortable at young children's playgroups, which they perceive as dominated by suspicious, unwelcoming mothers. Men's less intimate and more competitive friendship styles may also curtail fathers' willingness to discuss with other men their insecurities about fathering or focus on children's needs. Although Doucet doesn't emphasize typical health issues, her findings (and those of others) suggest fathers are more likely to be excluded from parental networks in which social support and children's health and childcare information are meaningfully shared. Moreover, fathers are less apt than mothers to take on the "community responsibility" tasks of engaging with adults involved with caring for children. Notably, some of these adults monitor children's physical, emotional, and mental health.

PROMOTING HEALTH-CONSCIOUS FATHERING

For far too long, many men have been ignorant of or ignored how their poor-health habits jeopardize their children's well-being. With the U.S. Department of Health and Human Services report *Healthy People 2020* on the horizon, now is the time to challenge men to foster positive health outcomes for their children. So what can be done to improve these patterns?

Generally speaking, men engage in more unhealthy behaviors and are less attentive to their self-care than women. The emerging evidence tells us, too, that men's exposure to health risks prior to their children's conception can contribute to prenatal problems. Moreover, men's poor health habits are related to children being more likely to smoke, abuse drugs, and eat poorly as well as be overweight and experience other negative health outcomes.

To understand and alter these patterns we must fully grasp fathers' lives as men and the diverse decisions affecting them. In other words, as the constrained choice theory implies, choices and priorities about health exist as part of a larger context and compete with other decisions about income, work, housing, partner/family, and personal image. Thus, we must commit to a multilevel approach to promoting social change that incorporates national and state policies, community-based strategies, workplace agendas, family support, and individual commitments.

Various social circumstances make it difficult for boys and men, some more than others, to forge and sustain healthy lifestyles and transmit similar values and benefits to their children. Constraints come in many forms, including conventional masculine discussion and business cultures that glorify stereotypes of the macho athlete or worker, inner-city and rural planning that limits recreational facilities for youth and adults alike, inadequate supports to educate men about reproductive health care and the consequences of paternity, workplace conditions and economic realities that expose men—especially those from economically disadvantaged backgrounds—to work-related health hazards, and peer pressure that extols a masculinity grounded in body toughness and risk-taking. Unfortunately, too many boys and men navigate their social networks, leisure, and work lives in ways that reinforce this less-than-ideal approach toward health.

Despite the constraints, men do have choices. Some recent research suggests, in fact, that men are capable of looking out for their own health and caring for others effectively if they put their minds to it. Ideally, as feminist values promoting gender equity inside the home gain wider appeal, and research accumulates to document the connections between fathers' and children's health, definitions of "good fathering" will summon fathers to pursue a healthier lifestyle while cultivating the same for their children.

Getting large numbers of men to adopt such a mindset requires broad public support and will require an intense public health service campaign—one that an Obama administration might be well-suited to launch. Realistically, though, concrete progress in altering individual commitments and choices will come when men regularly encourage each other to be more attentive to their own and their children's health. The seeds for this shift can be sewn most visibly in places that traditionally have been instrumental in discouraging health consciousness: locker rooms, fraternities, many work sites, and other places where male respect holds sway. Of course, these efforts also need to be augmented by men's partners.

The harsh reality for some men is that their chances to perceive and pursue healthy choices for themselves and their children hinge on politically sponsored national and state initiatives. In other words, for many men living in poor inner-cities and rural areas, structural and legislative changes are needed to improve access to fresh food markets, healthcare facilities and substance abuse programs, recreation sites, and organized sports so that more fathers and their children have viable, healthy options. It makes sense to expand Head Start programs by providing poor fathers with information, screenings, and referrals for a broad range of father-child health matters.

A less direct but critical step is to have schools, other youth-oriented organizations, and public health programs do a better job of providing teenage and young adult males comprehensive instruction on reproductive health. Because men have the capacity to influence fetal and infant life directly via their sperm quality and indirectly by how they treat the mother, they need to be educated at a young age about how their smoking, drinking, and drug use as well as their readiness to become fathers can affect their offspring's health. Although all males deserve these services, those living in poor neighborhoods are most vulnerable because they typically are the least prepared to assume many of the responsibilities associated with providing and caring for children.

Work sites are another place where men can be afforded opportunities to make better decisions about eating, exercise, stress management, and substance use, and receive medical check-ups and education about how these experiences matter for their children's health. Workplace policies, reinforced by more father-friendly corporate cultures, can ensure fathers have increased access to flexible schedules and nonstigmatized family-leave time. These benefits can make it easier for fathers to accompany their children to medical visits as well as provide hands-on care for their sick children.

Health-care institutions, in addition to educating men, can promote men's greater participation by providing more convenient evening and weekend hours while making their operations more male-friendly. Like the decisive trend beginning in the 1970s that saw fathers participating in childbirth preparation classes and being present when their children were born, the medical community must find innovative ways to integrate more fathers into their children's pediatric care. Today's creative childbirth preparation classes might produce promising results by coordinating new fatherhood programs with interventions to curb smoking, drinking, and drug use.

Ultimately, fathers must answer the call to communicate proactively with their children and monitor their well-being in consultation with healthcare providers and others who have a vested interest. Perhaps most importantly, if men adopt healthier behaviors for themselves and reduce their stress, they can more readily model such behaviors and authentically encourage their children to do the same.

RECOMMENDED RESOURCES

C. E. Bird and P. P. Rieker. *Gender and health: The effects of constrained choices and social policies* (Cambridge University Press, 2008). An overview of how diverse layers of social life are interconnected, contributing to health disparities between men and women.

W. H. Courtenay. "Constructions of masculinity and their influence on men's well-being: A theory of gender and health," *Social Science & Medicine* (2000) 50: 1385–1401. A critical review of how various conditions shape the kind of masculinity men construct and how those practices contribute to differential health risks.

A. Doucet. *Do men mother? Fathering, care and domestic responsibility* (Toronto University Press, 2006). Grounded in qualitative data, this book highlights fathers' opportunities to be more nurturing and engaged parents.

C. Garfield, E. Clark-Kauffman, and M. M. Davis. "Fatherhood as a component of men's health," *Journal of the American Medical Association* (2006) 296: 2365–2368. A thoughtful interdisciplinary essay that conceptualizes the relationship between fatherhood and men's health.

C. L. Menning and S. D. Stewart. "Nonresident father involvement, social class, and adolescent weight," *Journal of Family Issues* (2008) 29: 1673–1700. Provides a quantitative analysis of two waves of the well-respected National Longitudinal Study of Adolescent Health survey.

38

American Preschoolers on Ritalin

JONATHAN LEO

*People in the contemporary world, Americans especially, are consuming pre-
scription drugs at a rate never seen before. We have drugs to combat baldness,
erectile dysfunction, and sadness. These drugs serve people's desire to conform to
appearance, behavior, and emotional norms. One of the biggest booms in the
prescription drug field has been the rise of drugs to combat ADD and ADHD.
People are less willing than ever before to tolerate behaviors such as fidgeting,
not paying attention, not sitting still, and not concentrating, especially in
school. The obsession of middle-class Americans with getting ahead has caused
many parents to try almost anything they can to give their children an advantage in the competitive atmosphere of achievement and success. At the same
time, teachers and school psychologists reap the benefits of drugs such as Ritalin
and Adderol because they cause normally unruly children to become better
behaved. What began as an aid to troubled individuals has turned into a
wholesale control mechanism, where people are channeled into the mainstream
with drug therapy. In this chapter, Leo discusses the rise and prevalence of
Ritalin as a treatment for children with ADD or ADHD and the social pressure promoting its use. He explores both the practical and ethical issues involved
with altering consciousness to attain conforming behavior. Have you or any of
your friends taken these prescription drugs? What are their benefits, and what
are their side effects? Do you know any people who take them without having
them prescribed? How do you feel about prescribing them to adults as opposed
to children? Can you think of drugs that have similar issues? How would you*

SOURCE: Springer and Transaction Publishers, *Society*, 39(2), 2002, pp. 52–60, "American
Preschoolers on Ritalin," by Jonathan Leo. Used with the kind permission of Springer Science
and Business Media.

contrast the benefits and dangers of the level of drugs being prescribed in America today? What groups and/or forces are promoting higher levels of prescription drug use? What and/or who is opposing this trend?

At the start of the twenty-first century, it has become apparent that the American medical community's most controversial legacy to the science of child development and child rearing is a potent psychotropic drug. Ritalin is the drug of choice for Attention-Deficit Hyperactivity Disorder (ADHD) and every year, more of our children are taking it. American physicians, who prescribe 90% of the Ritalin produced worldwide, believe this is all based on "science." As an example of a child who the ADHD experts think should be medicated, take Sarah. Sarah's story is on the web site at the Department of Psychiatry at New York University (1/01).

"Sarah chooses to sit in the back of the classroom and much of the time she's doodling in her notebook or staring out of the window. She seldom completes assignments and often forgets to bring the right books to class. Her desk is a mess and she generally can't find what she's looking for. Then she gets weepy and says that nobody understands her." According to the experts at NYU, her diagnosis is Attention-Deficit Hyperactivity Disorder and the treatment of choice for her is Ritalin. This little girl is crying out, "Please understand me" and the American medical community's response is, "Medicate her." Sarah is a fourteen-year-old but we are currently prescribing Ritalin for children as young as two.

The ADHD experts are quick to point out that ADHD is one of the most thoroughly investigated and well-studied pediatric diseases. It is certainly true that millions of dollars, countless hours, and tremendous resources have all been consumed in an enormous effort to investigate ADHD. Yet, fundamental questions about ADHD are still vigorously debated. There is no proof of any underlying neurobiological deficit, it is not clear what the proper treatment should be, and it is not clear that the label "ADHD" is even valid. Even the *American Psychiatric Press Textbook of Psychiatry,* which overwhelmingly supports the idea that ADHD is a biological disease, has statements such as, "With unclear diagnostic boundaries, it is difficult to define or even conceptualize a unitary concept of ADHD or its etiology (p. 838)," or "there remains considerable uncertainty about the validity of ADHD as a diagnostic entity (p. 827)."

The issue of medicating children has recently taken on new importance because the National Institute of Mental Health (NIMH) has just started an unprecedented study on the use of medications to treat ADHD in preschoolers (three-year-olds). This is the single most important moral question the scientific community will face in the next decade. The ethics of genetically altered tomatoes, health care rationing, organ transplants, prescription drug reimbursements, and even assisted suicide are just a walk in the park compared to the ethics of exposing a developing brain to a psychotropic drug. Giving three-year-olds medications to help them be better nursery school students is a giant leap across an ethical threshold that will have profound consequences for our society. The responsibility, or as some would say, the irresponsibility, is enormous.

TREATMENT IN SEARCH OF JUSTIFICATION

To examine the rationale for these experiments a good place to start is with an article titled, "Trends in Prescribing Psychotropic Medications to Preschoolers." This article received a tremendous amount of coverage in the mainstream press because of one simple straightforward statistic. According to the authors, the number of preschoolers taking medications for ADHD increased 300% from 1990 to 1995. "Shocked," "concerned," and "surprised" were just some of the reactions from the American medical community. In fact, while both sides of the Ritalin debate expressed significant concern about this statistic, the reasons for their concern could not have been more different.

The problem, even for those who endorse the use of Ritalin, is that Ritalin has never been officially approved for children under six. Although prescribing these medications to children in nursery school may be legal, doctors are entering new territory when they do so because there have never been any experiments on the effects of Ritalin in such young children. According to Dr. Steven Hyman, the director of NIMH, "Without good clinical data, every child who receives this medication represents an uncontrolled experiment—that is entirely unacceptable." The White House, Hillary Clinton, and NIMH immediately stepped into the fray, and amidst great fanfare, announced the allocation of five million dollars to investigate the safety and efficacy of these drugs for preschoolers.

According to those who oppose the use of Ritalin, handing out more money to investigate the safety of Ritalin for even younger children is exactly the kind of thinking that has created the current mess, and more money will only make more of a mess. The response by the White House, the pro-Ritalin advocates, and the drug companies is nothing but a face-saving move to deflect a potential public relations nightmare. In light of the fact that there seems to be a correlation between the amount of time, effort, and money that NIMH devotes to ADHD and the rising numbers of American children using Ritalin, it is easy to see why a study estimating that a quarter-million American preschoolers are on Ritalin could be a public relations disaster. The political reasoning coming out of Washington goes something like this: a group of doctors has been prescribing medications to very young children; this group of doctors is now exposed; so, now we are going to give money to this same group to investigate the safety of what they have been doing. Why? So they can continue doing what they are already doing.

For those who oppose the escalating use of Ritalin in such young children, the appropriate response to statistics documenting toddlers on Ritalin is, "Stop—Enough is enough." As Drs. Michael McCubbin and David Cohen put it, "That public education and health systems permit the massive drugging of children despite the lack of knowledge should set off alarm bells." There is no reason for any child under six, much less three, to be taking any kind of medication for hyperactivity. Investigating the safety and efficacy of these drugs in toddlers will have one result: ten years from now even more three-year-olds will be taking medication for ADHD.

THE MTA STUDY

If one had to pick a "landmark" study in the history of ADHD research it would certainly be "Treatment Strategies for Attention-Deficit/Hyperactivity Disorder," written by the "MTA Cooperative Group." The fanfare surrounding the publication of this article was nothing short of extraordinary. In a segment titled, "Ritalin's Redemption," ABC News declared, "The early results of a large national study indicate that drug therapy for children with attention deficit and hyperactivity disorder is not only effective, but much better than psychological counseling alone." Note that ABC News did not quote an expert; they just declared it as fact.

In another article announcing the results of the study, one of the authors, James Swanson, was quoted as saying, "Treatment can mean the difference between a kid ending up at Berkeley or ending up in prison." In the ADHD literature, you would be hard pressed to find a single scientific study that has been more responsible for the huge number of Ritalin prescriptions written in this country. Any budding sociologist out there who would like to investigate why American physicians lead the world in passing out Ritalin should look no further than the MTA study.

According to the authors, the most important goal of the study was to answer the question, "How do long-term medication and behavioral treatments compare with one another?" It might seem like a fairly straightforward question, but it's loaded. A major point of contention in the Ritalin debate centers on answering the question of whether to treat an ADHD child with medication or psychosocial interventions.

The Ritalin proponents believe that a child with ADHD is "at risk" and will continue to fail at school, and if left untreated long enough the child will probably develop a more serious condition. Without some sort of medication, the child is doomed to a life of frustrations, failures and possibly even incarcerations. Talk therapy might sound good, but in the view of the Ritalin advocates withholding medication from these kids is irresponsible and even verges on child abuse.

As an example, in one case in New York State the parents of a child on Ritalin became concerned about what the drug was doing to their child and wanted to discontinue the medication. However, child protective services and the courts got involved and would not allow the parents to take the child off Ritalin. When asked about the ethics of this, Dr. Peter Jensen, one of the authors of the MTA study, replied, "Certainly child-protection laws and the courts are not the best way for us as a society to see that our children receive appropriate care. But when a child's well-being is at stake, we cannot default on our responsibilities to ensure that he or she gets the necessary help."

The primary goal of the MTA study was to supposedly answer the question "to medicate or not to medicate." However, it needs to be mentioned that the study's authors never entertained any doubt about the answer to this question themselves. In their previous writings, most of the MTA investigators have made it very clear that they strongly favor the use of medications. Their goal in creating this study was obvious: To bring those skeptics who do not quite share the MTA philosophy of raising children into the pro-Ritalin fold.

To compare medication and behavioral treatments, the investigators divided children aged seven to nine into several different groups. One group received medications, a second group received behavioral treatments, and a third group received both. There was also a fourth group that received no treatments from the MTA investigators but instead received the standard treatment available in the community. The experiment continued for fourteen months and the children in the different groups were compared. The main thrust of this study is found in the very first paragraph of the section titled, "Results." In the words of the investigators, "Robust differences were found according to two different data sources, indicating the superiority of medication management over behavioral treatment of ADHD symptoms." But who are these two data sources that say medication is better than behavioral management?

The first data source is the parents, and the second source is the teachers. To determine whether Ritalin or other similar drugs were working the authors of the study had the parents and teachers fill out a simple set of questions about the children's behavior. Note that neither the parents nor the teachers were "blind" to the treatments the children were receiving. According to the parents, the children who received medication did better than children who received behavioral treatment in terms of both attention span and hyperactivity/impulsivity. But can these parents be considered a random sample of the typical parents in this country? No, of course not.

The investigators have pre-selected a group of parents who believe that it is acceptable to medicate children; in this lack of random sampling, we find the experiment's main shortcoming. According to the MTA investigators, this is how they found the parents: "In all instances, the child's parents contacted the investigators to learn more about the study, after first hearing about it through local pediatricians, other health care providers, elementary school teachers, or radio/newspaper announcements." The MTA investigators have ended up with a group of parents who accept the very idea that ADHD is a disease—which in and of itself is a biased group.

Even as the study progressed, built-in mechanisms guaranteed the formation of a biased group of parents. After the children were initially screened and examined, they were assigned to the various experimental groups. At this point, out of 289 children who were going to be receiving medication, eighteen parents refused the medication and pulled their children from the study. It would be very interesting to find out why these parents withdrew from the study. Although the MTA investigators do not supply this information, these parents very likely had a problem with putting their children on medications—just one more example of how the MTA investigators ended up with a group of parents who do not represent the entire population....

The other problem with the study is that while the parents said that medication improved both attention and hyperactivity/impulsivity the teachers said that it only improved attention but *not* hyperactivity. Since we have known for many years that Ritalin will improve anyone's attention, this is really nothing new. What is surprising is that the teachers did not find any lessening in the level of hyperactivity.

The problem with the built-in bias of the parents is also further highlighted by the study itself, because, in addition to the parent and teacher observations, there was a third group of raters who observed the children in the classroom. Unlike the parents and teachers, these raters did not know which children were receiving medication or behavioral treatments and these raters found *no* difference between medication and behavioral therapy. According to the three groups of raters we have the following conclusions: (1) the parents, who were the most biased, found Ritalin to be the winner; (2) the teachers, who have a broader background in child behavior than the parents, found that Ritalin did not help in terms of hyperactivity/impulsivity, and; (3) the outside raters, the only unbiased group, found no difference between Ritalin and behavior management. Since the results of this unbiased group did not deter the MTA researchers' enthusiasm for Ritalin, it is not clear why they used this group at all.

Furthermore, from reading the MTA study it is impossible to really know what the teachers or parents actually reported about inattention and hyperactivity/impulsivity. In the discussion (p. 1077) the authors say that according to the teachers—the children on medication were better off in terms of both hyperactivity/impulsivity and inattention. However, Table 5 (p. 1082) in the data section contradicts this statement and says that according to the teachers—the students were better off in terms of inattention but *not* hyperactivity....

Is Ritalin quick, easy, and cheap? Yes. Will it work? Probably, if all that is meant by *work* is that the children are easier to control. But the real question is, will it *help* children? No one is going to disagree that Ritalin will make kids easier to control or that it improves their ability to pay attention. What we don't know is how Ritalin affects a child from within. Children cannot tell us what it is like to live with Ritalin.

However, an insightful, firsthand account of Ritalin's effects was written by Walter Kirn, an editor for *GQ* magazine who started taking Ritalin when he was thirty-one. For the first several months, he thought it was a wonder drug that gave him incredible powers of attention, but when he started to notice that it was fundamentally changing his personality he took himself off Ritalin. Besides lamenting the fact that children will not have the luxury of taking themselves off the medication, Kirn is also concerned about what constitutes success when it comes to evaluating the effects of Ritalin. In Kirn's words, "The pills are a kick. They're uppers. Get it? Uppers. They act like downers on kids who truly need them, according to the experts, but what do they know? The experts are on the outside looking in, monitoring behavior, not emotion. All they see are rows of little heads sitting obediently at little desks."

In summary, the MTA investigators found a group of parents who believe in medicating children and then asked them if it worked. They then trumpeted the results of a survey taken from a group of biased parents as the solution to keeping more kids out of jail. In retrospect, it appears that the media coverage and marketing of the MTA study far outpaced the actual scientific value of the study. The MTA study is significant only for those who already believe in medicating children; for those who do not believe in medicating children to help them get through the school day, the MTA study does not provide much direction.

THE PATS STUDY

Based on the success of the MTA experiment, the Ritalin experts are now investigating the use of Ritalin in preschoolers. The study is referred to as PATS, which stands for "Preschool ADHD Treatment Study." A recent article in *Science* magazine entitled "Planned Ritalin Trial for Tots Heads into Uncharted Waters," addressed some of the ethical issues surrounding the upcoming experiments on the use of medication in such young children. The end of the article holds a surprising paragraph about the laboratory classroom that Dr. Lawrence Greenhill, with funding from NIMH, is planning. In response to a question about how researchers will know whether a three-year-old is functioning "on task" (one of the goals of giving Ritalin), Dr. Greenhill explains: *"We're going to set up a laboratory classroom, and we'll observe common tasks done in nursery school,* such as stacking blocks and stringing beads on a thread. Children will be asked to sit in a circle and take part in group events. The test will be whether the child is *compliant* and participates or *attends for a few seconds before drifting away and doing everything else in the room"* (Greenhill's words are in italics, the reporter's paraphrasing in the original article is not in italics).

Medications aside, the NIMH laboratory classroom, with its heavy emphasis on children's compliancy, is not the type of nursery school that many parents want for their children. Evidently, the PATS investigators have never heard of Montessori schools. In a Montessori school, there are several stations that are set up in the classroom and children are urged to visit whatever station they desire. The children are given the *choice* to stack blocks; they are not *forced* into stacking blocks. Maria Montessori was not overly concerned with making preschoolers "compliant"; instead she focused on creating an environment that was conducive to the child, not forcing the child to fit the environment. In Montessori's words, "A more just and charitable approach toward the child would be to create an 'adaptive' environment different from the repressive one in which he operates and which has already formed his character. The implementation of any educational system ought to begin with the creation of an environment that protects the child from the difficult and dangerous obstacles that threaten him in the adult world."

To say that Maria Montessori and the PATS investigators have entirely different views about the ideal classroom would be an understatement. The NIMH classroom seems to be about drawing lines; Montessori's classroom is all about expanding boundaries....

The PATS investigators are designing the best way to whittle away at square pegs (the students) so that they fit into a peg board (the classroom) with round holes, but there are still schools in this country that take the opposite tack, which is to adapt the school to the child.

When the PATS investigators see a child who does not want to stack blocks, they see a disease that needs to be medicated. When educators like... Montessori see a child who does not want to stack blocks they look to the environment. It needs to be mentioned that the people who believe that non-compliant three-year-olds have some sort of neurobiological disease are the same people who are largely responsible for the epidemic of Ritalin use in this country. These are the same people who for years have said the following: ADHD is a "disease"; these children have a malfunctioning cerebral cortex; 3% to 5% of our children have this disease; and the best treatment for these children is Ritalin....

In the book, *Punished by Rewards,* Alfie Kohn does not talk about Ritalin but he does talk about the issue of control. According to Kohn, "Before we resort to control, we should be absolutely certain that less intrusive, more respectful interventions cannot work. We should also think about how an act of control is exercised: Do we justify it with a reasonable explanation? Do we pause to ask whether what we are getting the child to do is really necessary? Are we thinking about how best to help the child become a responsible person (as opposed to just getting her to obey)?"...

Our society is leading the world in Ritalin consumption and we cannot ignore the fact that this is partially due to the value system in this country. Sure, Ritalin will help us control our kids, but as a society maybe we need to take a step back and reevaluate this issue of "control." The real ethical question is much more complex and complicated. If the American medical community is not treating a disease but is instead selling a performance-enhancing drug the ethical ramifications are overwhelming.

The major ethical question is not, "Is Ritalin safe?" but, "Is it right to drug little children because we don't like their behavior, or because they don't fit in?" If the entire approach to dealing with these children by drugging them is ethically bankrupt then any questions about Ritalin's safety are a non-issue. Who decides whether 2%, 5% or 10% of our children have this disease? Is it right to give a three-year-old a drug to control him? Why does America lead the world in Ritalin consumption, while the British are talking about banning Ritalin for children under five? These are the real ethical questions.

Besides objecting to the PATS investigators' somewhat simplistic philosophy of education, there are also plenty of objections to be made on a purely scientific level. For instance, research has shown that the dopamine receptor, which has been implicated in the formation of ADHD, reaches a peak density at about three years of age and then starts to taper off. Given that the long-term treatment of many psychotropic drugs has been shown to alter the numbers and sensitivity of dopamine receptors, a perturbation of the dopamine system at a critical developmental time period could have severe consequences. Considering that we know so little about the effect of Ritalin on the developing brain, even the developing rat brain, it seems odd that we are dispensing it to such young children.

THE DIAGNOSIS OF ADHD

To the news reading public, the diagnosis of ADHD is out of control. In the last several years, in addition to the legal consumption of Ritalin, the illegal use of Ritalin has risen sharply. When one examines the ADHD diagnosis and sees that it is essentially going to a doctor and saying, "I cannot pay attention" or "my child cannot pay attention" it is surprising that anyone would go through great lengths to get Ritalin illegally. In twenty-first century America it is easier to get a legal prescription for Ritalin than it is to get a library card.

Dateline NBC recently reported on the illegal use of Ritalin in our schools and colleges. On college campuses across the country, undercover reporters simply asked students studying in the library where they could get Ritalin illegally. In one case the students recommended just going to the student health service and getting a prescription because the ADHD test which the doctors give is so subjective.

What the *Dateline* report failed to recognize is that the major problem with Ritalin is not the *illegal* use but the *legal* use. Many of the students who *Dateline* interviewed were using Ritalin illegally for the very same reason that doctors prescribe it—to pay attention. It is disingenuous to criminalize these students who are seeking to improve their performance, and then turn around and use the performance enhancing aspect of the drug as the major reason to prescribe it. If anything, the fault lies more with the adults than with the students, because the adults should know better. Think about the hypocritical message we are sending to our children. On one hand we tell them it is acceptable to go to a doctor and get a prescription for Ritalin to improve one's ability in school. But on the other hand we tell them it is wrong to get the very same drug for the very same reason from a classmate. No wonder our children are confused about drugs.

CONTROLLING OR DIAGNOSING NORMAL KIDS?

When challenged with statistics documenting out of control Ritalin use it is common for the "experts" to fall back on the "little monster scenario." In defense of Ritalin, the advocates point out that even if Ritalin is over prescribed, there are children out there with a real disease who need Ritalin to function and that without their medication these children are out of control one-man wrecking crews. The organization, "Children and Adults with Attention Deficit Disorder," also known as CHADD, is a strong proponent of the disease model of ADHD. CHADD supports the idea that the best treatment for ADHD is medication. According to E. Clarke Ross, the CEO of CHADD, "The critics always spotlight a handful of children who have experienced side complications or side effects from medication. But what about the millions of children who have been helped by medication?" Okay. So let's spotlight a child who CHADD thinks is helped by medication. This case study appears in a pamphlet that CHADD distributes to schoolteachers.

"John, a third grade student, is often non-compliant and does not begin tasks when asked. During a two-week observation period, he exhibited the following behaviors on a routine basis: John sharpened his pencil three times before sitting down and working. John fell out of his chair when given an assignment with 50 problems. He pretended to be the class clown. The class laughed. After leaving his reading group, on the way back to his seat for independent work, John tripped Sally. He was sent to the corner of the room."

According to the Ritalin advocates: John has a neurobiological disease; his antics in the classroom are only a foreshadowing of bigger problems; he is destined to a life of frustrations and failures; his problem is biological and he needs medication to function. Granted, he would have been better off if he had been treated in preschool, but it's not too late to turn his life around with Ritalin.

But is John a "little monster" or is he the class clown? Is it possible that he is bored with school and needs more intellectual stimuli? Is he in a classroom with thirty students and one teacher? Is there really nothing else that will work for John other than medication? In the eyes of the Ritalin advocates, John has a disease that needs medication but in the eyes of many educators, John's behavior would be considered fairly normal. The problem is not that we are prescribing Ritalin to kids who apparently don't need it. The problem lies with the Ritalin advocates' definition of who needs it. According to many people in this country, the diagnosis of John with ADHD represents nothing less than a fundamental misunderstanding of children....It is wrong to blame the average physician when it is clear the fault lies with the creators and marketers of the diagnosis. They have given the American medical community an unworkable, unscientific, and unreliable diagnosis that could fit just about any kid in America....

RITALIN MEETS HUMAN GENOME TECHNOLOGY

The human genome project is going to change the way we treat disease. No one has discovered an "ADHD gene" and it is highly unlikely that an ADHD gene or even a set of ADHD genes will ever be discovered yet talk of gene therapy for these children is in the air. Several generations from now, heart disease, diabetes, Parkinson's disease, and ADHD will all be treated with gene therapy. Wait a minute. Rewind. Double take. Are we really ready to use gene therapy to treat ADHD? If we accept the logic of biological psychiatry that ADHD is just like any other disease, then the answer is yes. If ADHD is really a disease then the technology of the Human Genome Project holds great promise. Consider Dr. Alan Zametkin's closing remark in a discussion on the future of ADHD research: "Can pharmacological or gene manipulations lead to a cure?"

There are two barriers to using gene therapy—technology and ethics. The technological barrier will be overcome shortly, yet most people believe the ethical barrier will prevent us from altering the genome. But they are wrong, because as a society we have already embraced the belief that it is acceptable to chemically alter the neurobiology of a developing child. The logic of the Ritalin advocates has taken us across the ethical barrier; the technology of the human genome project will simply make the entire affair more efficient. Medicating three-year-olds in nursery school is just a harbinger of things to come.

PART V

Social Change

Throughout this book we have integrated social change into each section, but now we address it directly. So many changes are shaping the world in which we live, that we found it important to offer you a contemporary view of the current issues of our society and the controversies that arise from them. Social change occurs as new technologies arise to transform our lifestyles, both liberating us and capturing us with their advances. New ideas foster change as well, and these have led to corresponding transformations in the structures and forms of society, such as gender, the family, religion, and medicine. All of society's institutions, organizations, statuses, and roles, in fact, are subject to the influence of new trends and patterns that occur, some sweeping through rapidly, while others evolve slowly and gradually, almost imperceptibly.

Hierarchies may slowly and subtly shift, or they may suffer more sudden reversals and upheavals. Changes may occur as a result of forces outside of society or through developments that arise from within, altering social values, norms, institutions, social relationships, and self-identities. These may result from environmental events, warfare and invasion, cross-cultural and subcultural contact, and the diffusion of ideas from one society or group to another. New innovations may arise in people's lifestyles, inventions, marketing, healing, educating, spending, voting, and communicating that may radically transform the way they think and live. The rate of social change is astounding many people today. Less than a generation ago we did not have computers, cellular phones, ATMs, the Internet, or a host of other "modern conveniences" that now we could no longer imagine living without. These have all transformed our major social institutions and the way they operate.

In our first section we consider some of the changes that have occurred as the result of technological innovations. Graphs charting the rate of social change show that this has continued to accelerate at an increasing pace. People living in today's society take for granted a level of technological support that would have been unimaginable a generation ago. Your cohort of college students, often now referred to as the "Millenials," are the first to have grown up in the "microwave" age, when people expect things to happen in seconds. You cannot imagine doing the types of mechanically mundane tasks that your parents faced! These innovations have profoundly altered the way people live in relation to the natural and social world. Communication has become instantaneous, transforming the concepts of space and time.

Kaveri Subrahmayam and Patricia Greenfield discuss the electronic technologies of communication available through the Internet and cell phones, focusing especially on their use by adolescents. They offer a plethora of information about who uses these technologies, to what extent, and in what way, and analyze how these media have dramatically changed teens' lives and interrelations with friends, parents, and strangers. Along the way they discuss some of the possible dangers associated with this widely open access to the world and how to help manage them.

Monica T. Whitty further explores this cyber realm in her study of romantic cyber relationships and cyber-cheating. Online connections and the norms governing them are still rapidly evolving, so that behaviors considered uncertain or problematic today may be radically altered a decade hence. Young people (as well as adults) have moved into new frontiers of flirting and "sexting" through the social media, texts, and tweets now available. Having an open portal through which friends, acquaintances, and strangers can access people's lives offers huge opportunities for more visceral and immediate interactions with others, but can also lead to greater unwelcome intrusions. Whitty explores the social meaning of Internet "relationships" and the challenge they pose to their face-to-face counterparts. In her experiment, she offers a simple scenario and asks young men and women to assess the nature and role of online flirtatious and romantic connections. In general, "sexters" view sexting as harmless and fun, whereas non-sexters see it as dangerous. One thing is sure, sexting, like sex, is here to stay. This evokes the nebulous nature of liaisons that have no physical anchor, the meaning of these ephemeral bonds, and a new set of Internet and cell phone norms that are still evolving.

Another important development has been the shrinking of the world. Compared to any previous age, today's college students are more likely to study abroad, to communicate with people in different countries, to live and work with people from other lands, and to purchase goods and services that originated outside their country of origin than in any previous age. We have seen a global flow of tourism, labor, products, and culture. More than in any other era, we are aware of our lodging in an international arena. As a country we are struggling to find ways to adapt to the inherent advantages and disadvantages. We can buy cheap clothes, cameras, phones, and digital recorders made in China, but at what cost?

Phil Taylor and Peter Bain's selection addresses the issue of outsourcing. In the 1990s, as free trade agreements were signed, companies took their manufacturing plants to other countries where labor was cheaper and environmental regulations more lax, enabling them to produce goods at lower prices and undercut domestically made commodities. Large industries such as steel, rubber, textiles, clothing, energy production, and others were relocated offshore, robbing American inner cities of the core economic opportunities that had sustained generations of workers. Low-skilled or skilled blue-collar workers found themselves displaced, unable to retool or get a footing back into the job market. Inner-city families were fractured by the male breadwinners' inability to earn a living, and hundreds of thousands of men were demoralized as their wives gained jobs in the pink-collar service sector, effectively replacing them as the household providers. At first, the employment opportunities of the educated classes were largely unaffected, as the economy transformed from manufacturing into a service base. The twenty-first century has seen this security undercut, as we are now witness to the increasing outsourcing of service jobs. Slowly at first, and concentrated in the most low-skilled and repetitive kinds of jobs, employment for white-collar workers is flowing from first-world into developing countries. Taylor and Bain describe something most of us have experienced firsthand: calling a domestic company and being connected to a service provider in India or another country. Yet their depiction of offshore call centers shows the difficulties these organizations encounter in staffing their positions and the humiliations and challenges workers taking these jobs face abroad.

Another critical dimension of the globalized world can be found in its ugly underbelly: the international trafficking of slave labor, whether for manufacturing, household domestic, or sex work. David R. Hodge and Cynthia A. Lietz describe the global flow of labor, which has lured many young women to leave their countries of origin, where they are a drain on their families' inadequate resources, to seek a means of economic survival in the unknown. They represent ripe pickings for those who would exploit them, buying, selling, or stealing them into forced servitude. The plight of these (mostly) women and children is heart-wrenching, as they endure unbearable conditions, wretched treatment, and suffer disease and death due to beating, malnutrition, and sexually transmitted or other forms of disease.

Given these and other changes, what can we say about the social transformations they have wrought and what they portend for the future? In the section on *Visions of Society* we look at broad trends influencing the shape of society. The world that we encounter has been altered in ways that we interact with daily on a commercial basis. Our sociological odyssey takes us to look at George Ritzer's concept of the "McDonaldization," or bureaucratization, of society. As we travel the country we can see the disappearance of "mom and pop" stores, of personal relationships between workers and their customers, replaced by strip malls featuring the same set of stores, fast-food restaurants, big-box warehouses, banks, hotels, and commercial landscapes. These have homogenized the country, blending everything together into one standardized, sterile Americana. We can see these trends influencing the shape of our entire culture, its organizations, and the roles, relationships, and behavior people enact within them. McDonald's standardization imaging is now a pervasive force in nearly all aspects of social and commercial life. Look for these all around you when you shop, eat, select entertainment, or even in the way that you choose your college courses.

In some ways, the most important issue we face as a nation in the 21st century is the challenge of *community*. What once surrounded and nurtured us has eroded, and we worry about the quality of our lives and our society as a result. The small, traditional communities where people associated with each other in primary relationships and across generational borders (what the early sociologist, Ferdinand Tönnies, called *gemeinschaft*) has been replaced by the antiseptic and distanced relations of contracts, instrumental behavior, mass society, and urban areas where people do not have as much in common or even know each other (*gesellschaft*). The traditional forms in which we used to recognize community have changed. Just as the "Pleasantville" families from the 1950s have become rarer, so have the neighborhoods in which they lived. Developers still create suburban tracts and people flock to them, but the feeling of neighborliness is not the same. People are not rooted in their local social institutions the way they used to be. Robert Putnam expresses this when he writes of the decline of the traditional civic institutions and questions what has become of community. We are forced to wonder if community is a necessary environment to nurture people and their development, and what component parts, if any, are the most critical. We face the challenge of social change and the maintenance of our core values.

Yet counterposed to this fearful portrait of the decline of community is our own depiction of the rise of cyber communities, which have flourished and abound on the Internet. Looking specifically at the communities of self-injurers, our study of online groups and the relationships and identities they foster stands as a blueprint for most online groups. At the same time as geographically bounded communities have declined, virtual communities, forged around people's more central preoccupations and selves, are arising to take the place of what is gone. You must decide what you think of this trend, and how you evaluate chat rooms and other interest groups. What have we gained and what have we lost in this trade-off, and how has society been affected by this dramatic development? This is your society, and the decisions and values you adopt will help shape future generations.

Finally, we close with a few words about the role of sociological thinking in making the world a better place. Sociology is not just for sitting back and describing and analyzing what is happening. It can also be about jumping in, being active, and promoting social change. We are not here just to take what is offered to us and look out for ourselves, but to give back to our community, our country, and our environment in ways that matter. Brett Johnson, Ross Haenfler, and Ellis Jones draw on their experiences as sociologists to close the book as we opened it: with the use of their sociological imaginations. They show us how we can change things for the better all around us every day, just by applying our sociological insights and passions. So don't just take this course and read this book; go out there and make a difference!

39

Online Communication and Adolescent Relationships

KAVERI SUBRAHMANYAM AND
PATRICIA GREENFIELD

The term "online communication" refers to reading, writing, and communication via networked computers. It encompasses synchronous computer-mediated communication (CMC, whereby people communicate in real time via chat or discussion software, with all participants at their computers at the same time); asynchronous computer-mediated communication (whereby people communicate in a delayed fashion by computer, using programs such as e-mail); and the reading and writing of online documents via the World Wide Web. Growth in online communication, particularly through social networking, is rapidly impacting the media world. Youth are leading the transition from face-to-face (FTF) to CMC, with close to nine out of ten teenagers using the Internet. Not only has the wired share of the teenage population grown, but teens' use of the Internet has intensified. Teenagers now use the Internet more often and in a greater variety of ways than they did at the turn of the century. Half of all teenagers report going online daily, with similar numbers having access to broadband. Wired teens also make more use of instant messages, online gaming, and use the Internet to shop, to get news, and to get health information. Nearly half of all teenagers own a cell phone and one-third engage in texting. Texting on cell phones is particularly common among those who already go online frequently and use other Internet tools often. Emailing is still a fixture in teenagers' lives, but this is gradually giving way to the increasing use of IM and texting. The current generation of adolescents and young adults is also the most adept at multi-tasking, being able to do homework, listen to music, IM, email,

SOURCE: Kaveri Subrahmanyam Patricia GreenfieldOnline Communication and Adolescent Relationships, The Future of Children - Volume 18, Number 1, Spring 2008, pp. 119–146. From The Future of Children, a collaboration of the Woodrow Wilson School of Public and International Affairs at Princeton University and the Brookings Institution.

search the Web, and text to friends at the same time. With the proliferation of tweeting, or micro-blogging, youth are increasingly becoming accustomed to producing and to digesting information bits that are limited to 140 characters.

Subrahmayam and Greenfield assess the parameters of this phenomenon by reviewing recent findings about teenage online communication. Which of the modes of online communication they discuss do you currently use? How would your life be different without these electronic media? Have you or your friends personally experienced any of the problems they discuss? Do you think that teenagers need to be protected from some of the dangers lurking on the Internet? Did your folks put any restrictions or give you any guidance regarding your Internet use? If you were a parent, how would you treat your child's relation to these media? How do you think social networking is affecting the nature of friendships? Do people spend more time online interacting with people who simply respond to them compared to FTF friends in real life? How would you assess the nature of online versus FTF relationships? Do online relationships take away from FTF ones, or serve as staging grounds helping people learn to navigate real world relationships?

The communication functions of electronic media are especially popular among adolescents. Teens are heavy users of new communication forms such as instant messaging, e-mail, and text messaging, as well as communication-oriented Internet sites such as blogs, social networking, photo and video sharing sites such as YouTube, interactive video games, and virtual reality environments, such as Second Life. Questions abound as to how such online communication affects adolescents' social development, in particular their relationship to their peers, romantic partners, and strangers, as well as their identity development, a core adolescent developmental task.

In this article, we first describe how adolescents are using these new forms of electronic media to communicate and then present a theoretical framework for analyzing these uses. We discuss electronic media and relationships, analyzing, in turn, relationships with friends, romantic partners, strangers, and parents. We then explore how parents and schools are responding to adolescents' interactions with electronic media. Finally, we examine how adolescents are using electronic media in the service of identity construction.

Adolescents have a vast array of electronic tools for communication—among them, instant messaging, cell phones, and social networking sites. These tools are changing rapidly and are just as rapidly becoming independent of a particular hardware platform. Research shows that adolescents use these communication tools primarily to reinforce existing relationships, both friendships and romantic relationships, and to check out the potential of new entrants into their offline world. But while the Internet allows teens to nourish existing friendships, it also expands their social networks to include strangers.

The newly expanded networks can be used for good (such as relieving social anxiety) or for ill (such as sexual predation). Although researchers have conducted no rigorous experiments into how adolescents' wide use of electronic communication may be affecting their relationships with their parents, indications are that it may be reinforcing peer communication at the expense of communication with parents. Meanwhile, parents are increasingly hard-pressed to stay aware of exactly what their children are doing, with newer forms of electronic communication such as social networking sites making it harder for them to control or even influence their children's online activities. Schools too are now, amidst controversy and with difficulty, trying to control the distracting uses of the Internet and other media such as cell phones while children are at school. The challenge for parents and schools alike is to eliminate the negative uses of electronic media while preserving their significant contributions to education and social connection....

ELECTRONIC MEDIA AND RELATIONSHIPS

Establishing interpersonal connections—both those with peers, such as friendships and romantic relationships, and those with parents, siblings, and other adults outside the family—is one of the most important developmental tasks of adolescence. As electronic media technologies have become important means of communicating with others, it is important to consider them in the context of the interpersonal relationships in adolescents' lives. Two themes have framed discussions of adolescent online communication and relationships. One is concern about the nature and quality of online and offline relationships. The other is how online communication affects adolescents' relationships and well-being and whether the effects are positive or negative.

Electronic Media and Relationships with Friends

We first examine the role of electronic media in youth's existing friendships. One study of detailed daily reports of home Internet use found that adolescents used instant messaging and e-mail for much of their online interactions; they communicated mostly with friends from offline lives about everyday issues such as friends and gossip.[1] Another study found that teens use instant messaging in particular as a substitute for face-to-face talk with friends from their physical lives.[2]...

To study the communicative purposes of text messaging, one study asked ten adolescents (five boys and five girls) to keep a detailed log of the text messages that they sent and received for seven consecutive days. Analysis of the message logs revealed three primary conversation threads: chatting (discussing activities and events, gossip, and homework help), planning (coordinating meeting arrangements), and coordinating communication (having conversations about having conversations). The teens ended most text conversations by switching to another setting such as phone, instant messaging, or face-to-face.[3]

Effects of Electronic Communication on Friendships   How does adolescents' electronic communication with their friends affect their friendship networks and, in turn, their well-being? According to a 2001 survey by the Pew Internet and American Life Project, 48 percent of online teens believe that the Internet has improved their relationships with friends; the more frequently they use the Internet, the more strongly they voice this belief. Interestingly, 61 percent feel that time online does not take away from time spent with friends....[4]

Whereas survey participants who used instant messaging communicated primarily with existing, offline friends, those who visited chat rooms communicated with existing friends less often. This pattern makes sense because chat is generally a public venue providing wide access to strangers and little access to friends, whereas instant messaging is primarily a private medium....

Even when adolescents are communicating with their friends, social networking sites such as MySpace may by their very nature be transforming their peer relations. These sites make communication with friends public and visible. Through potentially infinite electronic lists of friends and "friends of friends," they bring the meaning of choosing one's social relationships to a new extreme. They have thus become an essential part of adolescent peer social life while leading to a redefinition of the word "friend." A recent focus group study of MySpace on a college campus found that most participants had between 150 and 300 "friends" on their MySpace site.[5] Friends' photos and names are displayed on users' profiles, and each profile includes a list of "top" friends, ranging from a "top four" to a "top twenty-four." Such public display of best friends seems a potentially transformative characteristic of a social networking site. But how does making (and not making) someone's "top" friends list affect adolescent relationships and self-esteem?...

Electronic Media and Bullying   The news media are increasingly reporting that adolescents are using electronic technologies such as cell phones, text messages, instant messages, and e-mail to bully and victimize their peers. In a 2005 survey conducted in the United Kingdom, 20 percent of the 770 respondents, aged eleven to nineteen, reported being bullied or receiving a threat via e-mail, Internet, chat room, or text, and 11 percent reported sending a bullying or threatening message to someone else. Text bullying was most commonly reported, with 14 percent reporting being bullied by mobile text messaging. Bullying in Internet chat rooms and through e-mails was reported by 5 percent and 4 percent of the sample, respectively. A new form of harassment appears to be emerging through cell phone cameras: 10 percent reported feeling embarrassed, uncomfortable, or threatened by a picture that someone took of them with a cell phone camera. The majority of the respondents reported knowing the person who bullied or threatened them.[6]...

A large-scale online survey conducted at a popular teen Internet site in 2005 found a much higher rate of harassment—72 percent—using two different methods of estimating prevalence for the previous year.[7] The online recruitment probably yielded relatively heavy Internet users for whom the risk of cyberbullying would be greater. The online anonymity of the questionnaire may also have fostered greater honesty.

Research on cyberbullying has tried to create profiles both of youth who are likely to perpetrate harassment and of those who are likely to be the victims of harassment. Cyberbullies are more likely to report poor parent-child relationships, substance use, and delinquency. Youth (aged ten to seventeen) with symptoms of depression are more likely to report being harassed. Among boys, those reporting major depression were three times more likely to be harassed than those reporting mild to no depression.

Cyberbullying illustrates how traditional offline adolescent issues are moving to the electronic stage. A questionnaire study of eighty-four thirteen- to eighteen-year-old teens found that text messages were the most common form of electronic bullying. Most important, the findings suggest that students' role as victim and perpetrator of bullying in the offline world predicted their role in electronic bullying.

Respondents who had experienced repeated school-based bullying were seven times more likely to be subjected to repeated online bullying. Heavy use of the Internet also increased the risk, as did the use of particular Internet tools, specifically, instant messaging and webcams. These latter factors, however, were much less powerful than was school-based bullying.

Electronic Media and Relationships with Romantic Partners

Given that adolescents are using electronic media to interact with peers, it is important to see how they use them in the area of romantic relationships. Finding a romantic partner and establishing a romantic relationship are important adolescent developmental tasks. Related to these tasks are adolescents' developing sexuality and their construction of their sexual selves.

Adolescents appear to use electronic media to reinforce existing romantic relationships, just as they do friendships. According to a recent online survey by Teenage Research Unlimited, nearly a quarter of teens in a romantic relationship have communicated with a boyfriend or girlfriend hourly between midnight and 5 a.m. using a cell phone or texting. One in six communicated ten or more times an hour through the night.

Online communication forms that allow for anonymity offer adolescents a new avenue to practice partner selection. Using a sample of 12,000 utterances from adolescent chat rooms, researchers have found that the search for partners was ubiquitous online, with approximately two partner requests each minute.[8] Participants who declared they were older searched more actively for a partner and were also more likely to specify the gender of the partner they were seeking. Also participants who stated that they were females were more likely to make partner requests. The gender difference indicates that the online environment provides a safe space for females to initiate romantic relationships.

Research also suggests that anonymous online contexts provide a forum for sexual exploration, another major task of adolescence. Of the 12,000 utterances in the chat study just noted, 5 percent were sexual ones (about one sexual remark a minute). Participants who self-presented as older were more likely than younger ones to make explicit sexual utterances. Gender was also related to modes of sexual expression: utterances by users with masculine screen names were more sexually explicit; those by feminine screen names, more sexually implicit. Adolescents also use online bulletin boards to learn about sexuality. Finally, they use the Internet to engage in cybersex. In one study of 692 Czech secondary school students, 16 percent of twelve- to twenty-year-olds reported having tried virtual sex. A significant number reported having their first sexual experience online.[9] The study also found that 43 percent of the boys and

8 percent of the girls admitted to viewing pornographic materials. Although adolescents' exposure to online sexual content can be either intentional or unsolicited, more research is necessary to assess how this early exposure may affect sexual identity and intimacy during emerging adulthood. Studies have found that inadvertent exposure to sexual media in childhood and adolescence often has negative emotional effects, such as shock, disgust, or embarrassment, and that these effects can be enduring. Online forums may also provide sexual minority adolescents with a safe haven for sexual exploration without the prejudice and harassment that gay, lesbian, and bisexual adolescents sometimes face at the hands of peers and adults.

Much less is known about adolescents' use of electronic communication for romantic relationship formation. The 2001 Pew survey on teenagers and instant messaging reported that among teens who used instant messaging, 17 percent used it to ask someone out and 13 percent, to break up with someone.[10] One recent study of romantic relationships among college students explored the use of Facebook, a social networking site, among 1,440 first-year students at Michigan State University.[11] According to the study, the lowest-ranked use was finding casual sex partners; the next-lowest was finding people to date. The students may, however, have been using Facebook to check out people they had met as prospective dates. Uses such as checking out people they have met socially or in class or others who live in their dorm are all ranked relatively highly.

Relationships with Strangers and Acquaintances

Because online interactions lack important features of face-to-face communication, such as gestures and eye contact, they are believed to be less rich than offline ones. When the communication is with strangers or individuals not part of one's offline life, it is believed to represent weak ties, which have been characterized as relationships that have superficial and easily broken bonds, infrequent contact, and narrow focus. Questions about the relative richness of online communication have raised concerns about the extent of adolescents' online interactions with strangers and about the social impact of such weaker interactions and relationships.

Trends in Relationships With Strangers The potential for online stranger contact varies depending both on the particular technology used and the time period under consideration. In the earlier years of the Internet, when chat rooms were the rage, teens were more likely to be in contact with strangers; once instant messaging became popular, teens seemed to be using it to connect mostly with offline friends. With the advent of today's popular social networking sites, video and photo sharing sites, and blogs, adolescents may again connect and interact with people who are not a part of their offline lives....

Trends suggest that although adolescents may be using online communication forms as a way of extending their interaction with peers from their offline lives, the potential for interactions with strangers is high and therefore merits further exploration....

Do Online Relationships Move Offline?   Another question is whether relationships with strangers that begin online move offline. In a national survey of 1,501 youth, 256 respondents reported close online relationships and 41 percent of them reported face-to-face meetings with their online friend.[12] It appears that relationships move from online to offline only occasionally; however, given that the newer friendship forms of networking center on making "friends," this issue needs further exploration.

Who Forms Online Relationships With Strangers?   It is also important to consider the characteristics of adolescents who are more likely to interact with strangers and to form relationships with them. Such interactions can compromise the safety and well-being of the adolescent if the strangers are not peers but, rather, older, unscrupulous adults. Early research on this question found that more troubled adolescents were more likely to have formed close online relationships. Girls who had high levels of conflict with their parents and boys who had low levels of communication were more likely to have formed close relationships. Troubled adolescents have similarly been found to be more likely to visit chat rooms, where users usually encounter strangers rather than friends or family....

With the newer generation of online communication forms and the greater privacy controls they offer, youth now have the choice to interact online both with strangers and with people from their offline lives. Researchers have compared adolescents who primarily talk online with strangers and those who talk online both with strangers and with friends. Participants who communicated more frequently were less likely to communicate with a stranger, whereas those who communicated at more length were more likely to talk with strangers. Adolescents were also more likely to talk to strangers if they communicated online to meet people to assuage boredom and to compensate for their lack of social skills. Those who communicated online to maintain relationships were less likely to talk to strangers.

Benefits of Talking to Strangers   Online communication with strangers may offer some benefits for adolescents. One study using detailed daily diaries found that contact with unknown peers in forums such as chat rooms and social networking sites might help adolescents cope with threats to "belonging" in their offline lives.

Positive Content in Online Stranger Interaction   The Internet is filled with anonymous discussion groups and bulletin boards devoted to all kinds of topics of interest to youth, from music groups and bands, television shows, and fan fiction to sports, health, sexuality, and even college admissions.

One reason why teens might like to get their health-related information online is the anonymity of such communication. Young people may feel more comfortable asking strangers sensitive health-related questions than they would asking a parent or physician in person. Another advantage of online bulletin boards and discussion groups is their full-time availability. They also make it possible to get information passively (by looking at other people's questions and the responses they received) and to get advice and suggestions from far more sources than would be possible from one's circle of face-to-face friends.

The interpersonal connections with strangers made possible by electronic media may be particularly valuable for youth suffering from illnesses, such as AIDS, eating disorders, and self-injurious behavior, about which they may not feel comfortable talking with their friends in person. Online bulletin boards and chat rooms allow youth to form such connections. A study of the personal Web pages of adolescent cancer patients found that they often expressed a strong desire to help other young cancer patients through providing information, sharing personal experiences, and giving advice. The guest books found on most of the Web pages (which are analogous to electronic bulletin boards) indicated that the pages were producing cyber communities providing patient-to-patient support for cancer victims.[13]...

Negative content in online stranger interaction.   Although the anonymous and public natures of these online forums may provide benefits to youth, they may also disinhibit users and lead to negative content in their online interactions. Racial slurs and comments were much more common, for example, in unmonitored chat rooms frequented by older adolescents than in the monitored chat rooms frequented by younger adolescents.

The most dramatic instances of young people engaging in racist behavior online occur on hate sites targeted to children and teens. Websites, chat rooms, multi-user domains, discussion boards, music, audio- and videotapes, games, and literature are some of the most common tools used to disseminate online hate. Hate groups reach out to young people online by a number of means, including the creation of Web pages specifically geared to children and teens. Ideas may be worded to be more understandable to young people. The sites may even feature messages by youth directed *to* other youth.

Online stranger contact and sexual solicitation.   Online contact with strangers also puts adolescents at risk for sexual solicitation and sexual exploitation by predators, though such risks were far higher in the earlier days of the Internet before the widespread recognition of the potential dangers inherent to online stranger contact. Most online communication forms today have privacy controls that, if used, can greatly reduce the risks for sexual victimization.

Youth who engaged in a pattern of risky online behaviors in their interactions with strangers were more at risk for unwanted sexual solicitation or harassment. These behaviors included aggressive behavior in the form of rude or nasty comments, embarrassing others, meeting people in multiple ways (for example, on an online dating site or when instant messaging), and talking about sex with strangers. Youth who are victims of unwanted sexual solicitation also report emotional distress, depressive symptoms, and offline victimization.

Electronic Media and Family Relations

Two major questions on the topic of electronic media and family relations warrant further study. First, to what extent do youth use electronic media to communicate with their parents, siblings, and other family members? Second, how has adolescents' use of electronic communication affected their relations with their parents and other family members? Numerous media reports, as well as anecdotal observation, suggest that more and more parents are turning to text messaging and instant messaging to communicate with their adolescents; text messaging in particular can be very useful to parents trying to keep tabs on their teen.

Concern is growing that adolescents' extensive use of electronic communication to interact with their peers may impair their relations with their parents, siblings, and other family members. There is some evidence that electronic media may enhance peer relations at the expense of family, especially parent-child relations. An intense four-year video study of thirty dual-earner families with children provides a glimpse of the role of technology in modern family life.[14] When the working spouse, usually the father, came through the door at the end of the day, the other spouse and children were often so absorbed in what they were doing that they greeted him only about one-third of the time, usually with a perfunctory "hi." About half the time, children ignored him and continued multitasking and monitoring their various electronic gadgets. Parents had a hard time penetrating their children's world and often retreated. Electronic multitasking has become pervasive, sometimes at the expense of face-to-face family interaction, among siblings as well as with parents....

The role of cell phones in adolescent life and family relations is also worthy of attention. A series of focus groups with teenagers, young adults, and parents in Norway found that teens used the cell phone to establish generational boundaries (for example, screening calls from parents into voice mail) and also that cell phone use undermined family rituals, such as mealtimes and vacations.[15] Perhaps the most powerful way in which the mobile telephone undermined family interaction in favor of peer communication was through the individualization of communication. When peers called one another through a mobile telephone, they knew that they could talk directly with their friends, without any filtering or monitoring from parents or others in the household. In the words of the authors of the study, "Adolescents control the people with whom they talk and have more room into which they can share thoughts and messages that might not be [socially] acceptable. This plays on the peer group's ethos that their inner communications be shielded from non-members, and particularly parents." One of the authors found further qualitative evidence of such undermining in a focus group in which one participant told about a girl whose boyfriend had secretly given her a cell phone so she could stay in touch with him against her parents' wishes....

Have Social Relationships Been Altered by Electronic Media?

... Teens now conduct a higher *proportion* of their communication through writing in an electronic medium rather than face-to-face or voice-to-voice—in effect, relatively depersonalizing the process of interpersonal communication. It is also clear that electronic communication expands adolescent social networks. For example, for a teen to have 150–300 "friends" would have been unheard of before social networking. It is also evident that electronic communication brings together—for both good and ill—common-interest groups whose uniting characteristic, such as adolescent cancer or self-mutilation, may be rare in anyone's group of friends or family. The quasi-experimental monitoring studies in teen chat have also indicated that the anonymity of the Internet produces a disinhibiting effect on both sexual and racist behavior. The daughter of an *L.A. Times* reporter told her mother that MySpace had become necessary for her social life....[16]

ELECTRONIC MEDIA AND SCHOOLS

How have schools responded to the increasing presence of electronic media in the lives of today's youth? News reports suggest that some schools and school districts have responded by blocking the use of electronic media in schools, in particular text messaging, cell phones, iPods, and video games. Many school computer systems also block access to websites popular among teens such as those that provide access to instant messaging, e-mail, blogs, and social networking utilities. School authorities argue that these media are distracting, isolating, and disruptive and that they facilitate cheating (as when cell phone cameras are used to copy exams) and other illegal activity (as when cell phones and pagers are used in drug and gang activity)....

The ban against cell phones in high schools is perhaps the most controversial restriction. Parents and youth alike favor cell phones as invaluable tools for everyday planning and coordinating that can be critical in the event of emergency. But in a case brought by parents, the New York State Supreme Justice ruled in favor of New York City's ban on cell phones in the schools. Partial or complete cell phone bans have now been put in place in Toronto, Los Angeles, Detroit, and Milwaukee.[17] It remains unclear how effective such bans are in preventing the behaviors they are designed to target. Researchers need a better understanding of what teachers and school administrators know about adolescent use of electronic media and how such technologies might be integrated in school settings.

ELECTRONIC COMMUNICATION AND
IDENTITY DEVELOPMENT

According to Erik Erikson, the German developmental psychologist, establishing a coherent identity is the fundamental psychosocial task of adolescence.[18] Adolescents must establish a clear sense of who they are, what they believe in, and where they are headed. Early on, some observers saw the Internet, with its potential for anonymity and disembodied interaction, as a perfect venue for such identity exploration and experimentation. Online, it was thought, people could be whoever they chose to be and could slip in and out of various identities. But over time concerns were raised that such identity play may hinder, not help, adolescent development.

In fact, the evidence is mixed as to whether adolescents engage in extensive pretense and identity play online. In one study of twelve- to fifteen-year-olds, of the 175 participants who responded to questions about online pretense, 49 percent had never pretended to "not be yourself," and 41 percent reported pretending a couple of times.[19] Seven participants reported pretending often and two reported that they pretended all the time. Most common was pretending to be older, and was often done in the company of a friend and as a joke. Only 2 percent reported that they pretended to explore a new self or identity.

By contrast, in a study of Dutch adolescents, 246 out of a total of 600 participants reported having experimented online with their identity at least sometimes.[20] Pretending to be someone older was most commonly reported, especially among girls. The most common motives for identity experiments were self-exploration (to observe others' reaction), social compensation (to make up for shyness), and social facilitation (to form relationships). Taken together, the findings of both studies suggest that although youth do pretend to be someone else online, they do not do so frequently, and when they do, they may simply pretend to be older. Given that many online sites have age restrictions, it is quite possible that such pretense is not a true form of identity exploration but more a way to sidestep age-related restrictions. Although youth do not seem to be using electronic media to experiment with different roles and identities in the manner envisioned by Erikson, nonetheless these media afford them opportunities to explore as well as to practice self-disclosure and self-presentation, which are both important steps toward constructing a coherent identity. Anonymous forums such as chat rooms, in particular, enable such exploration and self-presentation.

CONCLUSION

Society's traditional adolescent issues—intimacy, sexuality, and identity—have all been transferred to and transformed by the electronic stage. Among the hallmarks of the transformation are greater teen autonomy, the decline of face-to-face communication, enhancement of peer group relations at the possible expense of family relations, and greater teen choice. Given the connectedness between the physical and virtual worlds, the challenge is to keep adolescents safe (both physically and psychologically) while at the same time allowing for the explorations and interactions that are crucial for healthy psychosocial development. This conflict is nicely illustrated by instant messaging, which helps teens stay in touch with friends, but is also widely used for electronic bullying. Meeting strangers on social networking sites such as MySpace offers another example. Although such virtual contacts can endanger adolescents, research has found that interactions with strangers may also help alleviate the negative effects of social rejection in the physical world. The benefits of exploring identity and intimacy online must also be weighed against the harmful effects of viewing sexual content and being bullied online. The thrust of the research at present suggests that real-world relationships and adolescent issues influence adolescents' electronic communication at least as much as electronic communication influences their real-world relationships and developmental outcomes.

NOTES

1. Gross, "Adolescent Internet Use" (see note 3).

2. Bonka S. Boneva and others, "Teenage Communication in the Instant Messaging Era," in *Information Technology at Home*, edited by Robert E. Kraut and others (Oxford University Press, 2006), pp. 612–72.

3. Rebecca E. Grinter and Margery A Eldridge, "Wan2tlk?: Everyday Text Messaging," *Proceedings of the SIGCHI Conference on Human Factors in Computing Systems* (New York: ACM Press, 2003), pp. 441–48.

4. Amanda Lenhart, Lee Rainie, and Oliver Lewis, "Teenage Life Online: The Rise of the Instant-Message Generation and the Internet's Impact on Friendships and Family Relationships" (Washington, D.C.: Pew Internet and American Life Project, 2001) (http://www.pewinternet.org/pdfs/PIP_Teens_Report.pdf/ [August 9, 2007]).

5. Adriana A. Manago and others, "Self-Presentation and Gender Differences on the MySpace Network," Department of Psychology, UCLA, 2007.

6. NCH, "Putting U in the Picture: Mobile Bullying Survey 2005" (http://www.nch.org.uk/uploads/documents/Mobile_bullying_%20report.pdf [August 9, 2007]).

7. Jaana Juvonen and Elisheva F. Gross, "Extending the School Grounds? Bullying Experiences in Cyber-space," University of California—Los Angeles, 2007.

8. David Šmahel and Kaveri Subrahmanyam, "Any Girls Want to Chat Press 911: Partner Selection in Monitored and Unmonitored Teen Chat Rooms," *CyberPsychology and Behavior* 10, no. 3 (2007): 346–53.

9. Zbynek Vybíral, David Smahel, and Radana Divínová "Growing Up in Virtual Reality: Adolescents and the Internet," in *Society, Reproduction, and Contemporary Challenges*, edited by Petr Mares (Brno: Barrister & Principal, 2004), pp. 169–88.

10. Lenhart, Rainie, and Lewis, "Teenage Life Online" (see note 4).

11. Lampe, Ellison, and Steinfeld, "A Face(book) in the Crowd: Social Searching vs. Social Browsing" (see note 1).

12. Wolak, Mitchell, and Finkelhor, "Close Online Relationships" (see note 45).

13. Lalita K. Suzuki and Ivan I. Beale, "Personal Home Web Pages of Adolescents with Cancer: Self-Presentation, Information Dissemination, and Interpersonal Connection," *Journal of Oncology Nursing* 23, no. 3 (2006): 152–61.

14. Elinor Ochs and others, "Video Ethnography and Ethnoarcheological Tracking," University of California—Los Angeles, 2007.

15. Rich Ling and Brigitte Yttri, "Control, Emancipation, and Status: The Mobile Telephone in Teens' Parental and Peer Relationships," in *Computers, Phones, and the*

Internet: Domesticating Information Technology, edited by Robert Kraut, Malcolm Brynin, and Sara Kiesler (Oxford University Press, 2006), pp. 219–34.

16. Catherine Saillant, "Testing the Bounds of MySpace," *Los Angeles Times*, April 8, 2006, p. 1 ff.

17. City News, "Cellphone Use Banned in all Toronto Public Schools" (www. citynews.ca/news/news_9977.aspx [August 3, 2007]).

18. Erik Erikson, *Identity and the Life Cycle* (New York: W. W. Norton, 1959).

19. Gross, "Adolescent Internet Use" (see note 3).

20. Patti M. Valkenburg, Alexander Schouten, and Jochen Peter, "Adolescents' Identity Experiments on the Internet," *New Media & Society* 7, no. 3 (2005): 383–402.

40

The Realness of Cybercheating

MONICA T. WHITTY

We noted in the last chapter that college students have grown up in an Internet age. They relate through chat rooms, listservs, bulletin boards, and a multitude of networking opportunities. Online interaction differs from FTF interaction, offering people the freedom to represent themselves in ways previously unavailable. The anonymity of Internet interactions combines with their relatively intense and focused nature to foster bond formation of a greater speed and intensity than that found in the offline world. Norms governing Internet interactions and relationships are still evolving, and vary greatly from one context to another and between individuals. The extent to which people invest themselves in these cyber relationships varies enormously as well. Whitty explores a fascinating dimension of this arena in her chapter that uses a hypothetical scenario to evoke people's feelings about the seriousness of Internet relationships. To what extent do you agree with the boys' versus the girls' responses to the questions she asks? Why do you think they have such marked differences? What do these reveal about the underlying feelings people have about what "makes" a relationship real? How are these scenarios and people's responses to them affected by the tendency to misrepresent or "bend" their self-presentations? How seriously should we take Internet self-presentations? Is this an arena in which people should be free to just have fun or is this serious business? One of the patterns we have witnessed by adults, even politicians, is an increase in cyberflirting, cybercheating, and sexting. It is estimated that anywhere from 20% to 50% of adult men cheat and 15% of them have online sex. Why do you think this has grown so pervasively? Does opportunity fuel behavior or is there something more going on in our sexual norms and

SOURCE: Monica T. Whitty, "The Realness of Cybercheating: Men's and Women's Representations of Unfaithful Internet Relationships," *Social Science Computer Review,* 23(1), pp. 57–67. Copyright © 2005 SAGE Publications. Reprinted by permission of SAGE Publications, Inc.

values? Should significant others treat such behaviors as serious violations? How does cybercheating compare to hooking up in real life? If someone sent you a flirtatious cyber message, might you be inclined to respond in kind?

Currently, little is known about whether certain interactions that occur online are perceived by some as a threat to an offline romantic relationship. Shaw (1997) suggested that "Internet infidelity is, of course, behaviorally different from other kinds of infidelity; however, the contributing factors and results are similar when we consider how it affects the way partners relate" (p. 29). She did not, however, qualify how online and offline infidelities are behaviorally different. To investigate this question further, Whitty (2003a) surveyed people about their attitudes toward offline and Internet infidelity. Her study considered acts such as sexual intercourse, cybersex (describing the sexual act while typically masturbating), hot chatting (a type of erotic talk that moves beyond light-hearted flirting), emotional disclosure, and various types of pornography online and offline. It is interesting to note, the research revealed that individuals do believe that some interactions that occur online are acts of betrayal. Some of these behaviors, such as cybersex, posed a greater threat than other behaviors, such as downloading pornography. Of further importance, the study (Whitty, 2003a) found that there are separate components of infidelity that we need to consider, including sexual infidelity, emotional infidelity, and pornography. This is consistent with previous research on offline infidelity that has purported that infidelity should not be reduced to simply sexual infidelity, but that mental exclusivity is also an important component of fidelity. However, what is unique to Whitty's (2003a) study is that the factor analysis she performed revealed that online acts of betrayal do not fall into a discrete category of their own. For example, sexual intercourse, hot chatting, and cybersex all combined to make one factor. Therefore, we might conclude from such a study that people hold similar attitudes toward online and offline infidelities.

The findings from Whitty's (2003a) study challenge the notion that acts that occur in cyberspace cannot have a so-called real impact on an individual's life. Perhaps this is because although there are no physical bodies present online, this, in turn, does not mean that the action is unreal. Instead, as Whitty (2003a) and Whitty and Carr (2003) argued, Internet relationships are better understood if we focus on the reconstruction of the body online, which is imperative to the success of many online interpersonal interactions. There are a few more reasons why acts, such as cybersex and hot chatting might be considered as acts of betrayal. For instance, Yarab and Allgeier (1998) claimed that when considering sexual fantasies, the greater the threat of the sexual fantasy to the relationship, the more likely the fantasy is considered to be unfaithful. It is probable that participants perceived that sexual acts such as cybersex and hot chatting were more of a threat than pornography because individuals were more likely to meet face-to-face with their cyberloves than they are with porn stars. A further explanation might be that although many sexual encounters do initiate online, it has been suggested that a large proportion of individuals continue these relationships offline. Hence, it is likely that some individuals see their partners' erotic interactions with another on the Internet as a so-called real threat to their relationship.

This current research is another step forward in trying to understand how people might experience Internet infidelity. Rather than ask participants directly about what they believed were acts of Internet betrayal, the current study employed a qualitative method to investigate people's representations of Internet infidelity. Drawing from Kitzinger and Powell's (1995) story completion method, this article presents data generated in response to a cue relating to Internet infidelity....

METHOD

Participants were given one of two versions of a story completion task based on the task devised by Kitzinger and Powell (1995).... In Kitzinger and Powell's (1995) study, participants were asked to write a story to a cue story exercise that stated the following:

VERSION A: "John and Claire have been going out for over a year. Then John realizes that Claire is seeing someone else."

VERSION B: "Claire and John have been going out for a year. Then Claire realizes that John is seeing someone else."

In the current study, the instructions were slightly changed to read:

VERSION A: "Mark and Jennifer have been going out for over a year. Then Mark realizes that Jennifer has developed a relationship with someone else over the Internet."

VERSION B: "Jennifer and Mark have been going out for over a year. Then Jennifer realizes that Mark has developed a relationship with someone else over the Internet."

The term *seeing* was substituted for *relationship* as people do not typically refer to any forms of relationships on the Internet as "seeing" someone. The term *Internet* was explicitly used so that only Internet relationships could be explored. Writing to a cue story using the third person was maintained for a couple of reasons. It has been argued that using the third person allows one to reveal more socially undesirable information than using first-person cues, and it allows individuals to distance themselves so as to not warrant or justify their own behavior and motivations.

For the current study, 3rd-year students who were enrolled in a psychology subject at the University of Western Sydney, Australia, were invited to participate in the study. As pointed out by Kitzinger and Powell (1995) university students are particularly appropriate for research of this kind, as they are fairly literate and reasonably fluent writers, who are accustomed to requests to express their ideas in writing. The current study was passed by the university ethics committee, students were assured anonymity and were not given credit for participating, nor were they penalized for not participating in the study. Nonetheless, 100% of students agreed to participate....

RESULTS

Although Kitzinger and Powell (1995) found that 90% of their sample interpreted their cue story, which was developed in respect to offline infidelity, to be an act of sexual involvement, this was not the case in the current study. Instead, the stories produced for the current study painted a more complex picture. Although all of the participants understood this to be a dilemma about infidelity, some were divided as to whether the betrayer believed he or she was committing an act of infidelity, whereas others wrote that the partner was not certain that he or she had been betrayed. Moreover, unlike Kitzinger and Powell's study, when participants interpreted the cue story as a story about sexual involvement, this was not necessarily about sexual relationships but, in many cases, was exclusively an emotional involvement.

Of the sample, 51% wrote that the betrayer believed that he or she had been unfaithful, 27% wrote that the betrayer believed he or she had not been unfaithful, whereas 22% either did not represent the betrayer's point of view or were unclear. In contrast, 84% of the sample wrote that the partner felt that he or she had been betrayed, 9% wrote that the partner had not, and 7% did not represent their perspective or were unclear.

There were several reasons given in the stories for why the perpetrator or the aggrieved did not consider the Internet interaction as an act of infidelity....

The most common explanation given for why the scenario should not be considered an act of infidelity was that the interaction was "just a friendship," as illustrated in the following extracts:

> Jennifer explains to Mark that the relationship is not romantic and that
> they are only friends. (24 female Jenny)
> He did not think he had done anything wrong, after all this girl was
> only a friend who existed in text. (38 female Mark)

Another explanation was that the interaction was "merely flirtation or just a bit of fun," as illustrated below:

> Jennifer, on the other hand, thinks she's doing nothing wrong. So what
> if she flirts a bit with someone who lives far away. (13 female Jenny)
> Mark, at first, brushes it off thinking that it's "only the Internet, no
> harm in having fun." (12 male Mark)

It is interesting to note, there were others who pointed out that this was not infidelity, as the relationship was with an object (computer) in virtual space, rather than with a real human being.

> She tried to explain that he was just a faithful companion and the only
> feelings she had were not real as this man was just words on a screen.
> (55 female Jenny)
> When she confronts him about it one night over dinner, he denies
> everything saying that they were just friends. And that she should not
> take it so seriously and worry about it because it was not a real

relationship, but a net relationship. That net relationships mean
nothing because everyone lives in a virtual reality. (6 female Mark)

There were others who emphasized that the interaction could not be considered an act of betrayal as the two had never met, nor did they intend to meet.
Moreover, often the stories had the two cyberlovers interacting from different
countries, making it unlikely that they could potentially ever meet.

Jennifer retaliates and says, "How can I be having an affair, without
even meeting this guy." (19 female Jenny)

Mark tells her to calm down, and says that although he chats to her
regularly he has never offered to meet her, and she hasn't suggested it
either. (65 female Mark)

There was a significant result obtained for the explanation that "it cannot be
infidelity if there is not any physical sex occurring." Men and women were more
likely to write this when Mark was the betrayer.

"No I'm not cheating. It's not like I'm bonking her anyway."
(51 female Mark)

Although Mark believes that because there was no physical contact, he
has not cheated, Jen disagrees. (14 male Mark)

However, Mark said that it is not cheating at all. He said he just enjoyed
an imaginative relationship that is only made through computers. He
said he never met the girl he had been seeing on the net and his point
is that he doesn't think it is cheating unless he has a sexual relationship with someone else. (2 female Mark)

Finally, there were three participants who explained that this was not an act
of infidelity, but more sexual experimentation, as the cyber affair was between
two people of the same sex.

Finally he asks, "So how serious is it?" Jennifer replies, "Well it's just
really a bit of fun, you have nothing to worry about!" In fact, she
goes on to say, "It's actually a female!!" ARRGHH Shock Horror!!
(38 female Jenny)

Although some of the stories (as demonstrated above) focused on explaining
away why the online act was not an act of betrayal, many more took it for granted
that this was a scenario about infidelity. Quite a number of the stories provided
reasons for why this was an act of infidelity, either by admissions of guilt from
the perpetrator or as justifications for why the aggrieved felt they had been
betrayed. It should also be noted that sometimes the perpetrator, the aggrieved,
or both were uncertain as to whether this was actually a form of cheating and
discuss with each other why they think this might be an act of betrayal.

The most frequently stated reason for why this was an act of infidelity was
that participants reasoned that one should not have a romantic relationship with
more than one person. Sometimes this was because the online relationship was
considered to be as real as the offline relationship, or that it had the same effect as

being in another face-to-face relationship, or because the individual intended to meet up with the cyberlover, as demonstrated in the exemplars below:

> Jennifer can't live with Mark's betrayal and gives him an ultimatum—she is not prepared to "share" him with someone else. (54 female Mark)

One of the more interesting results obtained in the analysis was that emotional infidelity was stressed as much as sexual infidelity. Consistent with previous research on offline infidelity, women stressed the problems they had with emotional infidelity more than the men did. Emotional infidelity is perhaps best illustrated in the following extract:

> "It is cheating," she said rather calmly.
> "No I'm not cheating. It's not like I'm bonking her anyway. You're the one I'm with and like I said I have NO intentions of meeting her." He hopped into bed.
> "It's 'emotional' cheating," she said getting annoyed. "How so?" he asked, amusement showing in his eyes.
> "Cheating isn't necessarily physical. That's one side of it." He pulled the sheets over him and rolled over.
> "Well … I know you have not met her yet that's why, but I'm still a little annoyed, Mark." She sat on the edge of the bed.
> "Don't be mad. You're the one I love. So *how is it* emotional cheating." He sat up.
> "You're keeping stuff from me. Relationships are about trust! How can I trust you if you keep stuff from me about the 'Internet girl?" (51 female Mark)

When participants discussed sexual infidelity, sometimes they referred to cybersex, and on other occasions they wrote about flirting online or hot chatting. Rarely did they refer to these participants having offline sex with their cyberlovers. Examples of erotic encounters online that were believed to be acts of betrayal are provided below:

> He sneaks behind her and sees that his girlfriend is, in fact, flirting with a man by the name of Buzzy. He screams at her and tells her that their relationship is over. She begs him to stay, but he tells her that all this time he feels like she has been cheating on him and she can never gain his trust again. (50 female Jenny)
>
> Mark's obsession with the Internet is sure to cause a break-up in the relationship. Jennifer discovered, late one night after Mark had left his computer that he was partaking in "cybersex" with a woman by the name of "Buxom Blonde Bombshell." Jennifer confronted Mark who admitted that he was having an online relationship with this woman and that she was satisfying his sexual fantasies. This caused a relationship breakdown between Jennifer and Mark. (4 female Mark)

A further explanation given for why this cyberact must be an act of betrayal is because it was kept a secret. Had the perpetrator been an innocent player he or she would not have concealed the Internet activities from the partner, as shown in the extracts below:

> Mark follows this statement with why was he not informed of this relationship and had to find out for himself. (33 male Jenny)

Stories were also analyzed considering the kind of impact the cybercheating had on the offline relationship. Of the stories, 65% mentioned that the aggrieved had indeed been hurt or upset by this virtual encounter. The Internet infidelity did have a real impact on the aggrieved offline relationship, including in many cases leading to a break-up of the relationship.

The most frequently reported impact on the offline relationship was that the aggrieved felt upset by the incident. Women were significantly more likely than men to write about the aggrieved being upset. Sometimes the aggrieved expressed deep hurt, whereas others were considerably angry about the betrayal, as demonstrated below:

> Mark is shocked, upset, and hurt. He feels betrayed by Jennifer that she does not view him as being important enough to confide in him. Mark's hurt quickly turns to anger. He becomes defensive to cover his hurt. He doesn't understand why he feels this way. (22 female Jenny)
> Mark is not happy with Jen. He says, "Not happy Jen." (11 male Jenny)

Of the sample, 46% wrote that the offline couple broke up as a consequence of this Internet affair. In some cases, the offline relationship was already on rocky grounds and the Internet affair was sought out because of the dissatisfaction with the relationship, and in other instances it was simply the cyber affair that was the cause of the break-up. Women wrote about the couple's breaking up more than the men did in their stories.

> Next day, Jennifer told him everything about the affair. She thought that coming clean will be the best option, but she was wrong. Mark just asked her to leave the house and never come back. She couldn't do anything and was unable to say anything. She packs her bags and left. (3 female Jenny)
> She breaks up with him, giving him no explanation, stalks him until he develops signs of paranoia, then marries his best friend, who is a doctor and earns easily $500,000 p.a. at minimum, and looks like a god. (24 female Mark)

Of the sample, 42% wrote that the cybercheating leads to a loss of trust in the offline relationship. Women wrote this more than men, and individuals wrote this more when Mark was the perpetrator.

> From that day forth she began to question who was on the phone, who he stays out for drinks with, and who the girls are he works with. Jennifer tried to guess his e-mail password and checked his voice-mail messages. Eventually it was mistrust, the belief of deceit, and the obsession that ended the relationship. (43 female Mark)

There were some interesting and often rather cruel ways that the aggrieved sought revenge. On occasion, this involved logging on to pretend to be their partner to destroy the cyber relationship, sometimes this involved getting even by having their own affairs, whereas others wrote about psychologically or physically harming their partner or the cyberlover.

> Jennifer then decides to play a little game … through a little bit of deception and assumed identity, she manages to assume the identity of Mark's lover when he is chatting and assumes the identity of Mark when his lover is online. Jennifer then convinces them to meet each other, assuming they both love each other. Mark and his Internet lover agree…. When Jennifer met the Internet lover, she bludgeoned her to death with a keyboard, shoved a mouse up her arse, and then replaced her head with a monitor. Transporting her body to the meeting with Mark was next. Mark walks in to find the defaced body. Due to his shock, Jennifer was able to capture him. Keeping him as a human punching bag, whenever she returned home after a bad day she would kick the shit out of him. She would never forget or allow herself to be betrayed, and Mark had no choice but to remember what he had done. (38 male Mark)

Other effects on the offline relationship included the following: the perpetrator feeling upset by the affair and the impact it had on his or her partner, the cyber affair meant that there was less time spent with his or her offline partner, the aggrieved was shocked when he or she learned of the affair, the aggrieved felt sexually inadequate after learning that his or her partner would prefer to have virtual sex with a stranger rather than him or her, and the aggrieved felt a loss of self-esteem after learning about the affair. Women wrote more than men did that the cyber affair led to less time spent with the aggrieved.

DISCUSSION

The results of the current study suggest that when individuals are presented with a hypothetical scenario of cybercheating they do, in the main, consider this to be a real form of betrayal that can have just as serious an impact on a relationship as offline betrayal. Similar to Kitzinger and Powell's (1995) study and other studies on offline infidelity, the participants wrote that the aggrieved expressed upset and anger over the affair. In addition, akin to Kitzinger and Powell's study were the revenge stories that were elicited. As with the previous researchers work, the participants here wrote of seeking revenge by having their own affairs or by hurting, even murdering, the perpetrator or the lover. Moreover, trust was broken as a consequence of the affair. However, what is perhaps the most significant indication that this was a real affair was the amount of participants who wrote that the couple broke up as a result of the affair.

Although the results from the current study suggest that cybercheating can have a real impact on a relationship, it also needs to be recognized that some of the participants were not convinced that forming a virtual relationship is a real form of betrayal. Those who were not convinced stressed that the virtual relationship was not a threat as it was only fun and flirtatious and that the relationship could only be understood as a friendship. Some participants even went as far as saying that this is a relationship with an object rather than a person, whereas others wrote that if there was no intention of face-to-face contact then it cannot be real betrayal. Perhaps this can be explained by Yarab and Allgeier's (1998) research, which found the greater the threat of a sexual fantasy the more likely the fantasy is considered to be an act of betrayal. If the relationship is depersonalized as not being with a real person or if the person never intends to encounter this individual face-to-face then this could be perceived as a harmless sexual fantasy.

An important finding was the equal weighting given to emotional and sexual betrayal. Unlike previous studies on offline infidelity and in contrast to Whitty's (2003b) study on Internet infidelity, the participants in the current study did not consider sexual infidelity as having a more serious impact than emotional infidelity. Such a result suggests that cyber affairs could create problems for an offline relationship for very different reasons than an offline affair might. This result has some important therapeutic implications and warrants more attention in future studies....

REFERENCES

Kitzinger, C., and Powell, D. (1995). Engendering infidelity: Essentialist and social constructionist readings of a story completion task. *Feminism and Psychology, 5*(3), 345–372.

Shaw, J. (1997). Treatment rationale for Internet infidelity. *Journal of Sex Education and Therapy, 22*(1), 29–34.

Whitty, M. T. (2003a). Cyber-flirting: Playing at love on the Internet. *Theory and Psychology, 13*(3), 339–357.

Whitty, M. T. (2003b). Pushing the wrong buttons: Men's and women's attitudes toward online and offline infidelity. *Cyber Psychology and Behavior, 6*(6), 569–579.

Whitty, M. T., and Carr, A. N. (2003). Cyberspace as potential space: Considering the web as a playground to cyber flirt. *Human Relations, 56*(7), 869–891.

Yarab, P. E., and Allgeier, E. (1998). Don't even think about it: The role of sexual fantasies as perceived unfaithfulness in heterosexual dating relationships. *Journal of Sex Education and Therapy, 23*(3), 246–254.

41

Indian Call Centers

PHIL TAYLOR AND PETER BAIN

The American economy, and that of other developed nations, has been seriously
hurt by the movement of manufacturing jobs to countries where labor and other
costs are cheaper, as noted in Chapter 36 by Legerski and Cornwall on industrialized job loss. Blue collar workers, often in inner cities but also in suburbs,
have suffered as their traditional lines of work, followed generation after generation, have evaporated. Cites, states, and regions have suffered economic blight as
one industry after another has moved offshore. Large domestic and multinational
corporations have sought to enhance their profits by squeezing their expenses,
busting unions, and relocating their headquarters to other countries in an effort to
avoid taxes. This trend pushed middle- and lower-middle class Americans
downward in the structure of society, thrust many into unemployment, and
tempted youth to increasingly enter the illicit economy. Developed countries found
themselves increasingly transitioning from a manufacturing to a service economy.
While this trend has become widely recognized, we have more recently witnessed
the movement of service jobs overseas as well. Technology now permits people
to drive up to a McDonald's and place their order to an individual located in
another country, another continent. More service jobs are becoming mechanized,
with machines replacing human beings. Service labor, an industry once thought to
be secure, now faces the prospect of eroding, with potentially dire consequences for
the economies of developed nations. Taylor and Bain focus on one of the most
common and growing forms of outsourced service: call centers. Indeed, resentment
has brewed for several years over people calling familiar companies and finding
themselves talking to people abroad. In this chapter, Taylor and Bain discuss
some of the financial logic that drives companies to move certain divisions to
developing countries, yet they face unanticipated problems there. At the same

SOURCE: Phil Taylor and Peter Bain, "'India Calling to the Far Away Towns': The Call Centre
Labour Process and Globalization," *Work, Employment and Society*, 19(2), 2005. Copyright © 2005 BSA
Publications Ltd. Reprinted by permission of SAGE Publications Ltd.

time, workers taking these jobs labor under difficult conditions that put their social class, educational attainment, and job satisfaction at odds. How do you feel about this trend? What should be done about it? Are we better off over the long term controlling our markets and jobs or opening them to global competition and adapting to face this? How will the global nature of travel, communication, labor, and manufacturing affect your generation more than any other?

In recent years the overseas migration of call centre services has generated widespread media coverage, as a string of prominent UK-based companies (e.g. Prudential, Norwich Union, British Telecom, Lloyds-TSB and Abbey National) have relocated operations to India. To many, this process of offshoring, to use the term that increasingly entered business parlance in the United States in the 1990s, will ravage call centre employment in the developed countries. For example, one UK MP predicted that "there would not be a call centre in Scotland in five years" (Brian Donoghue, *Herald* 14 March 2003). This common-sense perception that massive cost savings will be delivered in India, that call centres can be readily relocated and that offshoring will devastate the service sector, just as manufacturing disappeared *en masse* to the Far East, resonates with the arguments of the "hyperglobalists." The essence of this perspective, that the contemporary "global economy is genuinely borderless" and "that information, capital and innovation flow all over the world at top speed" in search of lower costs, seems to apply to call centre offshoring. After all, the information and communication technologies, integral to the call centre's functioning, can connect customers in real time to remote servicing centres. Surely call centres exemplify Castells' informationalism (2000), which, in the context of a networked, globalized economy, replaces the "space of places" with the "space of flows," and represents the "death of distance."

A vital component of this paradigm of inexorable offshoring is the substitutability of labour in the so-called developing world for that in the advanced industrial countries. For UK offshorers the ease with which work organization can be transplanted to India, and the particular qualities of that country's labour force, are seen as central to the success of remote servicing. A succinct illustration of the widely-held conviction that call centre migration is a seamless process is to be found in the words of the CEO of the Hong Kong and Shanghai Bank (HSBC), who lauded the virtues of Indian workers, at the moment when his company embarked upon its first major wave of offshoring.

> They're quicker at answering the phones, highly numerate and keen to come to work every day.... Staff are hugely enthusiastic. A lot have degrees. I wouldn't say in all cases the performance of the UK is inferior. However, the quality of work overseas is exceptionally high.
> (Whitson, *Financial Times,* 10 August 2002)

This article challenges the assumption that the offshoring of voice services from the UK is unproblematic and questions its theoretical underpinning, principally through an investigation of the Indian call centre labour process....

METHODS AND SOURCES

Indian evidence comes from several sources. In January–February and June 2003 the authors undertook fieldwork in Delhi, Mumbai and Bangalore. Access to seven call centres was facilitated by the National Association of Software and Service Companies (NASSCOM). All call centres were third-party operations, but varied in company size (from 50–2000+ employees), establishment size (50–400 employees) and in the sectors and markets served. Potential case study organizations derived from our request to Nasscom that they should be representative of Indian call centre operations. Prior to final selection we evaluated company and establishment profiles as presented in Nasscom's (2002b) Directory.

Extended semi-structured interviews with senior management were based upon schedules of questions covering necessary contextual factors including company history and structure, infrastructure and connectivity, the relationship between call centre and back-office work, and the factors driving, facilitating and inhibiting company and industry growth. However, our central concerns were with the nature of work organization, the labour process and workforce management. Key questions focused on recruitment, selection and retention, training practices, linguistic and cultural capability, workflow complexity, working time and its effects, and the consequences of growth. While we were given tours of all the facilities and permitted to observe training sessions in one case, direct access to interview workers on-site was denied for reasons of corporate confidentiality. This deficiency was partly overcome by group discussions, serendipitously organized and conducted outside the workplace, with Mumbai call centre workers employed in two case study companies....

INDIAN WORKFLOWS

Primary and secondary data confirms that work organization in India approximates to, and may even constitute an extreme form of mass production. The types of calls handled tend to be highly standardized, simple in content, tightly scripted and of short duration. At Dell work largely consisted of rigidly structured credit card collections calls lasting three to four minutes. Four of the five Mumbai centres visited undertook basic transactional services for US financial sector clients. At Mum4, typical calls lasted 30–180 seconds. Agents, engaged in debt recovery, collections, outbound telesales, telemarketing and credit card campaigns, confirmed managers' accounts (Group discussion, 2 February 2003).

Nasscom recommends (2002b: 64) that third-party suppliers should bid for low-value, low-skill work. If they are successful in securing contracts, and then delivering on cost and to prescribed quality standards, operators hope to persuade clients to offshore more valuable, complex work. "Moving up the value chain" is the widely-used phrase which captures the essence of this strategy, whereby suppliers aim to capture "domain expertise," either deep knowledge of a specific industry, or greater expertise of a generic process. However, "moving up the value chain" remains largely an aspiration....

The practices of Scottish/UK companies confirm these assessments. Despite some variety, it is possible to identify the call types offshored; overflow, out-of-hours calls, basic customer services (e.g. change of address), routine insurance claims, directory enquiries, ticket fulfilment, debt recovery/reminders/collections, telemarketing/telesales, technical support/IT helpdesk. With the exception of the latter more complex tasks, the majority consist of standardized, low-value processes. An interesting statistical finding is that offshoring companies operate call centres whose mean workforce size (437) is more than double the industry average. It is not size per se that is significant, but the fact that larger call centres process higher volumes of standardized calls that lend themselves to both outsourcing and offshoring. Thus, most susceptible to migration are high-volume, low-value, routinized, short cycle-time workflows, mainly in financial services, in large workplaces, by companies which may have recently merged or undertaken significant rationalization....

WORK ORGANIZATION AND WORKER EXPERIENCES

The technological aspects of work organization in India are essentially the same as those in the advanced countries. Thus, ACD systems distribute calls, electronic wallboards display queue numbers and waiting times, and agents in Mumbai receive calls through headphones plugged in to the same turrets as found on workstations in Melbourne or Manchester. Agents navigate familiar screens and menus, entering data on globally-branded computers. Formally, the technical architecture of the labour process is similar, if not identical. Further, the forms and techniques of control manifest in call centres in developed countries, the subject of much research activity and debate, are also found in Indian establishments, although it is necessary to distinguish both subtle and substantial differences in their application and effects. Generally, these are the result either of culturally-specific distinctions and/or the functional relationship between the offshored process and its company and country of origin.

Agents' performance is extensively monitored in these highly routinized workflows. As with UK outsourcers, Service Level Agreements (SLAs), based on contracts between third-party providers and clients, prescribe quantitative measures (including call volumes, abandonment rates, call-handling times) and specify qualitative criteria. Both forms are strictly imposed in India; for example, contractual requirements dictate that agents adhere rigidly to scripts. In all case studies SLAs were translated downwards into agents' targets. One manager explained how his centre was more stringent than others.

> ... because we not only have targets, we grade agents into three categories: A—Excellent, B—Satisfactory and C—Not meeting standards.

With Personal Improvement Plans, the Cs are taken out, and if they do
not become As or Bs after a short time, then unfortunately they have to
go. (Mum4)

Diverse evidence suggests that Indian agents experience this work as demanding,
pressurized and frequently stressful. Nasscom, though committed to boosting the
industry, has even referred to "high-volume induced burnout and the tedium of
some of the tasks" (2002a: 52). High turnover is perhaps the single most serious
human resource problem and, while patently the product of rapid sectoral
expansion, is also the consequence of working conditions that many find intol-
erable. For "despite rather lucrative pay packets" (Nasscom, 2002a: 26) and limited alternative employment opportunities, 50 percent of those who exit a call
centre leave the industry altogether....

Given our previous discussion of call centre work outcomes, where call
throughput is prioritized and strict monitoring prevails, it is unsurprising that, in
India, where standardization occurs *in extremis,* employee exit and burnout
appear to be commonplace. However, distinctive characteristics of the Indian
industry exacerbate these generic problems. First, call-handling for overseas customers takes place at night, or during evenings, on shifts that can last more than
eight hours. In Datta's cases 9-hour shifts were the norm, but often were
extended "by another 3 to 4 hours for a 'reward' of $6" (2004: 7–8), while
Indians employed on British Telecom's contracts work two overlapping nine-
hour shifts between 2.30 pm and 2.30 am (Deb, 2004: 17). "Graveyard" shifts
commencing at one, two or three o'clock in the morning are common (Mum1;
Mum5). One experienced manager understood that "taking 150 different calls a
night is very difficult and there is high burnout," responsible even in "good"
centres for 30–35 percent attrition (Mum2). Several managers identified the *combination* of night-time working and repetitive call-handling as stressful, taking its
toll on agents' health and social and family life (Mum1; Mum2; Mum5). Long
commuting distances and travelling times compound these negative effects. Deb
(2004: 20) reports his, and fellow-workers', experiences of inhabiting "a self-
contained world of 13-hour days, taking into account the travelling time of up
to two hours each way, with space for little else to penetrate one's existence."
Some managers well understood the deleterious consequences: "A worker leaves
home at 2 pm, starts work at 5.30 pm and after work reaches home at 5 am. This
is not a good thing, but a sure recipe for health problems" (Mum3).

Agents confirm the prevalence of ill-health, with one typically commenting,
that "many cases have arisen where people have lost weight ... usually because
the cycle of eating and sleeping is disrupted. Symptoms include dark circles
under the eyes ... in most of the people" (Discussion, 2 February 2003). International research has long established the relationship between night-shift working and ill-health, but when combined with repetitive call-handling and
negligible control, it makes Indian agents' experiences even more pressurized
than their UK counterparts. Thirty-three percent of Datta's (2004: 13–14)
respondents reported difficulties in getting "normal sleep," a problem intimately
related to working long atypical shifts involving little task variation and

autonomy. At a session at Nasscom's ITES-BPO Conference devoted to HR issues, participants expressed concerns at the effects of extended unsocial working on health and work-life balance. The General Manager of Hughes BPO was emphatic in his identification of the problem.

> Working the night shift … is not something that you can willingly adjust to easily. It does lead to a high level of stress.… It isolates you from your family, friends, social life. When there are birthday celebrations in the evening, you find you can't go because you have to work. This is a very big area of concern, and why you see a lot of people getting out of the industry. Stress is accentuated by bad eating habits—eating food at all hours. At least one in two people is a smoker, which is not healthy.
> I think, as this industry grows, we need to ask questions, "what are the effects of this working routine at a physiological level, psychological level, social level, family level?" (Goyal speech, 13 June 2003)

Women, who constitute half the workforce, are particularly affected by conflicts between working hours and the demands of task performance, and domestic, family and social responsibilities and expectations.

Second, although the built environment bears favourable comparison with UK facilities, the fact that buildings are sealed and that temperature, air conditioning and humidity are controlled by Building Management Systems (BMS), means occupants are likely to experience problems. Lacking fresh air and opening windows, these artificially created micro-climates have been linked by the World Health Organization to "sick building syndrome" (Health and Safety Executive, 1995). Symptoms could be exacerbated by temperature and humidity extremes during the monsoon period, which BMS may be unable to counteract, particularly in circumstances of 24-hour building occupancy (Nasscom, 2003: 63). Research (Baldry et al., 1997) has shown that widespread ill-health results from routinized labour processes taking place within unsupportive ambient and ergonomic environments.

Third, there is evidence that customary Indian hierarchical workplace cultures have been transposed to call centres (Outsourcing Insight, 2001: 114). One associate in Datta's study revealed how when call volumes are high, "one has to raise a blue flag to attract the attention of the team leader for breaks and sometimes one cannot go to the loo" (2004: 7). Certainly, top-down methods dominate companies' staff communication systems (Nasscom, 2003: 149). Despite strong union traditions in telecoms, banking and insurance, there is neither recognition, nor even embryonic union organization in Indian call centres. This absence of employee voice constitutes a democratic deficit, which contrasts with widespread, if uneven, union recognition in the UK, and denies Indian workers opportunities to channel grievances and improve working conditions.

Fourth, the deleterious effects of task performance are aggravated by frustration, role confusion, a crisis of expectations and psychological tension experienced by Indian agents as they negotiate the contradictions between their culture, identity and aspirations, and the requirements of service provision for western customers. The widespread adoption of anglicized pseudonyms, of having to conceal their Indian locations, and the obligation to speak in "neutral" accents, or even emulate their customers' dialects, contribute greatly to a pressurized working experience. One manager (Mum2) was openly critical of these normative practices, which, originating in GE's call centres, were spread throughout the expanding industry through a managerial diaspora.

> If agents are taking 150 calls a night from the east coast, from the west coast [of the US], it puts even more pressure on them if they have to change from call to call to speak in the accent the customer uses. Customers can see through false accents.

Undoubtedly, the compulsion to conceal location—locational masking to use Mirchandani's term (2003: 16)—can increase tensions during encounters with customers. What motivates this deception is the assumption, shared by Indian providers and offshoring companies alike, that providing services from India produces "huge consumer opposition" (Mitial, 2002: 43–4). Agents report how perceptions that Indian call centres give inferior service can lead to customer frustration, and even aggression, directed towards them as front-line staff (Datta, 2004; Group discussion, 2 February 2003). When customers believe they are being lied to over location, anger can become overtly racist. In more subtle ways the performance of emotional labour is given an added, culturally mediated, twist, but this again intensifies work pressure, as in this instance where an agent recalled being congratulated by a satisfied customer,

> Well done! You really helped me, and he asks you your name. So I tell "Nick" (pause). I'm not Nick. I don't know who Nick is … what's the use of giving my alias, when I'm the one working. I'm solving the problem. I'm from India … no we cannot give my name. (Mirchandani, 2005, forthcoming: 9)

Also, the gulf between the stimulating work and prestigious career promised by employers, and the mundane nature of incessant call-handling can cause disenchantment and disengagement from job and organization. Articulating this conflict between career and status expectation on the one hand, and the reality of task performance on the other, these ex-graduates see work "as demeaning and inappropriate given their qualifications" (Mirchandani, 2005, forthcoming: 8). Directly contradicting the presumed loyalty of Indian workers, a male agent stated how, "We are all here, working, waiting for our opportunities to come," outside of current employment (Mirchandani, 2005, forthcoming: 10). This desire for more attractive, although less-remunerative employment, was also discovered by Deb (2004: 18).

Work and employment conditions generate both familiar and distinctive forms of resistance. Some workers clearly do know when supervisors monitor calls and adjust their behaviour accordingly, allowing them "to maximize their [performance] scores" (Deb, 2004: 19). Intriguingly, some are able to manipulate procedures for resolving customer service problems. Given that speed of resolution is essential, "where issues are complicated and customers are irate, workers sometimes decide to give customers a '*talla*'," a hoax solution. The extent to which

Indian workers employ creative methods to circumvent controls and maximize bonus payments is a fascinating question which future research hopefully will pursue. However, even these few examples show how far some Indian agents deviate from Whitson's (*Financial Times*, 10 August 2002) virtuous stereotype.

RECRUITMENT, TRAINING AND ATTRITION

Although Indian workflows are highly standardized, it does not follow that tensions between the cost-reduction and customer-oriented logics, the contradiction between quantity and quality, [have] been resolved satisfactorily in favour of the former. The *quality* of agent–customer interaction, albeit within the tightly-prescribed parameters of transactional, scripted calls, is of great importance....

Language and cultural training are central priorities, with one manager stating that "accent training is perhaps *the* most important criterion" (Mum4). The obligation to provide satisfactory standards of spoken English originates with the offshoring companies, who are sensitive to possible customer dissatisfaction, where migration causes perceptions of inferior service. The transnational dimension to service provision ensures two elements to language training. First, agents must be able to understand the national accents of English-speaking countries and customers' diverse regional accents. Second, customers must comprehend the speech of Indian agents. The universal practice is to "neutralize" accents, to "make sure that agents do not speak with Indian inflections" (Del1), a process of modification which varies according to the strength of regional accents.

During language training (Mum4) an expatriate Indian coaching agents how to relate to Canadian customers, commented on trainees' performances, following a role-play exercise aimed at improving pitch, empathy and pronunciation.

> There was a lack of empathy in the voice. Your speech tells us that you want to get off the phone as soon as you can ... you have to spend time with the customer and concentrate on the softness of the sounds. There are problems if you don't get modulation in your sounds. Also the "r" sounds are not coming out clearly and neither are the "v" sounds.

Mastering tongue twisters aimed to improve pronunciation in order to challenge western prejudices regarding Indians' "imperfect" speech. Having neutralized accents, most employers encourage, or train, agents to adopt their customers' accents. One manager's "ideal" was "to get agents to speak with accents, although to expect this immediately would be fanciful" since only by "prolonged exposure on the phone will agents get there" (Mum5). Modifying natural speech is combined with programmes designed to familiarize agents with the cultural background of customers' countries. One centre, providing two-day modules on US and UK culture, covered "typical customers, geography, currency, holidays, time-zones, slang and colloquialisms, and major sporting events" (Mum5).

A critical evaluation of these practices concludes that, notwithstanding agents' formal command of English and the high priority accorded to accent training, linguistic capability *generally* is not sufficiently developed nor flexible to ensure that more than routine call-handling can be conducted satisfactorily. That many in India share this judgement is illustrated by the comments of this senior official at the Ministry for Communications and IT: "On the quality side it is very important that we make moves from just felicity with the English language to higher communication skills and then on to superior domain knowledge" (Rajeeva Shah, 12 June 2003).

Further, truncated programmes cannot successfully bridge the cultural distance between India and the UK. In this respect, ill-informed commentary has appeared in the UK press, where linguistic and cultural assimilation is assumed to be non-problematic. For example, an article reporting a visit to an Indian centre was headlined "Greetings Jimmy ... great episode of Eastenders last night, by the way" (*Sunday Herald,* 9 March 2003). The implication here is that not only do Indian agents have a sophisticated understanding of English, but are conversant with Glasgow patois and are familiar with the most recent episode of a British soap opera. There is simply no evidence that this sort of interaction occurs, nor that training modules provide this depth of cultural understanding. Whether the objective is, as Mirchandani (2003) puts it, "Making Americans," or making English, Scots, or whatever nationality Indians are expected, chameleon-like to become, there are limitations on the ability of these neo-colonialist practices to succeed. Indeed, difficulties are acknowledged by UK companies. While leveraging the educated, English-speaking resource was cited as the second most important reason for migrating work, companies reported "accent and language problems/cultural differences/mistakes through misunderstanding" as the major disadvantage experienced (Taylor and Bain, 2003: 56–8)....

In sum, migrating call centre services to India has not enabled companies to escape from intractable labour management difficulties faced in the advanced countries. Indeed, it is remarkable how the Indian industry not only reproduces, albeit in exaggerated form, a labour process that has proved problematical for employers and employees alike in the west, but that it also displays, again through the lens of its distinctive national character, HR problems that are only too familiar.

CONCLUSION

While call centre offshoring to India certainly provides evidence of growing interconnectedness in the global economy, it is necessary to re-emphasize that globalization does not consist of a single pre-determined trajectory with inevitable outcomes. The perspective that globalization represents sets of tendencies, is confirmed by this analysis of the migration of voice services. Undoubtedly, powerful forces are driving offshoring, particularly the central capitalist imperatives of cutting costs, and maximizing profits and shareholder value, just as these were central to the emergence and widespread diffusion of the call centre as an organizational form in the English-speaking developed countries.... Fluency and clarity of speech, and depth of cultural understanding, combined with necessary tacit knowledge, are prerequisites for successful servicing and selling relationships.

However, it is difficult to make these qualities explicit and transmit them through formal mechanisms. Even allowing for the simplified, standardized nature of the bulk of Indian call-handling, it appears that cultural and linguistic differences are not readily overcome. Consequently, the call centre with its distinctive labour process may in the long run be offshored less readily than other non-customer facing, back-office activities.

REFERENCES

Baldry, C., Bain, P., and Taylor, P. (1997) "Sick and Tired? Working in the Modern Office," *Work, Employment and Society* 11(3): 519–39.

Castells, M. (2000) *The Rise of the Network Society*. Oxford: Blackwell.

Datta, R. C. (2004) "Worker and work—a case study of an international call centre in India," paper to 22nd International Labour Process Conference, University of Amsterdam, 5–7 April.

Deb, S. (2004) "Call Me," *Guardian Weekend,* 3 April: 14–23.

Health and Safety Executive (1995) *How to Deal with Sick Building Syndrome*. London: HSE Books.

Mirchandani, K. (2003) "Making Americans: Transnational Call Centre Work in India," Unpublished paper.

Mirchandani, K. (2005 forthcoming) "Webs of Resistance in Transnational Call Centres," in R. Thomas, A. Mills and J. Helms-Mills (eds) *Gender, Organization and the Micro-politics of Resistance*. London: Routledge.

Mitial (2002) *UK Call and Contact Centre Study, 2003*. London: Mitial.

Nasscom (2002a) *ITES: Background and Reference Resource*. New Delhi: Nasscom.

Nasscom (2002b) *Indian IT Enabled Service Providers: Directory 2002:* New Delhi: Nasscom.

Nasscom (2003) *Strategic Review: The IT Industry in India*. New Delhi: Nasscom.

Outsourcing Insight (2001) *Call for India*. London: Outsourcing Insight.

Taylor, P. and Bain, P. (2003) *Call Centres in Scotland and Outsourced Competition from India*. Stirling: Scotecon.

The International Sexual Trafficking of Women and Children

DAVID R. HODGE AND CYNTHIA A. LIETZ

The global flow of tourism, materials, and jobs is largely voluntary or, at worst, induced by economic hardships in some locations and opportunities in others. But globalization has also brought the resurgence of an ancient and despised practice: slavery. While it is difficult to accurately estimate the extent of the global slave trade, it flourishes in both entrepreneurial and highly structured organizations. Women are tricked, stolen, and purchased from poorer countries and transported into richer ones. This is a staggeringly lucrative industry, with women, and often children, sold or rented and then pimped out to hundreds of customers a week. The treatment of these victims and their life expectancy is abysmal. Hodge and Lietz offer us an excellent overview of some of the most compelling dimensions of this problem, discussing not only how this occurs but also what happens to the women and children who become victims of this exploitation. The United Nations is working to help reduce this problem, but it is severely underfunded and lacks the enforcement resources to be effective. What share of the blame should people from poor countries who traffic in human slaves bear versus those from rich countries who make this possible by consuming these illicit services? On which end of the problem should we focus? How can we stop the flow of people into slavery? How can we diminish the demand for it?

Human trafficking is the term that is commonly used to describe the present-day slave trade. Although slavery is often thought of as a relic of a bygone era, human trafficking is widely perceived to be a growing problem. According to the United Nations' Office on Drugs and Crime (2005), the trafficking of human beings has reached "epidemic proportions" over the past decade. Furthermore, trafficking is a problem in essentially every country; no nation is exempt.

Since human trafficking is part of the shadow economy, determining the number of individuals who are trafficked is difficult.... Among the more conservative estimates may be those the U.S. government provides. According to the U.S. Department of State (2004), 600,000 to 800,000 persons are trafficked across international borders annually. International trafficking tends to be easier to estimate than domestic trafficking. With this caveat in mind, on the basis of U.S. estimates, approximately 2 million to 4 million persons are trafficked within their home nations.... The largest subset of human trafficking is the sexual trafficking of young women and children. It is widely agreed that this subset of human trafficking has grown dramatically over the past decade....

THE GLOBALIZATION OF SEXUAL TRAFFICKING

Prior to the 1990s, most authorities believed that sex trafficking was largely concentrated in Asia. More recently, however, sex trafficking has globalized. In other words, trafficking is no longer primarily localized in one geographic region but has become increasingly transnational and fluid in character. A young girl may be recruited in Nigeria and sold and "trained" in Italy, with the Netherlands being the ultimate destination.

The United Nations mapped the emerging patterns in flows of international trafficking. The nations of origin, transit, and destination were assessed and classified into the following regions: Africa, Asia, the Commonwealth of Independent States (CIS—the countries of the former Soviet Union), Eastern Europe, Latin America, and the industrialized nations.... Destination countries tend to be wealthy nations in which large sex industries exist or where prostitution is legalized or broadly tolerated. For instance, the child pornography industry in the United States is one of the largest in the world and prostitution is legal in Germany and the Netherlands. Conversely, originating countries tend to be poor. Similarly, at the micro level, within regions and individual nations, it is often the most vulnerable who are victimized.

SEXUAL TRAFFICKING: HIGH REWARD–LOW RISK

Many factors contribute to the sexual trafficking of young women and children. Some observers have categorized them into push-and-pull factors. Poverty, war, and the lack of a promising future help to foster discontent, pushing individuals away from their originating countries. Conversely, media-constructed images of glamorous industrialized nations seem to offer the possibility of a better life, pulling individuals toward these countries.

While these factors help create an environment conducive to trafficking, Hughes (2000a) argued that the most critical element is criminal networks. Many areas are poor and chaotic, but not every area becomes a major supplier of women for the sex industry. It is the organized crime networks that manipulate these push-and-pull factors to recruit and traffic women. Criminal organizations have an interest in manipulating these factors because of the substantial profits that are involved in trafficking. After narcotics and arms sales, trafficking is the largest source of revenue for organized crime. The sale of trafficked young women and girls is profitable, although prices can vary significantly. In Italy, which serves as a major transit point for entry into the European Union, authorities uncovered a ring auctioning young women who were abducted from the former Soviet Union. These women were sold for an average price of approximately U.S. $1,000 a piece. Conversely, Asian women who are sold in the United States and Japan sell for up to U.S.$20,000 each.

Unlike narcotics, which are sold and used once, women who are sold into prostitution earn profits continually, year after year, for their exploiters or "pimps." The overhead is minimal and is typically paid for by the women. Russian women who are prostituted in Germany reportedly earn approximately U.S.$7,500 per month, of which $7,000 goes to the pimp. Many trafficked women retain little or none of the money they earn. Interpol reportedly estimated that sexually exploited women in Europe earn approximately U.S. $124,000 annually for their pimps.

In addition to high profits, the risks associated with trafficking are comparatively minimal. Prostitution is legal or semilegal in many places, complicating efforts to incarcerate traffickers. Even in settings where prostitution is clearly illegal, punishment is usually light. Furthermore, enforcement efforts typically focus on the women instead of the exploiters. In such settings, women often attempt to remain unnoticed for fear of being charged, particularly if they are trafficked internationally. The risk of incarceration is further reduced by the fact that many women are afraid of retaliation by their exploiters. Many traffickers use extreme violence to intimidate and control women. In Turkey, for instance, two trafficked women were reportedly murdered—thrown off a balcony—while six others were forced to watch....

RECRUITMENT STRATEGIES

Researchers have explored various aspects of trafficking in numerous locations.... Traffickers use a number of strategies to recruit women into the sex industry....

In the first strategy, seemingly legitimate organizations are frequently used to recruit young women and children. Employment, modeling, and marriage agencies purport to offer access to a better life in another, typically richer, nation. Those who are recruited usually have no idea of the fate that awaits them once they arrive at their destination. The second strategy is to approach healthy, non–drug-addicted women who are already engaged in prostitution in their country of origin. Women who work in nightclubs, for instance, may be approached and promised much higher earnings for doing similar work in wealthier nations. Women in this category are cognizant of the general type of work they will be required to perform but are unaware of the slavelike conditions in which they will be forced to work. The third strategy, which is less common, is kidnapping. Those who are approached to work in the sex industry but are unwilling to leave their country of origin may be kidnapped, along with completely unsuspecting individuals. To transport these victims to destination countries, traffickers often bribe officials to procure passports and/or safe passage. Finally, in some comparatively rare situations, recruiters may approach families or guardians who are living in poverty and seek to purchase girls or young women. The recruiters may point out how the money may be used to help the family and promise that the daughter will have access to a better life in a richer nation....

TRANSPORTATION AND ENSLAVEMENT

The costs that are involved in traveling to another country are typically borne by the women through the mechanism of "debt-bondage." The traffickers pay the costs associated with transportation, and the women incur the costs as debt, which they repay out of future earnings. A woman may be passed from organization to organization as she makes her way to her destination country, incurring further debt with each exchange.

When they arrive at their destination countries, the victims are commonly told what they are expected to do and how much money they must repay to regain their freedom. For many individuals, this is the first time they realize that they are in some type of bondage.

Their options, however, are limited. By this point, they must cope with a new setting, country, language, and culture without the assistance of family members, friends, or other supportive social support networks. In this context, they are largely dependent on the traffickers for their psychological and physical survival.

The traffickers use various strategies to increase the victims' isolation and dependence. Passports may be held. Psychological techniques may be used. Women may be involved in fraud or some other type of illegal activity during transportation. In such cases, the victims are often hesitant to consult with local officials out of concern about possible legal consequences.

In spite of the procedures, many women resist induction into the sex industry. To procure compliance, traffickers often use threats and violence. Women may be told that they have committed crimes and are threatened with arrest or that family members in their country of origin will be harmed. If the victim has any children, the children may be seized or threatened with harm. In addition to threats, violence may be used, particularly if law enforcement is ineffective.

Rape—vaginal, anal, and gang—is common. Young women and girls may be beaten; burned with hot irons, cigarettes, or other objects; or cut and salt poured into the wounds. The violent acts tend to be perpetrated upon nonvisible areas of women's and girls' bodies, however, so as to not reduce the marketability of the "product." Murder is not uncommon, although it is often used as a last resort, since future profits are lost. Consequently, murders are typically staged in a manner that helps to establish control over others.

In short, traffickers seek to break victims psychologically—to bring them to the point where they feel they no longer have control over their own safety. Rather, the traffickers do. Consequently, to survive, the women must become subservient to the traffickers' wishes. As one young respondent who was trafficked from Laos to Thailand put it, "After eight days of torture, I thought I will accept the job to save my life" (quoted in Zimmerman et al. 2003, p. 46).

EXPLOITATION IN THE SEX INDUSTRY

Once control has been established, the victims are exploited in the sex industry, with prostitution being perhaps the most common venue. The victims frequently have little control over the type of clients they will serve, the number of clients they will be required to see, or the hours they will be forced to work. Some trafficked women reported serving as many as 40 to 50 clients on any given night, and 18-hour days are common. In many cases, their freedom of movement is monitored, and the victims are essentially imprisoned in brothels.

Similarly, victims often have little control over the acts they will be required to perform. Young women and girls may be forced to engage in abusive, degrading, and violent sexual acts. Examples include urinating or defecating on women, tying women to bedposts and whipping them until they bleed, and cutting their breasts and genital areas.

It is not uncommon for it to take a thousand sex acts to pay off the accumulated debt. However, since many victims are innumerate—they have never been taught to count—they have no way to keep track of their wages and debts. When those who are able to count come close to achieving their freedom, they may be trafficked to another country, incurring new debts that must be worked off.

Many victims are prostituted in pornography. Videotapes depict children being raped, tortured, and even murdered on film. Pornography is also used to control victims. Traffickers may take shots in which the victims are clearly identified and then threaten to reveal particularly degrading portrayals to family members in the victims' country of origin.

The Internet has opened up new opportunities for traffickers to prostitute women. Young women and girls can be exploited in areas where law enforcement can be bribed or where prostitution is legalized or tolerated. One Cambodian-based pornographic Web site, "the rape camp," featured trafficked "Asian sex slaves" who were gagged, bound, and blindfolded while they were used in various sex acts. Viewers could reply to requests for sexual torture that would be fulfilled in real time, from anywhere in the world. Although this particular site was shut down by governmental authorities after it received wide-spread attention, many similar sites exist. One Russian-based Web site advertises itself as the "most violent rape site on earth," while a Denmark-registered Web site, the "slave farm," features images of "needle torture," "hanging bondage," "tits nailed to board," and "drunk from the toilet," with a live chat option where viewers can "command the bitches." Some sites use encryption technology so that the various ways in which young women and girls are exploited can be hidden....

CONCLUSION

As Annan (1999) noted, women and children are not property to be bought and sold, used and discarded. Rather, they are human beings with certain fundamental human rights that preclude being sold into slavery. The young women and children who are enslaved in the sex trade represent some of the most oppressed individuals in the world. We must not ignore these victims....

REFERENCES

Annan, K. (1999). *Secretary-general says persistence of slavery, in era of human rights progress, demands action by global community*. Retrieved from http://www.un.org/News/Press/docs/19991130.sgsm7242.doc.html

Hughes, D. M. (2000a). The "Natasha" trade: The transnational shadow market of trafficking in women. *Journal of International Affairs, 53*, 625–651.

United Nations Office on Drugs and Crime. (2005). *Fact sheet on human trafficking*. Retrieved from http://www.unodc.org/unodc/en/trafficking_victim_consents.html

U.S. Department of State. (2004). *Trafficking in persons report*. Retrieved from http://www.state.gov/g/tip/rls/tiprpt/2004/34021.htm

Zimmerman, C., Yun, K., Shvab, I., Watts, C., Trappolin, L., Treppete, M., et al., (2003). *The health risks and consequences of trafficking in women and adolescents. Findings from a European study*. London: Gender Violence and Health Centre, London School of Hygiene and Tropical Medicine.

43

The McDonaldization
of Society

GEORGE RITZER

Ritzer uses the success of fast-food chains as a metaphor for some general trends characterizing contemporary American society. We have become a nation driven by concerns for rationality, speed, and efficiency that are so well illustrated by the McDonald's style of operation. Food, packaging, and service are designed to move quickly and cheaply through these restaurants, giving customers the most modern eating experience. Speed, convenience, and standardization have replaced the flair of design and creation in cooking, the comfort of relationships in serving, and the variety available in choice. McDonaldization has become so pervasive that one can travel to nearly any city or town in America and find familiar chain-style restaurants, shops, hotels, and other avenues for commercial exchange. This has fostered the homogenization of American culture and life, streamlined along a set of rational, efficient, and impersonal principles. What have been the benefits of this sweeping sameness? What have we lost? Into what areas aside from fast food has this trend expanded?

A wide-ranging process of *rationalization* is occurring across American society and is having an increasingly powerful impact in many other parts of the world. It encompasses such disparate phenomena as fast-food restaurants, TV dinners, packaged tours, industrial robots, plea bargaining, and open-heart surgery on an assembly-line basis. As widespread and as important as these developments are, it is clear that we have barely begun a process that promises even more extraordinary changes (e.g. genetic engineering) in the years to come. We can think of rationalization as a historical process and rationality as the end result

SOURCE: George Ritzer, "The 'McDonaldization' of Society," The *Journal of American Culture* 6(1), 1983, pp. 100–107. Reprinted by permission of Blackwell Publishing Ltd.

of that development. As a historical process, rationalization has distinctive roots in the western world. Writing in the late nineteenth and early twentieth centuries, the great German sociologist Max Weber saw his society as the center of the ongoing process of rationalization and the bureaucracy as its paradigm case. The model of rationalization, at least in contemporary America, is no longer the bureaucracy, but might be better thought of as the fast-food restaurant. As a result, our concern here is with what might be termed the "McDonaldization of Society." While the fast-food restaurant is not the ultimate expression of rationality, it is the current exemplar for future developments in rationalization.

A society characterized by rationality is one which emphasizes *efficiency, predictability, calculability, substitution of nonhuman for human technology,* and *control over uncertainty.* In discussing the various dimensions of rationalization, we will be little concerned with the gains already made, and yet to be realized, by greater rationalization. These advantages are widely discussed in schools and in the mass media. In fact, we are in danger of being seduced by the innumerable advantages already offered, and promised in the future, by rationalization. The glitter of these accomplishments and promises has served to distract most people from the grave dangers posed by progressive rationalization. In other words, we are ultimately concerned here with the irrational consequences that often flow from rational systems. Thus, the second major theme of this essay might be termed "the irrationality of rationality."…

EFFICIENCY

The process of rationalization leads to a society in which a great deal of emphasis is placed on finding the best or optimum means to any given end. Whatever a group of people define as an end, and everything they so define, is to be pursued by attempting to find the best means to achieve the end. Thus, in the Germany of Weber's day, the bureaucracy was seen as the most efficient means of handling a wide array of administrative tasks. Somewhat later, the Nazis came to develop the concentration camp, its ovens, and other devices as the optimum method of collecting and murdering millions of Jews and other people. The efficiency that Weber described in turn-of-the-century Germany, and which later came to characterize many Nazi activities, has become a basic principle of life in virtually every sector of a rational society.

The modern American family, often with two wage earners, has little time to prepare elaborate meals. For the relatively few who still cook such meals, there is likely to be great reliance on cookbooks that make cooking from scratch much more efficient. However, such cooking is relatively rare today. Most families take as their objective quickly and easily prepared meals. To this end, much use is made of prepackaged meals and frozen TV dinners.

For many modern families, the TV dinner is no longer efficient enough. To many people, eating out, particularly in a fast-food restaurant, is a far more efficient way of obtaining their meals. Fast-food restaurants capitalize on this by being organized so that diners are fed as efficiently as possible. They offer a limited, simple menu that can be cooked and served in an assembly-line fashion. The latest development in fast-food restaurants, the addition of drive-through windows, constitutes an effort to increase still further the efficiency of the dining experience. The family now can simply drive through, pick up its order, and eat it while driving to the next, undoubtedly efficiently organized, activity. The success of the fast-food restaurant has come full circle with frozen food manufacturers now touting products for the home modeled after those served in fast-food restaurants.

Increasingly, efficiently organized food production and distribution systems lie at the base of the ability of people to eat their food efficiently at home, in the fast-food restaurant, or in their cars. Farms, groves, ranches, slaughterhouses, warehouses, transportation systems, and retailers are all oriented toward increasing efficiency. A notable example is chicken production where they are mass-bred, force-fed (often with many chemicals), slaughtered on an assembly line, iced or fast frozen, and shipped to all parts of the country. Some may argue that such chickens do not taste as good as the fresh-killed, local variety, but their complaints are likely to be drowned in a flood of mass-produced chickens. Then there is bacon which is more efficiently shipped, stored, and sold when it is preserved by sodium nitrate, a chemical which is unfortunately thought by many to be carcinogenic. Whatever one may say about the quality or the danger of the products, the fact remains that they are all shaped by the drive for efficiency....

One of the most interesting and important aspects of efficiency is that it often comes to be not a means but an end in itself. This "displacement of goals" is a major problem in a rationalizing society. We have, for example, the bureaucrats who slavishly follow the rules even though their inflexibility negatively affects the organization's ability to achieve its goals. Then there are the bureaucrats who are so concerned with efficiency that they lose sight of the ultimate goals the means are designed to achieve. A good example was the Nazi concentration camp officers who, in devoting so much attention to maximizing the efficiency of the camps' operation, lost sight of the fact that the ultimate purpose of the camps was the murder of millions of people.

PREDICTABILITY

A second component of rationalization involves the effort to ensure predictability from one place to another. In a rational society, people want to know what to expect when they enter a given setting or acquire some sort of commodity. They neither want nor expect surprises. They want to know that if they journey to another locale, the setting they enter or the commodity they buy will be essentially the same as the setting they entered or product they purchased earlier. Furthermore, people want to be sure that what they encounter is much like what they encountered at earlier times. In order to ensure predictability over time and place a rational society must emphasize such things as discipline, order, systemization, formalization, routine, consistency, and methodical operation.

One of the attractions of TV dinners for modern families is that they are highly predictable. The TV dinner composed of fried chicken, mashed potatoes, green peas, and peach cobbler is exactly the same from one time to another and one city to another. Home cooking from scratch is, conversely, a notoriously unpredictable enterprise with little assurance that dishes will taste the same time after time. However, the cookbook cannot eliminate all unpredictability. There are often simply too many ingredients and other variables involved. Thus the cookbook dish is far less predictable than the TV dinner or a wide array of other prepared dishes.

Fast-food restaurants rank very high on the dimension of predictability. In order to help ensure consistency, the fast-food restaurant offers only a limited menu. Predictable end products are made possible by the use of similar raw materials, technologies, and preparation and serving techniques. Not only the food is predictable; the physical structures, the logo, the "ambience," and even the personnel are as well.

The food that is shipped to our homes and our fast-food restaurants is itself affected by the process of increasing predictability. Thus our favorite white bread is indistinguishable from one place to another. In fact, food producers have made great efforts to ensure such predictability.

On packaged tours travelers can be fairly sure that the people they travel with will be much like themselves. The planes, buses, hotel accommodations, restaurants, and at least the way in which the sites are visited are very similar from one location to another. Many people go on packaged tours *because* they are far more predictable than travel undertaken on an individual basis.

Amusement parks used to be highly unpredictable affairs. People could never be sure, from one park to another, precisely what sorts of rides, events, foods, visitors, and employees they would encounter. All of that has changed in the era of the theme parks inspired by Disneyland. Such parks seek to ensure predictability in various ways. For example, a specific type of young person is hired in these parks, and they are all trained in much the same way, so that they have a robot-like predictability.

Other leisure-time activities have grown similarly predictable. Camping in the wild is loaded with uncertainties—bugs, bears, rain, cold, and the like. To make camping more predictable, organized grounds have sprung up around the country. Gone are many of the elements of unpredictability replaced by RVs, paved-over parking lots, sanitized campsites, fences and enclosed camp centers that provide laundry and food services, recreational activities, television, and video games. Sporting events, too, have in a variety of ways been made more predictable. The use of artificial turf in baseball makes for a more predictable bounce of a ball....

CALCULABILITY OR QUANTITY
RATHER THAN QUALITY

It could easily be argued that the emphasis on quantifiable measures, on things that can be counted, is *the* most defining characteristic of a rational society. Quality is notoriously difficult to evaluate. How do we assess the quality of a hamburger, or a physician, or a student? Instead of even trying, in an increasing number of cases, a rational society seeks to develop a series of quantifiable measures that it takes as surrogates for quality. This urge to quantify has given great impetus to the development of the computer and has, in turn, been spurred by the widespread use and increasing sophistication of the computer.

The fact is that many aspects of modern rational society, especially as far as calculable issues are concerned, are made possible and more widespread by the computer. We need not belabor the ability of the computer to handle large numbers of virtually anything, but somewhat less obvious is the use of the computer to give the illusion of personal attention in a world made increasingly impersonal in large part because of the computer's capacity to turn virtually everything into quantifiable dimensions. We have all now had many experiences where we open a letter personally addressed to us only to find a computer letter. We are aware that the names and addresses of millions of people have been stored on tape and that with the aid of a number of word processors a form letter has been sent to every name on the list. Although the computer is able to give a sense of personal attention, most people are nothing more than an item on a huge mailing list.

Our main concern here, though, is not with the computer, but with the emphasis on quantity rather than quality that it has helped foster. One of the most obvious examples in the university is the emphasis given to grades and cumulative grade point averages. With less and less contact between professor and student, there is little real effort to assess the quality of what students know, let alone the quality of their overall abilities. Instead, the sole measure of the quality of most college students is their grade in a given course and their grade point averages. Another blatant example is the emphasis on a variety of uniform exams such as SATs and GREs in which the essence of an applicant is reduced to a few simple scores and percentiles.

Within the educational institution, the importance of grades is well known, but somewhat less known is the way quantifiable factors have become an essential part of the process of evaluating college professors. For example, teaching ability is very hard to evaluate. Administrators have difficulty assessing teaching quality and thus substitute quantitative scores. Of course each score involves qualitative judgments, but this is conveniently ignored. Student opinion polls are taken and the scores are summed, averaged, and compared. Those who score well are deemed good teachers while those who don't are seen as poor teachers. There are many problems involved in relying on these scores such as the fact that easy teachers in "gut" courses may well obtain high ratings while rigorous teachers of difficult courses are likely to score poorly....

In the workworld we find many examples of the effort to substitute quantity for quality. Scientific management was heavily oriented to turning everything work-related into quantifiable dimensions. Instead of relying on the "rule of thumb" of the operator, scientific management sought to develop precise measures of how much work was to be done by each and every motion of the worker. Everything that could be was reduced to numbers and all these numbers were then analyzable using a variety of mathematical formulae. The assembly line is similarly oriented to a variety of quantifiable dimensions such as optimizing the speed of the line, minimizing time for each task, lowering the price of the finished product, increasing sales and ultimately increasing profits. The divisional system pioneered by General Motors and thought to be one of the major reasons for its past success was oriented to the reduction of the performance of each division to a few, bottom-line numbers. By monitoring and comparing these numbers, General Motors was able to exercise control over the results without getting involved in the day-to-day activities of each division....

Thus, the third dimension of rationalization, calculability or the emphasis on quantity rather than quality, has wide applicability to the social world. It is truly central, if not the central, component of a rationalizing society. To return to our favorite example, it is the case that McDonald's expends far more effort telling us how many billions of hamburgers it has sold than it does in telling us about the quality of those burgers. Relatedly, it touts the size of its product (the "Big Mac") more than the quality of the product (it is not the "Good Mac"). The bottom line in many settings is the number of customers processed, the speed with which they are processed, and the profits produced. Quality is secondary, if indeed there is any concern at all for it.

SUBSTITUTION OF NONHUMAN TECHNOLOGY

In spite of Herculean efforts, there are important limits to the ability to rationalize what human beings think and do. Seemingly no matter what one does, people still retain at least the ultimate capacity to think and act in a variety of unanticipated ways. Thus, in spite of great efforts to make human behavior more efficient, more predictable, more calculable, people continue to act in unforeseen ways. People continue to make home-cooked meals from scratch, to camp in tents in the wild, to eat in old-fashioned diners, and to sabotage the assembly lines. Because of these realities, there is great interest among those who foster increasing rationality in using rational technologies to limit individual independence and ultimately to replace human beings with machines and other technologies that lack the ability to think and act in unpredictable ways.

McDonald's does not yet have robots to serve us food, but it does have teenagers whose ability to act autonomously is almost completely eliminated by techniques, procedures, routines, and machines. There are numerous examples of this including rules which prescribe all the things a counterperson should do in dealing with a customer as well as a large variety of technologies which determine the actions of workers such as drink dispensers which shut themselves off when the cup is full; buzzers, lights, and bells which indicate when food (e.g., french fries) is done; and cash registers which have the prices of each item programmed in. One of the latest attempts to constrain individual action is Denny's use of pre-measured packages of dehydrated food that are "cooked" simply by putting them under the hot water tap. Because of such tools and machines, as well as the elaborate rules dictating worker behavior, people often feel like they are dealing with human robots when they relate to the personnel of a fast-food restaurant. When human robots are found, mechanical robots cannot be far behind. Once people are reduced to a few robot-like actions, it is a relatively easy step to replace them with mechanical robots. Thus, Burgerworld is reportedly opening a prototypical restaurant in which mechanical robots serve the food.

Much of the recent history of work, especially manual work, is a history of efforts to replace human technology with nonhuman technology. Scientific management was oriented to the development of an elaborate and rigid set of rules about how jobs were to be done. The workers were to blindly and obediently follow those rules and not to do the work the way they saw fit. The various skills needed to perform a task were carefully delineated and broken down into a series of routine steps that could be taught to all workers. The skills, in other words, were built into the routines rather than belonging to skilled [craftspeople]. Similar points can be made about the assembly line which is basically a set of nonhuman technologies that have the needed steps and skills built into them. The human worker is reduced to performing a limited number of simple, repetitive operations. However, the control of this technology over the individual worker is so great and omnipresent that individual workers have reacted negatively manifesting such things as tardiness, absenteeism, turnover, and even sabotage. We are now witnessing a new stage in this technological development with automated processes now totally replacing many workers with robots. With the coming of robots we have reached the ultimate stage in the replacement of humans with nonhuman technology.

Even religion and religious crusades have not been unaffected by the spread of nonhuman technologies. The growth of large religious organizations, the use of Madison Avenue techniques, and even drive-in churches all reflect the incursion of modern technology. But it is in the electronic church, religion through the TV screens, that replacement of human by nonhuman technology in religion is most visible and has its most important manifestation....

CONTROL

This leads us to the fifth major dimension of rationalization—control. Rational systems are oriented toward, and structured to expedite, control in a variety of senses. At the most general level, we can say that rational systems are set up to allow for greater control over the uncertainties of life—birth, death, food production and distribution, housing, religious salvation, and many, many others. More specifically, rational systems are oriented to gaining greater control over the major source of uncertainty in social life—other people. Among other things, this means control over subordinates by superiors and control of clients and customers by workers.

There are many examples of rationalization oriented toward gaining greater control over the uncertainties of life. The burgeoning of the genetic engineering movement can be seen as being aimed at gaining better control over the production of life itself. Similarly, amniocentesis can be seen as a technique which will allow the parents to determine the kind of child they will have. The efforts to rationalize food production and distribution can be seen as being aimed at gaining greater control over the problems of hunger and starvation. A steady and regular supply of food can make life itself more certain for large numbers of people who today live under the threat of death from starvation.

At a more specific level, the rationalization of food preparation and serving at McDonald's gives it great control over its employees. The automobile assembly line has a similar impact. In fact, the vast majority of the structures of a rational society exert extraordinary control over the people who labor in them. But because of the limits that still exist on the degree of control that rational structures can exercise over individuals, many rationalizing employers are driven to seek to more fully rationalize their operations and totally eliminate the worker. The result is an automated, robot-like technology over which, barring some *2001* rebellion, there is almost total control.

In addition to control over employees, rational systems are also interested in controlling the customer/clients they serve. For example, the fast-food restaurant with its counter, the absence of waiters and waitresses, the limited seating, and the drive-through windows all tend to lead customers to do certain things and not to do others.

IRRATIONALITY OF RATIONALITY

Although not an inherent part of rationalization, the *irrationality of rationality* is a seemingly inevitable byproduct of the process. We can think of the irrationality of rationality in several ways. At the most general level it can simply be seen as an overarching label for all the negative effects of rationalization. More specifically, it can be seen as the opposite of rationality, at least in some of its senses. For example, there are the inefficiencies and unpredictabilities that are often produced by seemingly rational systems. Thus, although bureaucracies are constructed to bring about greater efficiency in organizational work, the fact is that there are notorious inefficiencies such as the "red tape" associated with the operation of most bureaucracies. Or, take the example of the arms race in which a focus on quantifiable aspects of nuclear weapons may well have made the occurrence of nuclear war more, rather than less, unpredictable.

Of greatest importance, however, is the variety of negative effects that rational systems have on the individuals who live, work, and are served by them. We might say that *rational systems are not reasonable systems. As we've* already discussed, rationality brings with it great dehumanization as people are reduced to acting like robots. Among the dehumanizing aspects of a rational society are large lecture classes, computer letters, pray TV, work on the automobile assembly line, and dining at a fast-food restaurant. Rationalization also tends to bring with it disenchantment leaving much of our lives without any mystery or excitement.

Production by a hand craftsman is far more mysterious than an assembly-line technology where each worker does a single, very limited operation. Camping in an RV tends to suffer in comparison to the joys to be derived from camping in the wild. Overall a fully rational society would be a very bleak and uninteresting place.

CONCLUSION

Rationalization, with McDonald's as the paradigm case, is occurring throughout America, and, increasingly, other societies. In virtually every sector of society more and more emphasis is placed on efficiency, predictability, calculability, replacement of human by nonhuman technology, and control over uncertainty. Although progressive rationalization has brought with it innumerable advantages, it has also created a number of problems, the various irrationalities of rationality, which threaten to accelerate in the years to come. These problems and their acceleration should not be taken as a case for the return to a less rational form of society. Such a return is not only impossible but also undesirable. What is needed is not a less rational society, but greater control over the process of rationalization involving, among other things, efforts to ameliorate its irrational consequences.

44

Bowling Alone

ROBERT D. PUTNAM

The desire for community may be seen as one of the strongest themes in contemporary America. Putnam heralds this concern by articulating the decline of those social institutions and activities—the church, the family, labor unions, civic and political engagement, fraternal organizations, service clubs, parent–teacher associations—through which Americans traditionally fused together and created the fabric of community life. He whimsically focuses the title of this selection around an ironic shift in social behavior that he thinks reflects this trend toward the decline of traditional community: more Americans are bowling today than ever before, but bowling in organized leagues has plummeted. Putnam ponders this decline in "social capital" and what he calls the consequent erosion in good neighborliness and social trust. How likely do you think Americans are to know their neighbors, to participate in community and civic events, and to belong to social clubs? Is Putnam right or wrong in his fears about the lack of community involvement of Americans? If he is right, how might this trend continue to affect the individual and group life of Americans?

No one is left from the Glenn Valley, Pennsylvania, Bridge Club who can tell us precisely when or why the group broke up, even though its forty-odd members were still playing regularly as recently as 1990, just as they had done for more than half a century. The shock in the Little Rock, Arkansas, Sertoma club, however, is still painful: in the mid-1980s, nearly fifty people had attended the weekly luncheon to plan activities to help the hearing- and speech-impaired, but a decade later only seven regulars continued to show up.

The Roanoke, Virginia, chapter of the National Association for the Advancement of Colored People (NAACP) had been an active force for civil rights since 1918, but during the 1990s membership withered from about 2,500 to a few hundred. By

November 1998 even a heated contest for president drew only fifty-seven voting members. Black city councilor Carroll Swain observed ruefully, "Some people today are a wee bit complacent until something jumps up and bites them." VFW Post 2378 in Berwyn, Illinois, a blue-collar suburb of Chicago, was long a bustling "home away from home" for local veterans and a kind of working-class country club for the neighborhood, hosting wedding receptions and class reunions. By 1999, however, membership had so dwindled that it was a struggle just to pay taxes on the yellow brick post hall. Although numerous veterans of Vietnam and the post-Vietnam military lived in the area, Tom Kissell, national membership director for the VFW, observed, "Kids today just aren't joiners."...

The civic-minded World War II generation was, as its own John F. Kennedy proclaimed at his inauguration, picking up the torch of leadership, not only in the nation's highest office, but in cities and towns across the land. Summarizing dozens of studies, political scientist Robert E. Lane wrote in 1959 that "the ratio of political activists to the general population, and even the ratio of male activists to the male population, has generally increased over the past fifty years." As the 1960s ended, sociologists Daniel Bell and Virginia Held reported that "there is more participation than ever before in America ... and more opportunity for the active interested person to express his personal and political concerns." Even the simplest political act, voting, was becoming ever more common. From 1920, when women got the vote, through 1960, turnout in presidential elections had risen at the rate of 1.6 percent every four years, so on a simple straight-line projection it seemed reasonable, as a leading political scientist later observed, to expect turnout to be nearly 70 percent and rising on the nation's two hundredth birthday in 1976.

By 1965 disrespect for public life, so endemic in our history, seemed to be waning. Gallup pollsters discovered that the number of Americans who would like to see their children "go into politics as a life's work" had nearly doubled over little more than a decade. Although this gauge of esteem for politics stood at only 36 percent, it had never before been recorded so high, nor has it since. More strikingly, Americans felt increased confidence in their neighbors. The proportion that agreed that "most people can be trusted," for example, rose from an already high 66 percent during and after World War II to a peak of 77 percent in 1964.

The fifties and sixties were hardly a "golden age," especially for those Americans who were marginalized because of their race or gender or social class or sexual orientation. Segregation, by race legally and by gender socially, was the norm, and intolerance, though declining, was still disturbingly high. Environmental degradation had only just been exposed by Rachel Carson, and Betty Friedan had not yet deconstructed the feminine mystique. Grinding rural poverty had still to be discovered by the national media. Infant mortality, a standard measure of public health, stood at twenty-six per one thousand births—forty-four per one thousand for black infants—in 1960, nearly four times worse than those indexes would be at the end of the century. America in *Life* was white, straight, Christian, comfortable, and (in the public square, at least) male. Social reformers had their work cut out for them....

In recent years social scientists have framed concerns about the changing character of American society in terms of the concept of "social capital." By analogy with notions of physical capital and human capital—tools and training that enhance individual productivity—the core idea of social capital theory is that social networks have value. Just as a screwdriver (physical capital) or a college education (human capital) can increase productivity (both individual and collective), so too social contacts affect the productivity of individuals and groups.

Whereas physical capital refers to physical objects and human capital refers to properties of individuals, social capital refers to connections among individuals— social networks and the norms of reciprocity and trustworthiness that arise from them. In that sense social capital is closely related to what some have called "civic virtue." The difference is that "social capital" calls attention to the fact that civic virtue is most powerful when embedded in a dense network of reciprocal social relations. A society of many virtuous but isolated individuals is not necessarily rich in social capital....

As this array of independent coinages indicates, social capital has both an individual and a collective aspect—a private face and a public face. First, individuals form connections that benefit our own interests. One pervasive stratagem of ambitious job seekers is "networking," for most of us get our jobs because of whom we know, not what we know—that is, our social capital, not our human capital. Economic sociologist Ronald Burt has shown that executives with bounteous Rolodex files enjoy faster career advancement. Nor is the private return to social capital limited to economic rewards. As Claude S. Fischer, a sociologist of friendship, has noted, "Social networks are important in all our lives, often for finding jobs, more often for finding a helping hand, companionship, or a shoulder to cry on."

If individual clout and companionship were all there were to social capital, we'd expect foresighted, self-interested individuals to invest the right amount of time and energy in creating or acquiring it. However, social capital also can have "externalities" that affect the wider community, so that not all the costs and benefits of social connections accrue to the person making the contact. As we shall see later in this book, a well-connected individual in a poorly connected society is not as productive as a well-connected individual in a well-connected society. And even a poorly connected individual may derive some of the spillover benefits from living in a well-connected community. If the crime rate in my neighborhood is lowered by neighbors keeping an eye on one another's homes, I benefit even if I personally spend most of my time on the road and never even nod to another resident on the street.

Social capital can thus be simultaneously a "private good" and a "public good." Some of the benefit from an investment in social capital goes to bystanders, while some of the benefit redounds to the immediate interest of the person making the investment. For example, service clubs, like Rotary or Lions, mobilize local energies to raise scholarships or fight disease at the same time that they provide members with friendships and business connections that pay off personally.

Social connections are also important for the rules of conduct that they sustain. Networks involve (almost by definition) mutual obligations; they are not interesting as mere "contacts." Networks of community engagement foster sturdy norms of reciprocity: I'll do this for you now, in the expectation that you (or perhaps someone else) will return the favor. "Social capital is akin to what Tom Wolfe called 'the favor bank' in his novel *The Bonfire of the Vanities*," notes economist Robert Frank. It was, however, neither a novelist nor an economist, but Yogi Berra who offered the most succinct definition of reciprocity: "If you don't go to somebody's funeral, they won't come to yours."…

Of all the dimensions along which forms of social capital vary, perhaps the most important is the distinction between *bridging* (or inclusive) and *bonding* (or exclusive). Some forms of social capital are, by choice or necessity, inward looking and tend to reinforce exclusive identities and homogeneous groups. Examples of bonding social capital include ethnic fraternal organizations, church-based women's reading groups, and fashionable country clubs. Other networks are outward looking and encompass people across diverse social cleavages. Examples of bridging social capital include the civil rights movement, many youth service groups, and ecumenical religious organizations.

Bonding social capital is good for undergirding specific reciprocity and mobilizing solidarity. Dense networks in ethnic enclaves, for example, provide crucial social and psychological support for less fortunate members of the community, while furnishing start-up financing, markets, and reliable labor for local entrepreneurs. Bridging networks, by contrast, are better for linkage to external assets and for information diffusion. Economic sociologist Mark Granovetter has pointed out that when seeking jobs—or political allies—the "weak" ties that link me to distant acquaintances who move in different circles from mine are actually more valuable than the "strong" ties that link me to relatives and intimate friends whose sociological niche is very like my own. Bonding social capital is, as Xavier de Souza Briggs puts it, good for "getting by," but bridging social capital is crucial for "getting ahead."…

Is life in communities as we enter the twenty-first century really so different after all from the reality of American communities in the 1950s and 1960s? One way of curbing nostalgia is to count things. Are club meetings really less crowded today than yesterday, or does it just seem so? Do we really know our neighbors less well than our parents did, or is our childhood recollection of neighborhood barbecues suffused with a golden glow of wishful reminiscence? Are friendly poker games less common now, or is it merely that we ourselves have outgrown poker?…

So we are spending significantly less time nowadays with friends and neighbors than we used to do. What might we be doing instead that has implications for social capital? One common form of leisure activity is participation in sports. What can we learn about trends in social capital from an examination of Americans at play? Have we perhaps shifted the locus of our social encounters from the card table or the neighborhood bar to the softball diamond or the exercise class?…

Bowling is the most popular competitive sport in America. Bowlers outnumber joggers, golfers, or softball players more than two to one, soccer players (including kids) by more than three to one, and tennis players or skiers by four to one. Despite bowling's "retro" image, in 1996 even twenty-somethings went bowling about 40 percent more often than they went in-line skating. More recently, even greater numbers of young people have reportedly been attracted by a high-tech combination called "cosmic bowling" or "Rock 'N' Bowl." Moreover, participation in all other major sports is more highly concentrated among either young men, or the upper middle class, or both. Unlike health clubs, bicycling, jogging, exercise, swimming, tennis, golf, softball, and all other major sports, bowling is solidly middle-American—common among both men and women, couples and singles, working-class and middle-class, young and old.

Given population growth, more Americans are bowling than ever before, but *league* bowling has plummeted in the last ten to fifteen years. Between 1980 and 1993 the total number of bowlers in America increased by 10 percent, while league bowling decreased by more than 40 percent.... [T]he long-run trend in league bowling in America is a profile that precisely matches the trends in other forms of social capital that we have already examined—steady growth from the beginning of the century (except during the Depression and World War II), explosive growth between 1945 and 1965, stagnation until the late 1970s, and then a precipitous plunge over the last two decades of the century. At the peak in the mid-1960s, 8 percent of all American men and nearly 5 percent of all American women were members of bowling teams. Yet ... if the steady decline in league bowling were to continue at the pace of the last fifteen years, league bowling would vanish entirely within the first decade of the new century.

Lest bowling be thought a wholly trivial example, I should note that, according to the American Bowling Congress, ninety-one million Americans bowled at some point during 1996, *more than 25 percent more than voted in the 1998 congressional elections.* Even after the 1980s' plunge in league bowling, between 2 and 3 percent of American adults regularly bowled in leagues, although as we have seen, that figure was dropping fast. The decline in league bowlers threatens the livelihood of bowling lane proprietors, because according to the owner of one of the nation's largest bowling lane chains, league bowlers consume three times as much beer and pizza as do solo bowlers, and the money in bowling is in the beer and pizza, not the balls and shoes. The broader social significance, however, lies in the social interaction and even occasionally civic conversations over beer and pizza that solo bowlers forgo. Whether or not bowling beats balloting in the eyes of most Americans, bowling teams illustrate yet another vanishing form of social capital....

45

Cyber Communities of Self-Injury

PATRICIA A. ADLER AND PETER ADLER

In juxtaposition to Putnam's fears about the decline of civic participation and other face-to-face forms of community, the twenty-first century has witnessed the explosion of community in the World Wide Web. Forums such as MySpace, Facebook, and a host of textual and virtual media offer individuals the opportunity to go online and locate individuals like themselves. People who used to be isolated can now find others, bringing them together in communities of interest rather than communities of place. Online communities may include networking groups, support groups, interest groups, travel groups, deviant groups, or people looking to locate and to meet others like themselves. In this selection, we discuss our research into the hidden worlds of self-injury and discuss what draws these formerly isolated loners into contact with each other and the types of communities they find and build there. Compared to people your parents' age, how common do you think it is for college students to belong to Internet communities? To how many do you belong? When was the last time someone "Facebooked" you or vice versa? How important are these communities and relationships for you? Do you think that memberships in Internet communities and relationships serve as a staging ground preparing people to better present the selves they want to be in real life, or do they pull people away from the solid world and isolate them by drawing them more exclusively into cyber communities and cyber relationships?

SOURCE: Patricia A. Adler and Peter Adler, "The Cyber Worlds of Self-Injurers: Deviant Communities, Relationships, and Selves," *Symbolic Interaction*, Vol. 31, No. 1: 33–56. © 2008, Society for the Study of Symbolic Interaction and the University of California Press. Used by permission. All rights reserved.

The cyber world represents a new frontier of life, one that extends what have often been colloquially referred to as the fourth and fifth dimensions: time and space. Just as Melbin (1978) analyzed the night as a temporal frontier, the cyber world is a domain that occurs in a new form of space that is both "out there" and "in here"; it is simultaneously public and social, while remaining private and solitary. Unfolding and developing before our eyes, the limits of its untapped potential are yet to be realized. It is a postmodern form of space, created by technology and populated by disembodied people in a virtual universe that is detached from the physical locations known as place. These spaces are fertile locations for the rise of virtual communities, defined by Rheingold (1993:5) as "social aggregations that emerge from the Net when enough people carry on those public discussions long enough, with sufficient human feeling, to form webs of personal relationships in cyberspace." These challenge traditional notions of identity and community, with some suggesting that cyberspace is radically altering our conceptions of community and the nature of our communities.

We focus in this chapter on the way self-injury, the deliberate, non-suicidal destruction of one's own body tissue, incorporating practices such as self-cutting, burning, branding, scratching, picking at skin or re-opening wounds, biting, head-banging, hair-pulling (trichotillomania), hitting (with a hammer or other object), and bone-breaking, has been impacted by the Internet. Through the computer-mediated-communication (CMC) of the World Wide Web, these individuals, unconnected in real life, have constructed myriad cyber forums and communities, replacing what Bellah et al. (1985) called the "culture of separation," with a "culture of coherence."

Drawing on data gathered through analysis of self-injury bulletin boards and Usenet groups, in-depth, life history interviews with self-injurers, and email communications and relationships formed over a period of five years, we examine self-injury cyber communities and their characteristics.

METHODS

We have conducted over 100 in-depth interviews, in person and on the telephone, the largest qualitative data set with a non-clinical self-injuring population to date. Participants ranged in age from 16 to their mid-fifties, comprising more women than men, nearly all Caucasian, coming from the United States, Canada, Great Britain, Germany, Australia, New Zealand, Bulgaria, Holland, and France. We initially found participants through a convenience sample of individuals who heard, usually on one of our two campuses, that we were looking to talk with people who self-injured. Interested people requested interviews via email.

In addition, beginning in early 2002 we began to explore the Websites and public postings of self-injurers, subscribing to three public Internet self-injury groups that we refer to as Alpha, Beta, and Gamma, and regularly visiting the message boards and Websites of a few others, joining a growing group of Internet ethnographers. Since 2002 we have collected tens of thousands of Internet messages and emails including those posted publicly and those written to and by us. This CMC offered insight into the subcultures of self-injurers and their naturally-occurring conversations. Further, like other cyber-researchers, we used the Web as a means of recruitment, going to various sites, lists, groups, boards, and chat rooms to solicit subjects. We posted copies of our informed consent form and the complete range of interview topics on Patti's website, directing potential participants to view these before agreeing to be interviewed.

SELF-INJURY IN CYBERSPACE

During our early face-to-face (FTF) interviews, most of the people we encountered worked hard to hide their self-injury and felt the sting of social condemnation and shame. They were primarily loner deviants, unconnected to other self-injurers and lacking the social support and information diffusal prevalent in deviant subcultures. Around this time, information on this subject began to appear on the Internet, and since many of them were college students and computer literate, they easily could have gone there to discover it. But the majority chose not to, remaining isolated. When we expanded the focus of our research into cyber venues we found people's experiences to be vastly different. We describe here the ways that self-injurers' lives were dramatically changed by their cyber communication, compared to being real world isolates.

Beginning in the late 1990s and early 2000s, people began going to the Internet as a resource. Most felt confused and alone, unable to find real world counterparts, and sought help for themselves. Paula, a 38-year-old holistic massage therapist who had picked open wounds for years, described the frustration that led her to search the Web:

> Sometimes the picking episodes would be like three or four hours long and when I would use the needle, this wasn't a hugely bloody thing, but it was a little bloody. And I'd be in a position like this [*leaning the top of her head forward towards the mirror, but with her eyes peeking up*], in kind of a grimace, because you can't be in a position like this for three hours without being really physically just pshhh. And you know, I'm emotionally disconnected, so there'd be this sort of like insane look in my eyes, and I'm looking in the mirror and I'm not really seeing my reflection because I'm focusing on this. Blood is gathered on my hands, so I have caked blood all over my fingers, maybe some moments where more blood comes, and it really starts to drip. So I had one of these, "I'm here for hours," and then all of a sudden the veil comes up and I see myself like this, and I see the look in my face and I see the blood on my hands, and that's when I went to my computer and I said, "I need fuckin' help." And I know there's gotta be something out there, and I don't know what the hell it is, but I need help. And that's when I got online and just put in words.

Self-injurers participated in three of the four common modes of Internet engagement. They participated passively by going to Websites, reading others' postings and poetry, and viewing their images, some of which were rather graphic. They participated interactively in message boards, newsgroups, or Usenet groups that offered supportive communities. They found real-time communication in chat rooms, populated around the clock all over the world. We found no virtual spaces ("MUDs" or "MOOs"), the interactive, multi-user, online cyber games, cyber pubs, cybercafes, or other forums that go beyond mere words to offer visual representations of characters in "textual virtual reality," however, marked by or for self-injurers.

CYBER COMMUNITIES

Critical to becoming a regular participant in self-injury cyberspace was establishing membership in one of the many cyber communities.

Finding a Community

People searched until they found a community that seemed to fit their specialized needs. In this, they considered the size, level of activity, demographics, and orientation of the group. Some groups generated no more than a few postings daily, while others hosted a hundred or more communiques. People with the greatest communication needs gravitated towards those busier communities. Smaller groups often had greater levels of participation and were highly cohesive. Larger groups, despite their size, could feel relatively small because only a few people posted regularly. The demographics of the groups were also important to making people feel at home. Some specifically marketed themselves as oriented towards teens, others were for older people, and many invited a mixture. About half of the people we encountered were between the ages of 12 and 18. The next largest group, around 40 percent, was 19 to 30 years old. Those in their 30s represented a fading age category, with people also scattered around into 40s and 50s.

When they found a community that fit them well, it was rewarding. Paula, the holistic massage therapist, expressed the sense of community she got from her group:

> It's a good feeling to find a community that can accept your darkest shadows, but it's also a really scary thing to see those shadows. So it was double-edged. But it got to where, the same way I would look forward to coming home to pick, I would look forward to getting home and getting on the computer and reading all the emails, and I would go on the chats. It's a world, it is definitely a world.

Others used memberships in multiple communities to express different identities or aspects of their identities. For example, some men were revealed to be hiding behind female-sounding names, perhaps trying to fit into subcultures that were predominantly composed of women. In another case, Tim, a 21-year-old part-time college student who held various jobs and who moderated his own self-injury group, felt that he had to offer a hopeful self-presentation to the people he was trying to help on his site, so he proclaimed himself self-injury-free for two years, when in fact he had lapsed in and out of cutting. Tim used his membership in another group to discuss his ongoing problem, presenting a different self and identity there, although he used the same screen name.

Nature of the Community

As we see from the above remarks, communities differed in their policies, their norms, and their orientations towards self-injury. Some were highly regulated while others were fairly open; some were highly focused, while others were diffuse; some were stringently anti-self-injury, while others were more accepting of people's continuing practice. Alpha had a strict focus and policy. As one of the moderators posted:

> The reason that this group exists is to help people in recovery. All members are asked to identify the alternatives s/he tried to use to avoid using SI as a coping mechanism. For those who are not ready to embrace recovery, this is the wrong group.

Alpha moderators and members accepted people's slips into self-injury as long as they only discussed their feelings about it and not their injurious acts, and as long as they remain staunchly committed to quitting. Their rules not only prohibited any discussion of individuals' self-injury, but also any use of the actual term.

Other sites were more ambivalent about self-injury. They had everything, from people who loved cutting to those who had quit. If people wanted to stop, there were those who would support them. If people wanted to continue, others would accept them. Many fostered conversations among participants about their daily problems in dealing with friends and family members who did not understand them, about individuals' various psycho-medical conditions and medications, and about self-injurious problems and practices.

Finally, a few sites were more avowedly pro-self-injury. As we noted in an earlier article (Adler and Adler 2007), these sites approached self-injury in much the same way as the pro-Ana and pro-suicide movements. They treated it as a voluntary lifestyle choice and a long-term coping mechanism. Considering individuals' decision to injure themselves rather than injuring others constructive, they encouraged people to help themselves embrace their self-injury and, like others in a tertiary deviant stage (Kitsuse 1980), to shed the stigma. Along with this they offered practical suggestions for engaging in self-injury and dealing with the physical problems this generally engendered. Zoe, a member of the unregulated Beta Website, posted the following view of self-injury:

> I honestly dont see what is so wrong with cutting. I think I m kinda looking to see if anyone agrees. I mean, instead of punching a pillow, you

just take it out on yourself. As long as you dont do it too deep, whats the big deal??? Its better than abusing the people around you. The real problem with it is the emotions and the depression BEHIND the cutting, right? If it isnt "adversely affecting one's life," as is required for anything to be a legal disorder, then why does everyone else think it is wrong …
Am I making any sense to anyone???

She received the following response from Angie:

Hi there!! Nice to hear from you, welcome! As I was reading your posts, I couldn't help but feel as tho I was reading something that I had written!!! I don't see too much wrong with it either, it doesn't hurt anyone but myself.

Identification with the Community

Although people belonged to various sites and sometimes went for long periods between postings, when they found a community that fit well it gave them a sense of identity. They experienced this whether they were actively self-injuring or not. Jones (1997) noted that our sense of identity is not only derived from identification with the group, but from our understanding of the group identity. As Erica, an 18-year-old college freshman, noted:

You've been there; you know what it's like. I have traits in common with other members of the community: being sexually abused, being a perfectionist, having an ED. Always like, trying to help other people, doing community service, volunteer work, I'm really into that. Like everything they say on those Websites is completely me. I don't think it's all cutters; I think it's the majority of cutters. I just happen to fit. So it makes me feel more connected to the community as a member.

Identifying with members of the community was vitally important to most people we encountered, whether they had fully functioning work and social lives and hid their self-injury, or if they were trapped in their houses or bedrooms, unable to make contacts with people in the real world. McKenna and Bargh (1998) suggested that people with concealable stigma identify more strongly with these Internet support groups and consider them more important to their identities. As a result of this identification, they are also more likely to achieve greater self-acceptance, decreased estrangement from society, and decreased social isolation. Deshotels and Forsyth (2007) proposed that identities forged with the aid of Internet groups may help people disengage themselves from normative social control. Yet while people found these sites helpful, their identification with the community might also reinforce their self-injurious behavior, as Amber, a 20-year-old college junior, noted:

If you go to, like, the same chat room and stay there, you kind of get this group of friends, maybe. I guess you could get a sense of belonging or something. It's like you need to cut to stay in that group, you know?

Because that's what the chat rooms are for. It's a cutting chat room I guess, even though it says it's a no-cutting chat room. And so I think it just escalates people because we're kind of co-dependent in a way because, like say someone tells their friends the experience of it in that group, everyone will try it and they'll just keep on doing it and it'll just keep on escalating because, like, that's what's expected in that group and it just gets worse because there's no outside force preventing you from doing that, I guess.

When people identified with a community, that identity often transferred to them. Lemert (1967) discussed how primary deviants, who keep their deviance hidden from others, have the luxury of denying self-identification with their behavior. Becker (1963) echoed this theme, arguing that "secret deviants" are unlikely to conceive of themselves through the deviant lens. Erica, the college freshman, explained what it was about membership in her site that changed her identity:

Just the fact that there were other people doing it. Maybe like it really is, there's a group of people. I *am* part of this group, obviously. That helped me connect my identity to a self-abuser. Whereas before I was just, like, one of two people doing it so it wasn't really an identity, it was more of a problem. I didn't really think it was a problem, just a habit. Whereas on the Internet it's a lifestyle almost, the way you are, instead of just a habit. They were connected to it in a more long-term way. It was a more central focus of people's lives. It was the central focus of mine for quite a while.

Oscillating in and out of Communities

As many scholars have noted, a common feature of cyber communities, much more than in the physical world, was members' transience. People moved fluidly through groups looking for one that felt right. As their lives evolved and changed, what had once been a good fit might no longer suit them. People's support needs were dependent on the stage of their self-injury career. Bonnie, the bankruptcy coordinator, commented on how she was ready to move on to a different kind of group:

And I think [Gamma] is a good group and I've met a lot of really nice people there. But there's also a lot of that constant crisis, like help me right now, right now, right now kind of thing. And I'm trying to avoid being in that situation again. Because I was in that situation so I have a little harder time with that group now.

It was also common to see people announce that they were going to either leave or take a break from the group. Sometimes people indicated what was causing their departure, but other times they just said goodbye. Bob, a regular on the Alpha list, posted this in an upset message:

I must excuse myself for a while while i cope with this situation. i will look in on mail daily—please contact me directly. i will not open group mail for a while until i feel better about things. I sign off wishing everyone a safe weekend while i mull over whether or not i wish to remain a member of this group or not.

Although Bob rejoined the group, some people eventually left the community for good. Many held on long after they had (allegedly) desisted from self-injuring for months or years, enjoying the outreach they provided to those still in the throes of the behavior. As Brown (1991) noted for professional—exs, individuals often helped themselves by helping others. But both lay people and the therapeutic community viewed self-injury as a by-product of pain or unhappiness rather than a psychological syndrome unto itself, and when people found stability and joy in their real lives they left the cyber community behind. Cindy, the retail sales clerk, found a better job, got into therapy, met a boyfriend, and her life improved significantly. She no longer felt the need to self-injure, and although her group had been a huge part of her life for three years, she gradually faded out of the picture. At first she did not write as often but did read some of the posts, but eventually she found people's stories depressing and self-absorbed. She stayed with the group for as long as she felt strong urges to self-injure, but as these weakened, she was able to leave. Reflecting on her life after self-injury, she noted that if she had a problem or got upset, she was likely to turn to her boyfriend or find some other way of dealing with it.

CONCLUSION

The cyber world represents an ephemeral space of creation and destruction. It offers people who are dispossessed by mainstream society a place where they can form their own culture and communities. Schutz (1967) noted that space refers not only to physical, but to social proximity, to how far or close we feel with others, the connections between people. It includes everything with the character of "beside-each-otherness."

In an era where we have become concerned about the loss of community (Putnam 2000), the decline of civic participation and neighborhood communities has been replaced with membership in these virtual communities. Foster (1997) and Cerulo (1997) suggested that these represent enclaves of *Gemeinschaft* within a *Gesellschaft* world. Bellah et al. (1985) argued that hyper-individualism, rather than urbanization or a lack of morality, has obscured the idea of community in contemporary society, with the rhetoric of individualism as our primary language. The resultant "communities of interest" are simply aggregations of self-interested, self-seeking individuals who join together to augment each individual's good. Yet we find that these cyber self-injury communities offer a context within which self-awareness develops and personal identities are formed. Although they are ephemeral and transient, with many members lurking silently or flitting in and out, they display norms with sanctions, and they offer a forum that joins people together to form a social order, enabling cooperation and association.

These cyber communities vary by the type of media supporting them, with very different kinds taking shape in the passive websites, the more active bulletin board and group discussions, the interactive chat rooms, and the visual realities of the cyber cafes and pubs. As people move up the ladder of interactivity and virtuality, these venues become more compelling and engaging, progressively offering greater potential for a real community feel, for identity construction and self-lodging, and for cyber-absorption to the exclusion of the real world. Although these virtual self-injury communities lack authentic embodiment, spatial grounding, the commitment associated with the permanence and obligation of social roles and relationships, and some even temporal "telecopresence," they offer the freedom from risk of anonymity, the intimacy of privacy, the forgiveness of self-recreation, the ease of point-and-click interaction, and the multiplicity of overlapping simultaneous participation in different venues. For many people, especially those raised with the convenience of this type of interaction, the web of community offered in the cyber world is preferable to the real world.

REFERENCES

Adler, Patricia A. and Peter Adler. 2007. "The Demedicalization of Self-Injury: From Psychopathology to Sociological Deviance." *Journal of Contemporary Ethnography*, 36(5):537–70.

Becker, Howard S. 1963. *Outsiders*. New York: Free Press.

Bellah, Robert, Richard Madsen, William M. Sullivan, Ann Swidler, and Steven Tipton. 1985. *Habits of the Heart*. Berkeley: University of California Press.

Brown, J. David. 1991. "Preprofessional Socialization and Identity Transformation." *Journal of Contemporary Ethnography* 20:157–78.

Cerulo, Karen A. 1997. "Reframing Sociological Concepts for a Brave New (Virtual?) World." *Sociological Inquiry* 67:48–58.

Deshotels, Tina H. and Craig J. Forsyth. 2007. "Postmodern Masculinities and the Eunuch." *Deviant Behavior* 28:201–18.

Foster, Derek. 1997. "Community and Identity in the Electronic Village." Pp. 23–27 in *Internet Culture*, edited by D. Porter. New York: Routledge.

Jones, Steven G. 1997. "The Internet and its Social Landscape." Pp. 5–35 in *Virtual Culture: Identity and Communication in Cybersociety*, edited by S. Jones. Thousand Oaks, CA: Sage.

Kitsuse, John. 1980. "Coming Out All Over: Deviants and the Politics of Social Problems." *Social Problems* 28:1–13.

Lemert, Edwin M. 1967. *Human Deviance, Social Problems and Social Control*. Englewood Cliffs, NJ: Prentice-Hall.

McKenna, Katelyn Y. A. and John A. Bargh. 1998. "Coming Out in the Age of the Internet: Identity 'Demarginalization' through Virtual Group Participation." *Journal of Personality and Social Psychology* 75:681–94.

Melbin, Murray. 1978. "Night as Frontier." *American Sociological Review* 43:3–22.

Putnam, Robert. 2000. *Bowling Alone*. New York: Simon & Schuster.

Rheingold, Howard. 1993. *The Virtual Community: Homesteading on the Electronic Frontier*. Reading, MA: Addison-Wesley.

Schutz, Alfred. 1967. *The Phenomenology of the Social World*. Translated by George Walsh and Frederick Lehnert; with an introduction by George Walsh. Evanston, IL: Northwestern University Press.

46

Creating a More Just and Sustainable World

BRETT JOHNSON, ROSS HAENFLER, AND ELLIS JONES

Now that you've come to the end of this course, possibly your first sociology class, it's time to reflect on what you've learned and how it can apply to your everyday life. You may have thought about some of the problems and issues in the world over the course of this term, and we hope that you have done so from a different perspective than you are used to, a sociological perspective. In this chapter Jones, Haenfler, and Johnson offer some practical yet small suggestions for how you might use the concepts and the knowledge you've acquired this term to shift your view on the world and how you might take any inspiration you have developed and harness it. Feel free to pick and choose from the myriad creative ideas they toss out and find the ones that you might like. A lot of people want to make the world a better place, but they don't know how. After reading this chapter you will be an expert to your friends!

Sociology allows us to see the world in a powerful new way, exposing the inner-workings of our societies and helping us to uncover social patterns that most people take for granted. It teaches us to "zoom out" of our own limited personal experiences, revealing the complexity of the social world that shapes our lives. Unfortunately, we don't always like what we see. You may have learned about racial inequality, wealth and health care disparities, sexism, global poverty, wars, urban violence, and other social problems. On top of that, you are likely worn down from a semester of hard work and inadequate sleep. Though you may be ready to take your final exam and pack it in, we

SOURCE: Brett Johnson, Ross Haenfler, and Ellis Jones, "Creating a More Just and Sustainable World." Reprinted by permission of the authors.

hope you will spend some time looking past the end of the term to consider the place of sociology in your future. Beyond revealing the world's problems, thinking sociologically—using your sociological imagination—creates possibilities for change and opportunities to make a real difference in the world. In this chapter, our goals include outlining a vision for a more just and sustainable world, demonstrating the lessons sociology teaches to help us create that world, and suggesting ways you can make a difference, both as individuals and as part of social movements. By the time you are done reading, we hope that you will be able to put sociology into action.

LESSONS OF SOCIOLOGY AND SOCIAL CHANGE

Lesson: There Is Massive Inequality in Our Present World

It doesn't take a sociologist to tell you about the staggering amount of inequality and suffering in the world, and the prospects of change in the near term seem slim at best. Consider just a few examples:

- One *billion* people live on one dollar per day or less (Chen and Ravallion 2004).

- The richest 10% of humans make as much income as the bottom 90% combined (UNDP 2005, p. 4).

- Everyday in the world, 24,000 people die from preventable, hunger-related causes (UNDP 1999).

- Since World War II, 20 million people have been killed in 150 wars (Zinn 1997).

- The U.S. median household wealth for blacks is $6,000—for whites, $88,000 (14 times as much) (Associated Press 2004).

- In the U.S., one-and-a-half million women are raped or sexually assaulted by their intimate partner every year (Tjaden and Thoennes 2000).

- 18% of all babies in Angola die before their first birthday—67 times the rate in Sweden (CIA 2008).

- Average life expectancy in rich countries is almost 80 years but is only 50 in sub-Saharan Africa (and only 41 in Zambia!) (UNDP 2008, pp. 229–232).

This kind of information overwhelms most students, often leading to a culture of "doom and gloom" that hovers over sociology courses (Best 2001). Sociology teachers are often wary (rightly so) of telling students what they "should do" with their newly gained sociological knowledge so that courses end up bursting with information on problems, seemingly with no solutions. To many, these horrendous conditions are so foreign to their daily experience that they simply are hard to believe. These statistics represent real people—as real as you and me—living, breathing, human beings with dreams of a better life. Sociology enables us to confront uncomfortable realities of which we may rather have remained unaware. Unfortunately, too often in the sociology classroom, increasing awareness of human suffering perpetuates a cycle of cynicism in which you (1) learn about a particularly awful situation, (2) want to help but can't see how, (3) fail to act and become disheartened or frustrated, and finally (4) try to shut out the world's problems because they only serve to aggravate and depress you. Of course, ironically, such cynical thinking just perpetuates the problems even further! It is important that we change this cycle of cynicism into a cycle of active engagement that allows us to begin resolving these social problems.

Lesson: Social Problems Have Structural Foundations

By asking critical questions and conducting rigorous research, sociology equips us to understand the structural foundations or root causes of social problems, an important step to effective social change. Inequalities are perpetuated on a daily basis through our societal institutions: our governments, media organizations, religious institutions, economies, and families. These institutions tend to reflect the interests of those with more influence or power—those with higher levels of status, education, and income—resulting in many of the inequalities we've discussed. Social problems are embedded in institutions, not individuals. For example, violence toward women is not an anomaly confined to a few abusive men—it happens everyday. It is situated within patriarchal systems that devalue women and things feminine. Men's attempts to live up to a cultural ideal of being a "real man" drive much of this violence. This "tough guise," fostered in the media, glorifies certain qualities associated with manhood, such as power and control, while disdaining others, like vulnerability and compassion (Katz 2000). We must alter the cultural institutions that devalue women and glorify hypermasculinity.

Lesson: Social Change Requires a New Vision

It's easy to point out the world's problems, but more difficult to envision and create something better. Given the diversity and complexity of human societies, creating a one-size-fits-all prescription seems like a daunting task. Nevertheless, we need a vision for the future, something to spark our imaginations and creativity, something to inspire us. While sociology does not provide an "objective" formula for how a society should be structured, much less how one should live, it does offer some tools and insights useful for envisioning a better world.

One important sociological insight is each of us has been socialized by our families/media/religion to believe in a set of values. By learning about other cultures, we quickly realize that if we would have grown up in a different corner of the world, our values and beliefs would be very different than the ones we have now. That profound realization brings with it a destabilizing sense of doubt. Harvard education professor William Perry (1981) calls this process moving from *dualism*, where we think our values are right and others' are wrong, to *relativism*, where all values are considered of equal merit. Relativism is great for building tolerance and understanding across groups but it does not provide much guidance for a life philosophy. So what are we supposed to believe? Perry identifies the final stage (and most challenging) of critical thinking as *commitment*—something that involves balancing consistent openness to new ideas, reflective analytical thinking, and a deep and active commitment to core values. It involves putting ideas into action, taking risks, making mistakes, and learning from them. It also involves valuing the humanity in others as well as ourselves.

We encourage you to spend time contemplating your values. Based on our reflection, we envision a world built on the following seven foundations (Jones, Haenfler, and Johnson 2007):

Economic Fairness: A world dedicated to economic fairness would strive to meet every person's basic needs so that no one would lack food, shelter, clothing, or meaningful work. People's strength of character and passion should determine their opportunities rather than the economic circumstances into which they were born. The gains from economic growth should be widely shared.

Comprehensive Peace: A world committed to comprehensive peace would shift its creative energies toward cooperating rather than competing, resolving conflict rather than escalating it, seeking justice rather than enacting revenge, and creating peace rather than preparing for war.

Ecological Sustainability: A world committed to ecological sustainability would create a new vision of progress that recognizes that the future of humanity depends upon our ability to live in harmony and balance with our natural world.

Deep Democracy: A world built on deep democracy would empower citizens to participate in shaping their futures every day (not just on election day), provide broad access to quality information, and democratize our most powerful institutions.

Social Justice: A world dedicated to social justice is a place where everyone receives respect and equal access to jobs, education, and health care regardless of race, gender, ethnicity, sexual orientation, age, physical or mental abilities, or economic background.

Simple Living: A society that embraces simple living would encourage each person to find meaning and fulfillment by pursuing their true passions, fostering loving relationships, and living authentic, reflective lives rather than by seeking status and material possessions.

Revitalized Community: A revitalized community would create a healthy and caring environment for people to celebrate their many shared values while embracing individual differences, and would provide support for each person's physical, emotional, and spiritual needs.

Lesson: We Are All Connected

One of the broad, fundamental questions that sociologists tackle is "What is the relationship between the individual and society?" Put another way, how does society, or social context, influence individuals and how might individuals, in turn, influence society? Sociology reveals that forces outside of us shape our beliefs, values, and opportunities. You were not born knowing girls should play with dolls and boys should play with "action figures"—you learned, from your parents, peers, teachers, and the media a host of gender "rules." Not everyone has an equal chance to be the CEO of a Fortune 500 company—someone born to a poor rural family in Appalachia faces greater challenges than someone born to a wealthy family that can provide an elite education and powerful business connections. Recognizing we are not completely free can be a blow to the ego. On the other hand, the sociological imagination provides us an opportunity to increase our freedom and expand our options, even if we can never be totally free of society's influence. The first step is realizing our basic interconnectedness—what you do affects me, what I do affects you, even if we can't always see how.

American culture's focus on individualism challenges our ability to acknowledge our interconnectedness, especially the opportunities and constraints society poses for our lives, preferring to believe we shape our own destinies. We tend to think our achievements result solely from our own hard work. In reality, we all rely on each other for our daily existence. Ecology teaches us that we eat food that grows in soil nurtured by microscopic organisms and we drink water that has vaporized from the oceans. Sociology teaches us that we wear clothing made by people we will never meet who live across the planet. Tax dollars have paid salaries of those who have educated us. Our personal well-being is inextricably linked to the well-being of our families, our friends, our communities, and our planet.

Lesson: Your Actions Make a Difference

As you begin to understand the interconnected nature of the world, the relationship between individuals and social structures, you realize that you are both very powerful and yet very small—you influence everything around you, yet you exist as part of something much greater than you. You may be thinking, "I don't influence society—CEOs, famous people, legislators, and activists influence society, not people like me." Social systems are made up of patterns of action taken by millions of people. When you study the connections that bind us all together, you gain awareness of how each of your actions affects other people and the planet. Your political activity (or inactivity) perpetuates our political system. How you spend (and invest) your money props up our economic system. Your lifestyles encourage a set of values and behaviors and discourage others. Your participation can even transform a small group of people into the beginnings of a social movement. The key question is not *if* but *how* you are impacting society. Even by doing "nothing" you perpetuate the society in its current state. Recognizing that each of your actions sends little ripples throughout the rest of society creates a sense of personal responsibility for your role in creating a better world.

Lesson: It's Not Easy to Deviate from the Mainstream

Given the power of social structures to shape who we are and what we believe, it's not always easy to stray from the norm. As you leave this course, there are at least two paths for you to choose from. The first path, the "conventional path," involves living as if your sociology course never happened, shielding yourself from human suffering, staying away from people who are "different" from you, pulling back from the world and devoting yourself entirely to your own well-being and that of your immediate circle of friends and family. Programmed like a robot, you may find yourself working long hours and rewarding yourself with lots of stuff. Like the momentum of a river, our society will "naturally" carry you to this path. You don't have to choose this path, it will choose you.

In order to live the second path, the socially engaged path, you must intentionally choose it. This path is less well charted, though many before you have taken it. It involves actively engaging the world and creatively integrating your own desire for happiness with your desire for a better world. Instead of carrying out conventional daily routines (a.k.a. living on auto-pilot), this path requires a "discursive consciousness" (Giddens 1984) where you evaluate and reflect upon your values and intentionally choose actions that support them in the world. Despite the extra effort, this path is ultimately richer, deeper, more fulfilling, and it allows you to have your voice heard in this collective project of building a better world.

Lesson: Societies Are in a Perpetual State of Change

Sometimes social patterns seem so persistent, so natural that it's easy to forget how much things have changed in just a few generations. Society is in a constant state of flux—though the change is often imperceptible on a daily basis. Asking your grandparents how society has changed over their lifetimes reveals the shifting nature of society—more egalitarian gender roles, improved race relations, shrinking family size, and the globalization of the economy are just a few of the significant transformations over the last 50 years. Remember that our society is made up of human relationships and interactions and therefore can be recreated by human action. It is important to not *reify* societies, meaning to treat social realities as if they are unchanging and "natural." After all, where would we be if abolitionists hadn't challenged slavery? If suffragettes hadn't demanded the right to vote? If labor activists hadn't fought for safe working conditions and an end to child labor? Societies change—and it is people like you who can nudge it in the right direction.

TURNING INSIGHTS INTO ACTIONS

Even if you have been persuaded by this piece, you might be thinking: "I don't have the time or energy to 'change the world'." Almost all of us feel like we are too busy to take on something new. Creating social change requires varying levels of time commitments. Give the time that you have. You don't have to be a saint or give up your whole life for a cause to make a difference.

Alternatively, you may be thinking "I don't know enough about the issues" to get involved. Certainly, action without understanding is foolish but always needing to "learn more" can be a crutch that lets you off the hook from ever taking action. Get active and you will learn more along the way. Every journey begins with a single step. You might think "I can't make a difference, so why even try." These kinds of mental traps serve the interests of the status quo by convincing us we are powerless. Consider the following story of how college students can make a real difference:

> One spring at Luther College, a small group of friends got together and decided to attend a local conference on sweatshops—factories in poor countries where workers are abused and underpaid. They came back inspired and started a campus group committed to economic justice. The group decided to organize "Fair Trade Fest," a small gathering in the student union that featured student musicians, short speeches on fair trade, and a table full of "fair trade products" made by artisans through the world (from www.tenthousandvillages.org). The event helped educate and inspire students, faculty, and staff about building economic justice. The group grew slowly over the semester and the energy of the new people helped ignite a spark of creativity and action. The next campaign was to improve the lives of global workers who produce college apparel. After attending another conference and doing lots of research, they made an effective presentation to student leaders and administration to become a member of the Worker Rights Consortium <www.workersrights.org>, which monitors the working conditions in factories around the globe. Next the group held a silent auction of student artwork to raise $300 to invest in "micro-lending"—small loans given to the world's poor to break them out of the trap of poverty (learn more at Kiva.org). Few of these extraordinary, ordinary people considered themselves activists when they came to college. They are people who fit their activism in between exams and hanging out with their friends (actually they often hang out with their friends and do some of these things at the same time). They are people just like you.

PART I: CREATING STRUCTURAL CHANGE

Social change occurs through myriad forces including population dynamics, natural disasters, and technological and economic transformations. Pursuing our vision of a better world requires a more intentional approach to change. At the macro level, social change is most likely to occur through collective action such as political organizing and social movements. Social movements inspire people to reconstruct their notions of what is possible and what should be done to create that possibility. Movement participants then integrate movement values into their lives and spread the movement's message, as in the environmental movement. Movements also put collective pressure on decision-makers in corporations and government to change the way they operate. Fundamental social change occurs when social institutions are transformed.

Actively Participate in Our Political System

Political organizing and involvement help create structural social change. Politicians get to set an agenda for the society. They can jumpstart research into solar energy or give tax breaks to oil and coal companies. They can peacefully negotiate with other countries or invade them. They can help improve the lives of the poor or ignore them and give tax breaks to the most well-off, and they can restrict or expand women's and gay rights.

For beginners, it's important first to register and then vote on a regular basis (every two years). It may seem like such a small thing, but it is essential to our democracy. The idealism so prevalent in young people could inject some much-needed change into our politics, yet voter turnout among students is generally quite low.

Advanced political engagement involves providing support to candidates who share your vision of a better world (giving money, canvassing, phone banking) and supporting organizations that lobby political leaders to implement public policy that reduces inequality, builds peace, and promotes sustainable use of natural resources.

- Register for an absentee ballot so that you can vote from your home/dorm/apartment.
- Check out Project Vote Smart (www.vote-smart.org) and League of Women Voters (www.lwv.org) for some valuable, unbiased information.
- Start a registration drive on campus. Take a look at Rock the Vote (www.rockthevote.com) for resources.

Join Social Movements that Create Long-Term Social Change

Social movements consist of groups of people dedicated to pursuing a shared vision of a better community or world. Participating in social movements requires building solidarity with others who share common goals. It includes having conversations, going to meetings, spreading your message, and writing letters to the editor. The best place to start is to join an organization that is working hard on an issue you really care about.

Be sure to support organizations that address the root causes of social and environmental problems, in addition to ones that meet your community's immediate needs. For example, volunteering at your local homeless shelter has an immediate impact on people's lives (micro-level change), while working with the National Low Income Housing Coalition will help eliminate the need for homeless shelters (macro-level change). Donating money to a developing country suffering from famine is important, but so is giving to a group such as Oxfam <www.Oxfam.org> that helps develop self-sufficient, long-term, sustainable economic development throughout the world.

Top 7 Organizations to Join/Start at Your Campus

1. ONE Campaign [www.one.org]
2. Student Global Aids Campaign (SGAC) [www.fightglobalaids.org]
3. Student Environmental Action Coalition (SEAC) [www.seac.org] or Sierra Student Coalition [www.ssc.org]
4. Gay/Straight Alliance [www.gaystraightalliance.org]
5. Student Peace Action Network [www.studentpeaceaction.org]
6. Students Against Sweatshops (SAS) [www.studentsagainstsweatshops.org]
7. Amnesty International [www.amnesty.org]

Top Campus Collective Action Campaigns

Organizing social change events takes time but can be fun and empowering. There is power in numbers! You may only spend a few years at your college but your actions while you are there can impact that campus for many years into the future. Lots of students are ready to get involved and are just looking for an opportunity to make a difference. Form coalitions with students groups—human rights, environmental, and diversity activists share many common values. Try mobilizing your campus faith communities—many students involved in such groups already put their faith into action, undertaking volunteer projects.

Celebrate "Buy Nothing Day" on Campus   "Buy Nothing Day" is an international event on the Friday after Thanksgiving (the most popular U.S. shopping day of the year) to publicize the effects of over-consumption on our families, culture, and the planet. Never underestimate the impact of "symbolic politics"— that a fun, thought-provoking event can effectively challenge dominant cultural notions. Learn more at <www.adbusters.org/bnd>.

Introduce "The Graduation Pledge."   Encourage your fellow students to take the voluntary "graduation pledge": "I pledge to explore and take into account the social and environmental consequences of any job I consider and will try to improve these aspects of any organizations for which I work." The pledge is a great way to start conversations about how students want their work to contribute to a better world. Learn about how colleges/universities have integrated the pledge into their campus culture at <www.graduationpledge.org>.

Promote Fair Trade Coffee   Campuses serve a lot of coffee to their hardworking, sleep-deprived students. Take action to ensure that coffee farmers are paid a fair wage. Ask your campus cafes and cafeterias to carry "fair trade certified" coffee. Learn more about "fair trade" from United Students for Fair Trade <www.usft.org> and Co-op America's "Guide to Fair Trade" <www.fairtradeaction.org>. To find out what other colleges are doing visit: www.transfairusa.org/content/support/campus.php.

Work to End Sexual Violence on Campus   Sexual assault has long been a terrible problem on college campuses. You can help prevent rape by joining (or starting) your campus anti-rape group—your student health center should be able to point you in the right direction. Organize a Take Back the Night rally [www.takebackthenight.org] or a V–Day event [www.vday.org] to help spread awareness. If you are a man, consider how you might get involved (and involve other men) in this important work.

Promote Tolerance and Diversity   Chances are your campus has a variety of student groups that promote diversity, respect, and social justice. You can help by joining their organizations, attending their events, supporting their causes, and helping organize against bias and intolerance. Celebrate National Coming Out Day on October 11 [www.hrc.org]. Support targeted goals for your institutions about enrollment by first generation college students and domestic students of color. Teaching Tolerance [www.tolerance.org] offers "10 Ways to Fight Hate on Campus" and "101 Tools for Tolerance."

Green Your Campus   From small projects such as getting recycling bins conveniently located in each dorm to big projects such as installing a wind turbine to produce "green energy," there are endless ways for you to make your campus more sustainable. As an easy starting point, consider getting your college to set up double-sided printers in your campus computer labs to default to print double-sided automatically. This one-time action can reduce paper usage nearly in half, saving forests worth of trees for years to come!

The Association for the Advancement of Sustainability in Higher Education (AASHE) <www.aashe.org> has amazing resources for environmentally friendly initiatives at colleges/universities including "green" building designs, recycling, alternative transportation, energy efficiency, and green energy. Encourage your institution to become a member of AASHE and lobby your president to sign onto the American College & University President's Climate Commitment to fight global warming <www.presidentsclimatecommitment.org>. Other great resources for greening college campuses include the National Wildlife Federation's "Campus Ecology" website: <www.nwf.org/campusecology> and The Campus Climate Challenge <www.climatechallenge.org>.

Become a Sweatshop-Free Campus   Student activists have effectively mobilized at universities across the country to demand fair working conditions for the workers who make products (particularly clothing) with their college's logo. One of the most powerful steps you can take to spread economic justice is to get your institution to become a member of the Worker Rights Consortium <www.workersrights.org> which monitors working conditions at college apparel factories. Over 180 colleges and universities are members. Check to see if your campus is one of them. If not, you can be the one to start the effort and make human rights the priority it should be. Learn more about sweatshops from: United Students Against Sweatshops <www.studentsagainstsweatshops.org> and Co-op America's excellent "Guide to Ending Sweatshops" <www.sweatshops.org>.

Help Make College More Affordable   To promote equality of opportunity, people of all races and social classes must have access to higher education. Collective action is necessary to increase government financing for higher education and to reduce the debt burden of graduating college students. Supporting the United States Student Association <www.usstudents.org> will help their lobbying and organizing efforts across the country.

Make Your College's Endowment Socially Responsible   Help integrate your institution's mission and its investment practices. Consider a goal to invest 1% of your college/university's endowment in domestic and international "community investments" such as microlending and community development loans funds (e.g., Calvert Social Investment Foundation). You will unleash hundreds of thousands or millions of dollars to fight poverty throughout the globe. Learn more at: The Responsible Endowments Coalition <www.endowmentethics.org> and the Sustainable Endowments Institute <www.endowmentinstitute.org>.

Sociologists who study social movements know that before people will participate they must feel a sense of efficacy—they must believe that what they do could make a difference. Unfortunately, with any political or social movement participation you are never guaranteed the results you desire. Sometimes your action will appear to make little difference but in reality you can never accurately assess the positive effects of your actions. What we can accurately predict is that cynicism and inaction is a self-fulfilling prophecy. If you believe you can't make a difference, you won't.

PART II: INDIVIDUAL ACTIONS

Engaging in social movements and politics is not the only way to create change. There are many actions you can take in your "private life" that will contribute to the creation of a more peaceful, just, and sustainable world. As we've indicated, social structure is comprised of actions taken by individuals. Social systems rest on the cumulative actions—and cooperation—of a majority of people in society. Similar to Gandhian non-cooperation, you can withdraw your support for aspects of the status quo and redirect your actions to support alternatives that are in-line with your values. For example, to challenge how children are socialized into gender roles, parents could avoid buying Barbie dolls (for girls) and violent toys (for boys), replacing them with gender-neutral and nonviolent alternatives.

One of the biggest challenges is to figure out how to integrate making a better world and living a meaningful, enjoyable life. First off, you don't have to follow some pre-designed path for making the world better. Living a reflective, vital life means making a commitment to living out your core principles. This commitment is a lifelong challenge to integrate your desire to create a better world with your daily actions—leading to a truly meaningful life. As Mohandas K. Gandhi, the renowned practitioner of nonviolence, famously expressed, "you must be the change you wish to see in the world."

Buy Less Stuff In essence, *live simply*. Find pleasure in your relationships, work, and passions instead of in the make-believe worlds that Madison Avenue creates. The present advertising barrage promotes the idea that satisfying each of our desires is more important than contributing to the welfare of others and seeking deeper fulfillment—a very destructive message that ads convey 24 hours a day, 7 days a week. Seek to liberate your consciousness from the impulse to consume. This act of mental self-defense will open up time and energy to more fully engage the world. For things you feel you must buy, try to buy used, durable, and reusable products rather than new, cheap, and disposable junk. Check out the Center for a New American Dream's "Simplify the Holidays" Campaign for more ideas (www.newdream.org/holiday).

Watch Less TV It is time to renegotiate our relationship with television. The average American watches over four-and-a-half hours of TV per day (Nielsen Media Research 2006). Watching less TV is one step toward resisting commercialism and connecting with your community. It also opens up time for fulfilling activities like: spending quality time with friends and family, catching up on some reading, pursuing hobbies, volunteering, playing music, enjoying nature, exercising, playing with your dog, or just getting outside. Consider dropping your cable subscription.

Vote with Your Dollars Every dollar you spend can be a vote for the world you want to live in. It's important to support farmers' markets, food co-ops, and local, independent businesses that spread wealth throughout your community and make it an interesting place to live. Whether it's groceries, electronics, or fast food, college students make powerful impacts as consumers. When you buy clothing from more socially responsible companies like Patagonia or Levi's and boycott harmful companies like Wal-Mart and Dillard's you cast a vote for a more just world. Look for the "fair trade" certification logo and check out Better World Shopper at www.betterworldshopper.org to find report cards that rate the social responsibility of many different companies.

Study and Volunteer Abroad Over and over again our students tell us about the transformative power of studying abroad. Living in another culture will teach you things that no classroom experience ever could. We encourage you to seek out different cultures, beyond your national borders, to gain new insights into the meaning of life and the multiplicity of realities that define people's lives. The cross-cultural interaction that comes from studying and volunteering abroad encourages understanding, builds compassion, and decreases the likelihood of destructive conflict including war.

    Council on International Educational Exchange [www.ciee.org]
    Volunteers In Asia [www.via.org]
    The Peace Corps [www.peacecorps.gov]

Volunteer In-Country We also strongly encourage you volunteer within your own country, as a way of experiencing the diversity that makes up this vast nation. You'd be amazed at the difference you can make, the personal growth you can experience, and the strong bonds you can form when you work with others across the nation on projects dedicated to building stronger communities.

Americorps [www.americorps.org]

Habitat for Humanity [www.habitat.org]

Teach for America [www.teachforamerica.org]

Create a Giving Budget   Even though as a student money may be tight, now is exactly the time to create a semester-based budget for supporting those organizations that are working to create the kind of world you want to live in. Here is the key. The amount you give is not important—$50, $20, $10, or $5—it doesn't really matter. What matters is that you begin creating this budget now and give it to a good organization every semester. You'll have plenty of time when you're older to consider giving $100 or more. Your participation, like voting in an election, may seem small but it is, in fact, essential to the long-term well-being of these organizations.

If you really want to try something powerful, try being a microfinancier. You can fight global poverty by making loans as small as $25 to the world's poor at <www.kiva.org>.

Stay Informed   It's not always easy to stay informed with high quality information about what's going on in the world when you're in a constant cycle of working, studying, eating, learning, and (just maybe) sleeping in college. Public radio, alternative magazines, and a few, select TV shows can provide the solid investigative journalism you need. Make it a habit to learn about the pressing issues of the day. This knowledge will provide a foundation for informed action. Magazines such as *Mother Jones* offer hard-hitting stories often absent from mainstream news coverage. PBS programs like Frontline produce fascinating documentaries about a variety of contemporary social issues.

National Public Radio [www.npr.org]

Public Radio International [www.pri.org]

The New York Times [www.nytimes.com]

Mother Jones [www.motherjones.com]

Ode [www.odemagazine.com]

Frontline [www.pbs.org/frontline]

Challenge Your Stereotypes   Given the power of social context over our beliefs, it is virtually impossible to avoid being infected by some degree of racism, sexism, and homophobia. Stereotypes about racial and ethnic minorities, lesbians and gay men, women, immigrants, and others abound, even if many people know such views are unfair. This is less a personal failing than a public issue. Examining your own prejudices requires courage but can transform your relationships. Avoid slang that denigrates other groups—"fag," "gay," "homo," "bitch," and racial slurs may be the vernacular of the day, but using such words belittles women and men alike. Along similar lines, complaining about "illegal" immigrants without understanding the sociological forces at work demonstrates ignorance.

Eat Green Tweaking eating habits is an easy way to make a daily difference. Eating less fast food, buying some organic food, and eating less meat are all ways to make environmentally friendly food choices. Not only is fast food bad for your health, the industry promotes intensive, destructive farming practices. Certified organic food may be a bit more expensive but ensures that producers have not used chemical pesticides, hormones, genetic modification, steroids, or antibiotics. Meat production, especially beef, requires tremendous amounts of land, water, and fossil fuels.

Fast Food Nation: The Dark Side of the All-American Meal by Eric Schlosser. New York: Harper Perennial, 2005.

Be a Mentor Helping kids develop into responsible, caring young women and men is one of the most powerful (and most fulfilling) actions you can take. Attention from compassionate adults is one of the best ways to keep kids out of trouble and help them become successful young adults. Volunteer at an after-school program, be a Big Brother or Big Sister, or help out at your local Boys or Girls Club.

Big Brothers Big Sisters [www.bbbsa.org]
Boys and Girls Clubs [www.bgca.org]

Work for Justice As you look for employment consider that there are many opportunities to work with organizations whose sole purpose is to make the world a better place. Nonprofit organizations desperately need the skills you are developing in college. They need computer programmers, marketing services, administrative staff, and social service workers. You can focus virtually any interest you have into a job that promotes justice.

Idealist <www.idealist.org>
Environmental Career Opportunities <www.ecojobs.com>
Nonprofit Career Network <www.nonprofitcareer.com>

Choose a Fuel-Efficient Car—and Drive It Less People across the political spectrum finally acknowledge the threat of global warming. Sports utility vehicles and gargantuan trucks may be in vogue, but 10–15 miles per gallon isn't doing the atmosphere any favors. When you pick out a car, add fuel-efficiency to considerations like cost and preference and try to get a car that gets at least 30 miles per gallon. Not only will it improve air quality and reduce carbon emissions but your fuel-efficient car will save you thousands and thousands of dollars in reduced gasoline bills. To compare cars and find tips on driving efficiently check out www.fueleconomy.gov.

Green Your Home When you finally settle in one place for a bit, take steps to make your home energy efficient. Many ideas are relatively easy and inexpensive. Compact fluorescent bulbs, water heater jackets, and low-flow showerheads are simple ways to save energy and water and can be found in most hardware stores. NiMH rechargeable batteries for your remote controls, smoke alarms, toys, clocks, and so on help reduce hazardous waste. Energy Star rated appliances save energy, and nontoxic cleaners reduce toxic chemicals. Composting your food waste recycles organic material back into the soil. Buying green power invests in renewable energy sources. Perhaps most importantly, living in the smallest home that meets your needs and living close to work if possible are crucial opportunities to save energy and reduce pollution.

Nontoxic cleaners: www.seventhgeneration.com

Energy Star: www.energystar.gov

Green power: www.eere.energy.gov/greenpower

Share Housework and Childcare Equitably and Model Flexible Gender Roles   Despite the fact that women increasingly work outside the home they still find themselves responsible for a disproportionate amount of housework and childcare. This "second shift" means women have far less leisure time than their male partners (Hochschild and Machung 2003). Men need to step up. Furthermore, given that we teach (or socialize) kids what it means to be boys and girls, try modeling flexible gender roles so they understand that women and men are equally capable at nearly any task. Trade off on various household tasks, show that it's OK for both women and men to express a variety of emotions, and teach boys how to cook and girls how to fix the sink.

Bank and Invest in a Socially and Environmentally Responsible Manner   Socially responsible investing (SRI) has grown significantly overly the past two decades. There are now a wide variety of community development banks, such as ShoreBank <www.sbk.com>, dedicated to economic development in low-income communities. Just switch your bank accounts and your money is doing good—24/7. On-line banking now makes it easy to use a bank in another town. There are also many mutual fund companies committed to social responsibility. Learn more about SRI at <www.communityinvest.org> and <www.socialinvest.org>.

CONCLUSION

Years from now as you reflect upon your college experience we hope you fondly remember your introductory sociology course. More importantly, we hope you use your sociological imagination in whatever you pursue. Like many students, you may have chosen Soc 101 because it fulfilled a general education requirement or fit your schedule, not because of some intrinsic interest in the subject. At times, the course might have seemed abstract and theoretical, in other words not "useful." Nothing could be further from the truth. Not only have you learned to think sociologically, you now have a few tools to put your knowledge to good use. You understand that problems embedded in social institutions require structural solutions—we've armed you with some ideas and resources to engage in social movements working for long-term change. Just as importantly, you've learned that social institutions rest upon the actions of millions of individuals and that the path you take in your daily life is more than a personal choice—it is an opportunity for change. We hope you grasp that opportunity by joining together with others who share your vision for a better world and that you integrate this vision into your daily life.

REFERENCES

Associated Press. 2004. "Study Says White Families' Wealth Advantage has Grown." *New York Times*. October 18, 2004, p. A13.

Best, Joel. 2001. "Social Progress and Social Problems: Toward a Sociology of Gloom." *The Sociological Quarterly* 42: 1–12.

Central Intelligence Agency (CIA). 2008. *The World Fact Book*. [online] https://www.cia. gov/library/publications/the-world-factbook/

Chen, Shaohua and Martin Ravallion. 2004. "How Have the World's Poorest Fared Since the Early 1980's?" *World Bank Policy Research Working Paper* 3341, June 2004.

Giddens, Anthony. 1984. *The Constitution of Society: Outline of the Theory of Structuration*. Berkeley: University of California Press.

Hochschild, Arlie Russell with Anne Machung. 2003. *The Second Shift*. Penguin: New York.

Jones, Ellis, Ross Haenfler, and Brett Johnson. 2007. *The Better World Handbook: Small Changes That Make a Big Difference*. Gabriola Island: New Society Publishers.

Katz, Jackson. 2000. *Tough Guise: Violence, Media, and the Crisis of Masculinity* [film]. Media Education Foundation.

Nielsen Media Research. 2006. "Nielsen Media Research Reports Television's Popularity Is Still Growing" [press release]. September 21. Viewed at: www.nielsenmedia. com/nc/portal/site/Public/menuitem.55dc65b4a7d5adff3f65936147a062a0/?vgnex toid= 4156527aacccd010VgnVCM100000ac0a260aRCRD

Perry, William. 1981. "Cognitive and Ethical Growth: The Making of Meaning." pp. 76–116 in A. W. Chickering, ed., *The Modern American College*. San Francisco: Jossey-Bass.

Tjaden, P. and N. Thoennes. 2000. "Extent, Nature, and Consequences of Intimate Partner Violence: Findings from the National Violence Against Women Survey." July 1. US Department of Justice, Publication No. NCJ 181867, 2000. www.ojp. usdoj.gov/nij/pubs-sum/181867.htm

United Nations Development Programme (UNDP). 2008. Human Development Report 2007/2008.

United Nations Development Programme (UNDP). 2005. The 2005 Human Development Report. New York: Oxford University Press.

United Nations Development Programme (UNDP). 1999. The 1999 Human Development Report. New York: Oxford University Press.

Zinn, Howard. 1997. *The Zinn Reader*. New York: Seven Stories Press.

Glossary

achieved status position that is attained through individual effort, such as one's education or occupation

agent of socialization the person or group that provides information about social roles

aggregates large groups of people who actually have no relationship to one another except that they might happen to be in the same place at the same time

alienation term first used by Marx, that refers to the separation of workers from the product or result of their work, which can result in feelings of powerlessness

anticipatory socialization the process by which we prepare ourselves for future roles through thinking about and rehearsing the actions, emotions, and skills that may be involved in these new roles

ascribed status position that is attained through circumstances of birth and that cannot be changed, such as one's race-ethnicity or gender

backstage Goffman's term for the setting, or frame, in which impression management is not needed; contrast to *frontstage*

bourgeoisie term used by Marx and Marxian scholars to describe capitalists, people who own the factories and mills; contrast to *proletariat*

bureaucracies highly structured and formalized organizations that are governed by laws and rules

capitalism the free-enterprise economic system in which private individuals or corporations develop, own, and control business enterprises; contrast to *socialism*

case study a research technique that involves an in-depth look at one case, such as one person, one group, or one organization

causal model graphic device that illustrates sociological relationships between variables

causal relationship an association between variables in which one influences or causes the other

class consciousness Marx's term for people in a social class who are aware of their common interests and concerns and of the fact that these interests conflict with those of another class group

closed-ended questions survey questions that give respondents only certain possible options from which to choose their answers

collective consciousness Durkheim's term describing the common beliefs, values, and norms of people within society

community a group of people who have frequent face-to-face interactions and common values and interests, relatively enduring ties, and a sense of personal closeness to one another

comparative research research method that involves the comparison of data from a variety of groups or settings, such as nations or historical eras

conformity the process of maintaining or changing behavior to comply with the norms established by a society, subculture, or other group

control variable variable that is added to an analysis to see if it affects the relationship between an independent and a dependent variable

correlation the ways in which two variables may be related to each other in a predictable pattern

counterculture a group that strongly rejects dominant societal values and norms and seeks alternative lifestyles

crime actions that a society explicitly prohibits and that are sanctioned through official means

cultural capital Bourdieu's term describing how people behave, dress, and talk and how these manners and styles differentiate those in one class group from those in another

cultural diffusion process whereby elements of one culture or subculture spread from one society or culture to another

culture a common way of life; the complex pattern of living that humans develop and pass on from generation to generation

data factual information that is used as the basis for making decisions and drawing conclusions; the plural form of the word *datum*

deductive reasoning logical process of reasoning that moves from general theories or ideas to specific hypotheses or expectations; the opposite of *inductive reasoning*

demography the scientific study of populations and their effects

dependent variable variable that is said to be influenced or caused by another variable (the independent variable)

deviance behaviors that violate social norms

discrimination differential treatment accorded to a group of people based solely on ascribed characteristics such as race-ethnicity

domination form of power in which one party controls the behavior of others through sanctions; compare with *influence*

dramaturgical theory theory derived from the work of Goffman that uses the metaphor of a drama to explain how individuals play social roles and thus produce social structure

dysfunctions negative consequences of a structure of society for the whole of society

economy social institution that includes all the norms, organizations, roles, and activities involved in the production, distribution, and consumption of goods and services

education social institution that includes all the norms, organizations, roles, and statuses associated with a society's transmission of knowledge and skills to its members

elites powerful people who are able to influence the political process

elitism view that political power and influence are dominated by a small handful of people who are relatively unified and form a comparatively small, tight-knit social network

empirical based on experience and observation rather than pre-existing ideas

endogamy marriage rule requiring people to select partners from within their own tribe, community, social class, or racial-ethnic or other such group

ethnic group group of people who have a common geographical origin and biological heritage and who share cultural elements, such as language; traditions, values, and symbols; religious beliefs; and aspects of everyday life, such as food preferences

ethnocentrism the belief that one's own culture or way of life is superior to that of others

exclusion term describing attempts to entirely remove lower-paid groups from a labor market, as, for example, through restrictive immigration laws

exogamy marriage rule requiring people to select partners from outside their own tribe, community, social class, or racial-ethnic or other such group

experiment research method that uses control groups and experimental groups to assess whether a causal relationship exists between an independent variable and a dependent variable

experimental variable see *independent variable*

extended family family that includes relatives besides parents and children; contrast to *nuclear family*

falsification the logic that underlies the testing of hypotheses; we can never prove that a theory is true, but can only say that it has either been falsified (shown to be untrue) or not yet been finished

feminism an ideology that directly challenges gender stratification and male dominance and promotes the development of a society in which men and women have equality in all areas of life

field experiment experiment that takes place in a real-life setting as opposed to a laboratory

field research research method in which a researcher directly observes behaviors and other phenomena in their natural setting

folklore myths and stories that are passed from one generation to the next within a culture

folkways norms that govern the customary way of doing everyday things

frontstage Goffman's term for the setting, or frame, in which behaviors are designed to impress or influence others and in which impression management is important; contrast to *backstage*

function the part a structure plays in maintaining or altering the society

Gemeinschaft Tönnies's term describing relationships that might appear in small, close-knit communities in which people are involved in social networks with relatives and long-time friends and neighbors, much like those that appear in primary groups

gender-based division of labor rules about what tasks members of each sex should perform

gender identity one's gut-level belief that one is a male or a female

gender roles the norms and expectations associated with being male or female

gender segregation of the labor force the gender-based division of labor in the occupational world, or the phenomenon of men and women holding very different jobs; also called *occupational gender segregation*

gender socialization learning to see oneself as a male or female and learning the roles and expectations associated with that sex group

gender stratification the organization of society in a way that results in members of one sex group having more access to wealth, prestige, and power than members of the other sex group

generalized other Mead's term for the conception people have of the expectations and norms that others generally hold; the basis of the "me"

Gesellschaft Tönnies's term describing relationships that come about through formal organizations and economic relationships rather than kinship or friendship, similar to those that appear in secondary groups

globalization the process by which all areas of the world are becoming interdependent and linked with one another

group in sociological terms, two or more people who regularly and consciously interact with each other

through engaging in some common activity and having some relatively stable social relationship

health maintenance organizations (HMOs) organizations that provide, for a set monthly fee, total care with an emphasis on prevention to avoid costly treatment later

hegemony a hidden but pervasive power involving such extreme domination of social life that we seldom recognize it or question its legitimacy

historical-comparative research research that uses historical data to compare two or more societies

historical research research method that involves the examination of data from the past, often written artifacts and records

hypotheses statements about the expected relationship between two or more variables; often derived from theories

ideal types concepts or descriptions of phenomena that may not exist in a pure form in the real world but that define basic aspects of a given situation

ideologies complex and involved cultural belief systems

impression management Goffman's term describing how individuals may manipulate the impression or view that others have of them and give out cues to guide interactions in a particular direction

income how much money a person receives in a given time, such as $20,000 a year

independent variable variable that is said to cause or influence another variable (the dependent variable); also called the *experimental variable* in experiments

inductive reasoning logical process of reasoning that moves from specific ideas and observations to more general hypotheses and theories; the opposite of *deductive reasoning*

influence form of power whereby providing information and knowledge leads others to take different actions; compare with *domination*

institutional norms norms that prescribe appropriate structures and behaviors or organizations and other aspects of social institutions

interest groups political organizations that concentrate their activities on specific policy issues or concerns

internal validity the extent to which the conclusions of a study are true. Specifically in an experiment, the extent to which the observed changes in a dependent variable are caused by the introduction of an independent variable

intervening variable variable that comes between a dependent variable and an independent variable in a causal relationship

labeling theory theory which suggests that definitions of deviant behavior develop from social interactions and that the key element in becoming deviant is how others respond to people's behavior, rather than how they actually behave

latent functions functions that are less obvious and often unintended and that generally are unnoted by the people involved; contrast to *manifest functions*

longitudinal study study which involves data that have been collected at different times

macrolevel theories and analyses that deal with relatively broad areas of society rather than with individuals

male dominance cultural beliefs that give greater value and prestige to men and their roles and activities

managed care any system of cost containment that closely monitors and controls health care providers' decisions about medical procedures, diagnostic tests, and other services that should be provided to patients

manifest functions functions that are easily seen and obvious; contrast to *latent functions*

mean statistical average; computed by simply adding up all values and dividing by the number of cases

means of production term used by Marx and Marxian scholars to refer to the way in which people produce their living, such as by farming or manufacturing or hunting and gathering

median the midway point in a distribution; the point at which 50 percent of the cases are larger and 50 percent are smaller

medicalization the process whereby nonmedical problems become defined and treated as illnesses and disorders

methodology the rules and procedures that guide research and help make it valid

microlevel theories and analyses that deal with relatively narrow aspects of social life, such as individuals' day-to-day activities and relations with other people

middle-range theories theories that focus on relatively limited areas of the social world, as opposed to grand theories; often incorporate aspects of grand theories but are much more directed and applied toward specific research problems and can thus be more easily tested

modal the most common or frequently occurring category or case

monogamy the marriage of one man and one woman

mores norms that are vital to society, and violation of which is seen as morally offensive

neo-Marxist term used to describe recently developed theories that are in the Marxian tradition although they may depart from Marx's thought in certain ways

nonparticipant observation type of field research in which a researcher studies a group through observations without actually participating

nonverbal communication all the ways in which we send messages to others without words, including posture and movements, facial expressions, clothes and hairstyles, and manner of speaking

norms cultural rules defining behavior that is expected or required within a group or situation; includes folkways, mores, and laws

nuclear family family group consisting of a mother, a father, and their children; contrast to *extended family*

open-ended questions survey questions that allow respondents to give whatever responses they desire

panel study a longitudinal study that includes information on the same people over a long period of time

patriarchy a hierarchical system of social organization in which cultural, political, and economic structures are controlled by men

participant observation type of field research in which a researcher studies a group or event while actually participating in it

placebo a possible, simulated treatment of the control group, which is designed to appear authentic

pluralism view that the political power structure involves a number of powerful groups and individuals, all of which can potentially influence the decision-making process

population the entire group or set of cases that a researcher is interested in generalizing to; see also *sample*

post-test the measurement of the dependent variable that occurs after the introduction of the stimulus or the independent variable

power the ability of one social element, either a group or a person, to compel another social element to do what it wants

prejudice preconceived hostile attitudes toward a group of people simply on the basis of their group membership

pre-test the measurement of the dependent variable that occurs before the introduction of the stimulus or independent variable

primary groups groups that include only a few people and that are characterized by intimate, face-to-face interaction

probability sample sample that can be generalized to a larger group, typically chosen through some type of random selection process

proletariat term used by Marx and Marxian scholars to describe the workers; contrast to *bourgeoisie*

qualitative data measures of data that cannot be assigned real numbers; contrast to quantitative data

quantitative data measures that may be assigned real, or meaningful, numbers—for example, income or age

racial–ethnic group subculture that can be distinguished on the basis of skin color and ethnic heritage

random selection process that gives each member of a population an equal chance of being included in a sample

reliability the extent to which a measure yields the same results when used by different researchers on the same subject at different times

religion the social institution that deals with the area of life people regard as holy or sacred; it involves the statuses, roles, organizations, norms, and beliefs that are related to humans' relationship with the supernatural, including shared beliefs, ethical rules, rituals and ceremonies, and communities of people with common beliefs and standards

replication repetition of an earlier study to see if the same results occur and if they hold in other settings

role conflict situation in which a person holds roles with incompatible norms or obligations

role theory perspective that social structure is created and maintained because people generally act in ways that conform to social roles

role transition moves, or transitions, from one role to another during the life course

sample subset of a larger group or population; see also *probability sample*

sanctions social reactions to an individual's behavior, generally reflecting attempts to control the behavior; rewards and punishments

segregation the division or separation of neighborhoods in ways that lead to the inclusion of some groups and the exclusion of others

self one's view of oneself as a distinct person with a clear identity

self-concept the thoughts and feelings we have about ourselves

self-identity a set of categories used to define the self; the way we think about ourselves

significant others people with whom you interact and who are emotionally important to you

social capital resources or benefits people gain from their social networks

social change the way in which societies and cultures alter over time

social class groups of people who occupy a similar level in the stratification system

social control efforts to help ensure conformity to norms

social institutions the complex sets of statuses, roles, organizations, norms, and beliefs, that meet people's basic needs within a society

socialism an economic system that involves public rather than private ownership of the means of production; contrast to *capitalism*

socialization the way in which we develop, through interactions with others, the ability to relate to other people and to play a part in society

social mobility movement between social class groups

social role expectations, obligations, and norms that are associated with a particular position in a social network

social status positions that individuals occupy within the social structure

social stratification the organization of society in a way that results in some people having more and some people having less; divisions in a society based on social class

social structure relatively stable patterns that underlie social life; the ways in which people and groups are related to each other, and the characteristics of groups that influence our behavior

society a group of people who live within a bounded territory and who share a common way of life

sociological imagination Mills's term describing the ability to discern patterns in social events and view personal experiences in light of these patterns

sociology the science of society; the scientific study of the social world and social institutions

spurious correlation correlation between two variables that only occurs because of the influence of a third variable

status term used in the Weberian tradition to designate one dimension of stratification, that involving communities or social networks of people with similar lifestyles and viewpoints; synonymous with *prestige*

status characteristics the statuses that people hold and the evaluations and beliefs (characteristics) that are attached to these statuses

stigma according to Goffman, any physical or social attribute or sign that so devalues a person's social identity that it disqualifies that person from full social acceptance

stimulus the experimental condition of the independent variable that is controlled or introduced by the researcher in an experiment

structural discrimination discrimination that results from the normal and usual functioning of the society (the social structure) rather than from prejudice or from laws and norms that promote segregation or exclusion

structural functionalism sociological theory that tries to account for the nature of social order and the relationship between different parts of society by noting the ways in which these parts or structures function to maintain the entire society

survey research method of data gathering that involves asking people questions, through either interviews or written questionnaires

symbolic interactionism theory that social interaction involves a constant process of presenting and interpreting symbols through thinking about what another person is trying to communicate through the use of symbols

symbols anything that people use to represent something else; for example, language uses the symbols of words to represent objects and ideas

theories broad systems of ideas that help explain patterns in the social world

typology a classification of a group or phenomenon into discrete categories

unobtrusive research research method in which a researcher obtains data without directly talking to or watching people

upward mobility social mobility that involves movement to a social class position that is higher than one's parents occupied

urbanization the process of societal change that involves the movement of people from rural areas or small towns to metropolitan areas

validity the extend to which a measure actually represents the concept it is said to be measuring; when applied to a research design, indicates that we can trust the conclusions

values general standards about what is important to a group

variables logical groupings of attributes; literally, things that vary or have more than one value

wealth assets resulting from the accumulation of income, such as houses, cars, real estate, and stocks and bonds

white-collar crime nonviolent crimes that generally involve fraud and deception and are committed in the workplace